Infectious Diseases and Your Health

Prati Pal Singh
Editor

Infectious Diseases and Your Health

Springer

Editor
Prati Pal Singh
Centre of Infectious Diseases, Department of Pharmacology and Toxicology
National Institute of Pharmaceutical Education and Research
S. A. S. Nagar, Punjab, India

ISBN 978-981-13-1576-3 ISBN 978-981-13-1577-0 (eBook)
https://doi.org/10.1007/978-981-13-1577-0

Library of Congress Control Number: 2018957272

This Springer imprint is published by the registered company Springer Nature Singapore Pte Ltd.
The registered company address is: 152 Beach Road, #21-01/04 Gateway East, Singapore 189721, Singapore

In loving memory of late Prof. B. N. Dhawan.
You left fingerprints of wisdom on my life.

October 11, 1932–June 16, 2017
Former Director, Central Drug Research Institute, Lucknow, India

Foreword

I am very happy to write this foreword for this book entitled *Infectious Diseases and Your Health*, edited by Professor Prati Pal Singh, a distinguished colleague from NIPER, Mohali. This book has important contributions by several eminent experts on aspects relevant for better management of important diseases caused by parasites, bacteria, viruses, and fungi. Many among these topics are of immense clinical and public health importance to India and the developing world. It has topics like satranidazole in which Dr. Singh has himself contributed significantly; this drug has important role in the management of amoebic liver abscess, giardiasis, and trichomoniasis. Topics like cerebral malaria, drug resistance, autophagy, antioxidants, and epigenetics will be of immense interest to both scientists and medical professionals. Enterococci, staphylococci, chikungunya, candidiasis, HIV-AIDS, sexually transmitted diseases, and vulvovaginal infections continue to be relevant today and will have relevance for a long time to come. Nontuberculous mycobacterial diseases are gaining attention all over the world including our own country; this chapter has thus relevance today and more so in the future.

Considering the importance of the topics covered in such a comprehensive manner, I am optimistic that this book will be of immense interest to a wide spectrum of users including medical students, doctors, and scientists. On the whole, this will be useful to both medical professionals and scientists interested in finding new methods/compounds for the management of these infections of immense public health importance to India. I compliment the contributors as well as Prof. P. P. Singh in bringing out this book. I expect that in coming years this will be a popular book among a wide range of readers. I convey my best wishes to the contributors, editor, and potential readers.

Vishwa Mohan Katoch
MD, FNASc, FASc, FAMS, FNA

NASI-ICMR Chair on Public Health Research
at RUHS, Jaipur
Former Secretary, Department of Health
Research, Govt. of India, and
Director-General, Indian Council
of Medical Research
B-16, Govind Marg, Raja Park, Jaipur-302004
E-mail: vishwamohan_katoch@yahoo.co.in

Preface

> *In low-income countries, the main problems you have is infectious diseases.*
>
> Bill Gates

Lest I skip, I must mention that as I start to pen down this preface today, one hundred years ago, the first case of the *Spanish flu* was recorded at Camp Funston, Kansas, USA. Globally, nearly 500 million people were infected and 50–100 million died; in India, 17 million died. The pandemic killed in every part of the world. It turned out to be the "greatest medical holocaust in the history."

The success of Millennium Development Goals for infectious diseases, between 2000 and 2015, has been quite commendable. Nevertheless, the achievements do not gainsay the stark reality of an estimated 15 million deaths due to infectious diseases reported in 2010 and World Health Organization's prognostication of 13 million deaths by 2050. Every day, nearly 5000 children die due to diarrheal diseases. And these figures are just not statistics or sound bites, but of lost human lives. These lives could have been saved. Strangely, out of nearly 1400 recognized human pathogens, only a few of them caused a large number of these deaths. Today's world is a different place to live in. The reduced mortality due to infectious diseases, which are mainly childhood diseases, has ensued in epidemiological transition: reduced child mortality has translated in smaller family sizes and has ended up in increasing adult and elderly populations, which, in turn, has been reflected in terms of reduced deaths due to infectious diseases. Then there have been several other changes in the global health scenario, viz., massive flow of funds between 1990 and 2010 by the developed nations, especially, for infectious diseases like malaria, HIV/AIDS, and TB. These funds have helped in great achievements in HIV/AIDS treatments, restoration of life expectancy, and prevention of the transmission of infectious diseases. The Global Polio Eradication Initiative, between 1988 and 2013, has cut down by 99% the number of cases and has made India free of the disease by 2011. However, our war against infectious diseases is far from being completely won. And, thus, we must continue to fight this war on several fronts like antimicrobial resistance and the development of safe, effective, and affordable vaccines for malaria, TB, and HIV/AIDS. Our focus should be to build up upon several recent breakthrough research achievements.

I am inspired to conceptualize and bring out this book by the commands of the above narrative and several other considerations. The world has changed and will be a different place for both humans and infectious diseases. The global finances, scientific and technological innovations, the social changes in the health scenario, the impact of noninfectious and lifestyle diseases on the patterns and outcomes of infectious diseases, the urban-rural divide, the emergence of new and improved living conditions, and the impact of various global health initiatives will all have a bearing on the infectious diseases, in one way or the other. Through this book, I intend to connect the readers to the emerging dynamic relationships among different age groups of human populations, infectious pathogens and diseases, vectors and their transmission nuances, and, in turn, their poignant effects on various financial, scientific, technological, and social factors. Therefore, this book will be immensely beneficial to common citizens, public health workers, policy makers for health and economics, administrators, thought leaders, educationalists, researchers, and students.

I have edited this book due to my deep interest, emotional connect, and nearly 40 years of educational and research experience in infectious diseases. Additionally, the grand success of my earlier three related books, viz., *Proceedings of International Conference on Biotechnological Approaches to Neuroimmunomodulation and Infectious Diseases* (2009), *Human Parasitic Infections of Pharmaceutical and National Health Importance* (2009), and *Water and Health* (2014), has also provided me the enthusiasm and forte to edit this book.

Infectious Diseases and Your Health comprises of 22 articles assigned to 4 infection groups, viz., parasitic, bacterial, viral, and fungal. Most of the infectious organisms belong to one or the other of these groups. Because it is not possible to include all the human infectious pathogens in this book, I have limited myself only to some of the important ones. Each article has been authored by eminent experts in their own field and adds a special purpose to the book. Nonetheless, Chapter 1 is a bit different than the others and has a special significance. I have written this success story of the new anti-amoebic drug "satranidazole" that has been discovered, developed, and marketed in India. The article is an interesting must-read, especially for young new drug researchers, is of archival importance, and expounds on various complexities of human behavior, push and pulls of scientific and technological innovations, and hugger-muggers of decision-making.

I thank Dr. Nitya Anand, late Prof. B. N. Dhawan, late Prof. V. P. Sharma, Dr. K. Nagarajan, Dr. G. P. Dutta, Prof. R. C. Mahajan, and Prof. A. P. Dash for their encouragement and support, and with whom, I have had intense and long discussions on various aspects of infectious diseases. I am grateful to late Prof. P. C.

C. Garnham, CMG, FRS, for encouragement during my early research years. My interactions with Prof. Wallace Peters, Prof. Julius P. Kreier, Prof. Richard F. Mortensen, Prof. Robert M. Donahoe, Prof. Howard E. Gendelman, Prof. Trinad Chakraborty, Prof. Dirk Schlüter, Prof. Donatella Taramelli, and Prof. Bernhard Ryffel, over several years, have always been stimulating and inspiring. I acknowledge the support of Prof. Raghuram Rao Akkinepally, Director, National Institute of Pharmaceutical Education and Research, SAS Nagar.

Once again, I quote Bill Gates who told *Business Insider* "We are coming up on the centenary of the 1918 influenza pandemic. We have been fortunately spared anything on that scale for the past 100 years, but it is inevitable that a pandemic of equal virulence will emerge."

S. A. S. Nagar, Punjab, India — Prati Pal Singh
March 04, 2018

Contents

Contributors

S. K. Arora Department of Immunopathology, Postgraduate Institute of Medical Education and Research, Chandigarh, India

Upma Bagai (deceased) Department of Zoology, Panjab University, Chandigarh, India

Anamika Battu Laboratory of Fungal Pathogenesis, Centre for DNA Fingerprinting and Diagnostics, Hyderabad, India

Graduate Studies, Manipal Academy of Higher Education, Manipal, India

Priyanka Bhakt Laboratory of Fungal Pathogenesis, Centre for DNA Fingerprinting and Diagnostics, Hyderabad, India

Graduate Studies, Manipal Academy of Higher Education, Manipal, India

Trinad Chakraborty Institute of Medical Microbiology, Justus Liebig University, Giessen, Germany

Sidharth Chopra Division of Microbiology, CSIR-Central Drug Research Institute, Lucknow, India

Arunava Dasgupta Division of Microbiology, CSIR-Central Drug Research Institute, Lucknow, India

Swetarka Das Division of Microbiology, CSIR-Central Drug Research Institute, Lucknow, India

Priya Datta Department of Microbiology, Government Medical College Hospital, Chandigarh, India

Hemlata Dwivedi Division of Parasitology, CSIR-Central Drug Research Institute, Lucknow, India

Academy of Scientific and Innovative Research, New Delhi, India

Abhiruchi Galhotra Department of Community and Family Medicine, All India Institute of Medical Sciences, Raipur, India

Ritu Garg Department of Microbiology, Maharishi Markandeshwar Institute of Medical Sciences & Research, Mullana, India

Tanu Garg Division of Microbiology, CSIR-Central Drug Research Institute, Lucknow, India

Varsha Gupta Department of Microbiology, Government Medical College Hospital, Chandigarh, India

Namarta Kalia Department of Molecular Biology and Biochemistry, Guru Nanak Dev University, Amritsar, India

Amarjeet Kaur Department of Microbiology, Guru Nanak Dev University, Amritsar, India

Jasleen Kaur Department of Microbiology, Guru Nanak Dev University, Amritsar, India

Manpreet Kaur Department of Human Genetics, Guru Nanak Dev University, Amritsar, India

Rajvir Kaur Department of Microbiology, Guru Nanak Dev University, Amritsar, India

Rupinder Kaur Laboratory of Fungal Pathogenesis, Centre for DNA Fingerprinting and Diagnostics, Hyderabad, India

Sukhraj Kaur Department of Microbiology, Guru Nanak Dev University, Amritsar, India

Sumanpreet Kaur Department of Microbiology, Guru Nanak Dev University, Amritsar, India

Sumeeta Khurana Department of Medical Parasitology, Postgraduate Institute of Medical Education and Research, Chandigarh, India

Gurleen Mehta Department of Immunopathology, Postgraduate Institute of Medical Education and Research, Chandigarh, India

Abhisek Mishra Department of Community and Family Medicine, All India Institute of Medical Sciences, Patna, India

Alka Mital Department of Medicinal Chemistry, National Institute of Pharmaceutical Education and Research, S. A. S. Nagar, India

Helena Pillich Institute of Medical Microbiology, Justus Liebig University, Giessen, Germany

Kashi Nath Prasad Department of Microbiology, Sanjay Gandhi Postgraduate Institute of Medical Sciences, Lucknow, India

Madhu Puri Institute of Medical Microbiology, Justus Liebig University, Giessen, Germany

School of Life Sciences, Jawaharlal Nehru University, New Delhi, India

Nitu Saha Laboratory of Chromatin Biology, Department of Biological Sciences, Indian Institute of Science Education and Research, Bhopal, India

Ruchika Saroa Department of Zoology, Panjab University, Chandigarh, India

Preeti Sharma Department of Microbiology, Guru Nanak Dev University, Amritsar, India

Manjulika Shukla Division of Microbiology, CSIR-Central Drug Research Institute, Lucknow, India

Jatinder Singh Department of Molecular Biology and Biochemistry, Guru Nanak Dev University, Amritsar, India

Prati Pal Singh Centre of Infectious Diseases, Department of Pharmacology and Toxicology, National Institute of Pharmaceutical Education and Research, S. A. S. Nagar, Punjab, India

Satyendra Kumar Singh Department of Microbiology, Sanjay Gandhi Postgraduate Institute of Medical Sciences, Lucknow, India

Shreya Singh Department of Medical Parasitology, Postgraduate Institute of Medical Education and Research, Chandigarh, India

Lipika Singhal Department of Microbiology, Government Medical College Hospital, Chandigarh, India

Isha Soni Division of Microbiology, CSIR-Central Drug Research Institute, Lucknow, India

Qudsia Tahseen Department of Zoology, Aligarh Muslim University, Aligarh, India

Raghuvir Singh Tomar Laboratory of Chromatin Biology, Department of Biological Sciences, Indian Institute of Science Education and Research, Bhopal, India

Renu Tripathi Division of Parasitology, CSIR-Central Drug Research Institute, Lucknow, India

Academy of Scientific and Innovative Research, New Delhi, India

Ganesh Yadagiri Centre of Infectious Diseases, Department of Pharmacology and Toxicology, National Institute of Pharmaceutical Education and Research, S. A. S. Nagar, Punjab, India

About the Editor

Professor (Dr.) Prati Pal Singh, an internationally renowned educationalist and scientist, works at the National Institute of Pharmaceutical Education and Research (NIPER), SAS Nagar, India. For his Ph.D. (1980), he worked at the Central Drug Research Institute, Lucknow, and did his postdoctorate at Stanford University Medical Centre, CA, USA (1986), and Ohio State University, OH, USA (1984–1985). Prof. Singh was subsequently a visiting fellow at Montreal General Hospital, Canada (1985). He is an eminent biotechnologist, parasitologist, microbiologist, immunologist, and neuroimmunopharmacologist. He was involved in the preclinical development of *satranidazole*, a new anti-amoebic drug which is now on the market. He first reported on opiates as a new class of drugs for the treatment of parasitic/microbial infections and has made important contributions to the control and treatment of malaria, tuberculosis, leishmaniasis, amoebiasis, and trichomoniasis. Prof. Singh has co-edited three books: *Water and Health* (2014), *Human Parasitic Infections of Pharmaceutical and National Health Importance* (2009), and *Biotechnical Approaches to Neuroimmunomodulation and Infectious Diseases* (2009). He has over 250 research publications including 1 in *Nature Medicine*, 5 editorials, and 9 patents including 1 US patent to his credit. Prof. Singh has been editor-in-chief of the *Journal of Parasitic Diseases* and serves on the editorial board of the US journal *J. Neuroimmune Pharmacology* and several other international journals. Prof. Singh was elected Fellow of the National Academy of Sciences,

India (2004) and Fellow of the Indian Academy of Microbiological Sciences (2008), and was nominated for the fellowships of the Indian National Science Academy and Indian Science Academy. He was elected an at-large councilor (non-US) for the Society on Neuroimmune Pharmacology, USA, in 2010. Prof. Singh was awarded the prestigious Bill and Melinda Gates Foundation Global Health Award in 2008, the National Institute on Drug Abuse, USA Award in 2006, and the Tulasbai Somani Educational Trust Award of the Indian Academy of Neurosciences in 1992. Prof. Singh was also a co-investigator in an Indo-European Union 7th FP (2008) project "New Approaches to Target Tuberculosis." He has been invited for lectures/meetings in the USA, the UK, Canada, Switzerland, France, Germany, Austria, Italy, Belgium, and China. Prof. Singh is a senator of NIPER and its former associate dean (Academic Affairs). E-mail ID: drppsingh2016@gmail.com

Part I
Parasitic Infections

Chapter 1
Satranidazole and My Pharmaceutical Research Odyssey: A Success Story

Prati Pal Singh

Abstract The start of my nearly 40 yearlong pharmaceutical research odyssey almost coincides with the beginning of my association with Hindustan Ciba-Geigy Limited, Research Centre, Goregaon, Bombay (now Mumbai), and satranidazole. Satranidazole, an excellent and relatively superior molecule, is the first and the only antiamoebic drug that has been discovered, developed and marketed from India. In future, it may find therapeutic applications for many more indications. It is a product of long-drawn, very expensive and intense scientific and technological efforts, often marred with uncertainties and serendipity, of the dedicated scientists and technicians. Satranidazole had to survive spates of several squabbles but, in the end, has emerged as a champion and seen the light of day. Satranidazole will definitely go a long way to improve and invigorate the quality of human life. Nonetheless, for the new and budding drug researchers, the success story of satranidazole, full of different hues and shades of human complexities, will be a source of distinctive inspiration and has several important lessons to offer.

Keywords Amoebiasis · Antiamoebic drugs · Ciba-Geigy · *Entamoeba cricetti* · *Entamoeba histolytica* · *Entamoeba muris* · Experimental models · Luminal · Satranidazole

My association with Hindustan Ciba-Geigy Limited, Research Centre, Goregaon, Bombay (now Mumbai), and satranidazole (compound no. GO 10213) started on January 1, 1983. Ciba, much ahead of times, believed in research-based global pharmaceutical business. To commemorate the 25th anniversary of Ciba, Summit, NJ, USA (operations started on June 21, 1937), on May 25, 1962, the company organized a dedication ceremony for the nation's \$2,700,000 most modern research centre (CIBA 1962). Dr. Robert Kaeppeli, Chairman, Board of Ciba Ltd., Basel, Switzerland, was one of the lead speakers in the ceremony. And just after nearly 1 year, Ciba established another research centre in Goregaon, Bombay, India. The very idea for this research centre was conceived, set in motion and nurtured by Dr.

P. P. Singh (✉)
Centre of Infectious Diseases, Department of Pharmacology and Toxicology, National Institute of Pharmaceutical Education and Research, S. A. S. Nagar, Punjab, India

P. P. Singh (ed.), *Infectious Diseases and Your Health*,
https://doi.org/10.1007/978-981-13-1577-0_1

Kaeppeli. The basic objective behind the project was to enhance the profitability from the Ciba units in India by boosting their productivity driven by scientific and technological advancements. And soon the centre started evolving from an inchoate notion into one of the world-class pharmaceutical research organizations. The centre, christened as Ciba Research Centre, Goregaon, was officially inaugurated by our first Prime Minister Pt. Jawaharlal Nehru on March 21, 1963 (Figs. 1.1 and 1.2), who, on that occasion, made the following statement:

> *Altogether, if I was very much younger and inclined to do scientific research work, I would rather like to do the work in such surroundings and under the conditions which will no doubt prevail here.*

Fig. 1.1 (**a, b**) Prime Minister Pt. Jawaharlal Nehru at the inauguration of the Ciba Research Centre, Goregaon, Bombay (now Mumbai)

Fig. 1.2 (Clockwise) The inaugural plaque, library, research buildings and a corridor of Hindustan Ciba-Geigy Limited, Research Centre, Goregaon, Bombay (now Mumbai)

In the cool morning of January 3, 1983, I entered my new work place for the first time. More than the laboratories and the facilities, I was greatly moved by the scenic and picturesque surroundings. I and my predecessor had an overlapping period of 1 month to smoothen the transition. I was amazed by the lab management and meticulous record keeping. All the scientists in the lab were highly trained and quite experienced, and a reasonable degree of lab citizenship among them was clearly visible. By the time I joined the centre, it had already been dedicated to tropical disease research. I was to head the protozoan diseases research programme, to be more pointed, to modernize, strengthen, infuse dexterity, accelerate and give a much-needed new dimension(s) to the ongoing new antiamoebic drug discovery programme; and to add new related protozoan diseases to the programme.

Before I go any further, a few words about satranidazole. Briefly, nearly 45 years ago, it all began in Basel, Switzerland, with the merger of Ciba and J. R. Geigy Ltd., in 1971, which resulted in the emergence of a composite identity Ciba-Geigy. This merger, due to several reasons, affected the functioning of Ciba Research Centre, Goregaon, only in the dedication of its research priorities to tropical diseases. This prioritization of the drug discovery programmes was done after careful deliberations and later proved to be a boon to the centre. In October 1967, in a lavish international symposium in New York, Ciba launched its new antischistosomal drug niridazole that was initially marketed with lots of fanfare but eventually had to be withdrawn. Taking the strength from niridazole's good antiamoebic activity, and inspired by metronidazole, a Basel scientist synthesized another compound CGP 291, a potent antiamoebic drug. As the centre in Goregaon now was to concentrate

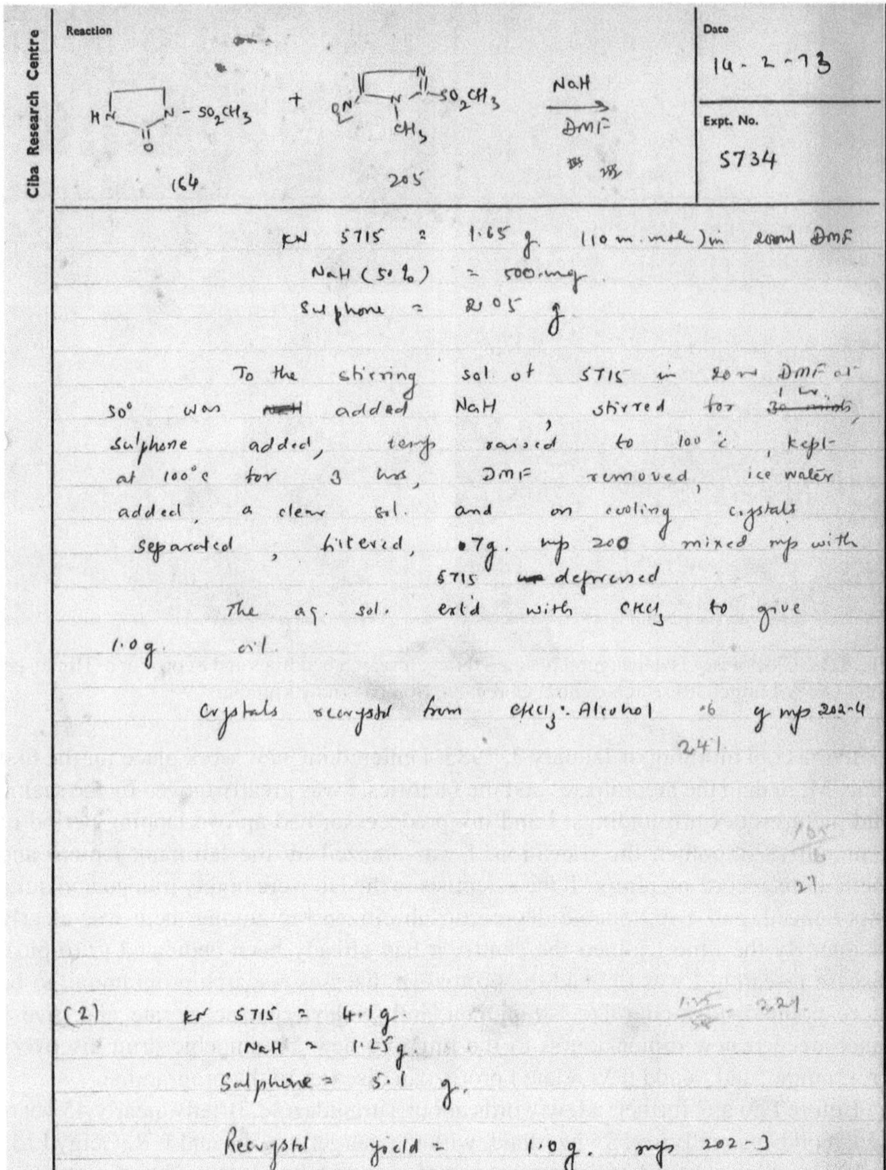

Fig. 1.3 IAB record of first synthesis of GO 10213

on tropical diseases, CGP 291 was shifted to Goregaon for its bulk production for clinical development. Unfortunately, its synthesis yielded a complex mixture of compounds, which warranted an entirely different synthetic approach (Fig. 1.3) leading to GO 10213 (satranidazole) that was synthesized in the laboratory of group leader Dr. K. Nagarajan, along with several other new analogues. GO 10213 stood superior as compared to CGP 291 and even metronidazole in standard animal models for antiamoebic screening (Nagarajan et al. 1982). Some other analogues of GO 10213 also showed superiority sufficient enough to warrant their consideration. However, after lots of arguments related to patent authorships and several other considerations including short-term toxicity studies, finally, GO 10213 was chosen for further development. A bulk quantity of GO 10213 was synthesized, and after safety pharmacology, metabolic/pharmacokinetic and stability studies, tablets were formulated. Later, an IND was filed, and after phase I, II and III clinical trials, finally, an NDA was filed to the Drug Controller of India.

Soon after joining, to my utter surprise, I found that though the NDA for GO 10213 has been filed, it has only been tested for its antiamoebic activity against *Entamoeba histolytica*, the causative agent of human amoebiasis, either in mixed bacterial cultures, in vitro, or against hepatic and caecal amoebiasis in hamsters and mice, respectively. In these models, *E. histolytica* grows and multiplies along with various other living associates, in the absence of which it will perish. One inherent limitation of such models is that it is not possible to discern whether or not the antiamoebic activity observed is due to the direct antiamoebic activity of the test compound or it is indirect due to the killing of the living associates. Certainly, it was a very important and crucial question. No doubt, to a large extent, the future of GO 10213 was dependent on its direct amoebicidal activity. By now the centre had already gone quite far with GO 10213, and at this stage, even the thought of its failure as a direct antiamoebic drug was a dreadful one and nightmarish. Soon, after holding a series of discussions with the Director, Dr. Nagarajan and other colleagues, I mobilized my team and embarked on the initiation of a new screening activity, i.e. the evaluation of test compounds for their direct antiamoebic activity against axenically grown *E. histolytica* (NIH 200 strain). Well, it wasn't that easy. Besides making arrangements for necessary facilities, the most important thing was to train the scientists for this new screening methodology that required both immense sophistication and dexterity. After lots of efforts, at last, we were successful in establishing the methodology in the lab, and the new antiamoebic screening activity started with a big bang. My team of scientists, though highly efficient and motivated, was not immune to commit mistakes. We were testing several known antiamoebic drugs along with several batches and various analogues of GO 10213. Unfortunately, a mix-up occurred, and the concerned scientist conveyed me the results we were not expecting, and our headlights just went dim. Personally, I am not the one used to hide my own defalcations, I kept my cool and later went through all the procedural details, calculations and records, several times over, and then zeroed on to the lapse. We started again, firm and resolute, with meticulous precision and then came the eureka moment: one fine morning the concerned scientist informed me that GO 10213 (MIC 1.0 μg/ml) was almost twice as potent as metronidazole (MIC 1.95 μg/ml). This revelation came as a new feather in the cap of GO 10213 and was a big delight to all of us. Later, while in the USA, I published this data, something not a high priority to the centre (Singh et al. 1985) (Figs. 1.4 and 1.5).

Fig. 1.4 A photograph of Dr. K. Nagarajan taken in 1974

Fig. 1.5 L to R: Dr. K. Heusler, Prof. Dr. E. F. Jenny, I (Dr. Prati Pal Singh) and Dr. J. A. McFadzean during the Symposium on "Recent Advances in Protozoan Diseases", November 28–29, 1983

Next, we developed a new animal model for screening new potential luminal amoebicides (Singh et al. 1987). Experience has shown that in new drug discovery research, experimental models of diseases play a major role and, at times, even change the entire course of investigations and their outcome. I should like to repeat that after the materialization of Ciba-Geigy as the composite entity, the Goregaon Research Centre having been assigned to concentrate on new antiamoebic drug discovery research also started to synthesize new compounds around the lead molecule diloxanide furoate, a known luminal amoebicide. On June 28, 1972, compound GO 9863 was synthesized but was found to be inactive in both hepatic and caecal models of experimental amoebiasis, and the same was true for the other known luminal amoebicides as well. Those days, at the Goregaon Research Centre, only these commonly used animal models of experimental amoebiasis were established and operational. No specific experimental model of luminal amoebiasis was available. However, in a US patent in 1976 and later in a research paper, Sterling Winthrop reported that the compound GO 9863 was highly active in a hamster/*E. cricetti* model for the screening of potential luminal amoebicides. Later, in 1984, at Mexico, they registered it as quinfamide, a new luminal amoebicidal drug. Thus, just because of the lack of an appropriate experimental model, Ciba-Geigy lost a drug or may be fortunes. I took this as a challenge and started thinking to come out with something to ensure that this doesn't happen again. Instead of establishing and using the known hamster/*E. cricetti* model, I decided to revisit the entire space and come up with something innovative, different and, of course, more meaningful. In literature, I found that akin to hamsters having natural *E. cricetti* infection, Wistar rats also have natural *E. muris* infection, albeit more intense. Wistar rats were easily available in the animal house of the centre. My scientists then got involved in a very labour-intensive and long-drawn project, and after several months of zeal, their efforts were met with success and ended up in the development and standardization of a new model for the identification and evaluation of potential antiamoebic agents for their luminicidal activity. Interestingly, compound GO 10213 turned out to be active, albeit weakly, in this in vivo model also. On my return from the USA, with financial support of the research centre, I published this work (Singh et al. 1987).

Lest I pass over, I should like to mention here some of the difficulties I and my team encountered while developing this new and novel model. In my opinion, difficulties in the execution of a research project are at times equally or even more important than its successful outcome. In the development of rat/*E. muris* model, besides scientific and technical difficulties, we had to face several human problems which were more challenging than the scientific work. When, for the first time, I proposed the very idea of this model, most of my biology colleagues almost ridiculed it. Nevertheless, unfazed, I continued with our work, and after surviving one onslaught after the other, we, over a few months, generated lots of data, which documented the nuances of this model. I remember very well that in one of our review meetings, where some senior scientists from Ciba-Geigy, Basel, were also present, I presented the entire data related to the development of this new model. And then all of a sudden, to my astonishment, one of the senior scientists (obviously, whom I can't name) started, repeatedly and abortively, to interrupt the flow of my presentation and distract the focused attention of other participants of the meeting. I could read the faces of several participants that they could decipher this attempted gimmick. Well, unabated, I continued my presentation and drove the main punch of my work home, quite convincingly. My presentation was well taken by all the participants. In fact one participant, who had been a regular partaker in such meetings, during the following tea break, told me that this time he was even more convinced of the novelty, authenticity, reproducibility and usefulness of this new experimental model of luminal amoebiasis. In the next few months, we routinely used this model, and several compounds including satranidazole were tested; satranidazole was active albeit at relatively higher doses and thus started a new activity for the testing of potential luminal amoebicides in an experimental model developed in-house.

On the other hand, this time around, Hindustan Ciba-Geigy was in a highly euphoric mood. Despite the fact that our centre had practically given all the relevant data and information to the Basel group, several of their scientific leaders used to visit the Goregaon Research Centre for "hard talks" on satranidazole. Additionally, I remember the visit of late Prof. M. G. K. Menon to the centre on those days, and I had the occasion to discuss with him regarding satranidazole. At that time, I had never imagined that nearly 32 years later, he will be the one to write the foreword of one of my books *Water and Health* (Singh and Sharma 2014). We the scientists at the centre and those from Basel used to have long and intense overhead projection presentation and discussions. And, at the end of the day, we all used to have several get-togethers in the form of sundowners and dinners in the exquisite hilltop guest house in the scenic beauty of the centre. One pleasant afternoon, I learned that the Director has planned for a meeting with all the scientists. Soon after he called the meeting to order, he informed us that it has been decided to organize a symposium on protozoan diseases to commemorate the 20th anniversary of the centre. The occasion was also planned to launch satranidazole. I was chosen as the secretary for the Scientific and Publications Committee. After some deliberations, the symposium was titled Symposium on "Recent Advances in Protozoan Diseases" and was scheduled for November 28–29, 1983. The President of the symposium was chosen to be the Vice-President of Glaxo company Dr. Paul Anand, and we the members of the core organizing committee used to meet in the President's office near the Mumbai TV tower at Worli. R&D heads of some other pharmaceutical companies were also taken as the members of the organizing committee, and we after our meetings used to join in the Band Box restaurant nearby. For some financial help, we also approached the Chairman, Unichem Ltd., who very courteously entertained us at the Racecourse at Mahalaxmi and also generously funded the symposium. I was involved mainly in the development of the scientific programme, preparation and receiving of the scientific material and selection, approach and formally extending the invites to the prospective speakers. I was also entrusted with the responsibility to invite four international scientists, and the entire cost of their participation was to be supported by Ciba-Geigy, Basel. I corresponded with several eminent scientists including Prof. Julius P. Kreier, OH, USA; Dr. A. Martinez Palomo, Mexico; Dr. J. T. Ponnampalam, Malaysia; Dr. R. D. Powell, IL, USA; Dr. J. A. McFadzean, UK; and several national scientists. A big team of relevant scientists from Ciba-

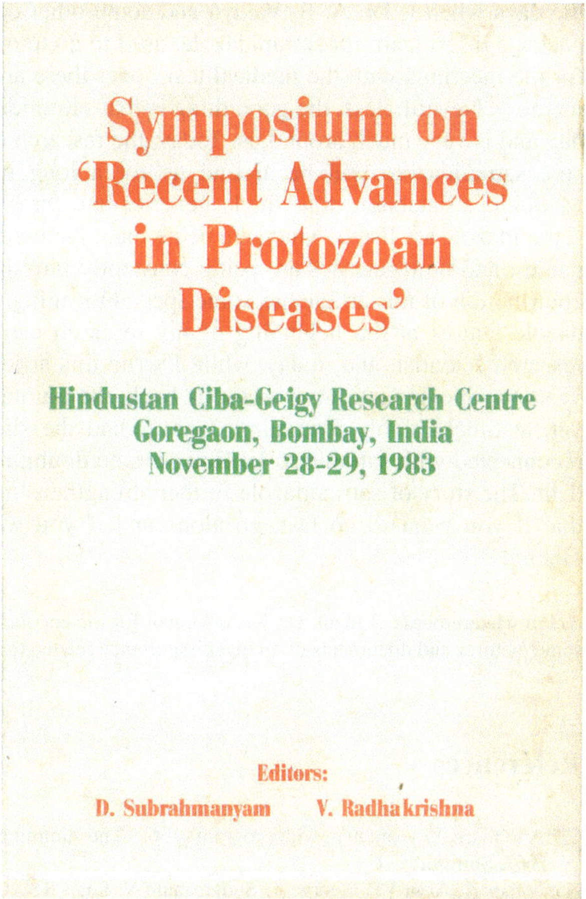

Fig. 1.6 The proceedings of the symposium held to commemorate the 20th anniversary of the centre and to launch satranidazole

Geigy Ltd., Switzerland, was also invited to the symposium. Almost all of them accepted the invite and contributed immensely to objectives of the symposium. Dr. K. Nagarajan and several other scientists involved with the discovery, development and clinical trials of satranidazole also participated and presented volumes of their data in the symposium. The symposium was a grand success. Satranidazole had now become a buzz word. Dr. K. Nagarajan, the brightest star in the constellation, nonetheless, kept his characteristic humility throughout. I was told that, perhaps, no other symposium was ever organized by the centre, at least of this magnitude. The proceedings of the symposium was published a couple of years later (Fig. 1.6; Subrahmanyam and Radhakrishna 1985).

Today, the Goregaon Research Centre with its immense research facilities, beautiful laboratory and housing buildings and lush green gardens and lawns has been subsumed somewhere. During my last visit to Mumbai a few years back, I saw that a huge commercial tower is all that can be seen at that place. All those scientists and research assistants have gone here and there, and some have even passed away. But the days when I, Dr. A. B. Vaidya and some other colleagues, as members of the package insert team for satranidazole, used to go to our Head Office at Churchgate for the meetings with the medical team over there are still very much alive in my memory. Nevertheless, the good thing is that satranidazole has seen the light of the day and is very much around us. Today, the research centre is not around, but I am sure satranidazole will be around us for a long time. Presently, according to Medindia's database, nine single generics and three combined generic brands of satranidazole are listed, which are being manufactured by five pharmaceutical companies, and new generics are being constantly introduced. For me, personally, the contribution of this article has some special meaning. I got associated with satranidazole almost at the beginning of my research career as a new drug discovery research scientist, and, today, while I write this article, I am left with only a few years to superannuate. And as I look back, the journey had been long and arduous yet, at times, highly rewarding. I never had the slightest comprehension to get reconnected with satranidazole. However, no doubt, now I deem it a pleasant accident. The story of satranidazole further strengthens my belief in the African adage that if you want to go fast, go alone and if you want to go far, go along with others.

Acknowledgements I thank Dr. K. Nagarajan for the encouragement and generously providing some pictures and documents of archival importance related to satranidazole.

References

CIBA to mark 25 years here with program (1962) The summit herald and summit record. May 24, 1962, Summit N. J

Nagarajan K, Arya VP, George T, Sudarsanam V, Shah RK, Goud N, Shenoy SJ, Kulkarni VS, Rao MK (1982) Nitroimidazole part 4.1-[methyl-5-nitroimidazolyl(2)]-2-oxo-3-sulphonyl (carbamoyl-thicarbamoyl) tetrahydroimidzoles. Ind. J Chemother 21B:928–940

Singh PP, Sharma VP (2014) Water and Health. pp i–xxiii + 404, Springer India. ISBN: 978-81-322-1028-3 (Print) 978-81-322-1029-0 (Online); https://doi.org/10.1007/978-81-322-1029-0

Singh PP, Tendulkar JS, Nagarajan K (1985) Amoebicidal activity of CG-10213 GO and other nitroimidazoles against axenically grown *Entamoeba histolytica*. Ind J Parasit 9:99–100

Singh PP, Srivastava VB, Talwalkar S (1987) Natural *Entamoeba muris* infection in rats: a new model for screening potential luminal amoebicides. Ind J Parasit 11:15–20

Subrahmanyam D, Radhakrishna V (1985) Symposium of the proceedings of the symposium on recent advances in protozoan diseases

Chapter 2
Amoebiasis Revisited

Alka Mital

Abstract Amoebiasis, a disease characterized by abdominal pain, loose motion, and blood-containing feces, is caused by a unicellular protozoan parasite *Entamoeba histolytica*. It afflicts nearly 500 million people worldwide and is the second leading cause of death from a parasitic infection in children under 5 years of age. The cysts are the infective stages, and the infection spreads through ingestion of mature cysts in fecally contaminated food and drinking water or by hands from an infected person. A majority of infected persons often remain asymptomatic; however, in some cases, infection can lead to severe clinical complications like dysentery and amoebic liver abscesses. Though metronidazole (a 5-nitroimidazole) continues to be a drug of choice for the treatment of amoebic dysentery, for its complete cure, a combination of a tissue amoebicide (usually a 5-nitroimidazole) and a luminal amoebicide (e.g., diloxanide furoate) is required. Unfortunately, the occurrence of *E. histolytica* strains resistant to metronidazole, its potential carcinogenicity, and a strong metallic taste after a few days of use are some of its big limitations. Satranidazole, a novel 5-nitroimidazole discovered and developed in India, is now available in several formulations. Nevertheless, the treatment of amoebiasis continues to remain far from satisfactory, and there is a strong and compelling need to discover and develop novel and improved antiamoebic drugs. Further, the availability of clean drinking water, proper sanitation, and good personal hygiene including handwashing will continue to remain important driving factors in tackling the problem of amoebiasis. This overview summarizes the history, life cycle, currently available drugs, and, more importantly, the progress on the discovery of novel antiamoebic agents and various preventive measures for the control of amoebiasis.

Keywords Amoebiasis · *Entamoeba histolytica* · 5-nitroimidazoles · Satranidazole

A. Mital (✉)
Department of Medicinal Chemistry, National Institute of Pharmaceutical Education and Research, S. A. S. Nagar, India

P. P. Singh (ed.), *Infectious Diseases and Your Health*,
https://doi.org/10.1007/978-981-13-1577-0_2

2.1 Historical Background

The earliest records of this infectious disease with symptoms of bloody, mucous diarrhea have been found in the Sanskrit document *Bhrigu Samhita* (1000 BC). Assyrian and Babylonian texts (600 BC) have also mentioned about amoebiasis which occurred in the Tigris-Euphrates basin region before the sixth century BC. Amoebiasis was first identified as a deadly disease by Hippocrates (around 300 BC), by describing a patient with symptoms of fever and dysentery (Imperato 1981). Around the sixteenth century, amoebiasis spread throughout Europe, Asia, Persia, and Greece due to the developing world trade. The prevalence of parasitic infection was mainly found in the rural and suburban communities of tropical countries with poor sanitation. It was more widespread among people living in the developing countries with poor sanitation and socioeconomic conditions.

The first accurate description of the disease was mentioned by James Annesley in his book *Researches into the Causes, Nature and Treatment of the More Prevalent Diseases of India and of Warm Climates Generally* written in the nineteenth century. More developments came in 1855 when it was suggested that the disease might have a parasitic origin. The first detailed description of a case of amoebic dysentery in 1875 by Fredrich Lósch, a Russian physician, was a milestone in the study of *E. histolytica* and amoebiasis. He discovered amoeba from the stool sample of a dysentery patient in St. Petersburg. He also provided drawings of amoeba and gave its detailed description, including structure, size, motility, nucleus, vacuoles, and such intracytoplasmic elements as red blood cells. Koch in 1883 described amoeba in stained tissue sections. Kartulis in 1886 published his observations on 150 cases of dysentery among Egyptians, describing the organisms he found in both stool specimens and in tissue sections and conducting a number of experiments on their survival time in sugar solutions, in salt water, and in hanging drop preparations. Kartulis had concluded that amoebae were present in every case of dysentery and coined the term "tropical dysentery" for the disease. Koch published his 1883 observations, and Kartulis also published his studies of amoebic liver abscess, a complication of tropical dysentery in 1887. Councilman and Lafleur in 1891 first recognized amoebiasis as a distinct clinical disease due to a specific pathogen, which they called *Amoeba dysenteriae*. They first used the common terms "amebic dysentery" and "amebic abscess of the liver." The major breakthrough work was carried out by Walker and Sellards in 1913, who distinguished the infective cyst form of pathogenic *E. histolytica* and non-pathogenic *E. coli*. They also demonstrated that *E. histolytica* did not always give rise to clinical disease.

In 1912 Leonard Rogers, a pathology professor at the Medical College Hospital in Calcutta, reported the first successful treatment of amoebiasis by injectable salts of emetine. Emetine, the principal alkaloid of dried root of *Psychotria ipecacuanha*, had been effective in killing amoebae in vitro, and it was a major breakthrough in the treatment of the disease (Rogers 1912). Emile Brumpt in 1925 had proposed the existence of two types of parasites, the invasive (*E. histolytica*) and noninvasive (*E. dispar*). He suggested that both *E. histolytica* and *E. dispar* were identical morphologically, but only *E. histolytica* was pathogenic for humans. Dobell in 1919 described the life cycle of *E. histolytica* (Imperato 1981). The axenic cultivation of *E. histolytica* was first achieved by Diamond in 1961, and it was an important step in understanding of the cell biology (Diamond 1961). The World Health Organization (WHO) in 1969 defined amoebiasis as an "infection with *E. histolytica*, with or without clinical manifestations" thus indicating that all the strains were potentially pathogenic. WHO in 1997 had recognized that *E. histolytica* was two species (Brumpt had earlier proposed in 1925) and had given clear guidelines for distinguishing both the entamoeba species.

2.2 Introduction

Parasitic diseases like malaria, amoebiasis, toxoplasmosis, trypanosomiasis, and leishmaniasis are caused by protozoan parasites and affecting approximately 25% of the world's population (Azam and Agarwal 2007). Amoebiasis is a parasitic protozoan infection caused by *E. histolytica.* It is the second leading cause of death worldwide, infecting over 50 million people per annum and causing 110,000 deaths annually mostly in developing countries (Hayat et al. 2016). Most morbidity and mortality due to amoebiasis occur in India, tropical regions of Asia and Africa, Mexico, and parts of Central and South America (Hayat et al. 2016). It has been reported throughout India, affecting about 15–20% of the Indian population. Poverty, overcrowded living places, poor sanitary conditions, and malnutrition favor the spread of the disease and increase burden. Amoebiasis is a common problem of the developed world, in persons with immunodeficiency, homosexual males, travelers, and immigrants from certain tropical countries.

There are three species of intestinal *Entamoeba* with identical morphological characteristics: *E. histolytica*, *E. dispar*, and *E. moshkovskii* (Peterson et al. 2011). Most symptomatic disease is caused by potent pathogenic *E. histolytica*, and *E. dispar* is generally considered non-pathogenic, which differ mainly in their cell surface phosphorylated glycolipids (Campos-Rodriguezp and Jarillo-Luna 2005). Infections with *E. moshkovskii* are also becoming more frequent (Heredia et al. 2012). The *E. histolytica* exists in two forms, the trophozoites, the causative agent, and the mature cysts responsible for transmission of amoebiasis. These two forms, from trophozoites to cysts and vice versa, are connected by "encystation" and "excystation" stages. The cysts are usually transmitted through fecal-oral route by ingestion of contaminated food and water or by dirty hands. Human beings are the only known host of the amoebiasis parasite, and all groups of people, regardless of age or sex, can become affected. Hence by developing novel methods to control encystation and excystation will potentially reduce the transmission of the disease by interrupting the life cycle of the parasite. The disease mechanism and the exact prevalence and incidence of infection caused by *E. histolytica* are still unknown. A large majority of infected persons do not show any symptoms, but in 10% cases, the infection can lead to severe complications of dysentery and amoebic liver abscess.

The naturally acquired immunity to *E. histolytica* infection is found in humans and is very much short lived. Hence development of a vaccine against amoebiasis by improving the human immune response conferred by natural infection is a significant challenge (Stanley 2006).

Several agents are used for the treatment of amoebic dysentery, which are classified as tissue and luminal amoebicides (Fig. 2.1). Tissue amoebicides include metronidazole (MTZ), tinidazole, ornidazole, and emetine, which kill the amoeba in the host tissue, whereas the luminal amoebicides such as iodoquinol, diloxanide furoate, and paromomycin are active in the intestinal lumen and generally given to eradicate the complete infection. Nitroimidazoles such as MTZ, tinidazole, ornidazole, secnidazole, and satranidazole are some of the main synthetic drugs given for invasive amoebiasis. Treatment with nitroimidazoles is not safe for humans, as they have several side effects and MTZ is potentially carcinogenic because of its cellular genotoxicity (Hayat et al. 2016). There are some reports where the in vitro generation of strains resistant to MTZ has been described (Orozco et al. 2002). Resistance to some of the current drugs such as MTZ, paromomycin, and nitazoxanide has been reported and poses a serious problem. Presently there is no complete safe treatment for amoebiasis, and therefore search for new compounds from synthetic and natural resources is still continuing to develop better and safer therapeutic agents. Various preventive measures such as improvement of basic sanitation and promotion of food hygiene, purification of drinking water supplies, and health education to the general public are essential to prevent the transmission of the disease.

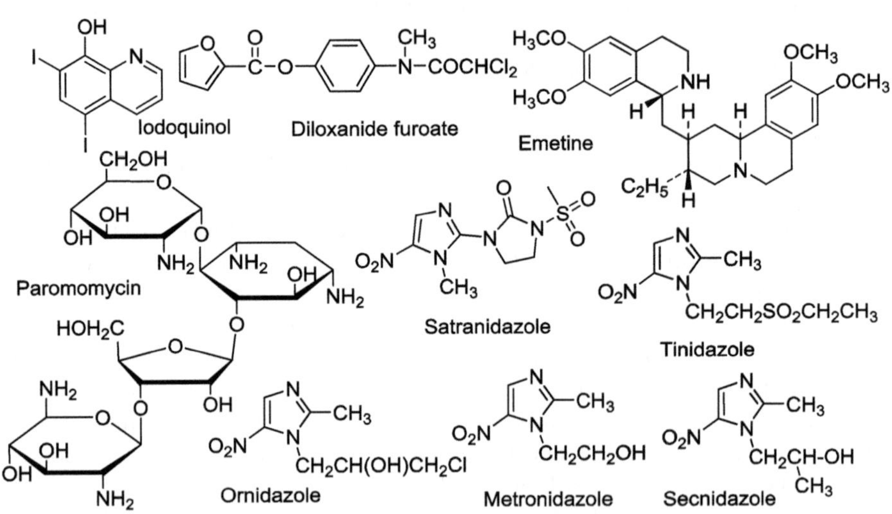

Fig. 2.1 Structures of some commonly used antiamoebic agents

2.3 Epidemiology

Amoebiasis is one of the most common parasitic infections occurring in humans. It has been reported worldwide, but its prevalence is disproportionately increasing in developing countries due to poor socioeconomic and sanitary conditions (Azam et al. 2015). An estimated 50,000–110,000 people die yearly from amoebiasis, making this disease the second leading cause of death from parasitic diseases (Hayat et al. 2016). The prevalence of the disease differs between countries and varies with the population of individuals affected mostly living in areas with poor socioeconomic conditions. Global statics on the prevalence of *E. histolytica* infection indicates that about 90% of individuals are asymptomatic and the remaining 10% develop clinically overt disease ranging from dysentery to liver abscesses. Amoebiasis cases are reported throughout India affecting about 15–20% of the Indian population. *E. histolytica* can be transmitted by a number of ways, but most commonly it is transferred from fecal matter, where *E. histolytica* cysts are present. It is also transmitted from contaminated food or water sources and from one person to another by oral or rectal contact with the infected individual.

A community-based study in Bangladesh highlighted that amoebiasis is a common diarrheal disease causing the death of 1 in 30 children before the age of 5. Rates of asymptomatic carriage in immigrants to the United States are reported to be between 17% and 33%. In travelers returning to Australia, the carriage rate is unknown and varies greatly depending on the countries visited (Hung 2007). Hepatic amoebiasis is endemic in Thailand, India, Egypt, and South Africa with high mortality rates. The results of a survey in Egypt indicated that 38% of individuals suffering with acute diarrhea had amoebic colitis (Stanley 2003). The *Entamoeba* infections are prevalent (4–21%) in developed countries such as Italy, Japan, and the United States and are due to the noninvasive species, *E. dispar*, which does not require treatment. It is mostly found in homosexual males who practice oral-anal sex, immigrants from or travelers to endemic areas, and HIV-infected patients. In Mexico, a national serosurvey had demonstrated that 8.4% of the population was infected with invasive amoebiasis, representing one million cases of the disease. In 2002, the prevalence of infection with *E. histolytica/E. dispar* was 0.78%, 3.9%, and 4.6% for central, northern, and southern part of Iran, respectively.

Human beings and some nonhuman primates are the only medium through which the *E. histolytica* spreads and multiplies. Most individuals with amoebiasis are infected by the ingestion of *E. histolytica* cysts through fecally contaminated food or water, by usually uncooked food and contaminated water supplies, or by contaminated hands. Some unusual modes of transmission also include oral and anal sex and contaminated enema apparatus. The spread of amoebiasis within families is due to poor hygiene and household contacts of infected patients. There was an increased risk for *E. histolytica* infection and invasive amoebiasis in HIV seropositive men and homosexual males in Taiwan (Hung et al. 2008). The two non-pathogenic *Entamoeba* species (*E. dispar* and *E. moshkovskii*) are morphologically identical to pathogenic *E. histolytica*, but the prevalence of human infections by these non-pathogenic species is not well studied.

2.4 *E. histolytica*: Structure and Life Cycle

The life cycle of *E. histolytica* parasite is simple and consists of two distinct forms, an infective cyst form and a multiplying trophozoite form (Stanley 2003; Figs. 2.2 and 2.3 are showing stages of life cycle and microscopic view of *E. histolytica*, sources: https://www.learnzoology.wordpress.com and https://www.cdc.gov/parasites/amebiasis/index.html). The main mode of transmission and incidence of amoebic dysentery and infection of the host occurs upon ingestion of fecally contaminated food or water containing the mature cysts of *E. histolytica*. Transmission can also occur through exposure to fecal matter during sexual contact, where trophozoites are also infective. The infective cysts exist outside the host only and are round in shape, usually 10–15 μm in diameter. They are non-motile and surrounded by a refractive wall containing chitin to help them survive the acids in the stomach

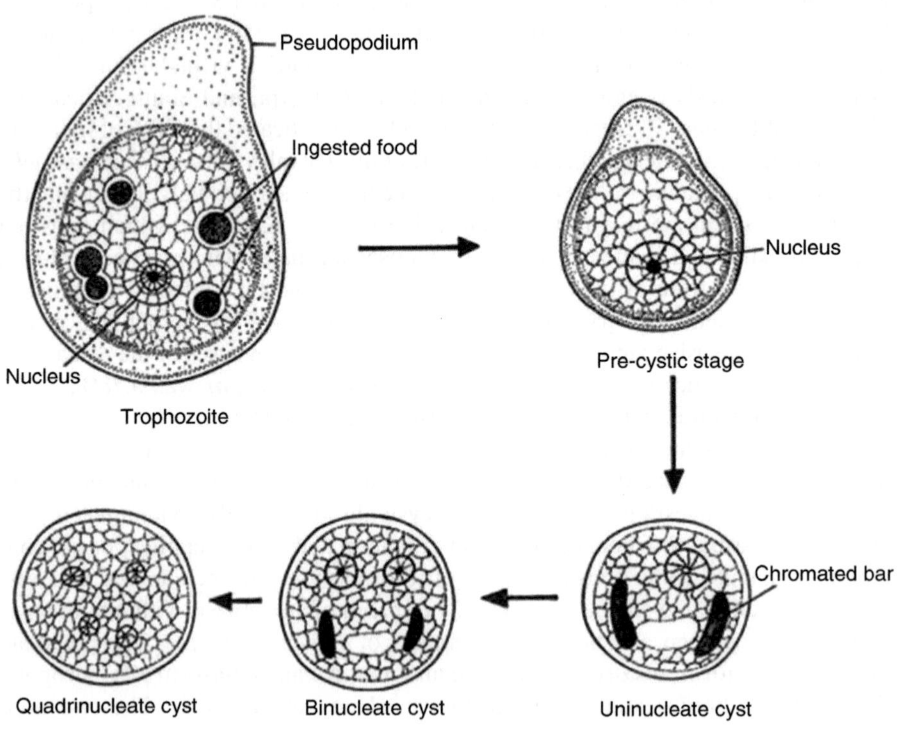

Fig. 2.2 Stages of the life cycle of *Entamoeba histolytica*. (Source: https://www.learnzoology.wordpress.com)

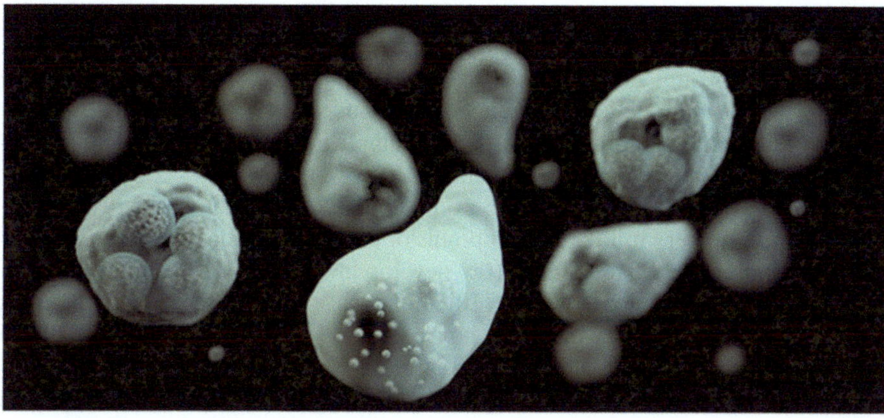

Fig. 2.3 Scanning electron photomicrograph of *Entamoeba histolytica*. (Source: https://www.cdc.gov/parasites/amebiasis/index.html)

and in the external environment for days to weeks. After ingestion, the cyst excysts in the small intestine and forms the amoeboid trophozoite, which then migrates to the large intestine of the host. *E. histolytica* trophozoites often remain confined to the intestinal lumen, and this results in an asymptomatic intestinal infection. Unlike the inert cysts, *E. histolytica* trophozoites are actively motile, having polymorphic shapes and sizes varying from 10 to 50 μm in diameter. Trophozoites exist inside the host and grow and multiply by binary fission in large intestine, ingest bacteria and food particles, and adhere and destroy the epithelial cells. The destruction of the epithelial tissue of large intestine causes the disease and symptoms. After penetration into the blood vessels, trophozoites can also migrate to other organs of the body such as the liver, lung, brain, and skin causing severe infections. Trophozoites transform into dormant and infectious cysts which are excreted into the environment through feces, where they are eventually ingested again. Cysts which are excreted in stools can survive for several months and continue their life cycle by further fecal-oral spreading. The asymptomatic person can also spread the infection by releasing new cysts into the environment through infected feces.

2.5 Symptoms and Clinical Diagnosis

Amoebiasis, also known amoebic dysentery, is an infection caused by any of the amoebas of the *Entamoeba* group. It can be present with no, mild, or severe symptoms. The symptoms commonly include frequent, watery, and/or bloody diarrhea with mild to severe abdominal pain, flatulence, fatigue, and loss of appetite. Liver abscess is the most common form of extraintestinal amoebiasis. Amoebic liver abscess is characterized by daily occurrence of fever and abdominal pain in most of the patients. The *Entamoeba* life cycle consists of either an infectious cyst stage or noninfectious trophozoite stage. Transmission occurs via ingestion of mature amoebic cysts, most commonly by consumption of fecally contaminated food or water and through direct anal-oral contact with an infected or asymptomatic carrier. Once the trophozoites have invaded the large intestinal walls, they can enter the bloodstream and migrate to various internal organs such as the liver, heart, lungs, brain, or other organs and cause more severe complications like severe colitis, intestinal perforation, liver abscess, and anemia.

Approximately 90% of people are asymptomatic when infected with *E. histolytica* and present very mild or no symptoms of the disease and spontaneously clear their infection, while the remaining 10% develop invasive disease. Clinical diagnosis of amoebiasis is sometimes very difficult because of the abdominal symptoms and signs of the infection which are nonspecific. They are easily confused with symptoms of noninfectious intestinal diseases (inflammatory bowel disease, ischemic colitis, etc.) and number of other bacterial dysenteries that are common in tropical countries. Accurate diagnosis is very much important for the asymptomatic infections, because infection can persist and develop to a symptomatic infection.

The most common symptom of the disease is intestinal amoebiasis and liver abscess. Any individual who presents the symptoms of diarrhea or liver abscess and who has traveled or lived in an endemic area in the previous 12 months can be considered suffering from amoebiasis. The incubation period of intestinal amoebiasis is 1–4 weeks, but it can vary, ranging from a few days to months or years. Diagnostic tests should be performed before starting the treatment for those suspected with amoebiasis. The laboratory diagnosis of amoebiasis is based on microscopy and serological methods (including enzyme-linked immunosorbent assay ELISA, indirect hemagglutination assay IHA, and latex agglutination LA) with detection of the parasite in stool samples and by biopsy of intestinal amoebic lesions or by drainage of liver abscess. The stool antigen detection tests offer a practical, sensitive, and specific way for the clinical laboratory to detect intestinal *E. histolytica*. Patients with asymptomatic intestinal infection generally have only cysts in the fecal sample. The most common means of diagnosis is microscopic examination of the motile trophozoites of *E. histolytica* in a saline preparation of fresh stool samples. The sample should be examined within 1 h of collection to search for motile trophozoites which may contain RBCs. The cysts and trophozoites lacking ingested RBCs are also visible on microscopy. Although quick and inexpensive, the test has limited sensitivity ranging from 60% to 70%, as it can give false-positive results due to misidentification of macrophages as trophozoites. All the current tests suffer from the fact that the antigens detected are denatured by fixation of the stool specimen, limiting testing to fresh or frozen samples. The estimation of the true prevalence of amoebiasis is not easy, as most diagnostic tests currently available do not reliably differentiate between the species of *Entamoeba* and are less sensitive and cumbersome to perform in developing countries due to poor sanitary conditions.

2.6 Treatment

Amoebiasis is a major public health problem in developing countries with poor sanitation, where entamoeba parasites can easily spread the infection. Treatment of amoebiasis includes pharmacologic therapy, surgical intervention, and various preventive measures, as required. Although the antiparasitic drugs currently available are highly effective against the trophozoite form of amoeba, the continuous usage or overdosing of such drugs could lead to the development of resistance in the targeted pathogen. Metronidazole is an effective drug for symptomatic, invasive disease, while paromomycin is the drug for noninvasive disease. Amoebiasis is treated first with a nitroimidazole derivative and then with a luminal agent to completely clear the infection. Paromomycin is safe, well tolerated, and effective in the treatment of intestinal amoebiasis, including in patients with HIV infection. Diloxanide is a dichloroacetamide derivative, which is amoebicidal against trophozoite and cyst forms of *E. histolytica.* Invasive amoebiasis (e.g., colitis, liver abscess) should be treated with MTZ for 10 days. The longer period of treatment with MTZ results in failures due to poor patient compliance. Although MTZ has some unpleasant side effects, such as headache, nausea, metallic taste, etc., its treatment has cure rates of approximately 90%. It is carcinogenic to humans, and resistance of *E. histolytica* strains to MTZ has been reported (Azam et al. 2015). An uncommon neurological side effect, such as vertigo or encephalitis, or neutropenia necessitates the discontinuation of treatment. MTZ kills trophozoites of *E. histolytica* in intestines and tissue but does not eliminate the cysts from intestines, therefore treatment with MTZ is to be followed with a luminal agent, and otherwise patients are at a risk of relapsing from residual infection. MTZ is the major drug of choice, and other nitroimidazole derivatives like tinidazole, satranidazole, secnidazole, and ornidazole are equally effective. Secnidazole is the newest among the nitroimidazole group and has a much longer half-life compared to MTZ, tinidazole, and ornidazole. Secnidazole appears to be superior due to the single-dose treatment of the drug and lack of side effects, as compared to other drugs with long duration of treatment and their side effects. Satranidazole (Satrogyl) is an antiamoebic, antiprotozoal, and antibacterial drug, first discovered in India, and is now being used in the treatment of various conditions, including trichomoniasis, giardiasis, liver abscesses, anaerobic infections, periodontitis, intestinal amoebiasis, and hepatic amoebiasis. The plasma elimination half-life of satranidazole is 1.01 h, which is significantly shorter than MTZ half-life of 3.62 h. Amoebicidal drugs and their recommended duration of treatment may be grouped under following categories:

Tissue amoebicides

- MTZ: 500 mg thrice daily for 5–7 days is the most effective treatment.
- Tinidazole: single dose of 2 g for 3–5 days.
- Ornidazole: 1.5 g once daily for 3 days or 500 mg twice daily for 5–7 days.
- Secnidazole: single dose of 2 g.
- Satranidazole: single dose of 300–500 mg, once or twice a day for 3–10 days.

- Nitazoxanide: 500 mg twice daily for 3 days.
- Chloroquine: 300 mg twice daily and continued by 300 mg daily for 21 days.

Luminal amoebicides

Prescribed after the course of tissue amoebicides:
Diloxanide furoate: 500 mg thrice daily for 10 days.
Quiniodochlor: 500 mg twice daily for 10 days.
Iodochlorhydroxyquin: 500 mg twice daily for 10 days.
Paromomycin: it is rarely used, given 30 mg/kg body weight thrice daily for 7 days.

2.7 Preventive Measures

The primary mode of transmission and infection of *E. histolytica* is through ingestion of fecally contaminated food and water. The effective prevention to reduce the transmission of cysts is extremely important. Thus eliminating the contamination can reduce the risk of infections, and it is the best method of prevention. It is possible by avoiding the use of contaminated water and food when traveling mainly to the endemic areas. Development of diagnostic methods for the identification of cysts even in asymptomatic carriers will be an effective preventive measure. General health measures for improving water supply and food safety, early detection and treatment of infections, educating people about improving good personal hygiene, avoiding unsafe sexual practices, handwashing, and sanitation will be effective in preventing the spread of many infections.

Handwashing is one of the most important things that can easily help in prevention of transmission. Global Handwashing Day (October 15) is being celebrated annually to raise awareness and to encourage people for washing their hands with soap. It was initiated by the Public-Private Partnership for Handwashing (PPPHW) in August 2008 at the annual World Water Week in Stockholm, Sweden. The UN general assembly suggested the date in 2008 (International Year of Sanitation). The 2017 Global Handwashing Day theme was "Our hands, our future!" The theme reminds us about the importance and practice of handwashing in protecting our own health in an easy, affordable, and effective manner against diseases. The simplest measure of washing and drying hands after defecation and before handling and consuming food can prevent the chances of developing many infections.

The parasite can also contaminate fruits and vegetables when they are grown in places where human feces are used as fertilizer. They can be transferred on the dirty hands of infected people who don't wash their hands properly. Therefore, thorough cleaning of the raw vegetables and other food items by washing and drying them should be practiced. Children should be taught about the importance of unhygienic habits like biting their nails, not washing their hands properly before touching or eating food and drinks, etc. The practice of hygienic habits such as avoiding contaminated drinking water and eating uncooked food, such as raw meat, peeled fruits, preserved salads, dairy products, ice creams, etc., can help in preventing the spread of the disease.

Avoiding the practice of open defecation can eliminate many severe health and environmental problems. Educating people about unsafe sexual activities that permit fecal-oral contact will reduce the risk of sexual transmission of infective cysts. Hence amoebiasis can be prevented worldwide by improving unhygienic living conditions, proper sanitation, and water treatment. Good sanitary practices and proper sewage disposal or treatment and government public health policies are necessary for the prevention and control of *E. histolytica* infection on an endemic level.

2.8 Antiamoebic Drugs

Amoebiasis is a contagious disease to treat because of its chronicity and inability of existing drugs to eradicate the cystic forms of the parasites completely. It is first treated with a tissue amoebicide (a nitroimidazole derivative) and then followed by treatment with a luminal amoebicide to eradicate the complete infection. There are numerous antiamoebic drugs used for treatment of amoebic dysentery and are classified as tissue amoebicides and luminal amoebicides (Fig. 2.1). Tissue amoebicides such as MTZ, tinidazole, and emetine kill amoeba in host tissue and organ, whereas the luminal amoebicides are poorly absorbed and are active only in the intestinal lumen. Antiamoebic drugs have been classified as luminal if their site of action is in the large intestine and extraluminal if their site of action is in other organs, mainly the liver. The luminal agents generally prescribed are iodoquinol, diloxanide furoate, nitazoxanide, and paromomycin (Azam and Agarwal 2007).

A large number of compounds possessing in vitro activity against *E. histolytica* parasite have been isolated and synthesized (Singh et al. 2009). They have been classified into azoles (MTZ, tinidazole, ornidazole, secnidazole), quinolines (iodoquinol), dichloroacetamides (diloxanide furoate), and carbamate derivatives and are recommended for treatment of different stages of amoebiasis (Fig. 2.1). Bioactive molecules showing significant antiamoebic activity are synthetic derivatives of imidazoles, alkaloids, furan, and quinolines (Sharma and Sharma 2001). The first effective drug emetine (a potent tissue amoebicide, Fig. 2.1) against *E. histolytica* infection was isolated from the root and rhizome of *Cephaelis ipecacuanha* (Wright and Phillipson 1990). Emetine and its synthetic derivative dehydroemetine were widely used to treat patients with severe amoebic dysentery and extraintestinal amoebiasis. However, because of cardiotoxicity, these drugs have largely been replaced by MTZ, and recent recommendations do not advise the use of emetine or dehydroemetine unless MTZ is ineffective or contraindicated. Luminal agents are generally recommended for the treatment of asymptomatic intestinal colonization with *E. histolytica* or after giving a tissue amoebicide to clear the infection. All patients with invasive amoebiasis require treatment with a tissue amoebicide followed by a luminal amoebicide in order to eliminate any surviving organisms in the colon and to prevent relapse.

The introduction of the nitroimidazole class of drugs had greatly improved the treatment of amoebic infection. The synthetic 5-nitroimidazole, MTZ, was first used in 1959 to treat chronic trichomonad infections. MTZ is the first-line medication for treating invasive amoebiasis, such as those with amoebic dysentery. Other synthetic nitroimidazole drugs, such as tinidazole, ornidazole, and secnidazole, have longer half-lives and shorter periods of treatment. They are better tolerated compared with MTZ, but they share same pharmacological profile and toxicity with MTZ. The highly selective effect of these drugs is due to reduction of these drugs by nitroreductase enzymes resulting in the formation of highly reactive free radical species (Leitsch et al. 2007). MTZ gets reduced to nitroradical anion and nitrosoimidazole by thioredoxin reductase (TrxR) or ferredoxin inside *E. histolytica* cell. The nitroradical anion formed further reduces O2 and generates reactive oxygen species, causing oxidative damage to cells. The highly reactive nitrosoimidazole species binds with proteins or nonprotein thiols and other biomolecules to inhibit the activity of *E. histolytica* (Leitsch et al. 2007).

A single high-dose therapy with 2 g MTZ is the preferred therapy, over multiple doses or therapy over an extended period. It is not well tolerated in some patients, including pregnant women and alcoholics. The Food and Drug Administration has classified MTZ as a class B risk factor for pregnancy (Ali and Nozaki 2007). Treatment with MTZ has several side effects that include headache, nausea, vomiting, loss of appetite, dry mouth, metallic taste, dizziness, vertigo, and some neurological complications. Other nitroimidazole drugs such as tinidazole, ornidazole, and secnidazole are recommended for better tolerance than MTZ and treatment for a shorter duration. Lumen-acting drugs, such as diloxanide furoate, iodoquinol, and paromomycin have been used to eliminate dormant luminal cysts and prevent reoccurrence of infection. Both the classes of drugs have been associated with one or more side effects including development of resistance.

Resistances to MTZ in many pathogenic bacteria and protozoa and their several side effects are well known (Wassermann et al. 1999). There is no carcinogenicity or mutagenicity of MTZ in human beings (Petri 2003), but it has been shown mutagenic in bacteria and carcinogenic in rodents (Legator et al. 1975; Rustia and Shubik 1972). Some reports also have described the in vitro generation of strains resistant to MTZ (Orozco et al. 2002). It is unknown whether it is the biological differences in the parasite, or differences in drug sensitivity and virulence, which are responsible for the treatment failure. The 5-nitroimidazole drug MTZ has been recommended as the most effective treatment of amoebiasis. Due to the high prevalence of these infections and due to its role as a second-line defense against *Helicobacter pylori* infections, MTZ has been included in the "essential medicines" list by the WHO (Leitsch et al. 2007). The MTZ complexes of Pd (II), Pt (II), Cu (II), Au (II), and Ru (II) have shown better in vitro results than MTZ (Bharti et al. 2002; Athar et al. 2005). The Cu-MTZ complexes have displayed highest in vitro activity than other metal complexes.

An oral synthetic broad-spectrum antiprotozoal agent, nitazoxanide, was found to be active against a number of intestinal helminthes and protozoans (Fig. 2.4). Nitazoxanide, a nitrothiazoyl-salicylamide derivative, and its active metabolite

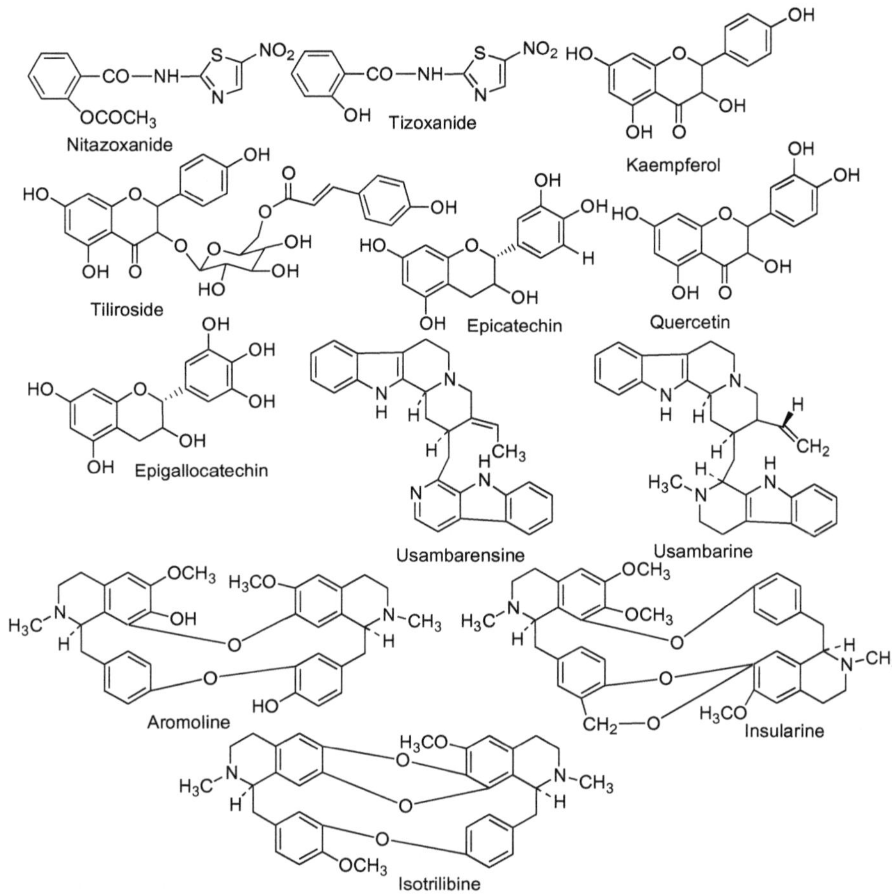

Fig. 2.4 Bioactive molecules from natural products having in vitro antiamoebic activity

tizoxanide have exhibited in vitro activity against *E. histolytica* and could be used as the first-line treatment against amoebiasis and other intestinal parasitic diseases in the future (Aslam and Musher 2007; Adagu et al. 2002). Nitazoxanide has broad-spectrum antiparasitic activity, against the protozoan's *E. histolytica*, *T. vaginalis*, *Cryptosporidium parvum*, *G. lamblia*, and *Isospora belli* and the helminths *Ascaris lumbricoides*, *Ancylostoma duodenale*, *Trichuris trichiura*, *Taenia saginata*, *Hymenolepis nana*, and *Fasciola hepatica* (Rossignol et al. 2007). Nitazoxanide is licensed in the United States for use in children as young as 12 months of age and is available as a pediatric suspension. The drug has few side effects and requires a shorter period of treatment. The molecular mechanisms of action, bioavailability, and drug interactions of nitazoxanide need to be further studied.

Satranidazole (Satrogyl) is an antiamoebic, antiprotozoal, and antibacterial agent, belonging to the nitroimidazole group of drugs (Fig. 2.1). It is a novel nitro-imidazole possessing a C-N linkage at C2 position of the imidazole ring. The nitro group on the imidazole ring is reduced by redox proteins present only in anaerobic organisms to the reactive nitro radical, which exerts its cytotoxic action by damaging DNA and other biomolecules. It is twice as effective as other nitroimidazoles. It is rapidly absorbed and exhibits higher plasma and liver concentration than MTZ. Satrogyl can be used to treat a wide range of anaerobic infections, such as trichomoniasis, giardiasis, periodontitis, liver abscesses, and intestinal and hepatic amoebiasis. The poor aqueous solubility of satranidazole results in low oral absorption and hence leads to low and variations in its bioavailability. This further gives rise to difficulties in the preparation of pharmaceutical formulations meant for oral and parenteral use. Therefore, the solubility of satranidazole needs to be explored further to improve its therapeutic efficacy and bioavailability. The main advantages of using satranidazole are better tolerability, no nausea, vomiting, or metallic taste, and absence of neurological and disulfiram-like reactions when used with alcohol. The generic drug satranidazole is manufactured in India by five pharmaceutical companies in the forms of tablet, capsule, syrup, cream, gel, ointment, liquid, or injection. Satrogyl and its combination with ofloxacin (Satrogyl-O) are available in the market as different brand names and are prescribed for the treatment of intestinal amoebiasis, intestinal infection, anaerobic infections, hepatic amoebiasis, trichomoniasis, and other conditions.

Some bioactive molecules from natural products that have shown significant in vitro antiamoebic activities are natural flavonoids like kaempferol, tiliroside, quercetin (Calzada et al. 1998), and alkaloids isolated from *Strychnos usambarensis*, usambarensine, and usambarine (Fig. 2.4) (Wright et al. 1991). Flavonoids (-)-epicatechin, (-)-epigallocatechin, and kaempferol (Fig. 2.4) were the most potent flavonoids against *Entamoeba*. Three bisbenzylisoquinoline alkaloids aromoline, isotrilobine, and insularine were also found to be active against *E. histolytica* (Fig. 2.4; Marshall et al. 1994). Carbamic acid derivative, ethyl-4-chlorophenylcarbamate, exhibited good activity against axenic culture of *E. histolytica* and reduced the development of amoebic abscess in hamsters (Fig. 2.5; Ordaz-Pichardo et al. 2005). Fatty acid derivatives such as alkylphosphocholines, oleyl-PC, octadecyl-PC, and nonadecenyl-PC exhibited the highest activity against two strains of *Entamoeba* SFL-3 and HM-1: IMSS with 50% concentrations for a 48 h treatment (Fig. 2.5; Seifert et al. 2001).

Another coumarinic acid derivative melilotoside isolated from *Teloxys graveolens* exhibited the most potent activity against *E. histolytica* (Fig. 2.5; Calzada et al. 2003a). Bisphosphonates (Ghosh et al. 2004), diphenyl bisamidine-liroldine (Venugopalan et al. 1996; Chatterjee et al. 1997), and benzyl glucosinolate isolated from the roots of *Lepidium virginicum* (Calzada et al. 2003b) showed in vitro activity against *E. histolytica* (Fig. 2.5; Azam and Agarwal 2007).

Some other compounds, such as cyclooctadiene ruthenium (II) complexes and copper (II) complexes of 2-nitrothiophene 2-carboxaldehyde thiosemicarbazones (**1**) (Singh et al. 2006, 2005b; Sharma et al. 2005a), copper (II) complexes of 5-nitrofuran 2-carboxaldehyde thiosemicarbazones (**2**) (Sharma et al. 2005b), quinoxaline derivatives of pyrazoline (**3**) (Budakoti et al. 2008; Abid and Azam 2006), Pd (II) complexes of 3-phenyl-2-pyrazoline thiosemicarbazones **(4), (5)** (Abid et al.

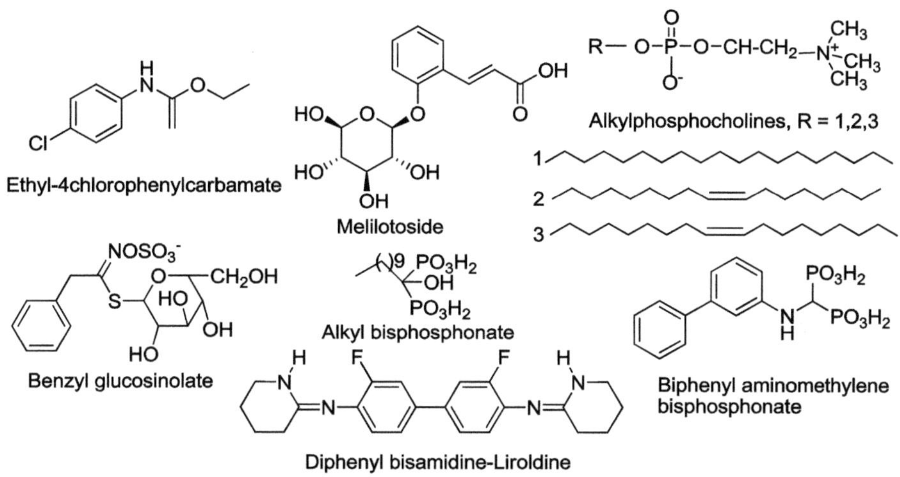

Fig. 2.5 Carbamic, fatty acid, coumarin, and bisphosphonate derivatives exhibiting in vitro antiamoebic activity

2009; Husain et al. 2008), bis-pyrazolines (**6**) (Bhat et al. 2009a), 3,5-bis-(2-chlorophenyl)-1,4,2-dioxazole (**7**) (Bhat et al. 2009b), triazines (**8**) (Singh et al. 2005a), and oxime ethers (**9**) (Abid et al. 2005) also showed promising in vitro activities against *E. histolytica* (Fig. 2.6).

A three-dimensional quantitative structure-activity relationship (3D-QSAR) and the comparative molecular field analysis (CoMFA) were performed on a set of 1H-benzimidazole derivatives (**10**) (Fig. 2.6) to study their activities against *E. histolytica* (Lopez-Vallejo et al. 2007). This was the first computational study aimed at exploring the structural requirements for the activity of benzimidazole derivatives and determining the tautomeric form that would probably fit a target receptor in *E. histolytica*. The antiamoebic activity was favored with steric bulk at position 5 of the benzimidazole ring and low electron density on the group at position 2. Compounds with a positive net charge had a high predicted biological activity (e.g., pIC_{50} >7.26), and compounds with a negative net charge had a low predicted biological activity (e.g., pIC_{50} <6.89). These CoMFA models would be very valuable in designing new and more potent compounds against *E. histolytica*.

The synthesis and antiamoebic activities of a wide range of molecules from different sources have been reviewed (Singh et al. 2009). Since then, several classes of heterocyclics with specific functional groups including azoles, hydrazones, chalcones, sulfonamides, pyrimidines, and metal complexes are being synthesized and were found to be potent amoebicidals. Chalcone bearing N-substituted ethanamine (**11**), MTZ-hydrazone conjugate (**12**), chloroquine-acetamide hybrid (**13**), N-acylhydrazone derived from substituted 4-piperazinyl-1-yl-quinoline (**14**), and MTZ-triazole hybrid (**15**) showed amoebicidal activity and were found to be more potent than MTZ in vitro. Hydrazone (**16**) and oxadiazoline (**17**) derivatives of 2-methyl-5-nitro-1H-imidazole were more potent against amoebiasis (Fig. 2.7)

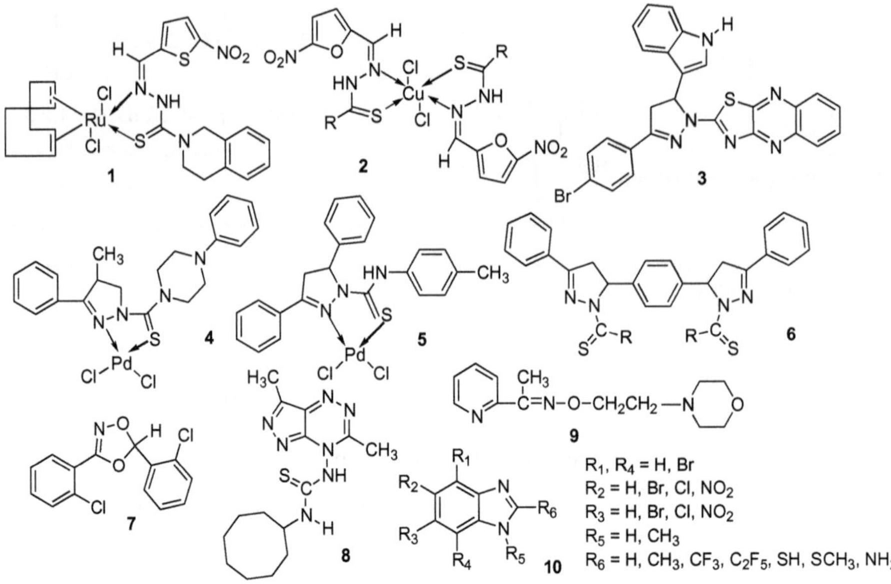

Fig. 2.6 Miscellaneous compounds and complexes showing potent in vitro antiamoebic activity

(Hayat et al. 2016; Azam et al. 2015). These compounds are effective therapeutic candidates and can be further explored for antiamoebic drug development.

The discovery of the amoebicidal activity of a US Food and Drug Administration (FDA)-approved drug, auranofin, for rheumatoid arthritis in 1985 (Fig. 2.7) offers a promising and completely new therapeutic option for the treatment of amoebiasis. Oral auranofin was effective both in vitro and in vivo against *E. histolytica*. It has been identified in a high-throughput drug screen as ten times more potent than MTZ against *E. histolytica*. The IC_{50} of auranofin against *E. histolytica* trophozoites was 2 μM. It had much higher cysticidal activity on *Entamoeba invadens* cysts than the standard amoebicide, MTZ, suggesting a most promising therapy for amoebiasis. Auranofin is an oral gold-containing compound with phosphine and thiol ligands in a linear arrangement, having IC_{50} significantly lower for *Entamoeba* and equivalent for *Giardia* to MTZ. Auranofin's antiparasitic activity is attributed to its monovalent gold molecule that readily inhibits the thioredoxin reductase, an antioxidant enzyme present in *E. histolytica*. The FDA has now given auranofin an orphan drug status for its use in the treatment of amoebiasis, which could serve as a potential new therapy to be advanced into phase IIa clinical studies more rapidly than other non-approved compounds (Debnath et al. 2012). Auranofin is prescribed for the treatment of rheumatoid arthritis and is being further investigated for potential therapeutic applications in a number of other diseases including cancer, neurodegenerative disorders, HIV/AIDS, parasitic infections, and bacterial infections (Roder and Thomson 2015).

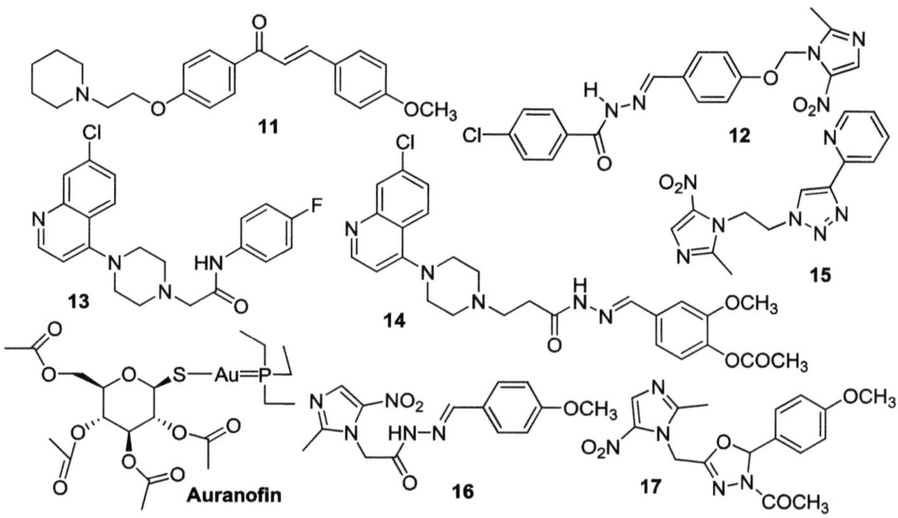

Fig. 2.7 Some heterocyclic compounds as potent amoebicidals

2.9 The Perspective

Amoebiasis is an important health problem in developing countries due to poor and adequate sanitation conditions. The basic approach to prevent amoebic infection is by improvement of living conditions and education in countries where invasive amoebiasis is prevalent. Various public health measures should be focused on (i) improving environmental sanitation including water supply and food safety, (ii) early detection and treatment of infections and/or disease, and (iii) health education. Preventive measures of improving personal hygiene are also vital for controlling the infections. Since transmission of the disease is mediated by cysts therefore, development of novel strategies to impair and control *E. histolytica* encystation is essential for achieving new preventive measures. A combination approach of interrupting the life cycle and eliminating the causative agent itself would be effective in controlling the spread of the disease.

The efficient therapy to combat drug resistance and methods for the treatment and prevention of the disease remains a major public health priority for the developing countries. One of the most important challenges presently facing the amoebiasis research program is to develop high-throughput, cost-effective, and highly predictive screening models. There is also an urgent need for the search of newer, safe, and efficacious chemotherapeutic agents for the diarrheal disease. The lack of new drugs and the resistance to different antimicrobial drugs have led to the search for active agents. A large number of synthesized heterocyclic scaffolds bearing different substituents have been identified as lead molecules exhibiting inhibitory activity on *Entamoeba* and are still being investigated. The major research work have to be focused on discovering the key biochemical pathways, identifying essential targets, in vivo testing models, and developing better assay methods that inhibit one or combination of these targets by a combination of molecules and to identify new chemotherapeutic agents. More studies with sensitive and specific diagnostic methods are needed to understand the relationship between HIV infection and *E. histolytica* infection in areas where both infections are prevalent. Thus development of new drugs, incorporation of new technologies into the diagnostic laboratory, development of improved control strategies, and preventive measures of hygiene achieved through national and international programs can help in controlling the spread of the disease.

References

Abid M, Azam A (2006) Synthesis, characterization and antiamoebic activity of 1-(thiazolo[4,5-b]quinoxaline-2-yl)-3-phenyl-2-pyrazoline derivatives. Bioorg Med Chem Lett 16:2812–2816

Abid M, Husain K, Azam A (2005) Synthesis and antiamoebic activity of new oxime ether derivatives containing 2-acetylpyridine/2-acetylfuran. Bioorg Med Chem Lett 15:4375–4379

Abid M, Bhat AR, Athar F, Azam A (2009) Synthesis, spectral studies and antiamoebic activity of new 1-N-substituted thiocarbamoyl-3-phenyl-2-pyrazolines. Eur J Med Chem 44:417–425

Adagu IS, Nolder D, Warhurst DC, Rossignol JF (2002) In vitro activity of nitazoxanide and related compounds against isolates of *Giardia intestinalis*, *Entamoeba histolytic*a and *Trichomonas vaginalis*. J Antimicrob Chemother 49:103–111

Ali V, Nozaki T (2007) Current therapeutics, their problems, and sulfur-containing-amino-acid metabolism as a novel target against infections by "amitochondriate" protozoan parasites. Clin Microbiol Rev 20:164–187

Aslam S, Musher DM (2007) Nitazoxanide: clinical studies of a broad-spectrum anti-infective agent. Future Microbiol 2:583–590

Athar F, Husain K, Abid M, Agarwal SM, Coles SJ, Hursthouse MB, Maurya MR, Azam A (2005) Synthesis and antiamoebic activity of gold (I), ruthenium (II) and copper (II) complexes of metronidazole. Chem Biodivers 2:1320–1330

Azam A, Agarwal SM (2007) Targeting Amoebiasis: status and developments. Current Bioactive compounds 3:121–133

Azam A, Peerzada MN, Ahmad K (2015) Parasitic diarrheal disease: drug development and targets. Front Microbiol 6:1–12

Bharti N, Shailendra, Coles SJ, Hursthouse MB, Mayer TA, Gonzalez Garza MT, Cruz-Vega DE, Mata-Cardenas BD, Naqvi F, Maurya MR, Azam A (2002) Synthesis, crystal structure and enhancement of the efficacy of metronidazole against *Entamoeba histolytica* by complexation with palladium (II), platinum (II) or copper (II). Helv Chim Acta 85:S2704–S2712

Bhat AR, Athar F, Azam A (2009a) Bis-pyrazolines: synthesis, characterization and antiamoebic activity as inhibitors of growth of *Entamoeba histolytica*. Eur J Med Chem 44:426–431

Bhat AR, Athar F, Azam A (2009b) New derivatives of 3,5-substituted-1,4,2-dioxazoles: synthesis and activity against *Entamoeba histolytica*. Eur J Med Chem 44:926–936

Budakoti A, Bhat AR, Athar F, Azam A (2008) Syntheses and evaluation of 3-(3-bromo phenyl)-5-phenyl-1-(thiazolo [4,5-b] quinoxaline-2-yl)-2-pyrazoline derivatives. Eur J Med Chem 43:1749–1757

Calzada F, Alanis AD, Meckes M, Tapia-Contreras A, Cedillo-Rivera R (1998) In vitro susceptibility of Entamoeba *histolytica* and *Giardia lamblia* to some medicinal plants used by the people of southern Mexico. Phytother Res 12:70–72

Calzada F, Velazquez C, Cedillo-Rivera R, Esquivel B (2003a) Antiprotozoal activity of the constituents of *Teloxys graveolens*. Phytother Res 17:731–732

Calzada F, Barbosa E, Cedillo-Rivera R (2003b) Antiamoebic activity of benzyl glucosinolate from *Lepidium virginicum*. Phytother Res 17:618–619

Campos-Rodriguezp R, Jarillo-Luna A (2005) The pathogenicity of *Entamoeba histolytica* is related to the capacity of evading innate immunity. Parasite Immunol 27:1–8

Chatterjee DK, Iyer SN, Venugopalan B, Raether W (1997) Antiamoebic activity of 3,3′-fluro-4,4′-di-(pyrrolidine-2-ylidene amino)-diphenyl (liroldine), against experimentally infected intestinal and hepatic amoebiasis. Indian J Exp Biol 35:765–770

Debnath A, Parsonage D, Andrade RM, He C, Cobo ER, Hirata K, Chen S, García-Rivera G, Orozco E, Martínez MB, Gunatilleke SS, Barrios AM, Arkin MR, Poole LB, McKerrow JH, Reed SL (2012) A high-throughput drug screen for *Entamoeba histolytica* identifies a new lead and target. Nat Med 18:956–962

Diamond LS (1961) Axenic cultivation of *Entamoeba histolytica*. Science 134:336–337

Ghosh S, Chan JM, Lea CR, Meints GA, Lewis JC, Tovian ZS, Flessner RM, Loftus TC, Bruchhaus I, Kendrick H, Croft SL, Kemp RG, Kobayashi S, Nozaki T, Oldfield E (2004) Effects of bisphosphonates on the growth of *Entamoeba histolytica* and Plasmodium species in vitro and in vivo. J Med Chem 47:175–187

Hayat F, Azam A, Shin D (2016) Recent progress on the discovery of antiamoebic agents. Bioorg Med Chem Lett 26:5149–5159

Heredia RD, Fonseca JA, López MC (2012) *Entamoeba moshkovskii* perspectives of a new agent to be considered in the diagnosis of amoebiasis. Acta Trop 123:139

Hung CC (2007) Amoebiasis: current status in Australia. Med J Aust 187:372

Hung CC, Dar-Der Ji HYS, Lee YT, Hsu SY, Chang SY, Wu CH, Chan YH, Hsiao CF, Liu WC, Colebunders R (2008) Increased risk for *Entamoeba histolytica* infection and invasive Amebiasis in HIV seropositive men who have sex with men in Taiwan. PLoS Negl Trop Dis 2:e175

Husain K, Abid M, Azam A (2008) Novel Pd (II) complexes of 1-N-substituted 3-phenyl-2-pyrazoline derivatives and evaluation of antiamoebic activity. Eur J Med Chem 43:393–403

Imperato PJ (1981) A historical overview of amebiasis. Bull N Y Acad Med 57:175–187

Legator MS, Connor TH, Stoeckel M (1975) Detection of mutagenic activity of metronidazole and niridazole in body fluids of humans and mice. Science 188:1118–1119

Leitsch D, Kolarich D, Wilson IB, Altmann F, Duchene M (2007) Nitroimidazole action in *Entamoeba histolytica*: a central role for thioredoxin reductase. PLoS Biol 5:e211

Lopez-Vallejo F, Medina-Franco JL, Hernandez-Campos A, Rodrıguez-Morales S, Yepez L, Cedillo R, Castillo R (2007) Molecular modeling of some 1H-benzimidazole derivatives with biological activity against *Entamoeba histolytica*: a comparative molecular field analysis study. Bioorg Med Chem 15:1117–1126

Marshall SJ, Russell PF, Wright CW, Anderson MM, Phillipson JD, Kirby GC, Warhurst DC, Schiff PL Jr (1994) In vitro antiplasmodial, antiamoebic, and cytotoxic activities of a series of bisbenzylisoquinoline alkaloids. Antimicrob Agents Chemother 38:96–103

Ordaz-Pichardo C, Shibayama M, Villa-Trevino S, Arriaga-Alba M, Angeles E, de la Garza M (2005) Antiamoebic and toxicity studies of a carbamic acid derivative and its therapeutic effect in a hamster model of hepatic amoebiasis. Antimicrob Agents Chemother 49:1160–1168

Orozco E, Lopez C, Gomez C, Perez DG, Marchat L, Banuelos C, Delgadillo DM (2002) Multidrug resistance in the protozoan parasite *Entamoeba histolytica*. Parasitol Int 51:353–359

Peterson KM, Singh U, Petri WA Jr (2011) Enteric Amoebiasis. In: Guerrant R, Walker DH, Weller PF (eds) Tropical infectious diseases: principles, pathogens and practice, 3rd edn. Saunders Elsevier, Philadelphia, p 614

Petri WA Jr (2003) Therapy of intestinal protozoa. Trends Parasitol 19:523–526

Roder C, Thomson MJ (2015) Auranofin: repurposing an old drug for a golden new age. Drugs in R & D 15:13–20

Rogers L (1912) The rapid cure of amoebic dysentery and hepatitis by hypodermic injections of soluble salts of emetine. Br Med J 1:14–24

Rossignol JF, Kabil SM, El-Gohary Y, Younis AM (2007) Nitazoxanide in the treatment of amoebiasis. Trans R Soc Trop Med Hyg 101:1025–1031

Rustia M, Shubik P (1972) Induction of lung tumors and malignant lymphomas in mice by metronidazole. J Natl Cancer Inst 48:721–729

Seifert K, Duchene M, Wernsdorfer WH, Kollaritsch H, Scheiner O, Wiedermann G, Hottkowitz T, Eibl H (2001) Effects of miltefosine and other alkylphosphocholines on human intestinal parasite *Entamoeba histolytica*. Antimicrob Agents Chemother 45:1505–1510

Sharma P, Sharma JD (2001) A review of plant species assessed in vitro for antiamoebic activity or both antiamoebic and antiplasmodial properties. Phytother Res 15:1–17

Sharma S, Athar F, Maurya MR, Azam A (2005a) Copper (II) complexes with substituted thiosemicarbazones of thiophene-2-carboxaldehyde: synthesis, characterization and antiamoebic activity against *E. histolytica*. Eur J Med Chem 40:1414–1419

Sharma S, Athar F, Maurya MR, Naqvi F, Azam A (2005b) Novel bidentate complexes of Cu (II) derived from 5-nitrofuran-2-carboxaldehyde thiosemicarbazones with antiamoebic activity against *Entamoeba histolytica*. Eur J Med Chem 40:557–562

Singh S, Husain K, Athar F, Azam A (2005a) Synthesis and antiamoebic activity of 3,7-dimethylpyrazolo[3,4-e][1,2,4]triazin-4-yl thiosemicarbazide derivatives. Eur J Med Chem 25:255–262

Singh S, Athar F, Azam A (2005b) Synthesis, spectral studies and in vitro assessment for antiamoebic activity of new cyclooctadiene ruthenium (II) complexes with 2-nitrothiophene-2-carboxaldehyde thiosemicarbazones. Bioorg Med Chem Lett 15:5424–5428

Singh S, Athar F, Maurya MR, Azam A (2006) Cyclooctadiene Ru (II) complexes of thiophene-2-carboxaldehyde-derived thiosemicarbazones: synthesis, characterization and antiamoebic activity. Eur J Med Chem 41:592–598

Singh S, Bharti N, Mohapatra PP (2009) Chemistry and biology of synthetic and naturally occurring Antiamoebic agents. Chem Rev 109:1900–1947

Stanley SL Jr (2003) Amoebiasis. Lancet 361:1025–1034

Stanley SL Jr (2006) Vaccines for amoebiasis: barriers and opportunities. Parasitology 133:81–86

Venugopalan B, Patel B, Karnik PJ, de Souza NJ, Chatterjee DK, Iyer N (1996) Synthesis of diphenyl bisamidines as potential amoebicides. Eur J Med Chem 1:485–488

Wassermann C, Hellberg A, Tannich E, Bruchhaus I (1999) Metronidazole resistance in the protozoan parasite *Entamoeba histolytica* is associated with increased expression of iron-containing superoxide dismutase and peroxiredoxin and decreased expression of ferredoxin 1 and flavin reductase. J Biol Chem 274:26051–26056

Wright CW, Phillipson JD (1990) Natural products and the development of selective antiprotozoal drugs. Phytother Res 4:127–139

Wright CW, Bray DH, O'Neill MJ, Warhurst DC, Phillipson JD, Quetin Leclercq J, Angenot L (1991) Antiamoebic and antiplasmodial activities of alkaloids isolated from *Strychnos usambarensis*. Planta Med 57:337–340

Chapter 3
Cerebral Malaria: Players in the Pathogenic Mechanism and Treatment Strategies

Hemlata Dwivedi and Renu Tripathi

Abstract Cerebral malaria (CM) is a major life-threatening disease caused by *Plasmodium falciparum* infection in humans. The complex pathogenic mechanisms underlying the fatal neurological complications of the disease are still not completely elucidated. The autopsy studies in fatal cases of human CM and advances in knowledge from various animal models have offered insight into the precise mechanism of the disease. The parasite sequestration in the brain microvascular endothelial cells and dysregulated host immune system together determine the pathophysiology of CM. Despite optimal treatment with antimalarials, 25% of the patients suffer from post-treatment neurological and cognitive deficits. In this review, we have discussed the components of the pathogenic mechanisms of CM and the current scenario of treatment.

Keywords Cerebral malaria · Blood-brain barrier · Cell adhesion molecules · Monocytes · Adjunct treatment · Antimalarial

3.1 Introduction

Malaria is a massive burden of morbidity and mortality, particularly to low- and middle-income countries. Asia ranks second to Africa in terms of malaria burden. The disease is known to be caused by the five species of *Plasmodium*, viz., *P. vivax*, *P. ovale*, *P. malariae*, *P. falciparum* and *P. knowlesi*. *P. falciparum* is the causative agent of the most severe form of malaria accounting for most of the mortalities. Plasmodium has a complex life cycle involving vertebrate and invertebrate host (female anopheles mosquito). It undergoes sexual stages in mosquito and traverses through liver stage in human before entering into the clinically active blood stage (Fig. 3.1). Sporozoite development takes place in the mosquito, and sporozoite transfer to human host occurs during the mosquito blood meal. Once in the human

H. Dwivedi · R. Tripathi (✉)
Division of Parasitology, CSIR-Central Drug Research Institute, Lucknow, India

Academy of Scientific and Innovative Research, New Delhi, India

P. P. Singh (ed.), *Infectious Diseases and Your Health*,
https://doi.org/10.1007/978-981-13-1577-0_3

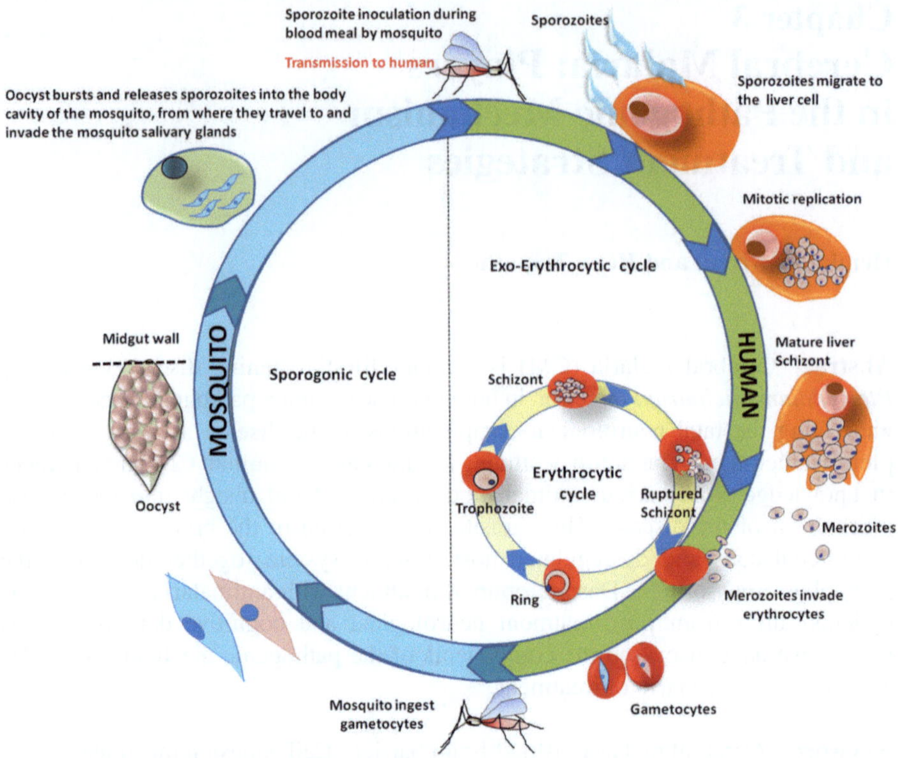

Fig. 3.1 Life cycle of *Plasmodium* spp.

bloodstream, the sporozoites travel rapidly to the liver and invade hepatocytes where they undergo exoerythrocytic schizogony. The infected hepatocytes rupture and release thousands of merozoites into the bloodstream. The merozoites then invade erythrocytes where they again develop into schizonts and rupturing schizonts release merozoites. The merozoites invade new erythrocytes to perpetuate the erythrocytic cycle, the stage of the parasite life cycle responsible for clinical symptoms. A small proportion of the merozoites differentiate into gametocytes ready to be taken up by the biting mosquito to complete the life cycle. The clinical features of the disease may range from mild symptoms including fever, shivering, excessive sweating, headache, vomiting etc. to the severe forms marked by deep anaemia, respiratory distress syndrome and acidosis, multiple organ failure or cerebral malaria.

3.2 The Burden of Cerebral Malaria

Falciparum malaria continues to be the massive burden of death and disease in tropical and sub-tropical countries. Cerebral malaria, a severe manifestation of *P. falciparum*, is characterized frequently by coma, fever, seizures and other neurological impairments (Fig. 3.1) (Warrell 1997). Severe cognitive deficit, behaviour disorder, loss of speech and hearing ability, spastic motor weakness and sensory disorders in CM survivors are collectively referred to as postmalaria neurological sequelae. The prevalence and risk of CM are greater in hyperendemic areas than in low-endemic areas (Reyburn et al. 2005). The intense malaria transmission in sub-Saharan Africa leads to natural acquisition of immunity during childhood. Thus, CM is uncommon in adults, but younger children under 5 are more vulnerable to the malaria infection. In Southeast Asia, malaria transmission is restricted to specific foci and not adequately intense enough to induce robust immunity in children. Hence, in this region CM predominantly occurs in grown-up children and adults (WHO 2015). The most recent estimates of CM mortality are 18–25% in pediatric populations in Africa and 30% in mainly adult populations from Southeast Asia (Seydel et al. 2015). Around 2–10% of CM survivors exhibit neurological complications lasting for more than 6 months after successful antimalarial treatment (Boivin et al. 2007). In an endeavour to better understand the factors leading to the progression of CM and subsequently improve the outcome for affected patients, a better understanding of the pathogenesis is crucial.

3.3 The Clinical Syndrome "Cerebral Malaria"

Cerebral malaria can be described as a potentially reversible clinical syndrome with unarousable coma (Glasgow coma score of 11/15 or less), diffuse encephalopathy often associated with convulsions, presence of asexual forms of *P. falciparum* on peripheral blood smears and absence of any other factors attributable to unconsciousness (Combes et al. 2010). CM, if untreated, can initiate vital organ dysfunction which can differ significantly in the African children (Fig. 3.2) and Southeast Asian adults (Wassmer et al. 2015). In adults, central nervous system dysfunction in concurrence with failure of renal and respiratory organ systems is frequent (Fig. 3.3), whereas neurological manifestation with rapid onset of coma, anaemia and seizures is frequent in children, but overt respiratory or renal compromise is generally absent (Miller et al. 2013).

3.4 Pathophysiology of Cerebral Malaria

Despite decades of research on cerebral malaria, the pathogenesis of the disease still remains a complex puzzle. There is still paucity of knowledge over why some people develop cerebral malaria, while others survive the peril of this disease despite being infected with same parasite, i.e. *P. falciparum*. Host factors and parasite interaction with host play the key role in the manifestation of the disease. Several hypotheses have been raised to explain the pathophysiology of CM. However, none of them can, per se, explain the pathogenesis. Until now there are three schools of

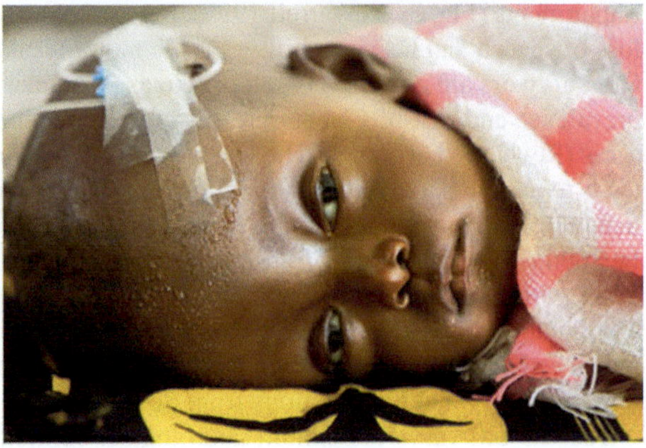

Fig. 3.2 A child suffering from CM. (Source: www.medimoon.com)

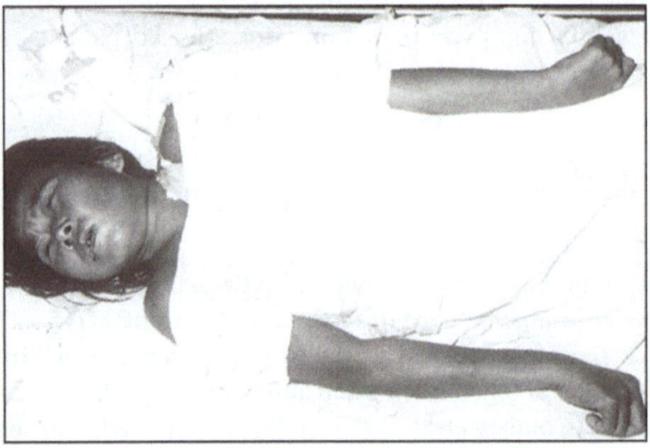

Fig. 3.3 Decorticate rigidity in a woman suffering from CM. (Source: http://www.nzdl.org)

thoughts to explain CM pathogenesis: the sequestration (or mechanical), the inflammation and the haemostasis hypothesis. However, the intricate interplay between them is responsible for the complexity of the disease (Sahu et al. 2015).

3.4.1 Sequestration Hypothesis

The adhesion of plasmodium-infected RBCs to endothelial cells of postcapillary venules as a mechanism to evade host immune response and splenic clearance is termed as sequestration. Sequestration happens in various organs, including the brain, liver, lung, heart, kidney, subcutaneous tissue and placenta, via parasite-derived proteins present onto the surface of red blood cells (Baruch 1999; Seydel et al. 2006). The parasite-derived multivariate *P. falciparum* erythrocyte membrane protein-1 (PfEMP-1) proteins that are expressed on the surface of iRBC bind to the endothelial cells via CD36, intercellular adhesion molecule-1 (ICAM-1), P-selectin and vascular adhesion molecule-1, E-selectin and chondroitin sulphate A (Ho and White 1999). Several post-mortem studies state the preferential binding of iRBCs in the cerebral vessels of CM patients and petechial haemorrhage as the hallmark features of human cerebral malaria (Fig. 3.5a, b) (Pongponratn et al. 2003; Sein et al. 1993). The contact point for parasitized RBCs to the endothelial cell surface receptor is a knob-like structure which are protrusions on the surface of *P. falciparum* pRBC caused by deposition of a knob-associated Histidine-rich protein (KAHRP) at the cytoplasmic side of the pRBC membrane (Howard et al. 1990), and containing several other parasite proteins, including PfEMP-1, RESA and MESA (Craig and Scherf 2001). More than 16 h after they infect red blood cells, malaria parasites secrete and export proteins onto the surface of pRBC and produce knobs, which contain a group of proteins including RIFIN, STEVOR, PfMC-2TM, PfEMP-1 and SURFIN. PfEMP-1 is thought to play a significant role in malaria pathogenesis through attachment to host endothelium (Crabb et al. 1997). Platelets and leukocytes also contribute to cytoadhesion in CM. Leukocytes (primarily macrophages/monocytes) interact with an inflamed/activated endothelium via binding to CAMs. Leukocytes are not as frequently seen as iRBCs but nevertheless present, and the presence has been associated with CM (Armah et al. 2005a, b). The microvascular congestion due to sequestration of pRBCs, platelets and leukocytes leads to severe endothelial damage causing breakdown of blood-brain barrier (Dorovini-Zis et al. 2011; Ponsford et al. 2012).

3.4.2 Inflammation Hypothesis

The immune response plays an essential role in curbing infections, but the imbalance in the delicate equilibrium of immune mediators may worsen the condition. During CM, according to the inflammation hypothesis, the dysregulated immune response leads to multi-organ failure and death. The inflammatory signaling cascade is multifaceted comprising myriad troupe of the innate and adaptive immune system which either sequentially or simultaneously acts in a coordinated fashion to deal with the disease. The chain of immune response events is still far from being completely elucidated in malaria progression. Studies have been focused on identifying the crucial regulators of the inflammatory state responsible for CM pathology. As soon as the schizont ruptures, parasite toxins release into the blood. Monocyte and macrophages are activated via toll-like receptor resulting in secretion of pro-inflammatory cytokines including TNF-α, IFN-γ, IL-1, IL-6, nitric oxide and LT-α which further recruits immune cells such as CD4+ and CD8+ cells (Gazzinelli and Denkers 2006; Martins et al. 2009). Though the immune response brings down parasitaemia and removes toxins like free heme, an overwhelming pro-inflammatory response leads to endothelial activation and worsens the CM outcome. The current trend, however, considers unified hypothesis of sequestration and inflammation to be responsible for CM pathogenesis (Fig. 3.4).

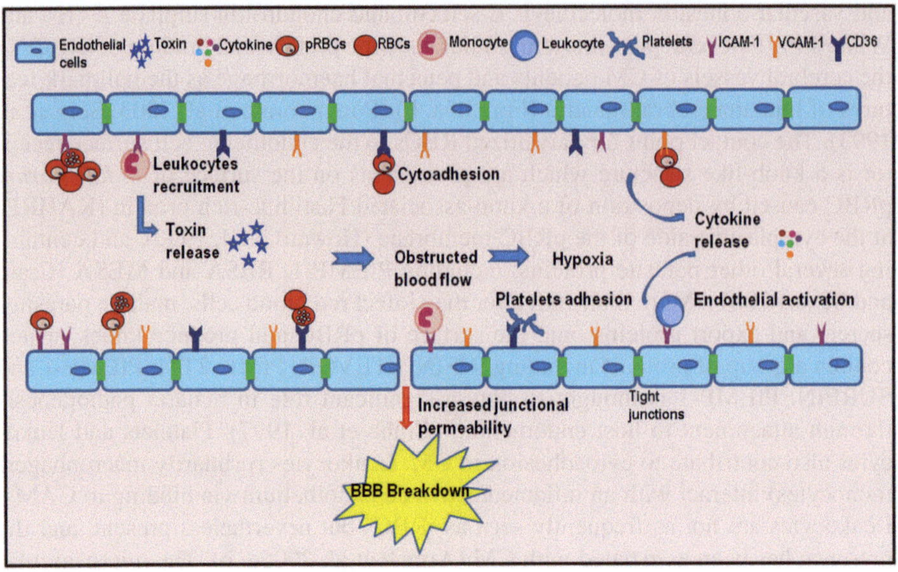

Fig. 3.4 Proposed mechanism of pathogenesis and BBB dysfunction in CM

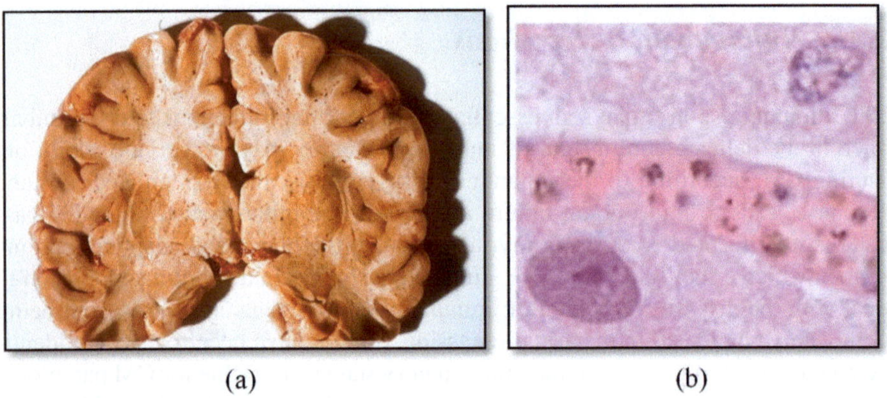

Fig. 3.5 Brain section of cerebral malaria patient showing (**a**) numerous petechiae and (**b**) obstructed blood vessel. (Source: http://itg.author-e.eu)

3.5 Players in the Pathogenesis of Cerebral Malaria

3.5.1 Blood-Brain Barrier and Parasitized RBCs

The BBB is a highly specialized structural and functional interface between the intraerythrocytic stages of *P. falciparum* and the human host. The sequestration of pRBCs in brain microvasculature through adhesion between pRBCs and endothelial cells (ECs) lining blood vessels is one of the salient feature of CM enabling the parasites to avoid splenic clearance. The adhesion is implicated to a diverse array of surface receptors expressed on ECs including intercellular adhesion molecule 1 (ICAM-1), CD36, platelet/EC adhesion molecule)/CD31 (PECAM-1), E-selectin, P-selectin and vascular cell adhesion molecule-1 (VCAM-1). The sequestration results in a cascade of intracellular signaling events that disrupt the tight junctions and increases the BBB permeability, focal disruptions at sites of PRBCs and localized hypoxia and haemorrhaging of surrounding brain parenchyma.

3.5.2 Cell Adhesion Receptors

ICAM1 ICAM1 (CD54) is a member of the immunoglobulin superfamily expressed on all cells contributing to CM development including lymphocytes, myeloid cells, platelets and endothelial cells. The interaction between ICAM-1 and pRBCs is through a distinct pair of domains found only in a subset of PfEMP1 variants (DBLb-C2 domains) (Chattopadhyay et al. 2004; Springer et al. 2004). The pRBC-binding site on ICAM1 has been mapped and is distinct from the binding site used by its natural ligand LFA-1 (lymphocyte function-associated antigen 1) (Ockenhouse et al. 1992; Tse et al. 2004). ICAM1 is known to synergise with CD36 to enhance static adhesion (Yipp et al. 2007). The isolates from patients with clinical malaria (severe and uncomplicated) have higher preference for adhesion to ICAM-1, but there exists no statistically significant correlation between ICAM1 binding and severe malaria in Africa (Heddini et al. 2001). In Asia, however, some studies show that ICAM1 binding is not necessarily associated with severe malaria in field isolate. ICAM1 is widely upregulated on microvascular endothelial cells in the presence of pro-inflammatory cytokines, such as tumour necrosis factor-α (TNF-α), which is raised significantly high in patients with severe malaria suggesting ICAM1-mediated cytoadherence to be the potential contributor during severe malaria.

CD36 CD36 (cluster of differentiation 36) is an 88-kDa cell surface class B scavenger receptor expressed on platelets, endothelial and epithelial cells, erythrocyte precursors, macrophages, monocytes and adipocytes making possible for the parasite to adhere at different organ sites. CD36 has been implicated in many diverse biological processes including angiogenesis, phagocytosis, inflammation and fatty acid metabolism. CD36 binding is a property of almost all *P. falciparum* isolates derived from malaria patients (Newbold et al. 1997a). The CD36 receptor binds to PfEMP-1 through cysteine-rich interdomain region (CIDR) domains (Hsieh et al. 2016).

Vascular Cell Adhesion Molecule-1 (VCAM-1) It is a member of the Ig superfamily of adhesion receptors and is upregulated on endothelium in response to inflammatory mediators. The field isolates from Thailand tether and roll on VCAM1, but no static adhesion was observed (Udomsangpetch et al. 1997). In African isolates, VCAM1 binding was extremely low and not associated with disease severity (Newbold et al. 1997b).

NCAM NCAM adhesion receptor has shown in vitro ability to adhere pRBC. It is known to be expressed on endothelium in two sites where sequestration is commonly observed in severe disease, skin and brain, tempting speculation that it may be an important receptor in severe disease (Pouvelle et al. 2007).

P-Selectin and E-Selectin Selectins are characterized by N-terminal domain of 117–120 amino acids that is homologous to Ca^{2+}-dependent animal lectins, and P-selectin has been shown to be involved in pRBC cytoadherence (Udomsangpetch et al. 1991). The binding site for pRBC on P-selectin is thought to be on the lectin domain but distinct from that of a natural ligand PSGL-1 (Ho et al. 1998). E-selectin is another member of the selectin family that has been suggested to be able to be involved in adhesion (Ockenhouse et al. 1992).

Platelets The vascular endothelium, red cells and platelets and the interactions of these three components play a key role in the pathogenesis of severe malaria. Thrombocytopenia driven by platelet activation is an essential feature of all forms of malaria. Increased P-selectin levels, platelet microaggregates (Scott et al. 2002), enhanced platelet activation and increased levels of platelet factor 3 (Emuchay and Usanga 1997) were identified in patients with malaria.

3.5.3 Monocytes/Macrophages

The monocytes and macrophages play an important role in protection against malaria by phagocytosis of IEs, antibody-dependent cell inhibition and regulation of cytokine production. Leukocyte sequestration, together with IE accumulation, can lead to mechanical blockage of brain microvasculature, and further downstream activation of host immune response augments the pathology. The platelet factor-4 (PF4), released from activated platelets, can facilitate the recruitment and retention of monocytes/macrophages in the brain vasculature and cause their excessive activation, contributing to CM pathology (Srivastava et al. 2010)

3.5.4 *Cytokines*

The complex malaria pathogenesis most likely entails both immunologic and non-immunologic mechanisms. The mounting evidence from experimental cerebral malaria convincingly ascertains the important role of inflammatory processes in the development of CM (Brian de Souza et al. 2002). The best-documented pro-inflammatory cytokines relevant for the development of CM are IFN-γ, TNF-α and IL-12. In some reports, the local production of TNF-α, IFN-γ, IL-1 and IL-10 is associated with hallmark features of human CM as determined by histology (Maneerat et al. 1999). Similar observations were made in primate model of CM (Tongren et al. 2000). However, some reports suggest that the mRNA of pro-inflammatory (TNF-α and IL-1β) cytokines, detected in post-mortem samples of patients suffering CM, does not correlate with the density of parasitized erythrocytes (Brown et al. 1999). In addition, a focal accumulation of TGF-β1, β 2 and β 3 during reorganization of brain parenchyma was found in patients with CM, suggestive of endothelial activation and immunologic dysfunction (Deininger et al. 2000).

3.6 Diagnosis of Cerebral Malaria

The initial symptom of cerebral malaria is usually fever (37.5–41 °C), vomiting and cough. The history of symptoms preceding coma may be as brief as 1 or 2 days. The severity of coma can be assessed according to the coma scale for children (Blantyre coma scale). Seizures are common before or after the onset of coma and are significantly associated with morbidity and sequelae. Important signs include intermittent nystagmus, salivation, minor twitching of a single digit or a corner of the mouth, an irregular breathing pattern and sluggish pupillary light reflexes. In children in profound coma, corneal reflexes and "doll's eyes" movements may be abnormal. The depth of coma in adult cerebral malaria patients is assessed by Glasgow coma scale. Convulsions, retinal changes, fixed jaw closure, tooth grinding (bruxism) and transient abnormalities of eye movement especially dysconjugate gaze are common.

3.7 Treatment of Cerebral Malaria

3.7.1 *Antimalarial Treatment*

Antimalarial drugs are the only interventions that unequivocally reduce mortality in patients. The cinchoids (quinine and quinidine) and artemisinin derivatives are most commonly used. To rapidly achieve the parasiticidal level of antimalarial drug for treatment of cerebral malaria, a loading dose is preferred. Quinine can induce hypoglycaemia by promoting insulin secretion. It kills parasites in the late stages of the

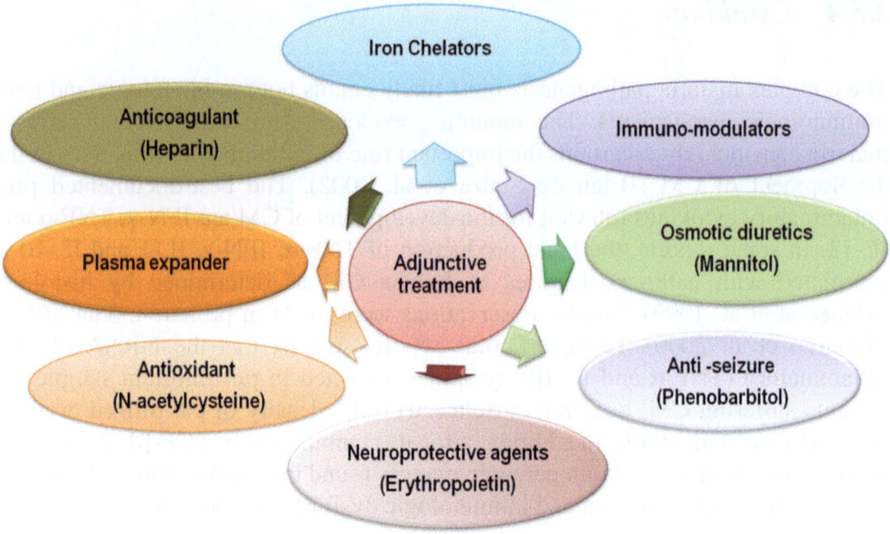

Fig. 3.6 Adjunct treatment of cerebral malaria clinically tested so far

erythrocytic cycle, thus having a narrow therapeutic window. The side effects of quinine treatment include complications such as dizziness and tinnitus after the patient regains consciousness. Quinine causes hypotension if given rapidly via the intravenous route and also slows ventricular repolarization. The artemisinin derivatives, artemether, arteether and artesunate, are the most rapidly acting and potent antimalarial compounds that kill all the parasitic stages. Artemisinin derivatives can be administered via the intramuscular route and have few local or systemic adverse effects.

3.8 Adjunct Therapy for Cerebral Malaria Treatment

Cerebral malaria mortality remains unacceptably high despite anti-plasmodial treatment. Therefore, adjunctive therapy for severe manifestations and complications is needed. Several agents had been studied over the past several years (Fig. 3.6).

3.9 Immunomodulators

3.9.1 Corticosteroids

Corticosteroids are anti-inflammatory agents that reduce intracranial pressure and curb the dysregulated immune response. It is also known to improve the integrity of the blood-brain barrier. These have been used as a standard treatment for increased intracranial pressure arising of vasogenic oedema from tumours or abscesses. These are the first compounds to be tested as adjunctive agents in randomized control trials of CM. Two independent trials of dexamethasone at different doses did not decrease mortality, but in one of the trials, dexamethasone was associated with increased complications including gastrointestinal bleeding, sepsis and prolonged coma resolution (Rampengan 1991).

3.9.2 Intravenous Immunoglobulin

The in vitro cytoadherence of *Plasmodium falciparum*-infected RBCs to melanoma cells is inhibited and reversed upon treatment with hyperimmune globulin (David et al. 1983). However, treatment with intravenous immunoglobulin as an adjunct to standard treatment for cerebral malaria in Malawian children resulted in increased mortality and neurologic sequelae (Taylor et al. 1992).

3.9.3 Anti-TNF-α Monoclonal Antibody

The pro-inflammatory cytokine TNF is extensively evaluated in cerebral malaria, and its inhibition attempted by administration of anti-TNF monoclonal antibody to children with cerebral malaria caused inhibition of fever but did not improve the survival in CM and was associated with excess neurological sequelae (Van Hensbroek et al. 1996; Looareesuwan et al. 1999).

3.9.4 Pentoxifylline

Pentoxifylline (PTX), a phosphodiesterase inhibitor, exhibits different beneficial effects including improved cerebral blood flow, increased RBC deformability, decreased platelet aggregation and decreased chemotactic movement and inhibition of pro-inflammatory cytokines. The clinical trials of PTX as adjunct CM treatment have yielded mixed results. In two clinical trials, PTX-treated group showed decreased coma resolution time and significant reduction in mortality, whereas in a small-scale study in Africa, the PTX group showed unexpectedly high mortality (Das et al. 2003; Lell et al. 2010). In another study conducted on complicated falciparum malaria patients in Thailand, administration of 20 and 40 mg/kg of PTX to the standard antimalarial regime did not produce any additional clinical benefit (Looareesuwan et al. 1998). The clinical data are intriguing, and a large-scale trial in patients with cerebral malaria is needed to reach to a conclusion.

3.10 Osmotic Agents

3.10.1 Mannitol

Osmotic agents lower elevated intracranial pressure and increase cerebral perfusion. The osmotic gradient build up by these agents drains water from brain parenchyma into the brain capillaries. Mannitol is an osmotic diuretic used extensively in clinical practice. The brain volume increases due to certain factors such as cytotoxic or vasogenic oedema, leading to elevated intracranial pressure which is promptly reduced upon administration of mannitol. A clinical trial of mannitol versus placebo in Ugandan children with CM exhibited no difference in mortality, coma duration or any other clinical outcome in either of the treated groups (Namutangula et al. 2007). In a recent study in India, adult cerebral malaria patients with brain swelling on CT scan were randomized to adjunctive treatment with intravenous mannitol or no adjunctive therapy. Mannitol-treated group showed prolonged coma recovery and 30% mortality vs 13% in without mannitol treatment (Mohanty et al. 2011).

3.11 Iron Chelators

3.11.1 Desferrioxamine

An iron chelator like desferrioxamine exerts its antimalarial property by withholding iron from being used by parasite for its metabolism and inhibits iron-induced free radical damage to cells and subcellular structures, which can render beneficial effect for patients with severe falciparum malaria (Gordeuk and Loyevsky 2002). A study undertaken in India showed that group of patients administered with oral deferiprone as supportive therapy showed improvement in the clinical course and final outcome (Mohanty et al. 2002). The drug was well tolerated and had no side effects. The parameters evaluated included the parasitaemia, fever and coma resolution. A randomized, double-blind, placebo-controlled trial of the iron chelator deferoxamine was conducted in Zambian children with cerebral malaria. Deferoxamine (100 mg/kg body weight/day, infused intravenously for 72 h) or placebo was administered as ancillary therapy along with the standard drugs quinine and sulphadoxine/pyrimethamine. The study concluded that iron chelation speeds up parasite clearance from the blood and augments recovery from deep coma in cerebral malaria (Gordeuk et al. 1992). However, another multicentre large-scale clinical trial of deferoxamine in Zambia showed a trend towards increased mortality with deferoxamine in one of the study sites (Thuma et al. 1998). There is insufficient data for any conclusions for both agents tested. There are non-significant trends towards harm (death) and potential benefit (fewer seizures) with DFO. With deferiprone, results suggest possible benefit (shorter coma recovery and parasite clearance). More rigorous trials with careful assessment of the adverse effects of iron chelators are needed to completely delineate the effect on clinical outcomes.

3.12 Oxidative Stress Reducing Agent

3.12.1 N-Acetyl Cysteine (NAC)

The increased oxidative stress is one of the contributors to malaria-related morbidity (Griffiths et al. 2001). N-acetyl cysteine is a well-tolerated antioxidant effective in reversing lack of deformability of parasitized as well as non-parasitized RBCs. It is the main treatment for acute paracetamol poisoning, acting through direct scavenging of free radicals and replenishment of glutathione and cysteine. In Thailand, a placebo-controlled trial was conducted with NAC which yielded positive results. The serum lactate level increased during malaria normalized to basal level sooner (twice as quicker) after NAC treatment as compared to placebo. There were no adverse effects of NAC (Watt et al. 2002). However, a succeeding larger study showed no difference in coma recovery, lactate clearance or mortality between groups, and parasite clearance time was slightly prolonged in those who received NAC (Charunwatthana et al. 2009).

3.13 Anticoagulant

3.13.1 Heparin

Rosetting of parasitized RBCs, a phenomenon observed in severe malaria, is related to pathogenesis of CM, and heparin, an anticoagulant, prevents the formation of rosettes of infected red blood cells (Rogerson et al. 1994). Heparin is a highly charged natural polymer which has abundant affinity for parasite proteins particularly MSP-1 and known to inhibit parasite egress (Glushakova et al. 2017). In a randomized study of falciparum malaria patients, patients receiving either heparin or aspirin showed no difference in mortality or any other outcome with either treatment (John et al. 2010). Heparin-induced bleeding is a real threat preventing in undertaking any large randomized trials in patients with severe malaria in whom incipient bleeding diathesis exists.

3.14 Exchange Blood Transfusion (EBT)

High blood parasitaemia in severe malaria is often associated with increased morbidity and mortality. Exchange blood transfusion in addition to antimalarial chemotherapy rapidly reduces high malaria parasitaemia, as it replaces blood with infected RBCs for blood with healthy RBCs. The transfused RBCs are uninfected and have normal deformability; therefore, EBT is considered to reduce antigenic stimulus, pro-inflammatory responses, RBC clearance and haemolysis. In the opinion of experts, if adequate amenities are accessible, then for very sick patients with parasitaemia >15%, EBT can be considered as an adjunct treatment (White 1996). EBT demands technical expertise and intense monitoring, which confines its suitability in resource-limited settings in developing countries. The clinical outcome of EBT in malaria has been diverse. A meta-analysis of studies in which individuals who received EBT had higher baseline parasitaemia and disease severity compared to those who did not receive EBT. However, there was no difference in mortality or other outcomes in both the groups (Riddle et al. 2002). A study conducted by Loutan et al. has reported three cases of severe falciparum malaria successfully treated by intravenous quinine and exchange transfusion. Though a speedy drop in parasitaemia was observed, serum levels of TNF remained markedly elevated during the first 48 h despite exchange transfusion (Loutan et al. 1992).

3.15 Plasma Expander (Albumin)

Plasma expanders such as albumin can improve microcirculation, correct hypoglycaemia and reduce lactic acidosis. In a study conducted in Kenya, the children with severe malaria who received albumin showed 3.6% mortality compared to 18% in those receiving saline (Maitland et al. 2005).

3.16 Neuroprotective Agent

3.16.1 Erythropoietin

Erythropoietin (EPO) downregulates pro-inflammatory response and possesses antioxidant and anti-apoptotic properties which merit it as a probable adjuvant therapy for CM. In a study, African children with high levels of EPO during a malaria episode exhibited improved clinical outcome than their counterparts with lower levels, suggesting that EPO provided some form of neuroprotection (Casals et al. 2008). In a recent study in India, a significant increase in circulating EPO during fatal CM calls for optimization of adjunctive treatments according to the targeted population (Dalko et al. 2016).

3.17 Conclusion

Malaria has plagued humanity since ancient times and is still one of the major causes of mortality and morbidity around the world. Cerebral malaria, the severe manifestation of falciparum infection, has the highest rates of complications and mortality despite treatment with effective antimalarial drugs. In falciparum malaria, 10% of all admissions and 80% of deaths are due to the central nervous system involvement. A number of agents have been or are being evaluated against animal model and in human trials. Adjunctive therapies that have been tested in patients with CM or severe malaria were disappointing with a very little success rate. Despite decades of intense research on cerebral malaria, the pathogenesis of the disease still remains unclear. Therefore, to gain insight into pathogenesis of CM and explore new adjunct treatment, a relevant model is the prerequisite. Besides, a careful interpolation of animal studies is needed when translating it to human.

References

Armah H, Dodoo AK, Wiredu EK, Stiles JK, Adjei AA, Gyasi RK et al (2005a) High-level cerebellar expression of cytokines and adhesion molecules in fatal, paediatric, cerebral malaria. Ann Trop Med Parasitol 99:629–647

Armah H, Wired EK, Dodoo AK, Adjei AA, Tettey Y, Gyasi R (2005b) Cytokines and adhesion molecules expression in the brain in human cerebral malaria. Int J Environ Res Public Health 2:123–131

Baruch DI (1999) Adhesive receptors on malaria-parasitized red cells. Best Pract Res Clin Haematol 12:747–761

Boivin MJ, Bangirana P, Byarugaba J, Opoka RO, Idro R, Jurek AM et al (2007) Cognitive impairment after cerebral malaria in children: a prospective study. Pediatrics 119:e360–e366

Brown H, Turner G, Rogerson S, Tembo M, Mwenechanya J, Molyneux M et al (1999) Cytokine expression in the brain in human cerebral malaria. J Infect Dis 180(5):1742–1746

Casals-Pascual C, Idro R, Gicheru N, Gwer S, Kitsao B, Gitau E et al (2008) High levels of erythropoietin are associated with protection against neurological sequelae in African children with cerebral malaria. Proc Natl Acad Sci U S A 105(7):2634–2639

Charunwatthana P, Abul Faiz M, Ruangveerayut R et al (2009) N-acetylcysteine as adjunctive treatment in severe malaria: a randomized, double-blinded placebo-controlled clinical trial. Crit Care Med 37:516–522

Chattopadhyay R, Taneja T, Chakrabarti K, Pillai CR, Chitnis CE (2004) Molecular analysis of the cytoadherence phenotype of a Plasmodium falciparum field isolate that binds intercellular adhesion molecule-1. Mol Biochem Parasitol 133(2):255–265

Combes V, El-Assaad F, Faille D, Jambou R, Hunt NH, Grau GE (2010) Microvesiculation and cell interactions at the brain-endothelial interface in cerebral malaria pathogenesis. Prog Neurobiol 91:140–150

Crabb BS, Cooke BM, Reeder JC, Waller RF, Caruana SR, Davern KM et al (1997) Targeted gene disruption shows that knobs enable malaria-infected red cells to cytoadhere under physiological shear stress. Cell 89:287–296

Craig A, Scherf A (2001) Molecules on the surface of the *Plasmodium falciparum* infected erythrocyte and their role in malaria pathogenesis and immune evasion. Mol Biochem Parasitol 115:129–143

Dalko E, Tchitchek N, Pays L, Herbert F, Cazenave PA, Ravindran B, Sharma S, Nataf S, Das B, Pied S (2016) Erythropoietin levels increase during cerebral malaria and correlate with heme, Interleukin-10 and tumor necrosis factor-alpha in India. PLoS One 11:e0158420

Das BK, Mishra S, Padhi PK, Manish R, Tripathy R, Sahoo PK et al (2003) Pentoxifylline adjunct improves prognosis of human cerebral malaria in adults. Trop Med Int Health 8:680–684

David PH, Hommel M, Miller LH, Udeinya IJ, Oligino LD (1983) Parasite sequestration in Plasmodium falciparum malaria: spleen and antibody modulation of cytoaderence of infected erythrocytes. Proc Natl Acad Sci U S A 80(16):5075–5079

Deininger MH, Kremsner PG, Meyermann R, Schluesener HJ (2000) Differential cellular accumulation of transforming growth factor-beta1, –beta2, and -beta3 in brains of patients who died with cerebral malaria. J Infect Dis 181(6):2111–2115
de Souza JB, Riley EM (2002) Cerebral malaria: the contribution of studies in animal models to our understanding of immunopathogenesis. Microbes Infect 4(3):291–300
Dorovini-Zis K, Schmidt K, Huynh H, Fu W, Whitten RO, Milner D et al (2011) The neuropathology of fatal cerebral malaria in Malawian children. Am J Pathol 178:2146–2158
Emuchay CI, Usanga EA (1997) Increased platelet factor 3 activity in *Plasmodium falciparum* malaria. East Afr Med J 74:527–529
Gazzinelli RT, Denkers EY (2006) Protozoan encounters with toll-like receptor signalling pathways: implications for host parasitism. Nat Rev Immunol 6:895–906
Glushakova S, Busse BL, Garten M, Beck JR, Fairhurst RM, Goldberg DE, Zimmerberg J (2017) Exploitation of a newly-identified entry pathway into the malaria parasite-infected erythrocyte to inhibit parasite egress. Sci Rep 7:12250. https://doi.org/10.1038/s41598-017-12258-x
Gordeuk V, Thuma P, Brittenham G, McLaren C, Parry D, Backenstose A et al (1992) Effect of iron chelation therapy on recovery from deep coma in children with cerebral malaria. N Engl J Med 327(21):1473–1477
Gordeuk VR, Loyevsky M (2002) Antimalarial effect of iron chelators. Adv Exp Med Biol 509:251–272
Griffiths MJ, Ndungu F, Baird KL, Muller DP, Marsh K, Newton CR (2001) Oxidative stress and erythrocyte damage in Kenyan children with severe Plasmodium falciparum malaria. Br J Haematol 113(2):486–491
Heddini A, Chen Q, Obiero J, Kai O, Fernandez V, Marsh K et al (2001) Binding of *Plasmodium falciparum*-infected erythrocytes to soluble platelet endothelial cell adhesion molecule-1 (PECAM-1/CD31): frequent recognition by clinical isolates. Am J Trop Med Hyg 65:47–51
Ho M, White NJ (1999) Molecular mechanisms of cytoadherence in malaria. Am J Physiol 276:C1231–C1242
Howard RJ, Handunnetti SM, Hasler T, Gilladoga A, de Aguiar JC, Pasloske BL et al (1990) Surface molecules on *Plasmodium falciparum*-infected erythrocytes involved in adherence. Am J Trop Med Hyg 43:15–29
Hsieh F-L, Turner L, Bolla JR, Robinson CV, Lavstsen T, Higgins MK (2016) The structural basis for CD36 binding by the malaria parasite. Nat Commun 7:12837
John CC, Kutamba E, Mugarura K, Opoka RO (2010) Adjunctive therapy for cerebral malaria and other severe forms of Plasmodium falciparum malaria. Expert Rev Anti-Infect Ther 8(9):997–1008
Lell B, Köhler C, Wamola B, Olola CH, Kivaya E, Kokwaro G, Wypij D, Mithwani S, Taylor TE, Kremsner PG, Newton CR (2010) Pentoxifylline as an adjunct therapy in children with cerebral malaria. Malar J 21:368–369
Looareesuwan S, Wilairatana P, Vannaphan S, Wanaratana V, Wenisch C, Aikawa M, Brittenham G, Graninger W, Wernsdorfer WH (1998) Pentoxifylline as an ancillary treatment for severe falciparum malaria in Thailand. Achieves Dis Child 58:348–353
Looareesuwan S, Sjostrom L, Krudsood S, Wilairatana P, Porter RS, Hills F et al (1999) Polyclonal anti-tumor necrosis factor-alpha fab used as an ancillary treatment for severe malaria. Am J Trop Med Hyg 61(1):26–33
Loutan L, Plancherel C, Soulier-lauper M, Pascual M, Subilia L, Chevrolet JC, Unger PF, Grau GE (1992) Serum TNF in patients with severe malaria treated by exchange transfusion. Trop Med Parasitol 43:285–286
Maitland K, Pamba A, English M, Peshu N, Marsh K, Newton C et al (2005) Randomized trial of volume expansion with albumin or saline in children with severe malaria: preliminary evidence of albumin benefit. Clin Infect Dis 40:538–545
Maneerat Y, Pongponratn E, Viriyavejakul P, Punpoowong B, Looareesuwan S, Udomsangpetch R (1999) Cytokines associated with pathology in the brain tissue of fatal malaria. Southeast Asian J Trop Med Public Health 30(4):643–649

Martins YC, Carvalho LJ, Daniel-Ribeiro CT (2009) Challenges in the determination of early predictors of cerebral malaria: lessons from the human disease and the experimental murine models. Neuroimmunomodulation 16(2):134–145

Miller LH, Ackerman HC, Su XZ, Wellems TE (2013) Malaria biology and disease pathogenesis: insights for new treatments. Nat Med 19:156–167

Mohanty D, Ghosh K, Pathare AV, Karnad D (2002) Deferiprone (L1) as an adjuvant therapy for Plasmodium falciparum malaria. Indian J Med Res 115:17–21

Mohanty S, Mishra SK, Patnaik R, Dutt AK, Pradhan S, Das B, Patnaik J, Mohanty AK, Lee SJ, Dondorp AM (2011) Brain swelling and mannitol therapy in adult cerebral malaria: a randomized trial. Clin Infect Dis 53:349–355

Namutangula B, Ndeezi G, Byarugaba JS, Tumwine JK (2007) Mannitol as adjunct therapy for childhood cerebral malaria in Uganda: a randomized clinical trial. Malar J 24:138

Newbold C, Warn P, Black G, Berendt A, Craig A, Snow B et al (1997a) Receptor-specific adhesion and clinical disease in Plasmodium falciparum. Am J Trop Med Hyg 57(4):389–398

Newbold CI, Craig AG, Kyes S, Berendt AR, Snow RW, Peshu N et al (1997b) PfEMP1, polymorphism and pathogenesis. Ann Trop Med Parasitol 91(5):551–557

Ockenhouse CF, Tegoshi T, Maeno Y, Benjamin C, Ho M, Kan KE et al (1992) Human vascular endothelial cell adhesion receptors for *Plasmodium falciparum*-infected erythrocytes: roles for endothelial leukocyte adhesion molecule 1 and vascular cell adhesion molecule 1. J Exp Med 176:1183–1189

Pongponratn E, Turner GD, Day NP, Phu NH, Simpson JA, Stepniewska K et al (2003) An ultrastructural study of the brain in fatal *Plasmodium falciparum* malaria. Am J Trop Med Hyg 69:345–359

Ponsford MJ, Medana IM, Prapansilp P, Hien TT, Lee SJ, Dondorp AM et al (2012) Sequestration and microvascular congestion are associated with coma in human cerebral malaria. J Infect Dis 205:663–671

Pouvelle B, Matarazzo V, Jurzynski C, Nemeth J, Ramharter M, Rougon G, Gysin J (2007) Neural cell adhesion molecule, a new cytoadhesion receptor for Plasmodium falciparum-infected erythrocytes capable of aggregation. Infect Immun 75(7):3516–3522

Rampengan TH (1991) Cerebral malaria in children. Comparative study between heparin, dexamethasone and placebo. Paediatr Indones 31:59–66

Reyburn H, Mbatia R, Drakeley C, Bruce J, Carneiro I, Olomi R et al (2005) Association of transmission intensity and age with clinical manifestations and case fatality of severe *Plasmodium falciparum* malaria. J Am Med Assoc 293:1461–1470

Riddle MS, Jeffery LJ, Sanders JH, Blazes DL (2002) Exchange transfusion as an adjunct therapy in severe *Plasmodium falciparum* malaria: a meta-analysis. Clin Infect Dis 1:1192–1198

Rogerson SJ, Reeder JC, Al-Yaman F, Brown GV (1994) Sulfated glycoconjugates as disrupters of *Plasmodium falciparum* erythrocyte rosettes. Am J Trop Med Hyg 51:198–203

Sahu PK, Satpathi S, Behera PK, Mishra SK, Mohanty S, Wassmer SC (2015) Pathogenesis of cerebral malaria: new diagnostic tools, biomarkers, and therapeutic approaches. Front Cell Infect Microbiol 5:75

Scott CS, Van Zyl D, Ho E, Ruivo L, Mendelow B, Coetzer TL (2002) Thrombocytopenia in patients with malaria: automated analysis of optical platelet counts and platelet clumps with the cell Dyn CD4000 analyser. Clin Lab Haematol 24:295–302

Sein KK, Maeno Y, Thuc HV, Anh TK, Aikawa M (1993) Differential sequestration of parasitized erythrocytes in the cerebrum and cerebellum in human cerebral malaria. Am J Trop Med Hyg 48:504–511

Seydel KB, Milner DA Jr, Kamiza SB, Molyneux ME, Taylor TE (2006) The distribution and intensity of parasite sequestration in comatose Malawian children. J Infect Dis 194:208–205

Seydel KB, Kampondeni SD, Valim C, Potchen MJ, Milner DA, Muwalo FW et al (2015) Brain swelling and death in children with cerebral malaria. N Engl J Med 372:1126–1137

Springer AL, Smith LM, Mackay DQ, Nelson SO, Smith JD (2004) Functional interdependence of the DBLbeta domain and c2 region for binding of the *Plasmodium falciparum* variant antigen to ICAM-1. Mol Biochem Parasitol 137:55–64

Srivastava K, Field DJ, Aggrey A, Yamakuchi M, Morrell CN (2010) Platelet factor 4 regulation of monocyte KLF4 in experimental cerebral malaria. PLoS One 5:e10413. https://doi.org/10.1371/journal.pone.0010413

Taylor TE, Molyneux ME, Wirima JJ, Borgstein A, Goldring JD, Hommel M (1992) Intravenous immunoglobulin in the treatment of paediatric cerebral malaria. Clin Exp Immunol 90(3):357–362

Thuma PE, Mabeza GF, Biemba G, Bhat GJ, McLaren CE, Moyo VM, Zulu S, Khumalo H, Mabeza P, M'Hango A, Parry D, Poltera AA, Brittenham GM, Gordeuk VR (1998) Effect of iron chelation therapy on mortality in Zambian children with cerebral malaria. Trans R Soc Trop Med Hygeine 92:214–218

Tongren JE, Yang C, Collins WE, Sullivan JS, Lal AA, Xiao L (2000) Expression of proinflammatory cytokines in four regions of the brain in macaque mulatta (rhesus) monkeys infected with Plasmodium coatneyi. Am J Trop Med Hyg 62(4):530–534

Tse MT, Chakrabarti K, Gray C, Chitnis CE, Craig A (2004) Divergent binding sites on intercellular adhesion molecule-1 (ICAM-1) for variant *Plasmodium falciparum* isolates. Mol Microbiol 51:1039–1049

Udomsangpetch R, Reinhardt PH, Schollaardt T, Elliott JF, Kubes P, Ho M (1997) Promiscuity of clinical Plasmodium falciparum isolates for multiple adhesion molecules under flow conditions. J Immunol 158(9):4358–4364

Van Hensbroek MB, Palmer A, Onyiorah E, Schneider G, Jaffar S, Dolan G et al (1996) The effect of a monoclonal antibody to tumor necrosis factor on survival from childhood cerebral malaria. J Infect Dis 174:1091–1097

Warrell DA (1997) Cerebral malaria: clinical features, pathophysiology and treatment. Ann Trop Med Parasitol 91:875–884

Wassmer SC, Taylor TE, Rathod PK, Mishra SK, Mohanty S, Arevalo-Herrera M et al (2015) Investigating the pathogenesis of severe malaria: a multidisciplinary and cross-geographical approach. Am J Trop Med Hyg 93(3 Suppl):42–56

Watt G, Jongsakul K, Ruangvirayuth R (2002) A pilot study of N-acetylcysteine as adjunctive therapy for severe malaria. Q J Med 95:285–290

White NJ (1996) The treatment of malaria. N Engl J Med 335(11):800–806

WHO (2015) World Malaria Report, WHO Bulletin

Yipp BG, Hickey MJ, Andonegui G, Murray AG, Looareesuwan S, Kubes P et al (2007) Differential roles of CD36, ICAM-1, and P-selectin in *Plasmodium falciparum* cytoadherence in vivo. Microcirculation 14:593–602

Chapter 4
Role of White Blood Cells in Immunopathogenesis of Cerebral Malaria

Ruchika Saroa and Upma Bagai

Abstract Cerebral malaria (CM) is a fatal complication caused by *Plasmodium falciparum* in humans claiming two million lives annually. The pathogenesis of this disease yet remains partially understood. Various studies have been carried out to understand the exact processes of CM which indicate towards the involvement of the immune response in the neurological complications. It has been hypothesized that CM occurs due to the over-vigorous immune response which originally evolved for the protection of the host against malaria. Some studies also examined immune-pathological responses occurring during CM and focused on reactions being carried out primarily in the systemic circulation. But these findings are not able to fully account for the development of neurological complications in malaria. There are multiple mechanisms which are involved in the induction of cerebral complications which contribute to the pathogenesis of CM. In the present study, results from human and mouse model demonstrating the contribution of various cells and cytokines in the development of CM and neurological complications have been summarized.

Keywords Cerebral malaria · Cytokines · Malaria · T cells

4.1 Introduction

Malaria, Tuberculosis and Acquired immunodeficiency syndrome (AIDS) are three most deadly diseases across the world causing millions of deaths every year. *Plasmodium falciparum*, *P. vivax*, *P. ovale*, *P. malariae* and *P. knowlesi* are five species of malaria parasites infecting humans. Among these species *P. falciparum* causes malignant malaria and is responsible for almost all malaria deaths worldwide (Kang et al. 2010).

In India, maximum mortality occurs in Orissa, West Bengal, Jharkhand and the central states of Chhattisgarh, Madhya Pradesh, Gujarat, Karnataka and Rajasthan.

R. Saroa (✉)
Department of Zoology, Panjab University, Chandigarh, India

U. Bagai (deceased)

P. P. Singh (ed.), *Infectious Diseases and Your Health*,
https://doi.org/10.1007/978-981-13-1577-0_4

P. falciparum infections alone are responsible for 88% of deaths which mark it as the main target for vaccine trials of the different initiatives and eradication programmes launched against malaria. In endemic areas, newborns are resistant to malaria due to the effect of mother's antibodies transferred through the placenta. When this immunity declines, the risk of contracting the disease increases. Therefore, young children are at high risk of malaria until they have developed their own protective immunity (Das et al. 2012).

Major problem in malaria control today is the increasing resistance of *P. falciparum* to various antimalarial drugs, lack of cost-effective drugs and non-availability of suitable vaccine against deadly parasite. Severe malarial anaemia, metabolic acidosis, placental-associated malaria and cerebral malaria are various complications associated with this disease (Medana et al. 2001a).

4.2 Life Cycle of *P. falciparum*

Anopheles mosquitoes are vectors of *P. falciparum*. Infected mosquito injects sporozoites into human skin during blood meal which travel to the liver for invading hepatocytes. Within 2 days, one merozoite transforms into a trophozoite and then into a schizont (Siciliano and Alano 2015).

After tissue schizogony, merozoites invade red blood cells (RBCs), and erythrocytic schizogony takes place. Inside the host, *P. falciparum*-infected red blood cells escape from immune surveillance and send adhesive proteins to the host's cell membrane. These proteins make the cells adhere to small blood vessels, which pose a threat to the human host since the clustered red blood cells create a blockage in the circulation system especially in the brain. Some schizonts develop into male and female gametes which are ingested by a mosquito when it feeds on infected blood (Fig. 4.1).

Inside the mosquito's midgut, male and female gametocytes fuse and develop into ookinete. The motile ookinetes penetrate the midgut wall and release sporozoites, which migrate to the salivary glands from where they are injected into humans during the next blood meal (Fig. 4.1; Siciliano and Alano 2015).

4.3 Signs and Symptoms of Cerebral Malaria

The major clinical symptoms of malaria include fever, malaise, splenomegaly, anaemia, convulsions, muscle pain and bloody stool due to the cyclic multiplication of the parasite into RBCs. During *P. falciparum* infection, the spectrum of severe pathology is broad and includes metabolic acidosis, cerebral malaria (CM) and severe malaria anaemia (SMA) accompanied by hypoxia, hypoglycaemia, lactic acidosis and multi-organ failure which may result in coma and death (McCall and Sauerwein 2010).

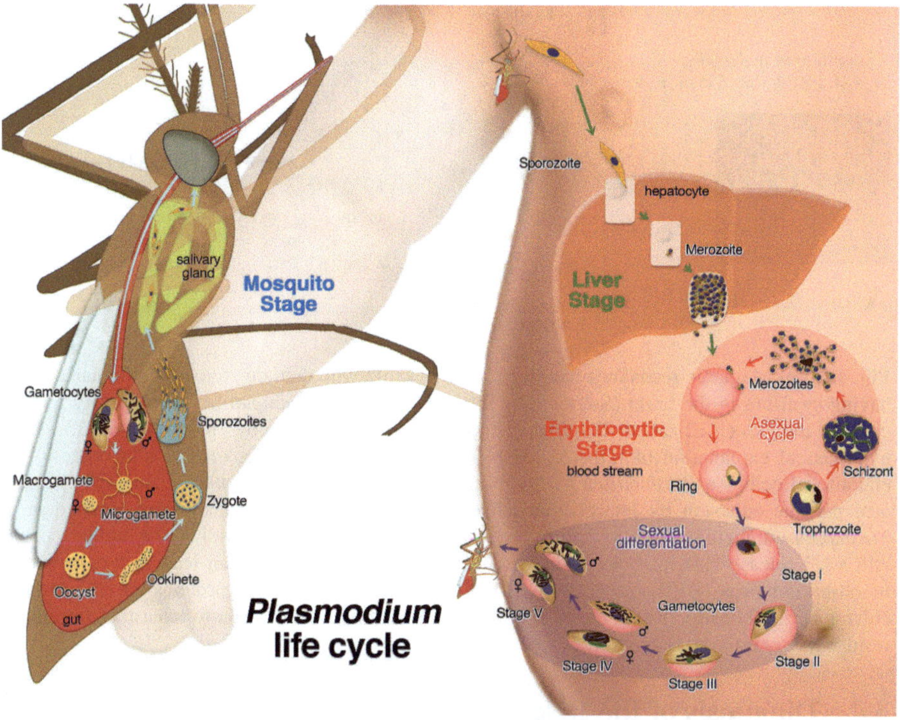

Fig. 4.1 Pictorial representation of the life cycle and different development stages of *P. falciparum* inside humans as well as mosquitos. (Source: Le Roch Laboratory, University of California, Riverside)

Fig. 4.2 African child suffering from cerebral malaria. (Source: Larry Johnson and Mike Urban: Insecticide-resistant mosquitoes challenging Gates malaria efforts)

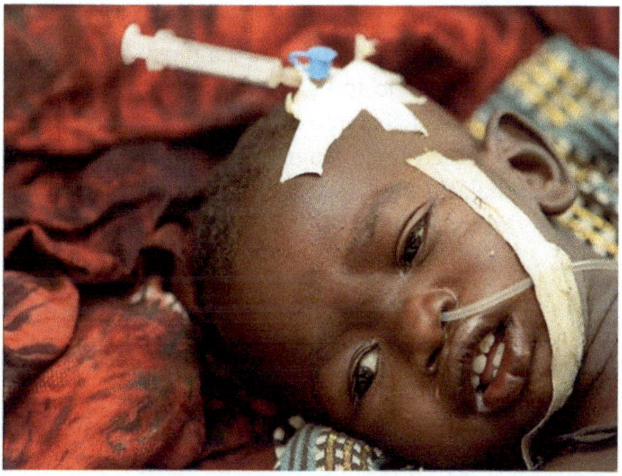

The clinical hallmark of cerebral malaria is impaired consciousness with coma being the most severe manifestation. Patients may develop coma following progressive weakness, brain swelling, intracranial hypertension and retinal changes (Figs. 4.2 and 4.3; Idro et al. 2010).

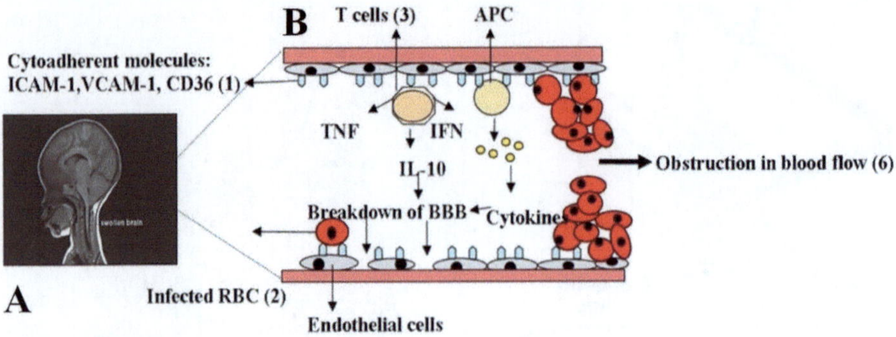

Fig. 4.3 (**a**) Magnetic resonance imaging (MRI) of child who died due to cerebral malaria (Left) having swelling in the brain (Source: Dr. Terrie Taylor from Michigan State University holds a child in the Queen Elizabeth Hospital in Blantyre, Malawi. Photo by Jim Peck, MSU.). (**b**) A hypothetical explanation of infected red blood cells (IRBCs) and leukocytes adherence to the endothelial cells via *Plasmodium falciparum* erythrocyte membrane protein-1 (PfEMP-1) mediated parasite adhesion by Intercellular adhesion molecule (ICAM-1) leading to attachment of *Plasmodium*-infected red blood cells to endothelial cells of the brain (2), adhered lymphocytes and APCs (3) release inflammatory cytokines (TNF-α, IL-10) (4) which can lead to breakdown of blood-brain barrier (BBB) permeability (5) allowing leakage of plasma proteins into the perivascular space and obstruction in blood flow (6) aggravated by platelets due to release of fibrinogen resulting in neurological syndrome

4.4 Diagnosis of the Disease

Various techniques are available for the diagnosis of malaria: Giemsa-stained blood smears, microhaematocrit centrifugation, fluorescent dyes, polymerase chain reaction (PCR), nucleic acid sequence-based amplification (NASBA) and ParaSight F test (dipstick test). Correct and well-timed treatments are important in malaria which require quick and valid diagnosis. Therefore, dipstick tests are the most important diagnostic tools which are rapid to perform and easy to read by even untrained persons. The most superior and sensitive method is histidine-rich protein II (HRP-2)-based rapid diagnostic test RDT. Some serological assays are also available for malaria detection including indirect immunofluorescence assay (IFA) or enzyme-linked immunosorbent assay (ELISA) (Batwala et al. 2010).

4.5 Mechanism of Cerebral Malaria

The key feature of this multifaceted, deadly manifestation is the cytoadherence of *Plasmodium falciparum* parasitized erythrocytes (PE) to endothelial cells (EC) and their sequestration in the cerebral microvasculature of the infected host. It has been reported that schizont-infected red blood cells block the brain capillaries causing obstruction in blood flow. Infected erythrocytes also adhere to normal erythrocytes (rosetting) and to other infected RBCs (autoagglutination) (Fig. 4.3). Vessels in the brain clogged with infected and normal RBCs result in insufficient supply of oxygen and nutrients to the brain leading to coma and death (Renia et al. 2006). Intravascular leukocyte sequestration has also been observed within brain venules from patients who died of CM (White et al. 2010).

4.6 Murine Model of Cerebral Malaria

P. yoelii and *P. berghei* ANKA (PbA) are the main sources of cerebral syndromes in rodent host which have various similarities to the human conditions. *P. berghei* ANKA infected murine erythrocytes sequester in the microvasculature of various organs via CD36 adhesion molecule, but sequestration is more in the lungs than in the brain (Engwerda et al. 2005).

A comparison within the types of immune responses generated in Balb/C mice and C57BL/6 mice infected with PbA led to the conclusion that Balb/C strain of mouse is resistant to the neurological conditions. So, the most widely used model for cerebral malaria is PbA infection in CBA or C57BL/6 mice (Franke-Fayard et al. 2005).

In human cerebral malaria, it is mainly infected RBCs that attach to the cerebral microvascular endothelium, whereas in rodent CM, it is chiefly leukocytes which exhibit adherence (Berendt et al. 1994).

4.7 Role of Adhesion Molecules in Cerebral Malaria

Sequestration of iRBCs into the brain microvasculature and other tissues is mediated by various receptors which are present on endothelial cells of venules and capillaries inside the host. These receptors include adhesion molecules such as CD36, intercellular adhesion molecule (ICAM)-1 and CD31, thrombospondin and chondroitin sulphate A (CSA) (Berendt et al. 1994).

Immunohistochemical staining of *P. berghei* ANKA infected host showed a strong ICAM-1 expression on vessels of the brain containing iRBCs (Silamut et al. 1999). There are many molecules which act as endothelial receptors for lymphocyte adhesion during inflammation and immune surveillance. The ligands for attachment of infected RBCs are produced by the *Plasmodium,* and the best is PfEMP-1 (*P. falciparum* erythrocyte membrane protein) (Favre et al. 1999). The parasite exports and inserts PfEMP-1 into the erythrocyte membrane at adhesive foci or knobs. Variants of PfEMP-1 expressed by different parasite stains have different binding affinities for the various endothelial receptors and influence the cell sequestration, e.g. CSA-binding parasites are implicated in placental malaria, whereas, ICAM-1 binding is important in CM (Favre et al. 1999).

PfEMP-1 is the major receptor for parasite-host interactions and is expressed on the surface of the parasitized RBCs. This is encoded by a multigene family of up to 60 var genes. Antigenic variations allow the parasite to evade the host's immune system promoting PfEMP-1 binding to a range of ligands including ICAM-1, vascular cell adhesion molecule-1 (VCAM-1), P-selectin, CD36, thrombospondin (TSP) and CSA. ICAM-1 on endothelial cells is markedly increased in response to various inflammatory mediators including pro-inflammatory cytokines such as TNF-α (tumour necrosis factor) and IL-1β (interleukin) (Chakravorty and Craig 2005).

Interaction of lymphocytes with ICAM-1 is mediated via receptors such as LFA-1 and Mac-1 which have a major role in the recruitment of various cells and their migration to tissues (Gardner et al. 2002). There seems to be a synergistic interaction between ICAM-1 and CD36 to bind parasitized RBCs because the blocking antibodies against either of the molecules reduce the binding (McCormick et al. 1997).

Histoarchitectural studies of infected rodent brain having CM showed sequestration of leukocytes into the cerebral parenchyma and PE accumulation intravascularly. The main brain-sequestered leukocytes are macrophages, T cells, few dendritic cells and natural killer (NK) cells (Schmutzhard et al. 2011).

4.8 T Helper and T Cytotoxic Cells

T cells consist of various sets of cells ($CD4^+$ and $CD8^+$ T cells) which fight against infection by secreting various cytokines and chemokines. $CD4^+$ T cells have been found to be crucial in malaria immunity. They not only act as helper cells for secreting antibodies but also act as effector cells in killing parasites. It is executed by secreting inflammatory cytokines and activating other cells like macrophages. They are also known as the major source of interferon-γ (IFN-γ) and TNF-α (Muxel et al. 2011) which are required for protection against the disease. Both induce nitric oxide synthase expression in the spleen to control parasite load, generate good antiparasitic response and mediate protection (Perez-Mazliah and Langhorne 2014).

In our laboratory, during lethal NK-65 infection in rodent host, an increase was observed in splenic $CD4^+$ and $CD8^+$ T-cell population during initial days of infection which declined with increase in parasite load. Whereas, mice immunized with parasite constituents exhibited a strong Th1/Th2 immune response and expansion of $CD4^+$ and $CD8^+$ T cells significantly leading to the clearance of the parasite from the host upon challenge (Kumar et al. 2015).

During cerebral complications in Anka-infected rodent host, a significant increase was observed in $CD4^+$ T-cell population leading to the secretion of IFN-γ and TNF-α in the brain which are responsible for the breaching of blood-brain barrier (Belnoue et al. 2002). $CD8^+$ T cells are cytotoxic cells which recognize pathogen via MHC-I molecule and destroy infected cells. During malaria, $CD8^+$ T cells are responsible for providing cell-mediated immunity against asexual stages of parasite in host (Belnoue et al. 2002). An elevated number of $CD8^+$ T cells were observed during self-clearing *P. chabaudi* infection, whereas animals depleted of these cells exhibited a significant delay in clearance of the parasite (Helmby et al. 2000).

Cytotoxic cells play a major role in the development of neurological complications. Flow cytometric analysis of leukocytes of the brain in mice having CM showed a selective increase in sequestered $CD8^+\alpha\beta$ T cells. This theory was confirmed by the transfer of splenic $CD8^+$ T cells from mice with CM into RAG-2-deficient mice. It was reported that these transferred cells migrate to the brain and induce neurological symptoms (Pais and Chatterjee 2005). A significant proportion of the $CD8^+$ T cells involved in CM development express the Vβ8.1,2 T-cell receptor (Claser et al. 2011).

In our laboratory, higher $CD8^+$ T-cell count was observed in *P. berghei* ANKA infected rodent host's brain with increase in infection as compared to the spleen. It points to the recruitment of cytotoxic $CD8^+$ T cells to the brain microvasculature with increase in parasitaemia (unpublished data).

Recruitment of $CD4^+$ T helper 1 (Th1) in infected host leads to generation of protective immune response and high levels of IFN-γ and pro-inflammatory cytokines like TNF and LT-α which are the major causes of cerebral complications (Hansen et al. 2003). Stimulated cytotoxic cells expand in the brain during the disease. Perforin-dependent cytotoxic mechanism and brain macrophages may be involved in the induction of neurological complications due to cerebral malaria (Hunt and Grau 2003).

4.9 Monocytes/Macrophages and Dendritic Cells

Macrophages recognize microbial product or pathogen via toll-like receptors leading to acute activation of these cells associated with the release of pro-inflammatory cytokines and chemokines (Chua et al. 2013).

Monocytes, dendritic cells and macrophages perform important roles during malaria infection in humans as well as in rodents. Monocytes are formed from haematopoietic lineage in the bone marrow and are released into the bloodstream upon maturation. Their further differentiation depends upon the downstream signals which trigger them to change into dendritic cells or tissue macrophages (Claser et al. 2011).

During *Plasmodium* infection, they are capable of reducing the parasite load in the blood via antibody-dependent cell inhibition (ADCI) which requires the acquisition of antibodies (IgG1 and IgG3 subtypes) against merozoite surface protein-1 (MSP-1) and glutamate-rich protein (GLURP). Antibody-opsonized merozoites further release the soluble mediators for the inhibition of parasite growth and multiplication (Pratt-Riccio et al. 2011).

It has been reported that macrophages are involved in complement-mediated phagocytosis which is necessary for providing protection against asexual stages of the parasite via complement receptor 1 (CR1 or CD35) (Silver et al. 2010).

Patients suffering from cerebral complications reported activation of platelet endothelial cell adhesion molecule -1 (PECAM-1). Activated platelets secrete various chemotactic molecules due to which the recruitment as well as retention of monocytes/macrophages is enhanced leading to complete occlusion of brain capillaries. These monocyte/macrophage sequestration together with attached parasitized RBCs ultimately lead to mechanical blockage of brain microvasculature (Dorovini-Zis et al. 2011).

Post-mortem histological analysis of infected brain expressed high amounts of stress proteins secreted by brain macrophages as well as showed demyelination and neuronal damage (Medana et al. 2001b). Changes in morphology of monocytes and macrophages indicating activation of retinal microglial cells during *Plasmodium falciparum* infection have also been reported (Hunt and Grau 2003).

Dendritic cells (DCs) are a variant type of antigen-presenting cells (APCs) which plays an important role in the initiation and maintenance of CMI (cell-mediated immune response). Based on the expression of their surface markers and response against pathogens, they can be divided into two types: plasmacytoid DC (pDC) and conventional DC (cDC) (Perry et al. 2004). It has been shown that macrophages are capable of initiating cDC population which can activate $CD4^+$ T cells. It has been shown that both $CD8^+$ and $CD8^+$ splenic cDC can activate parasite-specific $CD4^+$ T-cell responses during *P. chabaudi* infection in mice, whereas depletion of cDCs leads to protection of mice against cerebral malaria (Perry et al. 2004).

4.10 T Regulatory Cells (Tregs)

T suppressor cells are the subtypes of $CD4^+$ T cells which are known for their major involvement in generation of the immune response during various infections. Parasite can manipulate T regulatory cells by changing the T-cell immune response to an extent that could lower the parasite burden (Sakaguchi et al. 2006).

The well-defined markers for Tregs are $CD4^+CD25^{+high}$ and represent 10% of total peripheral $CD4^+$ T cells. They exhibit a high expression of Foxp3 which is necessary for differentiation and functioning of Tregs (Stevenson et al. 2011). They are mainly reported to suppress cellular immune responses through direct contact with immune cells via production of various cytokines like TGF-ß and IL-10 (Bacchetta et al. 2007).

During infection, Tregs reduce host immune responses through cell-to-cell contact, inhibitory cytokines or cytokine deprivation and prevent the effective generation of an immune response. Tregs are also known to downregulate Th2 responses such as IL- 5-dependent eosinophil activation which is required to kill parasites. During generation of immune response in host, the interplay and balance among Th1, Th2 and Tregs is crucial to fight against parasitic infection (Maizels and Yazdanbakhsh 2003). It has been observed that children with severe malaria were unable to control the inflammation during *P. falciparum* infection suggesting that this component may be rapidly overwhelmed by virulent infections (Torres et al. 2014).

Tregs have been reported to inhibit the pathogenic Th cells which are responsible for control of cerebral syndrome in resistant BALB/c mouse infected with ANKA strain. Whereas, in susceptible mouse strain, Tregs were found to be depleted which resulted in an increase in survival of mice with significant reduction in parasitaemia (Lee et al. 2011).

A comparative study on PbA-infected middle-aged CM-resistant and young CM-susceptible mice reported that this cell population has a regulatory involvement in the control of fatal pathogenesis and it worked in an IL-10-dependent manner (Shan et al. 2013).

$CD4^+$ T regulatory cells were found to be increased in *P. berghei* NK-65-infected rodent host, whereas, immunization of mice with parasite constituents showed a decrease in $CD4^+$ T regulatory cells leading to the generation of strong Th2 immune response. It resulted in the clearance of infection from the host upon challenge (Kumar et al. 2015).

4.11 Natural Killer (NK) Cells

During malaria, NK cells generate innate immune response with help of DCs and secrete various cytokines. They are found to be amplified during infection and are capable of lysing parasitized erythrocytes *in vitro*. In the bloodstream, NK cells produce IFN-γ in response to infected RBCs, leading to macrophage activation, and provide innate immunity to malaria. Pro-inflammatory chemokine IL-8 production by NK cells is also induced during malaria infection which leads to the recruitment and activation of other cells (Ariyasinghe et al. 2006).

NK cells are the strong inhibitors of liver-stage parasite and reported to regulate IgG antibody responses, which are significant for unrestricted malaria parasite control. These cells regulate not only cerebral malaria but also other complications such as pulmonary oedema and severe anaemia (Mitchell et al. 2005). In *Plasmodium berghei* ANKA infected mouse, CD1d-restricted NK cells induce early IFN-γ production and promote cerebral syndrome. IFN-γ production by NK cells has an effect on maintaining immune homeostasis in infected mice. It also influences parasite-specific antibody responses and the TH1/TH2 balance during infection (Brown et al. 1990).

4.12 Blood-Brain Barrier (BBB)

Free flow of molecules and ions in and out of brain parenchyma is regulated by the blood-brain barrier. The BBB along with blood-CSF (cerebrospinal fluid) barrier ensures a constant composition of the extracellular fluid in the brain. This is essential for normal neuronal function, and it also regulates the passage of immune cells into the central nervous system (Tsukita and Furuse 2000). Excessive production of various cytokines and chemokines like TNF-α, IFN-γ and interleukins can be toxic to the brain and may lead to the production of irreversible symptoms of coma during cerebral malaria. Significantly higher levels of TNF-α and IL-1β in the brain of children having cerebral malaria have been reported (Kwiatkowski et al. 1990).

Increase in levels of adhesion molecules during CM cause changes in endothelial cell junctional permeability. It leads to leakage of plasma proteins and fluids into the perivascular space and brain parenchyma causing cerebral oedema. This evidence supported the fact that functional BBB breakdown occurs in CM inducing systemic or local cytokine release from the blood through endothelial cells into the perivascular space and brain parenchyma (Adams et al. 2002).

4.13 Concluding Remarks

Cytokines secreted by white blood cells can be a major source of defence/inflammation to host against malaria. Elevated immune response generated during *Plasmodium* infection is responsible for creating several complications to host especially CM. It has been observed that immune cells providing protection to host are also responsible for secreting the adhesion molecules on the surface of the endothelial cells of capillaries and venules in the brain. Due to the activation of adhesion molecules in CNS, the infected red blood cells and leukocytes start adhering to the brain microvasculature and obstruct blood flow, creating cerebral complications.

This knowledge might lead to better understanding of the unclear mechanism of sequestration of cells in the brain during generation of strong immune response in malaria-infected host. It can be useful to device therapeutic approach to control the devastating pathogenesis during infection.

Acknowledgement Authors are thankful to UGC, New Delhi, for the financial support.

References

Adams S, Brown H, Turner G (2002) Breaking down the blood-brain barrier: signaling a path to cerebral malaria? Trends Parasitol 18:360–366

Ariyasinghe A, Morshed SR, Mannoor MK, Bakir HY, Kawamura H, Miyaji C et al (2006) Protection against malaria due to innate immunity enhanced by low-protein diet. J Parasitol 92:531–538

Bacchetta R, Gambineri E, Roncarolo MG (2007) Role of regulatory T cells and FOXP3 in human diseases. J Allergy Clin Immunol 120:227–235

Batwala V, Magnussen P, Nuwaha F (2010) Are rapid diagnostic tests more accurate in diagnosis of *Plasmodium falciparum* malaria compared to microscopy at rural health centres? Malar J 9:349

Belnoue E, Kayibanda M, Vigario AM, Deschemin JC, van Rooijen N, Viguier M et al (2002) On the pathogenic role of brain-sequestered alpha-beta CD8+ T cells in experimental cerebral malaria. J Immunol 169:6369–6375
Berendt AR, Turner GDH, Newbold CI (1994) Cerebral malaria: the sequestration hypothesis. Parasitol Today 10:412–414
Brown H, Turner G, Rogerson S, Tembo M, Mwenechanya J, Molyneux M, Taylor T (1990) Cytokine expression in the brain in human cerebral malaria. J Infect Dis 180:1742–1746
Chakravorty SJ, Craig A (2005) The role of ICAM-1 in *Plasmodium falciparum* cytoadherence. Eur J Cell Biol 84:15–27
Chua CL, Brown G, Hamilton JA, Rogerson S, Boeuf P (2013) Monocytes and macrophages in malaria: protection or pathology? Trends Parasitol 29:26–34
Claser C, Malleret B, Gun SY, Wong AY, Chang ZW, Teo P et al (2011) CD8+ T cells and IFN-gamma mediate the time-dependent accumulation of infected red blood cells in deep organs during experimental cerebral malaria. PLoS One 6:e18720
Das A, Anvikar AR, Cator LJ, Dhiman RC, Eapen A, Mishra N et al (2012) Malaria in India: the center for the study of complex malaria in India. Acta Trop 121:267–273
Dorovini-Zis K, Schmidt K, Huynh H, Fu W, Whitten RO, Milner D et al (2011) The neuropathology of fatal cerebral malaria in Malawian children. Am J Pathol 178:2146–2158
Engwerda CR, Mynott TL, Sawhney S, De Souza JB, Bickle QD, Kaye PM (2005) Locally up-regulated lymphotoxin alpha, not systemic tumor necrosis factor alpha, is the principle mediator of murine cerebral malaria. J Exp Med 195:1371–1377
Favre N, Da Laperousaz C, Ryffel B, Weiss NA, Imhof BA, Rudin W et al (1999) Role of ICAM-1 (CD54) in the development of murine cerebral malaria. Microbes Infect 1:961–968
Franke-Fayard B, Janse CJ, Cunha-Rodrigues M, Ramesar J, Buscher P, Que I (2005) Murine malaria parasite sequestration: CD36 is the major receptor, but cerebral pathology is unlinked to sequestration. Proc Nat Acad Sci USA 102:11468–11473
Gardner MJ, Hall N, Fung E, White O, Berriman M, Hyman RW et al (2002) Genome sequence of the human malaria parasite *Plasmodium falciparum*. Nature 419:498–511
Hansen DS, Siomos MA, Buckingham L, Scalzo AA, Schofield L (2003) Regulation of murine cerebral malaria pathogenesis by CD1d-restricted NKT cells and the natural killer complex. Immunity 18:391–402
Helmby H, Jonsson G, Troye-Blomberg M (2000) Cellular changes and apoptosis in the spleens and peripheral blood of mice infected with blood-stage *Plasmodium chabaudi chabaudi* AS. Infect Immun 68:1485–1490
Hunt NH, Grau GE (2003) Cytokines: accelerators and brakes in the pathogenesis of cerebral malaria. Trends Immunol 24:491–499
Idro R, Marsh K, John CC, Newton RJ (2010) Cerebral malaria: mechanisms of brain injury and strategies for improved neuro-cognitive outcome. Pediatr Res 68(4):267–274
Kang JM, Sung-Ung M, Jung-Yeon K, Shin-Hyeong C, Lin K, Woon-mok S, Tong-Soo K, Byoung-Kuk N (2010) Genetic polymorphism of merozoite surface protein-1 and merozoite surface protein-2 in *Plasmodium falciparum* field isolates from Myanmar. Malar J 9:131–136
Kumar V, Rakha A, Saroa R, Bagai U (2015) CD4+T cells expansion in *P. berghei* (NK-65) infected and immunized BALB/C mice. Journal clinical. Exp Pathol 5:229–234
Kwiatkowski D, Hill AV, Sambou I, Twumasi P, Castracane J, Manogue KR et al (1990) TNF concentration in fatal cerebral, non-fatal cerebral and uncomplicated *Plasmodium falciparum* malaria. Lancet 336:1201–1204
Lee JC, Hayman E, Pegram HJ (2011) In vivo inhibition of human CD19 targeted effector T cells by natural T regulatory cells in a xenotransplant murine model of B cell malignancy. Cancer Res 71:2871–2881
Maizels RM, Yazdanbakhsh M (2003) Immune regulation by helminth parasites: cellular and molecular mechanisms. Nat Rev Immunol 3:733–744
McCall MBB, Sauerwein RW (2010) Interferon gamma-central mediator of protective immune responses against the pre-erythrocytic and blood stage of malaria. J Leukoc Biol 88:1131–1143

McCormick CJ, Craig A, Roberts D, Newbold CI, Berendt AR (1997) Intercellular adhesion molecule-1 and CD36 synergize to mediate adherence of *Plasmodium falciparum*-infected erythrocytes to cultured human microvascular endothelial cells. J Clin Investig 100:2521–2529
Medana IM, Chaudhri G, Chan-Ling T, Hunt NH (2001a) Central nervous system in cerebral malaria: 'innocent bystander' or active participant in the induction of immunopathology? Immunol Cell Biol 79:101–120
Medana IM, Mai NT, Day NP, Hien TT, Bethell D, Phu NH et al (2001b) Cellular stress and injury responses in the brains of adult Vietnamese patients with fatal *Plasmodium falciparum* malaria. Neuropathol Appl Neurobiol 27:421–433
Mitchell AJ, Hansen AM, Hee L, Ball HJ, Potter SM, Walker JC et al (2005) Early cytokine production is associated with protection from murine cerebral malaria. Infect Immun 73:5645–5653
Muxel SM, do Rosario F, Castillo-Mendez SI, Sardinha LR, Rodriguez-Malaga SM et al (2011) The spleen CD4+ T cell response to blood-stage *Plasmodium chabaudi* malaria develops in two phases characterized by different properties. PLoS One 6:e22434
Pais TF, Chatterjee S (2005) Brain macrophage activation in murine cerebral malaria precedes accumulation of leukocytes and CD8+ T cell proliferation. J Neuroimmunol 163:73–83
Perez-Mazliah D, Langhorne J (2014) CD4 T-cell subsets in malaria: TH1/TH2 revisited. Front Immunol 5:671–676
Perry JA, Rush A, Wilson RJ, Olver CS, Avery AC (2004) Dendritic cells from malaria-infected mice are fully functional APC. J Immunol 172:475–482
Pratt-Riccio LR, Bianco C Jr, Totino PR, Perce-Da-Silva Dde S, Silva LA, Riccio EK et al (2011) Antibodies against the *Plasmodium falciparum* glutamate-rich protein from naturally exposed individuals living in a Brazilian malaria-endemic area can inhibit in vitro parasite growth. Mem Inst Oswaldo Cruz 106:34–43
Renia L, Potter SM, Mauduit M, Rosa DS, Kayibanda M, Deschemin JC et al (2006) Pathogenic T cells in cerebral malaria. Int J Parasitol 36:547–554
Sakaguchi S, Setoguchi R, Yagi H, Nomura T (2006) Naturally arising Foxp3-expressing CD25(+)CD4(+) regulatory T cells in self-tolerance and autoimmune disease. Curr Concepts Autoimmun Chronic Inflamm 305:51–66
Schmutzhard J, Kositz CH, Lackner P, Pritz C, Glueckert R, Fischer M et al (2011) Murine cerebral malaria: histopathology and ICAM 1 immunohistochemistry of the inner ear. Tropical Med Int Health 16:914–922
Shan Y, Liu J, Pan YY, Jiang YJ, Shang H, Cao YM (2013) Age-related CD4(+)CD25(+)Foxp3(+) regulatory T-cell responses during *Plasmodium berghei* ANKA infection in mice susceptible or resistant to cerebral malaria. Korean J Parasitol 51:289–295
Siciliano G, Alano P (2015) Enlightening the malaria parasite life cycle: bio luminescent *Plasmodium* in fundamental and applied research. Front Microbiol. https://doi.org/10.3389/fmicb.2015.00391
Silamut K, Phu NH, Whitty C, Turner GD, Louwrier K, Mai NT et al (1999) A quantitative analysis of the microvascular sequestration of malaria parasites in the human brain. Am J Pathol 155:395–410
Silver KL, Higgins SJ, McDonald CR, Kain KC (2010) Complement driven innate immune response to malaria: fuelling severe malarial diseases. Cell Microbiol 12:1036–1045
Stevenson MM, Ing R, Berretta F, Miu J (2011) Regulating the adaptive immune response to blood-stage malaria: role of dendritic cells and CD4(+)Foxp3(+) regulatory T cells. Int J Biol Sci 7:1311–1322
Torres KJ, Villasis E, Bendezu J, Chauca J, Vinetz JM, Gamboa D (2014) Relationship of regulatory T cells to *Plasmodium falciparum* malaria symptomatology in a hypoendemic region. Malar J 13:108
Tsukita S, Furuse M (2000) Pores in the wall: claudins constitute tight junction strands containing aqueous pores. J Cell Biol 149:13–26
White NJ, Turner GD, Medana IM, Dondorp AM, Day NP (2010) The murine cerebral malaria phenomenon. Trends Parasitol 26:11–15

Chapter 5
Chemotherapy and Experimental Models of Visceral Leishmaniasis

Ganesh Yadagiri and Prati Pal Singh

Abstract Visceral leishmaniasis (VL) is a neglected tropical parasitic disease in humans caused by protozoan parasite *Leishmania donovani* and transmitted to humans by the bite of an infected female sand fly, a haemoflagellate vector. According to WHO, every year 0.7–1 million leishmaniasis cases are reported globally, and over 20,000–30,000 deaths occur. Current anti-leishmanial drug (pentavalent antimonials, miltefosine, amphotericin B, pentamidine and paromomycin) therapy is fraught with several problems and causes serious adverse effects, which limit their clinical application. The emergence of drug resistance and non-availability of an effective vaccine(s) against leishmaniasis poses a serious challenge to leishmaniasis treatment and control. Environmental and socio-economic status of people like deforestation, global warming and poverty exacerbates both parasite survival and disease progression. Pentavalent antimonial-resistant strains of *L. donovani* are rampant in Bihar, a highly endemic zone of VL in India. Development of co-infections (HIV-VL and Malaria-VL) often leads to poor diagnosis and treatment. There are no proper prognostic and diagnostic markers for VL. Therefore, there is an urgent need for the development of new anti-leishmanial drugs for the treatment and control of devastating VL. Effective immunotherapy/immuno-chemotherapy is considered as a viable alternative to chemotherapy. Cytokines (granulocyte-macrophage colony-stimulating factor, interferon-γ and interleukin-12) both standalone and in combination with current anti-leishmanial drugs are being thought to reduce the drug resistance and useful in VL treatment. The development and availability of the reliable models for anti-leishmanial drug screening is very much warranted.

Keywords Co-infections · Cytokines · Drug resistance · Experimental models · Immunotherapy · Leishmaniasis · Vaccine

G. Yadagiri (✉) · P. P. Singh
Centre of Infectious Diseases, Department of Pharmacology and Toxicology,
National Institute of Pharmaceutical Education and Research, S. A. S. Nagar, Punjab, India

P. P. Singh (ed.), *Infectious Diseases and Your Health*,
https://doi.org/10.1007/978-981-13-1577-0_5

5.1 Introduction

Leishmaniasis is a neglected tropical disease caused by obligate intramacrophage protozoan parasites of genus *Leishmania* and is transmitted to humans by the bite of infected female phlebotomine sand flies. It affects mainly poor population of under-developed and developing countries. There are three main forms of the disease: cutaneous leishmaniasis (CL), visceral leishmaniasis (VL) or kala-azar (KA) and mucocutaneous leishmaniasis (MCL). Leishmaniasis overall (including VL, CL and MCL) carries the ninth highest disease burden of all the infectious diseases world-wide (WHO 2015). VL is the most dangerous form of leishmaniasis which is characterised by irregular fever, weight loss, enlargement of spleen and liver and decreased blood cell counts; if untreated, it causes the death of the patient. The causative agent for VL is *Leishmania donovani* in Indian subcontinent and *Leishmania infantum* in Europe, North Africa and Latin America. According to the World Health Organisation (WHO 2015), every year 0.7–1 million leishmaniasis cases are reported globally, over 20,000–30,000 deaths occur, and 50,000–90,000 new VL cases are reported globally; more than 90% of new VL cases are reported in Brazil, Ethiopia, India, Kenya, Somalia, South Sudan and Sudan. Leishmaniasis is believed to be the third most prevalent vector-borne diseases (the first two being malaria and lymphatic filariasis). CL is the most common form of leishmaniasis characterised by skin lesions (ulcers) on exposed parts of the body, leaving irremovable scars on the body. Most of the CL cases are caused by *L. major*. According to WHO, every year 0.6–1 million new CL cases are reported globally. The Americas, the Mediterranean basin, the Middle East and the Central Asia contribute about 95% of CL in the world, and more than two thirds of new CL cases are reported in Afghanistan, Algeria, Colombia, Brazil, Iran and Syrian Arab Republic. MCL is another form of leishmaniasis caused by *L. braziliensis*. MCL is characterised by destruction of mucous membranes in the nose, mouth and throat (larynx and pharynx). Plurinational States of Bolivia, Brazil, Ethiopia and Peru contribute about 90% of MCL cases, globally. Post kala-azar dermal leishmaniasis (PKDL) is a consequence of Kala-azar, characterised by the appearance of macular and nodular scars on face and body parts. PKDL cases are mainly reported in East Africa and India. In India, 5–10% of VL patients have been known to develop PKDL. Patients having PKDL can serve as a potential reservoir of VL (WHO 2015; Chappuis et al. 2007).

5.2 Life Cycle of Leishmania Parasite

In order to develop new drugs and combat the growing resistance in leishmania, one should thoroughly understand the life cycle of the parasites, so that viable drug targets can offer better classes of drugs (Fig. 5.1).

The leishmania parasite has a digenetic life cycle, as the parasite shuttles between mammalian host (as amastigotes) and vectors (as promastigotes). In some places,

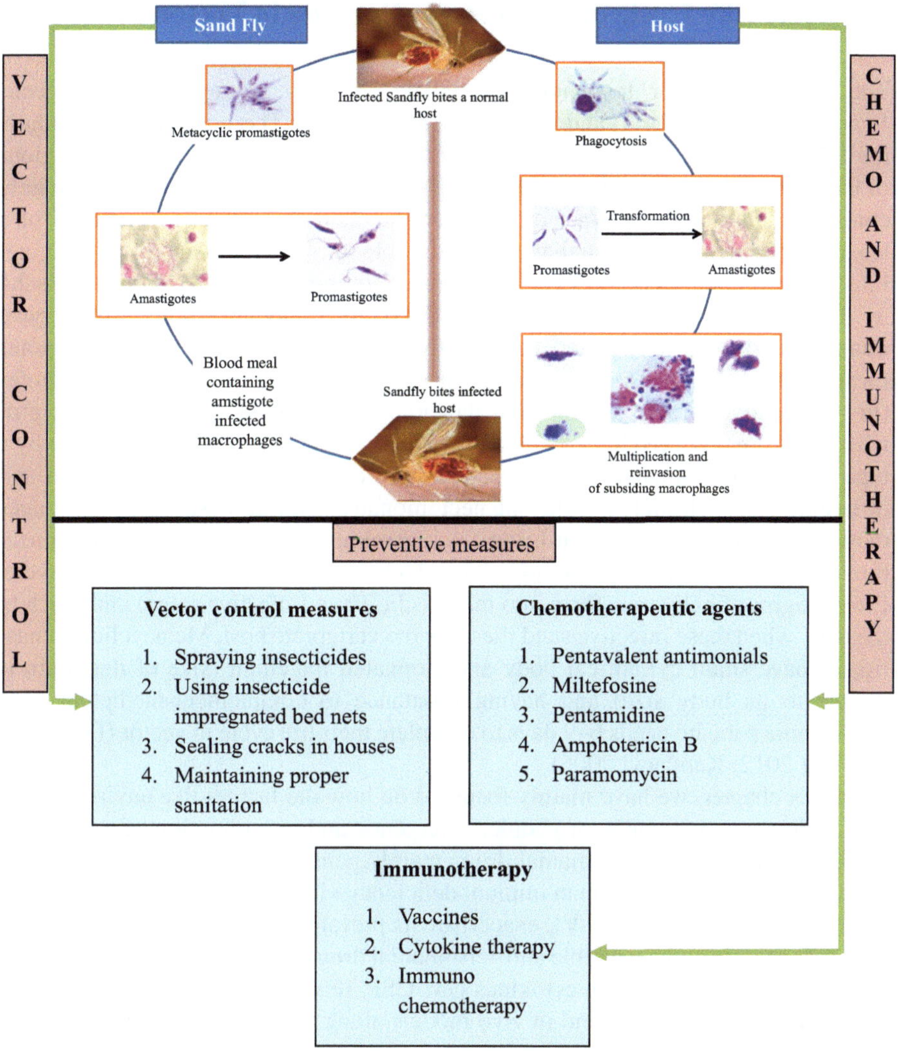

Fig. 5.1 Life cycle of *Leishmania* parasite and potential strategies for control of infection

domestic or wild animals are the reservoirs for the leishmania parasite. Based on this, leishmaniasis can be a zoo-anthroponotic (reservoir-vector-human) and anthroponotic (vector-human) disease. Infected female sand flies inoculate flagellate promastigotes in to the skin of humans; there the promastigotes are phagocytised by neutrophils and macrophages and form phagolysosomes. Within the phagolysosomes of macrophages, promastigotes are transformed into a flagellate amastigote stage which multiply rapidly and reinfect adjacent macrophages of the reticuloendothelial system.

5.3 The Life Cycle of Leishmania Parasite in Vector

The main vector for leishmania is female sand fly belonging to the genus *Phlebotomus* (Old World) and *Lutzomyia* (New World). These are tiny insects having 1.5–2 mm body length and are mainly found in tropical and subtropical regions of the globe. The development of leishmania parasite in vector (sand fly) is mainly confined to the digestive tract only. When the sand flies feed on the blood of a leishmanial-infected mammalian host, the blood meal containing amastigote-infected macrophages is ingested. These amastigotes reach to the midgut of the sand fly. The temperature and pH alterations in the vector's midgut favour morphological transformation of the ingested parasite from amastigote forms to promastigotes forms. The amastigote form transforms into weakly motile pro-cyclic promastigotes (first replicative forms), having short flagellum, and these sluggish pro-cyclic promastigotes are observed in early blood meal and are separated from the midgut by a type 1 peritrophic matrix. After 2–3 days, pro-cyclics slow their replication and differentiate into strongly motile long nectomonad promastigotes, and in later stages these nectomonads move towards the anterior midgut and transform into short nectomonad promastigotes (leptomonads), which enter another proliferative cycle. Promastigotes finally transform into metacyclic stage (infective stage) and reach to the host, when these infective sand flies bite the vertebrate host. Metacyclic promastigotes have small cylindrical body and elongated flagellum (size of flagellum is double to its body size) and having resistance to complement-mediated lysis. Leishmania parasite needs 6–9 days to complete their life cycle in vector (Dostalova and Volf 2012; Kamhawi 2006).

In this chapter, we have mainly focussed on how the factors like environmental and socio-economic status of people, emergence and consolidation of drug resistance to currently anti-leishmanial drugs mainly pentavalent antimonials, development of co-infections of human immunodeficiency virus (HIV-VL and malaria-VL) and diagnostic difficulties in VL exacerbate its prevalence. The strengths and weaknesses of the currently available anti-leishmanial drugs have been described, and the role of immunotherapy with cytokines GM-CSF, IFN-γ and IL-12 in the treatment of VL and several in vitro and in vivo models along with their pros and cons have been discussed.

5.4 Factors Influencing Leishmaniasis Infection

5.4.1 Environmental and Socio-economic Status

Environmental and socio-economic conditions of people are mainly responsible for the spreading of leishmaniasis. Environmental factors like global warming and deforestation favour the growth and survival of vector *Phlebotomus argentipes*. High temperatures in daytime and increased humidity in nighttime are the favourable conditions for vector growth and disease transmission. A small temperature change has greater effect on parasite growth in the midgut of the sand flies and enhances their infectivity. Due to deforestation there is a decrease in the boundaries between residential areas and forests. So, the people are highly exposed to vectors, and thus relatively more prone to vector-borne disease. American CL has become more prevalent in Central and South America after deforestation. Poor people are mainly affected by leishmaniasis compared to wealthy ones. Poverty, malnutrition, HIV and genetic susceptibility greatly influence the disease prevalence (Kolaczinski et al. 2008). Migration of people from their places to leishmania-endemic areas is one of the major causes of disease spread. Sand flies, mainly *P. orientalises* carrying *L. donovani* parasites, are mainly associated with black cotton soil and *Acacia balanite* forests in North West Ethiopia (Yared et al. 2014). This vector is actively present at nighttime in dried agricultural fields and causes leishmaniasis in agricultural labours, who reside nearer to this area. Poor sanitation, lack of personal hygiene and cracks in houses facilitate vector survival and lead to the spread of leishmaniasis. Spraying insecticides, using insecticide-treated bed nets and sealing of cracks in houses may decrease the chance of the occurrence of leishmaniasis. Goats and bamboo trees highly support the growth and survival of vector *Phlebotomus*; people who reside close to these areas are more prone to risk of infection. Ownerships of goats poses high risk of VL infection, compared to the ownerships of other livestock (Hasker et al. 2012).

5.4.2 *Drug Resistance, a Serious Obstacle for Leishmaniasis Treatment*

The emergence of drug resistance to currently available anti-leishmanial drugs leads to difficulty in the control of leishmaniasis. Low cost and ease of availability of anti-leishmanial drugs in India often leads to misuse of drugs which results in the development of resistance. Improper dosage forms, duration of intake and failure in completing course of treatment are the main reasons for the emergence of *L. donovani* strains resistant to anti-leishmanial drugs in clinical use. In HIV-VL co-infected patients, resistance emerges to anti-leishmanial therapy, and the rate of relapse is high. Pentavalent antimonials are the mainstay of therapy for the treatment of VL, but in India, especially in North Bihar (VL hyperendemic area), resistance has developed to these antimonials. Initially, in the early 1980s, very small doses and less duration (10 mg/kg for 6–10 days) of therapy to VL patients showed resistance. In later years, pentavalent antimonials at the dose of 20 mg/kg × 20 days also showed treatment failure in North Bihar. Thiol molecules cause oxidative stress inside the macrophages and reduce the formation of glutathione, which ultimately inhibits the conversion of pentavalent (inactive form) to trivalent (active form) antimonial, intracellularly. Increased intracellular thiol concentration leads to development of antimonial resistance. The main mechanism for the development of resistance to antimonials is thiol metabolism. The drug efflux transporters also have main roles in the emergence of drug resistance. Expression of drug efflux transporters like P-glycoprotein and multidrug resistance protein (MDR) results in the development of resistance to antimonials in laboratory strains of leishmania. Miltefosine, used to treat antimony-resistant VL patients, also suffers from the problem of drug resistance. Inactivation of miltefosine-internalising intracellular proteins *L. donovani* miltefosine transporter (LdMT) and *L. donovani* Ros 3 (LdRos3) lead to the development of resistance to miltefosine (Mohapatra 2014). *Leishmania* parasite has also acquired resistance towards lipophilic anti-leishmanial drug amphotericin B. Multidrug resistance (MDR) genes of ATP-binding cassette (ABC) family of promastigotes cause the efflux of amphotericin B which leads to the development of resistance. There are no molecular markers for the identification of drug-resistant strains of *L. donovani*. In vitro intramacrophage amastigote assay is a useful method for the monitoring of drug resistance in leishmaniasis. Therapeutic drug monitoring of VL patients, use of drug combinations with low dose of drugs for shorter time and use of immuno-chemotherapeutics may decrease the chances of the occurrence of anti-leishmanial drug resistance (Maltezou 2009). Identification and elucidation of mechanisms causing resistance is useful for development of newer anti-leishmanial drugs.

5.4.3 Nonavailability of a Proper Vaccine(s)

Presently available anti-leishmanial therapy is costly and causes serious adverse effects. Even though extensive research has been reported in the development of vaccines against leishmaniasis, till now there is no licenced vaccine(s) available around for human use. We are in urgent need for the development of proper vaccine for the prophylaxis and treatment of leishmaniasis. Earlier, leishmanization was a method that has been mainly practised in Soviet Union, Middle East and Israel. In this method, live virulent parasites are inoculated; however, the main drawback is the development of nonhealing cutaneous lesions and immunosuppression in some individuals. Whole-killed (autoclaved) promastigotes and autoclaved parasites along with adjuvant BCG also have been tested as vaccine. It has reduced the occurrence of leishmaniasis, but autoclaved parasites have shown reduced potency with time (Kumar and Engwerda 2014). Live attenuated, dendritic cell-based and DNA-based vaccines that have been tried against leishmaniasis are not fruitful.

5.5 Development of Drugs for Co-infections

Development of co-infections like HIV-VL and malaria-VL leads to the difficulty in the diagnosis and treatment of VL.

5.5.1 HIV-VL Co-infection

VL is a more commonly occurring opportunistic parasitic infection in immunocompromised patients affected by human immunodeficiency virus. It is very common in tropical, subtropical and Mediterranean regions of the globe. The HIV-VL co-infections are highly reported in Indian subcontinent, Western Europe and African regions, where VL is prevalent (Tavora et al. 2015). Both HIV and VL affect T-lymphocytes, macrophages and dendritic cells of the host. The immunosuppression caused by HIV facilitates uncontrolled multiplication of amastigotes in macrophages and accelerates the progression of VL disease and the replication of retrovirus (Lindoso et al. 2014). In vitro co-infection of monocyte-derived macrophages with HIV-1 virus and *L. donovani* promastigotes could hasten parasite growth as compared to the macrophages-infected with *L. donovani* promastigotes alone (Wolday et al. 1998). In HIV-infected patients, both T-cell proliferation and IFN-γ production are impaired, which supports the spread of the parasite in several locations. In HIV-VL co-infected patients, amastigote bodies have been observed in atypical locations including the digestive tract, skin, lungs and tonsils. These atypical symptoms have been more frequently observed in patients having a $CD4^+$ T-cell count of less than 50 cells/mm^3 (Rosenthal et al. 2000). In co-infected patients, HIV-mediated immunosuppression could lead to increased VL relapse rate in immunocompromised patients. Diarrhoea is the main symptom in HIV-VL co-infected patients. The choice of treatment is very difficult for this co-infection because both the disease accelerates the pathology of one another. Highly active antiretroviral treatment (HAART) is the main therapy for HIV-VL co-infected patients that could reduce the VL relapse (Okwor and Uzonna 2013).

5.5.2 Malaria-VL Co-infections

Malaria is a parasitic disease caused by *Plasmodium* species. Kala-azar and malaria are highly endemic in tropical countries like India and Nepal. The immunological status of patients in both the diseases is different. Low blood $CD4^+$ and $CD8^+$ T-cell ratio has been reported in chronic VL-infected patients, whereas in malaria patients this ratio remains unaltered. In co-infected patients, $CD4^+$ T-lymphocyte-mediated protective immunity against malaria was abolished by VL. Malaria cachexia is a condition in which the co-infection of malaria and VL leads to severe weight loss in children with malnutrition. In a case study of malaria-VL co-infected Nepal patient, fine needle aspiration of the left side lymph node showed *L. donovani* bodies in cytological diagnosis (Sah et al. 2002). Migration of VL patients to malaria endemic regions is one of the main reasons for acquisition of co-infections. Diagnosis and treatment of this condition is very difficult. In a case study, Nepalese VL patient was migrated to Terengannu in Malaysia where endemicity of malaria is high and there are patients co-infected with malaria. This co-infection led to difficulty in the diagnosis and treatment of VL (Ab Rahman and Abdullah 2011). In leishmania-malaria co-infected mice, *Plasmodium yoelii* and *L. mexicana* infections were significantly enhanced compared to mice infected with either parasite alone. Metastatic *L. mexicana* lesions were observed in co-infected mice, compared to *L. mexicana* alone infected mice model (Coleman et al. 1988).

5.6 Diagnosis of VL

The diagnosis of VL is difficult. The clinical features (irregular fever, weight loss, enlargement of the spleen and liver, decreased blood cell counts and anaemia) all resemble the clinical features of other diseases like malaria and tuberculosis, and currently available anti-leishmanial drugs are toxic. The lack of specificity in clinical features of VL needs highly sensitive and specific diagnostic tests (Table 5.1).

5.6.1 Microscopic Detection of Parasite

Microscopic detection of *L. donovani* parasites in the spleen, liver and bone marrow aspirates of VL patient is one of the best methods, but it is a painful procedure, and it requires skilful hand. Detection of *L. donovani* parasite in the spleen is a 95% more sensitive diagnostic method (Sundar and Rai 2002). Culture of bone marrow aspirates is, however, a more sensitive diagnostic technique than microscopy. Aspiration specimens are collected aseptically and cultured in Novy-MacNeal-Nicolle medium or in Schneider's Drosophila Medium supplemented with foetal calf serum. Cultures usually begin to show promastigotes in 2–5 days.

Table 5.1 Diagnostic methods for visceral leishmaniasis

S.No.	Method of detection	Characteristic observations
1.	Microscopic detection method	Identification *L. donovani* amastigotes in the liver, spleen and bone marrow aspirates of VL-infected patients. Parasite detection in spleen is 95% sensitive diagnostic method
2.	Culture method	Bone marrow aspirates of VL patients were cultured aseptically in NNN medium and observed for promastigotes in 2–5 days
3.	Haematological examination	Leucopenia, thrombocytopenia, pancytopenia, haemophagocytosis and anaemia are characteristic haematological features of VL. Estimation of total leucocyte counts, bone marrow cellularity and haemoglobulin content in VL patient is one of the diagnostic method
4.	Indirect fluorescent antibody test (IFAT)	Highly sensitive (96%) and specific (98%). Detects Abs which is present in early stages of infection
5.	rK39 ELISA	Specific diagnosis method for HIV-VL co-infected patients

5.6.2 Haematological Examination

Leucopenia, thrombocytopenia, pancytopenia, haemophagocytosis and anaemia are characteristic haematological features of VL. Estimation of total leucocyte counts, bone marrow cellularity and haemoglobulin content in VL patient is one of the diagnostic methods (Agrawal et al. 2013).

5.6.3 Indirect Fluorescent Antibody Test (IFAT)

IFAT is a specific test which detects antibodies, which are present in the early stages of infection, and these antibodies are not observed after anti-leishmanial drug therapy. It is highly sensitive (96%) and specific (98%), but the requirement of sophisticated laboratory conditions prohibits its application in the field. Other commonly used specific diagnostic tests are ELISA, direct agglutination test, immunochromatographic (ICT) strip test and PCR. rK39 ELISA is one of the best diagnostic method specifically for VL and HIV co-infected patients (Chappuis et al. 2007).

5.7 Biomarkers for VL

5.7.1 Direct Markers

Identification of kinetoplast DNA (kDNA) in clinical samples, and quantitative PCR for the detection of parasite load in VL patient's blood are direct markers of VL. Detection of VL-specific antigen is a predictive biomarker. KAtex is a urine-based latex agglutination assay, which detects a heat-stable low-molecular-weight carbohydrate antigen found in the urine of VL patients (Islam et al. 2004).

5.7.2 Indirect Markers

Indirect marker like macrophage-related marker neopterin, a catabolic product of GTP which belongs to heterocyclic pteridine compound, is synthesised by macrophages after interferon gamma (IFN-γ) activation. Neopterin is a marker for cellular immunity. Neopterin levels are elevated in infectious diseases like HIV, hepatitis B, hepatitis C, tuberculosis and malaria. Serum neopterin levels are elevated in VL patients and return back to normal levels after anti-leishmanial drug therapy (Schriefer et al. 1995; Hamerlinck et al. 2000). Adenosine deaminase (ADA) released from macrophages is mainly involved in purine metabolism. ADA levels are elevated in VL patients. In murine model of leishmaniasis, development of Th1 cytokine (IFN-γ, TNF-α and IL-12) response is associated with control of infection, and Th2 cytokine (IL-4, IL-10 and IL-13) response is associated with disease progression (Kip et al. 2015). IL-18 levels are raised in VL patients and reach to normal levels after chemotherapy (Hailu et al. 2004). Acute-phase proteins like C-reactive protein (CRP) and serum amyloid P component (SAP) are also elevated in active VL patients (Kip et al. 2015).

5.8 Current Chemotherapeutic Agents for VL

5.8.1 Pentavalent Antimonials

Antimony has been used as therapeutics from antiquity. Rai Bahadur Sir Upendranath Brahmachari, an Indian scientist, synthesised urea stibamine (carbostibamide), a pentavalent antimony compound used in the treatment for Indian KA-infected patients in 1920. Professor Brahmachari was nominated for Noble Prize in 1929 for this great discovery, which saved the lives of millions of poor Indian KA-infected patients. In 1936, Schmidt developed a stable, water-soluble pentavalent antimonial, solustibosan. Pentavalent antimonials were first introduced in 1945 as first-line drugs for the treatment of leishmaniasis. Sodium stibogluconate (Pentostam^{Tm}) developed by GSK and meglumine antimoniate (Glucantime) developed by Aventis are the two main antimonial drugs that are used for the treatment of both VL and CL. These drugs are mainly active against intramacrophage amastigote stages of *L. donovani* because these pentavalent antimonials (Sb^{V}) can be reduced to active and toxic trivalent antimonials (Sb^{III}) by amstigotes and not by promastigotes (Haldar et al. 2011). The mode of action of pentavalent antimonials is thought to be the specific inhibition of the DNA topoisomerise-1 of *L. donovani.* Sodium antimony gluconate (SAG)-treated *L. donovani*-infected macrophages releases reactive oxygen species (ROS) and nitric oxide (NO) by activating phosphoionositide 3-kinase (PI3K) and mitogen-activated protein kinase (MAPK), which are potentially involved in parasite-killing mechanisms (Basu et al. 2006). The treatment schedule is 20 mg/kg/day for 28 days. Long duration of treatment through painful intramuscular injections and tissue accumulation of antimonials causes treatment difficulty. The main serious side effects include vomiting, nausea, anorexia, myalgia, abdominal pain, headache, arthralgia, severe cardiotoxicity, pancreatitis and nephrotoxicity. Reportedly, resistance has developed to these antimonials in Indian subcontinents mainly in North Bihar (Croft et al., 2006b). The combination therapy of sodium stibogluconate (20 mg/kg) and allopurinol (20 mg/kg) for 14–54 days effectively cures the sodium stibogluconate-resistant VL patients; no amastigote bodies were observed in patient's splenic aspirates after 19 days of therapy (Chunge et al. 1985). Anti-leishmanial drugs like SAG, miltefosine and paromomycin have shown immunomodulatory effects like T-cell stimulating ability, enhanced production of IL-12,

TNF-α, NO and ROS on BALB/c mice peritoneal macrophages, directly (Ghosh et al. 2013). Emergence of resistance, poor bioavailability, painful daily injection schedule for long duration and severe side effects are some of the main drawbacks with pentavalent antimonials. Therefore, better improved pentavalent antimonial preparations like safe liposomal-encapsulated antimony-based formulations for targeting intracellular parasite and development of orally active pentavalent antimonials by improving their oral bioavailability with the addition of β-cyclodextrin may be useful in treating leishmaniasis (Frezard et al. 2009).

5.8.2 Miltefosine

Miltefosine, chemically hexadecylphosphocholine, is the first oral drug for the treatment of VL. In the late 1980s, it was developed as an experimental anticancer drug by German scientists. Miltefosine was registered in India in 2002 and has saved many people who were refractory to pentavalent antimonials and effectively treated childhood VL. Miltefosine is active on both the promastigote and amastigote forms of leishmania, in vitro. Miltefosine mainly affects the parasite lipid metabolism by inhibiting the synthesis of phosphotidylcholine and affects cell-signalling pathways and membrane synthesis and thus ultimately causes parasite apoptosis (Sundar and Olliaro 2007). Miltefosine shows immunomodulatory properties like enhancement of IFN-γ receptors, IFN-γ-induced STAT-1 phosphorylation, p38MAP kinase-dependent anti-leishmanial functions and IL-12 dependent Th1 responses in *L. donovani*-infected BALB/c mice (Wadhone et al. 2009). The combined therapy of miltefosine and amphotericin B/paromomycin is highly efficient against antimony-resistant Indian KA patients (Croft et al. 2006a). Miltefosine (25 mg/kg/ once or twice weekly, *p. o.*) could suppress the posttreatment recurrence in *L. donovani*-infected T-cell-deficient nude mice and can be useful in the oral maintenance therapy in T-cell-deficient patients mainly observed in HIV-VL co-infected patients (Murray 2000). The combination of antiretroviral drugs (protease inhibitors and reverse transcriptase inhibitors) and miltefosine has tested on *L. infantum*-infected BALB/c mouse peritoneal macrophages and promastigotes. In this, combination of miltefosine and efavirenz (non-nucleoside reverse transcriptase inhibitor) has shown better leishmanicidal activity and may be useful in the treatment of HIV-VL co-infected patients (Costa et al. 2016). The common side effects of miltefosine are vomiting, nausea, diarrhoea and loss of appetite. The main limitation for usage of miltefosine is teratogenicity (causes birth defects in pregnant women). Development of drug combination therapies and immuno-chemotherapy with miltefosine may be a better treatment strategy for decreasing the chance of the emergence of drug resistance.

5.8.3 Pentamidine

Pentamidine isethionate is the most commonly used treatment for *Pneumocystis carinii* pneumonia. After developing resistance to pentavalent antimonials, pentamidine which is a diamidine compound is used as second-line treatment for leishmaniasis. Pentamidine is used as secondary prophylaxis to prevent relapse in four immunocompromised patients (three with HIV-1 and one with idiopathic CD4+ lymphopaenia) with relapsing VL (Patel and Lockwood 2009). The combination therapy of pentamidine (300 mg/kg/day, *i.v*) and fluconazole (200 mg/kg/day, *p.o*) for 3 weeks could effectively reverse the relapse in HIV-VL co-infected patients, previously treated with liposomal amphotericin B and miltefosine (Rybniker et al. 2010). Pentamidine isethionate (two intramuscular injections of 4 mg/kg at 2 days interval) effectively treated *L. braziliensis guyanensis*-infected CL patients in French Guiana (Nacher et al. 2001). Pentamidine mainly acts on the AT bases of kinetoplastid DNA (kDNA) of the parasite. Gastrointestinal discomfort, bronchospasm and increased serum creatinine levels are commonly observed in pentamidine-administered patient. Pentamidine isethionate causes cardiotoxicity, renal toxicity and metabolic disturbances like diabetes mellitus (insulin-dependent diabetes in most cases) (Jha 1983).

5.8.4 Amphotericin B

Development of resistance to conventional pentavalent antimonial therapy in India, mainly in Bihar, where VL is hyperendemic, ensued in the introduction of amphotericin B as a first-line parenteral treatment for VL. In India, amphotericin B is introduced for the treatment of refractory VL due to the failure of existing drugs. Amphotericin B has the ability to sequester the cholesterol in host cell membrane, thereby inhibiting the macrophage-parasite interaction. The mechanism of action is mainly its ability to bind ergosterol in parasite cell membrane (Chattopadhyay and Jafurulla 2011). Amphotericin B effectively treats the *L. donovani*-infected euthymic and nude BALB/c mice and may be useful in the treatment of T-cell-deficient VL patients (Murray et al. 1993). Amphotericin B (0.75–1 mg/kg for 15–20 doses daily or on alternate days) through intravenous infusions showed more protection (nearly 100% cure rate) in VL patients. Painful intravenous route of drug administration is the main drawback for amphotericin B. Single dose of liposomal amphotericin B (5 mg/kg) and followed by 7–14 days of short-course oral miltefosine therapy effectively treated *L. donovani*-infected Indian VL patients, and this combination therapy may be useful in the reduction of development of drug resistance and duration of hospital stay of patients (Sundar et al. 2008). Amphotericin B causes severe adverse effects which include hypokalaemia, nephrotoxicity and myocarditis. These effects require close observation and hospitalisation of patients and raise the cost of therapy. These adverse effects are due to high exposure of free- drug to systemic circulation and deposition in organs. The lipid formulations of amphotericin B have improved pharmacokinetic properties and targeted delivery to organs like the liver, spleen and bone marrow macrophages, where leishmania parasites reside. Lipid-associated formulations including amphotericin B lipid complex (ABLC), liposomal amphotericin B (L-AmB) and amphotericin B colloidal dispersion (ABCD) were developed for the enhancement of therapeutic efficacy of amphotericin B (Hamill 2013).

5.8.5 *Paromomycin*

Paromomycin (aminosidine) is an aminoglycoside class of antibiotic with unique anti-leishmanial activity. Paromomycin was first isolated from cultures of *Streptomyces rimosus* having antibacterial activity for the treatment of intestinal infections like amoebiasis and giardiasis (Sundar and Chatterjee 2006). Aminosidine ointment (15% aminosidine and 10% urea in white paraffin) has been reported to cure CL effectively in *L. major*-infected patients in the Islamic Republic of Iran (Asilian et al. 2003). Aminosidine (11 mg/kg/day for 21 days) through intramuscular injections appproved as a first-line treatment for VL in Bihar (Moore and Lockwood 2010). Paromomycin involves with mitochondrial ribosomes and causes respiratory dysfunction in *L. donovani* promastigotes. Paromomycin interacts with 30S ribosomal subunit which leads to the inhibition of RNA and protein synthesis (Maarouf et al. 1997). Paromomycin acts synergistically with pentavalent antimonials. Combined therapy of paromomycin and antimonials has been shown to shorten the time course of therapy and reduces ototoxicity and nephrotoxicity caused by aminoglycoside antibiotics in VL patients in Kenya, India and Sudan (Sundar et al. 2009).

5.8.6 *Azoles*

Azoles were initially developed as antifungal drugs and after that were reintroduced in anti-leishmanial therapy by hampering the parasite ergosterol synthesis by inhibiting the enzyme lanosterol 14-α demethylase. Azoles actively inhibit leishmania culture growth in vitro. N-substituted azoles such as ketoconazole, fluconazole and itraconazole have more effect on leishmania parasite (Croft et al. 2006a, b). Fluconazole (200 mg capsule daily for 6 weeks) effectively treats CL caused by *L. major* in Saudi Arabian people (Zvulunov et al. 2002). The combined therapy of ketoconazole and allopurinol successfully treated a VL-infected renal transplant recipient who had developed pancreatitis due to the long-term use of sodium stibogluconate (Halim et al. 1993).

5.8.7 Sitamaquine in Clinical Trials: A Hope for Controlling VL

Sitamaquine (WR6026) is chemically (N,N-diethyl-N′-(6-methoxy-4-methyl-quinolin-8-yl)-hexane-1, 6-diamine) an 8-amino quinoline derivative developed by Walter Reed Army Institute of Research (WRAIR) and GlaxoSmithKline (GSK), collaboratively, as a potential compound for the treatment of leishmaniasis. Sitamaquine is an orally active anti-leishmanial agent like miltefosine. Sitamaquine reduces the risk of the development of resistance to leishmania parasite due to short elimination half-life (26 h) in humans (Loiseau et al. 2011). Sitamaquine could reverse miltefosine resistance by modulating LMDR1 (P-gp transporter in Leishmania ABC family involved experimental miltefosine resistance) in multidrug-resistant *L. tropica* line that overexpresses LMDR1 (Perez-Victoria et al. 2011). Sitamaquine binds to anionic polar head groups of phospholipids and accumulates in cytosol of *L. donovani* promastigotes in a sterol-independent manner. Lipid trafficking mechanism of leishmania could not be affected by sitamaquine and has been evidenced by ^{1}H NMR analysis of motile lipid (Loiseau et al. 2011). Sitamaquine induces oxidative stress and increases intracellular calcium levels by inhibiting succinate dehydrogenase in *L. donovani* promastigotes and ultimately causes apoptotic death of the parasite (Carvalho et al. 2011). In phase 2 clinical trial, *L. chagasi*-infected Brazilian patients are cured by the administration of sitamaquine, orally, at the dose of 2 mg/kg/day for 28 days (cure rate 67%; Dietze et al. 2001). In phase 2 clinical trials of sitamaquine in *L. donovani*-infected Kenyan patients showed different cure rates at different doses (1.75, 2.0, 2.5 and 3.0 mg/kg/day for 28 days show cure rates of 92%, 80%, 82% and 91%, respectively), and abdominal pain and headache are the common adverse effect; one patient showed severe renal failure at the doses of 2.5 and 3.0 mg/kg/day (Wasunna et al. 2005). Phase 2 clinical trials of sitamaquine in *L. donovani*-infected Indian VL patients showed 100% cure rate at 2 mg/kg/day dose for 28 days, and vomiting, dyspepsia and nephrotic syndrome are commonly observed adverse effects in Indian patients (Jha et al. 2005) (Fig. 5.2 and Table 5.2).

5.8.8 Cytokine Therapy for Leishmaniasis

The emergence of *Leishmania* parasites resistant to anti-leishmanial drugs and the non-availability of effective vaccine(s) against leishmania pose a serious challenge to leishmania control efforts. Effective immunotherapy/immuno-chemotherapy is considered as a viable alternative to the control of leishmaniasis. Cytokines (GM-CSF, IFN-γ and IL-12) stand-alone or in combination with current anti-leishmanial drugs may reduce the emergence of drug resistance.

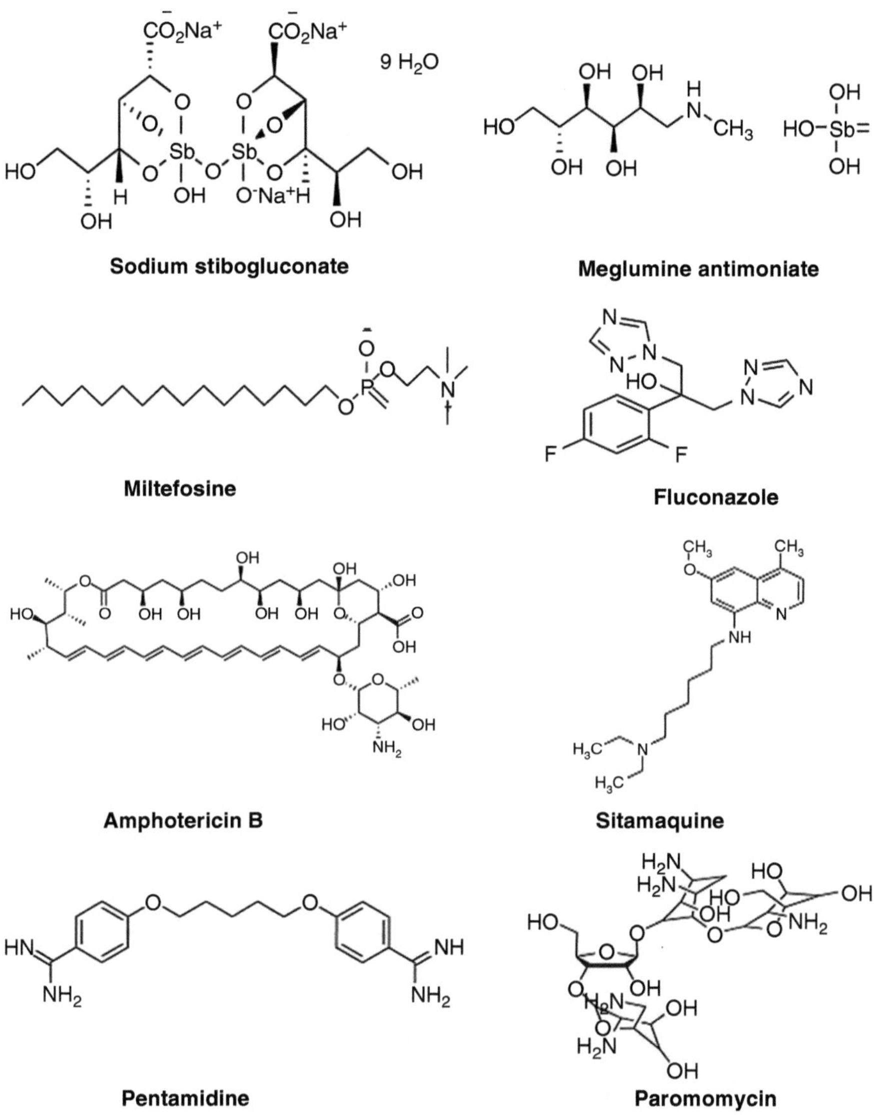

Fig. 5.2 Chemical structures of currently available anti-leishmanial drugs

GM-CSF GM-CSF is an immunoregulatory glycoprotein cytokine having a molecular weight of 23 kDa. In vivo, GM-CSF stimulates haematopoiesis, and in vitro, it stimulates differentiation and proliferation of committed progenitor cells. GM-CSF enhances the phagocytosis of macrophages and improves the host defence (Kaur et al. 2004). GM-CSF was first isolated and purified from a mouse lung-conditioned medium, which stimulated mouse bone marrow cells to proliferate into granulocytes or macrophages or both, by performing in vitro colony-stimulating factor activity

Table 5.2 Currently available anti-leishmanial drugs

S.No	Drugs	Mode of action	Dose	Advantages	Limitations
1.	Sodium stibogluconate	Acts as a prodrug. Converted Sb v to more toxic Sb III form, creates oxidative stress to parasite (macrophage stage)	20 mg/kg/day (i.m.) for 20–30 days	Low cost and easily available	Pancreatitis, acquired resistance in the Indian subcontinent
2.	Miltefosine	Parasite cell death by apoptosis (parasite cell membrane, inositol, PL activation)	50 mg/kg/day (p.o.)	First orally active, currently first-line drug in Indian subcontinent	Hepatic and renal toxicity
3.	Amphotericin B	Binds ergosterol in parasite membranes and creates pores which alter ion balance, increase membrane permeability and cell death	0.75–1.0 mg/kg for 15–20 infusions either daily or alternative days (i.v.)	Effective in antimony-resistant areas	High-cost, prolonged hospitalisation, fever with renal complications and hypokalaemia
4.	Liposomal amphotericin B (AmBisome)	Targeted delivery of drug to infected macrophages	3 mg/kg/day for 5 days (i.v.)	Highly effective, low toxicity, resistance is not documented	High cost
5.	Pentamidine	Accumulates in parasite mitochondria and inhibits kDNA	4 mg/kg/day for 3 times weekly for 15–20 dose (i.m. or i.v.)	Useful in combination of other drugs	Pancreatitis and irreversible IDDM
6.	Paromomycin/ Aminocidin	Protein synthesis inhibitor	11 mg/kg/day for 21 days (i.m.)	Low cost, effective in combination with antimonials	Reversible ototoxicity and lack of efficacy in East Africa

assay (Burgess et al. 1977). Recombinant GM-CSF has been used in the treatment of infectious diseases like malaria and tuberculosis. Immunoadjuvant therapy with rh GM-CSF has shown protective effects in pulmonary tuberculosis patients in phase 3 clinical trials. GM-CSF knockout mice are more susceptible to malaria than wild-type mice infected with *Plasmodium chabaudi* AS, as confirmed by higher peak parasitaemia, recurrent recrudescent parasitaemia and high mortality rate. Combination of rm. GM-CSF and enkephalin fragment peptide Tyr-Gly-Gly has shown protective effects in *Plasmodium berghei*-infected Swiss mice (Kaur et al. 2004). GM-CSF treatment enhances blood monocyte mobilisation, macrophage activation and ameliorates granulocytopenia in *L. donovani*-infected BALB/c mice. Anti-GM-CSF antibody treatment exacerbates the visceral infection, and liver parasite burden was more intensified in *L. donovani*-infected BALB/c mice. During leishmania infection in BALB/c mice, the splenic GM-CSF levels were greatly increased. A 12 kDa *L. donovani* amastigote antigen induced the elaboration of colony-stimulating factors (CSFs) by mouse macrophages, in vitro (Singal and Singh 2005). In leishmaniasis, increased number of GM-CSF cells mediates protection and causes resistance as seen in the case of leishmania-resistant mouse strain (C57BL/6) (Saha et al. 1999). rh GM-CSF has been reported to enhance the intracellular killing of *L. donovani* in human monocyte-derived macrophages in an LPS-independent manner, and the time required for activation of macrophages to show anti-leishmanial effect is very less (rh GM-CSF shows maximal activation at 36 h) as compared to rh IFN-γ, which requires 48–72 h to activate macrophages (Weiser et al. 1987). Purified GM-CSF, isolated from mouse lung-conditioned medium, enhances intracellular killing of *L. tropica* in mouse peritoneal macrophages (Handman and Burgess 1979). The intracellular amastigote killing by rh GM-CSF and M-CSF was more intensified by combining with rh IFN-γ. Combination of rh GM-CSF and sodium stibogluconate effectively treated the CL patients; the lesions were healed in very less time with an antimony dose of 20 mg/kg/day for 20 days. The use of rh GM-CSF in combination with sodium stibogluconate can reduce the dose and duration of antimony therapy and is useful in decreasing the drug toxicity. Combination of rh GM-CSF (10 mμg/ml in 0.9% saline is reapplied topically, and dressings were changed thrice/week for 3 weeks) and antimonials (20 mg/kg/day for 20 days) successfully treated the refractory CL patients (Almeida et al. 1999). A combination of GM-CSF (5 μg/kg/day for 10 days) and pentavalent antimony (10–20 mg/kg/day for 20 days) treatment can rapidly normalise the neutrophil, eosinophil and monocyte counts and reduced the viral and bacterial secondary infections in patients with acute VL (Badaro et al. 1994). Immuno-chemotherapy with GM-CSF and liposomal amphotericin B effectively treats the VL in VL-HIV co-infected patients (Mastroianni 2004). Transgenic recombinant L. major GM-CSF-expressing promastigotes infected BALB/c mouse peritoneal macrophages release GM-CSF can activate macrophages to release high levels of pro-inflammatory cytokines (IL-1β, IL-6 and IL-18) and chemokines (RANTES/CCL5, MIP-2/CXCL-2 and MCP-1/CCL2) involved in enhanced intramacrophage parasite killing in vitro and in vivo. These transgenic promastigotes delay the lesion development in BALB/c mice (Dumas et al. 2003).

IFN-γ IFN-γ is a pleiotropic glycoprotein cytokine having a molecular weight of 20–25 kDa, enhances host defence and is mainly used in the treatment of infectious diseases like toxoplasmosis, leishmaniasis and tuberculosis (Gallin et al. 1995). T-lymphocytes (Th1 cells, T_C cells and NK cells) are the source for IFN-γ production and activate macrophages for boosting defence against infectious diseases. IFN-γ enhances the intramacrophage leishmanicidal effect by activating macrophages to release Th1-type cytokines. Th1 cytokines are mainly involved in protection, and Th2 cytokines are involved in disease pathology (Sundar and Chatterjee 2006). Leishmania parasite control mainly involves increased levels of IFN-γ by $CD4^+$ T cells and exacerbates the disease condition with the absence of IFN-γ in *L. major*-infected C57BL/6 (resistant) and BALB/c (susceptible) mouse strains, respectively (Kima and Soong 2013). Treatment with monoclonal anti IFN-γ antibodies causes CL in C3H/HeN mice infected with *L. major*, which are naturally resistant to CL. Production of IFN-γ during initial period of host-parasite interactions in *L. major*-infected C3H/HeN mice is the major component of genetic control of natural resistance (Belosevic et al. 1989). Treatment with native human IFN-β and IFN-γ of *L. tropica* major-infected human mononuclear phagocytes has shown anti-leishmanial effects three times more than IFN-β, and IFN-γ released enhanced H_2O_2 levels in culture supernatants in a dose-dependent manner; the H_2O_2 release from monocytes was abrogated by the administration of monoclonal IFN-γ antibody. Leishmania-specific CD4 T cells produce IFN-γ which limits parasite replication in VL patients (Singh et al. 2014). In a pilot study, nine VL patients were treated with IFN-γ for 20 days and cleared 100% parasites in bone marrow aspirates of four patients, and five patients have shown reduced parasites in their bone marrow aspirates (Sundar et al. 1995). Combined therapy with rh IFN-γ and pentavalent antimony successfully treated the formerly untreated VL patients and refractory patients (Sundar and Chatterjee 2006). Immuno-chemotherapy with IFN-γ and pentavalent antimonials effectively treated the previously untreated VL patients and was helpful in reducing the duration of conventional therapy. The common side effects with IFN-γ are fever, body aches and flu-like symptoms. IFN-γ-induced fever can be reversed by the administration of antipyretics.

Interleukin-12 (IL-12) IL-12 is a 70 kDa heterodimeric immunoregulatory cytokine which is made up of two subunits of IL-12p35 (35 kDa) and IL-12p40 (40 kDa), linked by a covalent bond. IL-12 is mainly produced by antigen-presenting cells such as macrophages, dendritic cells, monocytes and neutrophils. IL-12 is helpful in the production of IFN-γ and Th1 type cytokines by inducing differentiation of naive $CD4^+$ T cells to Th1 cells and stimulating natural killer cells. IL-12 can be useful in the immunotherapy of diseases where Th1 response is desirable. Due to the induction of IFN-γ and Th1-type cytokines in phagocytes by IL-12, it can be useful in controlling infectious diseases like bacterial, viral and parasitic diseases (Hamza et al. 2010). IL-12 restores IFN-γ production and cytotoxic response in American VL *(L. chagasi)* characterised by the absence of lymphocyte proliferative response and IFN-γ production. Treatment with exogenous rh IL-12 of

L. donovani lysate-stimulated PBMC from active VL patients has been shown to enhance the production of IFN-γ and anti-leishmanial Th1-type response (Bacellar et al. 1996). IL-12 has been effectively shown to treat the established systemic intracellular infection in *L. donovani*-infected BALB/c mice (Murray and Hariprashad 1995). Recombinant murine IL-12 has been shown to increase protection against *L. major* in infected BALB/c mice (Heinzel et al. 1993). IL-12 regulates the leishmanicidal effects of pentavalent antimonials in experimental VL. Combination of IL-12 and pentavalent antimonials effectively treated the *L. donovani*-infected IL-12p35 knockout mice (Murray et al. 2000). Combination of IL-12 DNA and leishmanial recombinant open reading frame F (rORFF) protein induced protective immunity against experimental VL (Tewary et al. 2006). BALB/c mouse peritoneal macrophages pretreated with rm. IFN-γ (1 ng/ml) and rm. IL-12 (1 ng/ml) have shown resistance to *L. major* promastigotes at early phase of infection (Ota et al. 2008).

5.9 In Vitro and In Vivo Anti-leishmanial Screening Methods

The reliable in vitro and in vivo screening methods should have the good correlation with the clinical condition of the disease.

5.9.1 In Vitro Anti-leishmanial Drug Screening Models

Anti-leishmanial drug discovery requires potential drug screening models for testing leishmanicidal activities of newer drugs. In vitro anti-leishmanial drug screening models are advantageous in testing huge number of compounds within a short period of time. Drug testing is possible on parasitic stages like extracellular promastigote stage (survives and multiplies in sand fly's midgut) and intracellular amastigote stage (survives and multiplies in host macrophages). Requirement of less amount of the test compound, less number of animal usage and consistent and quick generation of results for high number of compounds are some of the main advantages of in vitro models.

Promastigotes as Drug Screening Model Promastigotes can grow in cell-free medium. Promastigotes are generally cultured in simple media like M199, RPMI-1640 and Leibovitz-15 and maintained at 22–26 °C in a BOD incubator. Testing of newer potential anti-leishmanial compounds on promastigotes is simple and a highly popular method. In new anti-leishmanial drug screening method, counted number of promastigotes, generally 1.0–2.0 × 10^6 cells/ml in culture medium in the presence and absence of appropriate concentrations of test compounds and kept in a BOD incubator maintained at 26 °C. After 3 days of incubation, promastigote growth inhibition by test compounds is determined and compared with controls,

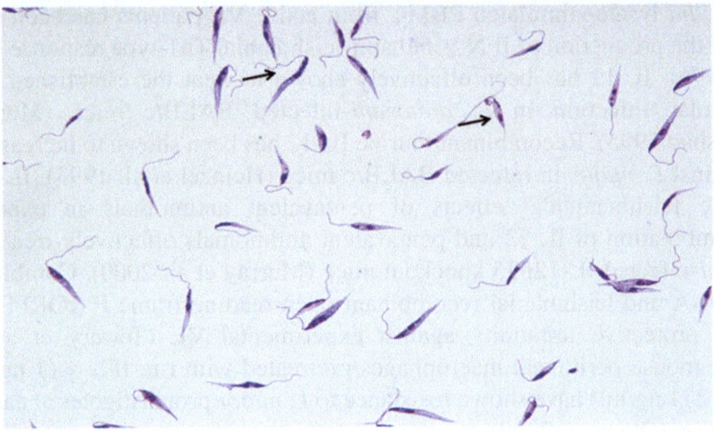

Fig. 5.3 Photomicrograph of *L. donovani* promastigotes (100X)

which can multiply three to six times. It is a rapid method of drug screening and requires very less time ease of maintenance in cell suspension form in vitro. But this flagellated promastigote stage is not present in the host, and thus it is not an appropriate target for anti-leishmanial drugs. Drug screening on promastigotes at 26 °C is of no or little value because in vivo temperature (37 °C) and the temperature at which the promastigotes are growing are different. Promastigotes can survive at 37 °C but there is no multiplication. The main limitation for this model is that the promastigote stage data do not correlate with that of amastigote stage data (Suman Gupta and Nishi 2011) (Fig. 5.3).

Axenic Amastigotes as Drug Screening Targets Axenic amastigotes permit in vitro primary screening of a large number of compounds in lesser duration and in an easy manner, very much like promastigotes. It represents the in vivo situation because this parasitic stage exists in hosts. Axenic amastigotes were maintained in Schneider's *Drosophila* medium supplemented with 20% FBS and 1% of penicillin streptomycin solution with acidic pH of 5.5 and are incubated at 32 °C. Amastigotes (2×10^5 cells/ml) are seeded and allowed to grow and multiply in the presence or absence of test compounds for 90 h. The number of amastigotes is calculated by using a haemocytometer and compared with that of controls, which can grow four to five times of the starting concentration (Callahan et al. 1997). High-throughput screening for new anti-leishmanial drugs by using luciferase gene expressing DNA-transformed *L. infantum* axenic amastigotes has also been developed. Metabolic processes of axenic amastigotes differ from that of intracellular amastigotes Sereno et al. 2001). Drugs are tested directly on amastigote stage, and the lack of the interplay of host cell-mediated effects (macrophages, phagolysosome formation and drug-induced toxic effects on host system) is the main limitation of this model.

Intracellular Amastigotes as Drug Screening Targets It is the most popular and widely used reliable method for new anti-leishmanial drug screening. In this method, amastigotes are allowed to infect cultured macrophages, and generally BALB/c mouse peritoneal macrophages (primary macrophages), J774A.1 macrophage cell-line (BALB/c mouse origin) and human monocyte transformed macrophage cell lines (THP-1, U-937, and HL-60) are used as the host cells. Macrophages are infected with promastigotes (multiplicity of infection 1:10 ratio), and after 2 h of incubation, extracellular (non-phagocytosed) promastigotes are removed and incubated at 37 °C in 5% CO_2 environment along with different concentrations of test and standard drugs. After 72–96 h of incubation, the activity of the test compound is determined by microscopic observation of the number of amastigotes/100 macrophages, and % inhibition is determined (Suman Gupta and Nishi 2011) (Figs. 5.4, 5.5 and 5.6).

$$\%\text{Inhibition} = 100 - (\text{AT} \times 100 / \text{AC})$$

AT – average number of amastigotes/100 macrophages in treated
AC – average number of amastigotes/100 macrophages in control

These differentiated non-dividing macrophages support the parasite division, and thus can be useful in the screening of new anti-leishmanial compounds. In vitro intramacrophage amastigote assay is a useful method for monitoring clinical resistance of leishmania parasites. New extension methods have been developed for this intramacrophage amastigote assay by applying quantitative real-time PCR, which can accurately determine the parasite DNA content in an amastigote-macrophage

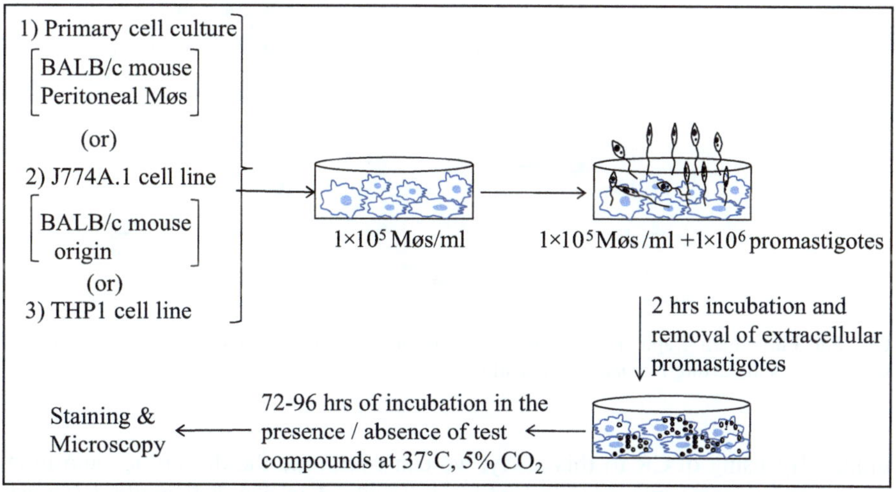

Fig. 5.4 Schematic representation of in vitro intramacrophage amastigote assay

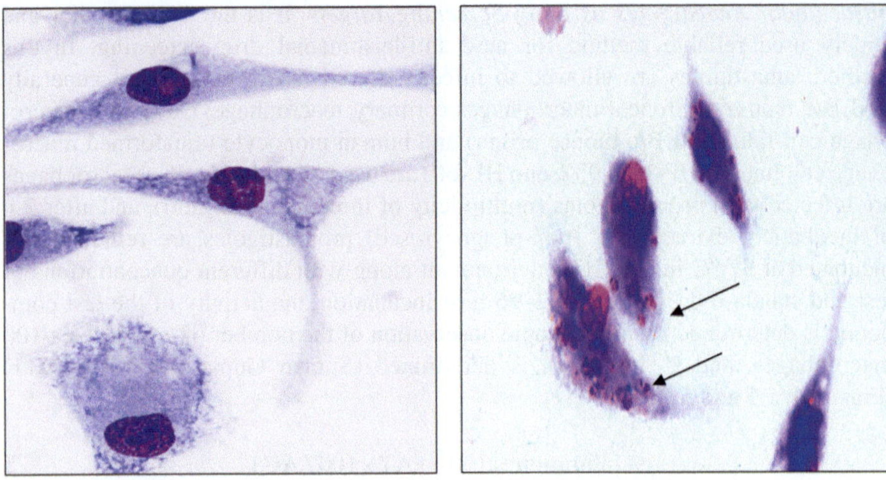

Control **Infected**

Fig. 5.5 Micro photographs of *L. donovani* amastigote-infected BALB/c mouse peritoneal macrophages (100X). Black arrows indicating *L. donovani* amastigotes

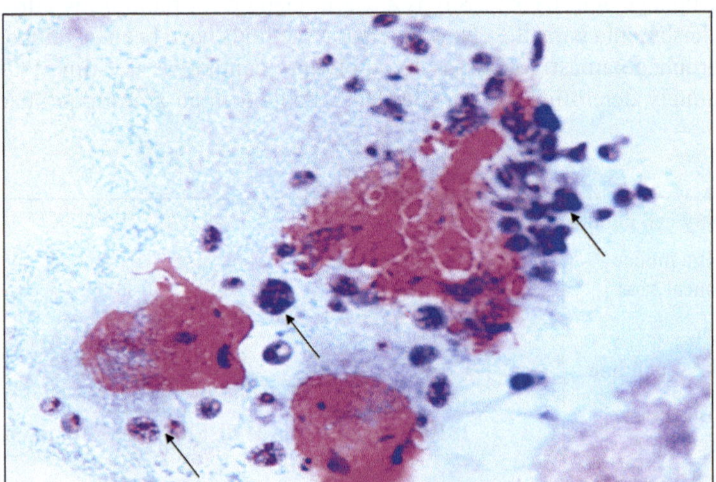

Fig. 5.6 Photomicrograph of *L. donovani* amastigote-infected J774A.1 macrophage (100X). Black arrows indicating *L. donovani* amastigotes

model. By using qPCR to this assay, we can overcome the difficulties and time-consuming microscopic counting which is involved in classical method (Gomes et al. 2012). Colorimetric assays have also been developed which can quantify the growth of intracellular amastigotes.

5.10 Reporter Gene Assays as Screening Models

Reporter genes generally encode a gene product that is a readily quantifiable phenotype and is easily distinguishable over endogenous cellular background. Based on their expression and existence, reporter genes can be classified as intracellular [chloramphenicol acetyltransferase (CAT), β-galactosidase, green fluorescent protein (GFP), firefly and bacterial luciferase and glucuronidase)] and extracellular reporter genes [(human growth hormone (HGH) and secreted alkaline phosphatase (SEAP)]. Reporter gene assays are advantageous as compared to conventional and currently available anti-leishmanial drug screening methods which are fraught with several problems like they are labour-intensive and time-consuming and lack automation (Dube et al. 2009). Genetically engineered recombinant leishmania parasites carrying reporter genes like GFP, CAT, β-galactosidase, firefly luciferase and alkaline phosphatases are mainly used reporter genes and facilitate monitoring of intracellular parasites. The main disadvantage of this method is the development of cross-resistance. In a high-throughput high-content intracellular *L. donovani*, assay in a 384-well plate has screened nearly 15,659 different compounds and has been proved better than the axenic amastigote screening method (De Rycker et al. 2013).

5.10.1 Green Fluorescent Protein (GFP) Assay

GFP, a bioluminescent protein, has been initially described as the green protein isolated from jelly fish *Aequorea victoria* (Chalfie 1995). Transfection of GFP in parasites like *Plasmodium* and *Leishmania* has been developed and used for drug-developing screening methods. Transgenic leishmania promastigotes which express GFP from episomal pXG vectors have proved the usefulness of GFP as a marker in the transfected leishmania (Ha et al. 1996). GFP is a cytoplasmic protein with low toxicity and has the possibility of continuous synthesis and ease of imaging and quantification. Introduction of GFP, as a marker in field strains of leishmania promastigotes by using leishmania expression vector pXG-GFP, has been developed, and in vitro anti-leishmanial compounds screening can be performed on these transgenic *L. donovani* promastigotes expressing GFP in their cytoplasm as a target in the cells. The relatively short duration of the screening experiments, possibility of automation, cost-effectiveness and the greater reliability are some of the main strengths of GFP assays and are more advantageous than classical drug susceptibility testing having the drawbacks of being labour-intensive and time-consuming and the requirement of macrophages (Singh and Dube 2004). GFP-tagged *L. donovani* promastigotes have been used for the observation of their developmental growth stages in the midgut of *Phlebotomus* vector easily as compared to non-tagged GFP (Guevara et al. 2001). *L. (Viannia) panamensis* promastigotes expressing GFP by transfection with p 6.5-egfp could retain its infection and are transformed into amastigote stages in U937 and J774 cell lines, in vitro, and the intracellular parasites expressing GFP can be easily identified by fluorescence microscopy, and flow cytometry can be useful in high-throughput screening of new potential anti-leishmanial compounds (Munoz et al. 2009).

5.11 In Vivo Anti-leishmanial Drug Screening Models

To understand the host-parasite interaction, immunobiology and pathology of VL needs a proper animal model. This knowledge is useful for the synthesis and development of new anti-leishmanial compounds. BALB/c mice and Syrian golden hamsters are the commonly used primary drug testing laboratory animal models for VL. Canines (dogs) and primates (monkeys) are commonly used as secondary drug testing models in VL drug discovery and development processes.

5.11.1 Mouse Model

The mouse model of VL has been extensively used for the development of vaccines and other related immunotherapeutics. Inoculation of *L. major* promastigotes in mice permits us to identify the immunological mechanisms involved in resistance (C57/BL6 strain) and susceptibility (BALB/c strain) to leishmaniasis. In-bred mouse strains are relatively more susceptible to leishmania infection. The susceptibility observed in BALB/c mice is due to the development of Th2 cytokines (IL-4, IL-10 and TGF-β), which can deactivate macrophages and favour intracellular parasite growth. Resistance in C57/BL6 strain is due to the development of Th1-mediated cytokine IL-12 which can activate macrophages to release IFN-γ, stimulates inducible NOS to release NO and kills the leishmania parasites (Matte et al. 2000). The mouse host, especially BALB/c mice, is used for studying organ-specific immunology. In *L. donovani*-infected BALB/c mice, amastigotes are highly replicated in the liver in the first few weeks of infection. Genetically resistant mice have natural resistance-associated macrophage protein 1 (NRAMP 1) gene, which is mainly involved in macrophage activation that can kill leishmania parasite by nitric oxide-mediated mechanisms (Loria-Cervera and Andrade-Narvaez 2014). Intravenous inoculation of 1×10^7 promastigotes has been reported to induce infection in mice. In the early stages of infection, the parasite burden is more in the liver as compared to the spleen. For routine new anti-leishmanial drug screening studies, BALB/c mouse strain has been extensively used. BALB/c mice are infected with 2×10^7 *L. donovani* amastigotes, intravenously, on Day 0 and randomly divided into groups (each group contains five mice). Treatment is given on day +7 to day +11 (five consecutive days), and the mice are sacrificed on day +14. Liver tissue touch prints are prepared on glass slides and stained for parasitological observation.

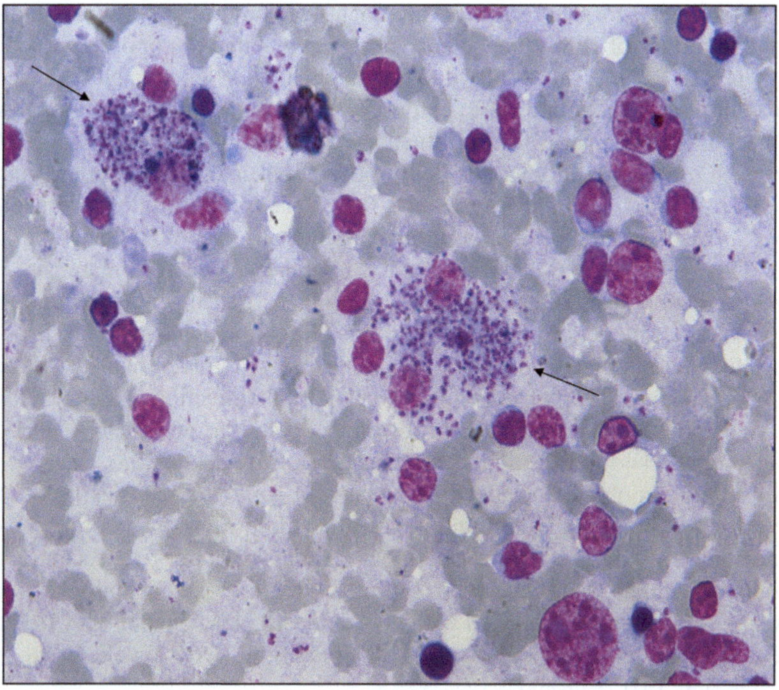

Fig. 5.7 Photomicrograph of *L. donovani*-infected BALB/c mouse liver tissue impression smears (100X). Black arrows indicate *L. donovani* amastigotes

The parasite burden is quantified as Leishman-Donovan Units (LDU; Suman Gupta and Nishi 2011) (Fig. 5.7).

$$\text{LDU} = \text{number of amastigotes} / 500 \text{host cell nuclei} \times \text{organ weight}(\text{mg})$$

5.12 Hamster Model

Syrian golden hamster (*Mesocricetus auratus*) is highly susceptible to infection by leishmania species like *L. donovani* and *L. infantum* and causes VL which absolutely resembles to human VL in clinical and pathological aspects like hepatosplenomegaly, pancytopenia, progressive cachexia, hypergammaglobulineamia, etc. It is extensively used as a screening model for new anti-leishmanial drug discovery (Nieto et al. 2011). An 8-day method for screening new anti-leishmanial drugs in golden hamster infected with 10^6–10^7 *L. donovani* amastigotes through intracardiac route has been developed for screening compounds in a short time (Stauber et al.

1958). *L. donovani*-infected hamsters are not able to control the replication of parasite due to failure of IFN-γ-mediated macrophage activation and decreased nitric oxide synthase-2 (NOS-2) activity due to defects in NOS-2 gene. NOS-2 is mainly involved in the production of nitric oxide (NO), a potent cytotoxic substance which kills the intracellular parasite. Leishmania-infected hamster macrophages do not generate NO, resulting in uncontrolled proliferation of intramacrophage parasites (Perez et al. 2006). The phagolysosome fusion has great implication in parasite survival, growth and multiplication in parasitophorous vacuole (Chang and Dwyer 1978). This knowledge is helpful in the chemotherapy of leishmaniasis. In VL patients haematopoiesis is severely affected which causes anaemia. Recently, it has been proved that induction of anaemia occurs due to the changes in erythropoiesis in the spleen and bone marrow of *L. donovani*-infected golden hamsters. Anaemia and leucopenia have been observed in 8 weeks of postinfection. Serum erythropoietin levels and BFU-E and CFU-E progenitor populations are greatly enhanced in the bone marrow and spleen of infected hamster (Lafuse et al. 2013). Because of the scarcity of immunological reagents, *L. donovani* -infected hamster model is not suitable for vaccination and immunotherapeutic studies. For overcoming the drawbacks of non-availability of immunological reagents for hamsters, recently, a new screening model has been developed. In this model, liver and splenic parasite burden was more in hamsters infected with 10^7 promastigotes, and not with 10^5 promasigotes, intracardially, after 155 days postinfection (Dea-Ayuela et al. 2007) (Figs. 5.8, 5.9, 5.10, 5.11 and 5.12).

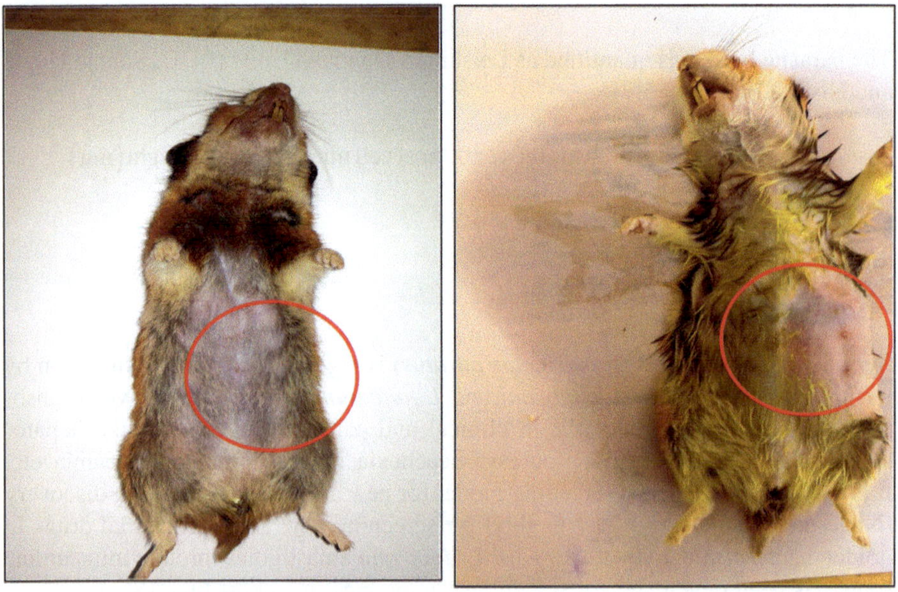

Fig. 5.8 Hepatosplenomegaly in *L. donovani*-infected hamster. The spleen and liver were enlarged in *L. donovani*-infected hamster compared to the control hamster

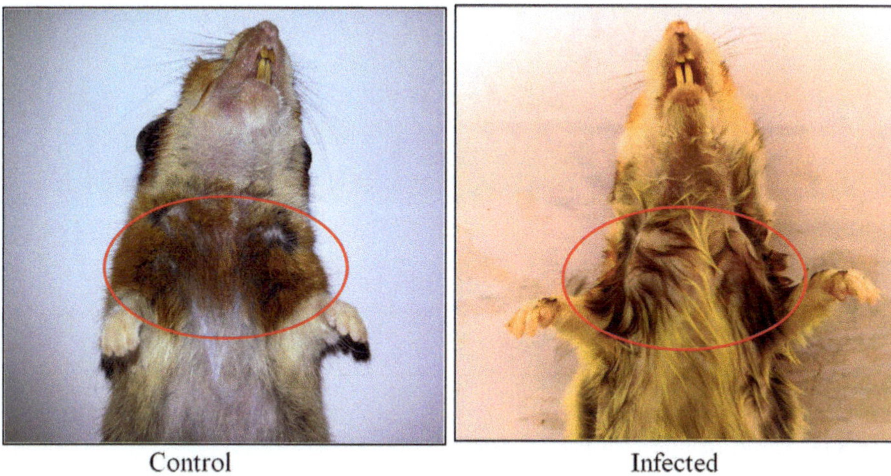

Fig. 5.9 Emaciation in *L. donovani*-infected hamster. Emaciation was clearly observed in *L. donovani*-infected hamster compared to the control hamster

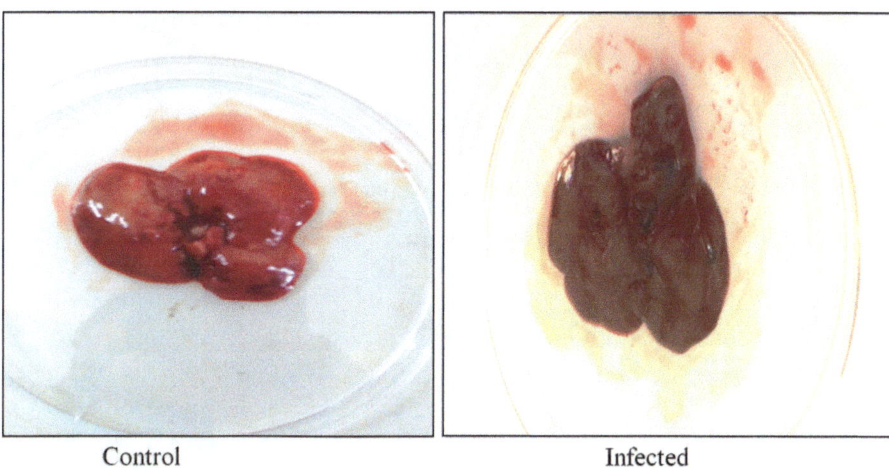

Fig. 5.10 *L. donovani*-infected hamster liver

5.13 Canine Model: A Secondary Drug Screening Model for VL

Canines are the best secondary drug screening models for VL. Drug metabolism and pharmacokinetic parameters in dogs, cats and monkeys are similar to human kinetic parameters. Mainly dog strains like stray, beagle and mongrel are more susceptible to VL infection and produce subclinical to fatal infection as like human VL. The

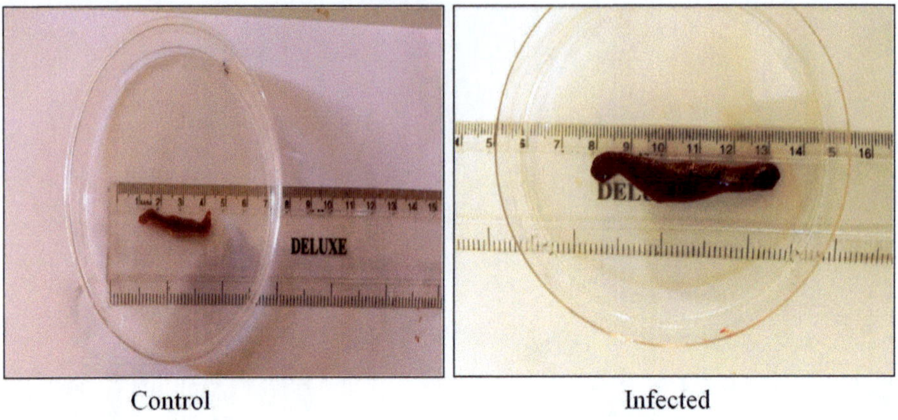

Fig. 5.11 *L. donovani*-infected hamster spleen

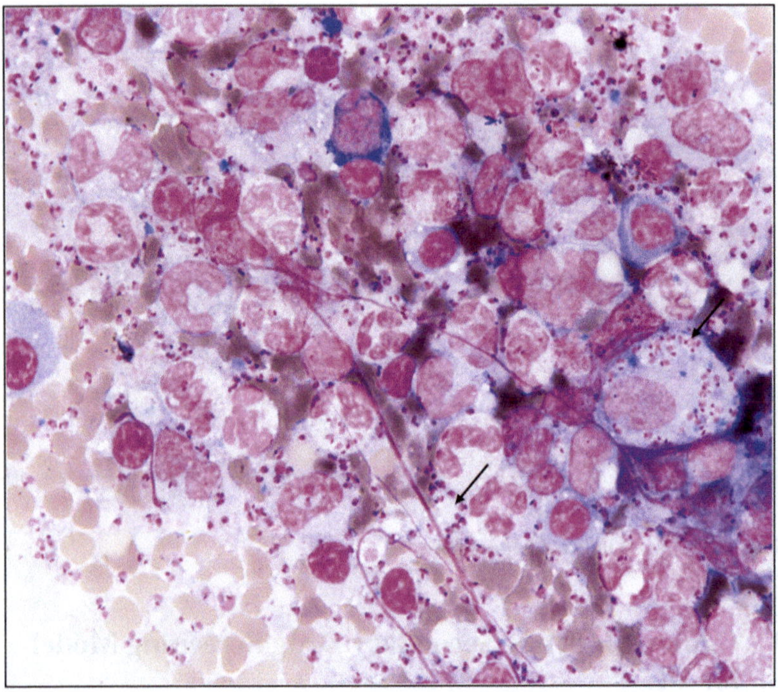

Fig. 5.12 Photomicrograph of *L. donovani*-infected hamster spleen tissue impression smears (100X). Black arrows indicate *L. donovani* amastigotes

dog is the major reservoir for *L. infantum* (in Middle East and Mediterranean region) and *L. chagasi* (in South America) (Loria-Cervera and Andrade-Narvaez 2014). Canine VL enhances the incidence of human VL in endemic regions where both the canine and human VL are prevalent, so development of vaccine against canine leishmania can reduce the incidence of human VL in both canine and VL endemic regions (Moreno and Alvar 2002). Anaemia, hypergammaglobulinaemia, fever, progressive weight loss with decreased appetite, swelling of lymph nodes, skin lesions and epistaxis (nose bleeding) are common pathological symptoms in canine leishmaniasis. In *L. infantum*-infected dogs, macrophages are known to be activated by IFN-γ and TNF-α to kill intracellular amastigotes through NO pathway; similar mechanism was observed in human VL. In *L. infantum*-infected dogs, m-RNA levels of IL-4 are greatly enhanced, and balanced production of Th1 and Th2 cytokines has been observed in infected dog spleen cells. Both CD4+ and CD8+ T-lymphocytes levels are to be diminished in *L. infantum*-infected dogs (Barbieri 2006). A transmission-blocking vaccine, FML vaccine (combination of FML antigen of *L. donovani* and saponin), has shown to protect against canine leishmaniasis (Saraiva et al. 2006). Because of the availability of huge immunological markers, the dog model of VL has become the best suitable experimental model for developing vaccine and immunotherapeutic agents for treating VL.

5.14 Primate (Monkey) Model: A Secondary Drug Screening Model for VL

Non-human primate models are phylogenetically close to human models. Physiology, pathology and immunology of non-human primates can mimic human VL. The monkey model is mainly useful for studying immunobiology of infection and host-parasite interactions. Non-human primate models like owl monkey (*Aotus trivirgatus*), squirrel monkey (*Saimiri sciureus*), marmoset (*Callithrix jacchus*), African green monkey (*Chlorocebus sabaeus*) and Indian langur monkey (*Presbytis entellus*) and vervet monkey (*Cercopithecus aethiops*) are commonly used preclinical models for anti-leishmanial drug screening studies (Olobo et al. 2001). Khartoum strain (WR378) of *L. donovani* amastigotes (3.25×10^7) that infected owl monkeys (total monkeys infected = 8) has shown the progressive weight loss, anaemia, hepatosplenomegaly, increased levels of serum hyperglobulinaemia, azotaemia and hyperalubinaemia, and high number of parasites have been observed in the liver, spleen, bone marrow and lymph nodes. Six owl monkeys died in 98 days of postinfection, and these findings support that owl monkeys are more susceptible to *L. donovani* infection and can be used as an animal model for VL drug discovery (Broderson et al. 1986). *L. infantum* (2×10^7 amastigotes/kg of body weight, *i.v.* route)-infected rhesus monkeys (*Macaca mulatta*) have shown clinical and immunopathological symptoms similar to human VL. Their findings suggested that rhesus monkey model is useful for preclinical screening of potential anti-leishmanial compounds and for the development of vaccine candidates for human VL (Porrozzi et al. 2006).

5.15 Conclusions

VL, a neglected tropical disease with high death toll every year, needs early diagnosis and proper treatment. Unfortunately, high cost, severe adverse effects of the currently available anti-leishmanials and emergence of leishmanial parasite resistant to pentavalent antimonials limit their clinical use. Development of highly sensitive and specific diagnostic methods for VL, HIV-VL and malaria-VL may be advantageous in leishmania disease control. Immunotherapy with cytokines (GM-CSF, IFN-γ and IL-12), both stand-alone and in combinations with current anti-leishmanials (immuno-chemotherapy), can emerge as a useful alternative therapy to halt the spread and consolidation of drug resistance. There is an urgent need for the discovery and development of new anti-leishmanial drugs which should be of low cost, require short-course therapy, have high oral bioavailability and are free from adverse effects, for the better control and management of VL. In vitro and in vivo drug screening methods are also available for testing of new anti-leishmanial compounds. In vitro intramacrophage amastigote method is a better reliable method which mimics human VL situation and is useful for the screening of potential new anti-leishmanials. In vivo animal models like rodent models, BALB/c mice, are useful in immunological studies, and Syrian golden hamsters are useful in chemotherapy studies. Higher models like canine model and non-human primate (monkey) models are useful secondary drug testing models for VL drug discovery and development.

Acknowledgements The author is thankful to Prof. R. R. Akkinepally, Director, National Institute of Pharmaceutical Education and Research (NIPER), S. A. S. Nagar for encouragement and help. The financial assistance from NIPER is greatly acknowledged.

References

Ab Rahman AK, Abdullah FH (2011) Visceral leishmaniasis (kala-azar) and malaria coinfection in an immigrant in the state of Terengganu, Malaysia: a case report. J Microbiol Immunol Infect 44:72–76

Agrawal Y, Sinha A, Upadhyaya P, Kafle S, Rijal S, Khanal B (2013) Hematological profile in visceral leishmaniasis. Int J Infect Microbiol 2:39–44

Almeida R, D'Oliveira A Jr, Machado P, Bacellar O, Ko AI, de Jesus AR et al (1999) Randomized, double blind study of stibogluconate plus human granulocyte macrophage colony stimulating factor versus stibogluconate alone in the treatment of cutaneous leishmaniasis. J Infect Dis 180:1735–1737

Asilian A, Jalayer T, Nilforooshzadeh M, Ghassemi RL, Peto R, Wayling S et al (2003) Treatment of cutaneous leishmaniasis with aminosidine (paromomycin) ointment: double-blind, randomized trial in the Islamic Republic of Iran. Bull World Health Organ 81:353–359

Bacellar O, Brodskyn C, Guerreiro J, Barral-Netto M, Costa CH, Coffman RL et al (1996) Interleukin-12 restores interferon-gamma production and cytotoxic responses in visceral leishmaniasis. J Infect Dis 173:1515–1518

Badaro R, Nascimento C, Carvalho JS, Badaro F, Russo D, Ho JL et al (1994) Granulocyte-macrophage colony-stimulating factor in combination with pentavalent antimony for the treatment of visceral leishmaniasis. Eur J Clin Microbiol Infect Dis 13:23–28

Barbieri CL (2006) Immunology of canine leishmaniasis. Parasite Immunol 28:329–337

Basu JM, Mookerjee A, Sen P, Bhaumik S, Sen P, Banerjee S et al (2006) Sodium antimony gluconate induces generation of reactive oxygen species and nitric oxide via phosphoinositide 3-kinase and mitogen-activated protein kinase activation in *Leishmania donovani*-infected macrophages. Antimicrob Agents Chemother 50:1788–1797

Belosevic MI, Finbloom DS, Van Der Meide PH, Slayter MV, Nacy CA (1989) Administration of monoclonal anti-IFN-gamma antibodies in vivo abrogates natural resistance of C3H/HeN mice to infection with *Leishmania major*. J Immunol 143:266–274

Broderson JR, Chapman WL, Hanson WL (1986) Experimental Visceral Leishmaniasis in the OwlMonkey. Vet Pathol 23:293-302

Burgess AW, Camakaris JA, Metcalf DO (1977) Purification and properties of colony-stimulating factor from mouse lung-conditioned medium. J Biol Chem 252:1998–2003

Callahan HL, Portal AC, Devereaux R, Grogl MA (1997) An axenic amastigote system for drug screening. Antimicrob Agents Chemother 41:818–822

Carvalho L, Luque-Ortega JR, Lopez-Martin C, Castanys S, Rivas L, Gamarro F (2011) The 8-aminoquinoline analogue sitamaquine causes oxidative stress in *Leishmania donovani* promastigotes by targeting succinate dehydrogenase. Antimicrob Agents Chemother 55:4204–4210

Chalfie M (1995) Green fluorescent protein. Photochem Photobiol 62:651–656

Chang KT, Dwyer DM (1978) *Leishmania donovani* hamster macrophage interactions in vitro: cell entry, intracellular survival, and multiplication of amastigotes. J Exp Med 147:515–530

Chappuis F, Sundar S, Hailu A, Ghalib H, Rijal S, Peeling RW et al (2007) Visceral leishmaniasis: what are the needs for diagnosis, treatment and control? Nat Rev Microbiol 5:873

Chattopadhyay A, Jafurulla M (2011) A novel mechanism for an old drug: amphotericin B in the treatment of visceral leishmaniasis. Biochem Biophys Res Commun 416:7–12

Chunge CN, Gacmra G, Muigai R, Wasunna K, Rashid JR, Chulay JD et al (1985) Visceral leishmaniasis unresponsive to antimonial drugs III. Successful treatment using a combination of sodium stibogluconate plus allopurinol. Trans R Soc Trop Med Hyg 79:715–718

Coleman RE, Edman JD, Semprevivo LH (1988) *Leishmania mexicana*: effect of concomitant malaria on cutaneous leishmaniasis. Development of lesions in a Leishmania-susceptible (BALB/c) strain of mouse. Exp Parasitol 65:269–276

Costa S, Machado M, Cavadas C, do Ceu Sousa M (2016) Antileishmanial activity of antiretroviral drugs combined with miltefosine. Parasitol Res 2016(115):3881–3887

Croft SL, Seifert K, Yardley V (2006a) Current scenario of drug development for leishmaniasis. Indian J Med Res 123:399–410

Croft SL, Sundar S, Fairlamb AH (2006b) Drug resistance in leishmaniasis. Clin Microbiol Rev 19:111–126

De Rycker M, Hallyburton I, Thomas J, Campbell L, Wyllie S, Joshi D et al (2013) Comparison of a high-throughput high-content intracellular *Leishmania donovani* assay with an axenic amastigote assay. Antimicrob Agents Chemother 57:2913–2922

Dea-Ayuela MA, Rama-Iniguez S, Alunda JM, Bolas-Fernandez F (2007) Setting new immunobiological parameters in the hamster model of visceral leishmaniasis for in vivo testing of antileishmanial compounds. Vet Res Commun 31:703–717

Dietze R, Carvalho SF, Valli LC, Berman J, Brewer T, Milhous W et al (2001) Phase 2 trial of WR6026, an orally administered 8-aminoquinoline, in the treatment of visceral leishmaniasis caused by *Leishmania chagasi*. Am J Trop Med Hyg 65:685–689

Dostalova A, Volf P (2012) Leishmania development in sand flies: parasite-vector interactions overview. Parasit Vectors 5:276–288

Dube A, Gupta R, Singh N (2009) Reporter genes facilitating discovery of drugs targeting protozoan parasites. Trends Parasitol 25:432–439

Dumas C, Muyombwe A, Roy G, Matte C, Ouellette M, Olivier M et al (2003) Recombinant *Leishmania major* secreting biologically active granulocyte-macrophage colony-stimulating factor survives poorly in macrophages in vitro and delays disease development in mice. Infect Immun 71:6499–6509

Frezard F, Demicheli C, Ribeiro RR (2009) Pentavalent antimonials: new perspectives for old drugs. Molecules 14:2317–2336

Gallin JI, Farber JM, Holland SM, Nutman TB (1995) Interferon-γ in the management of infectious diseases. Ann Intern Med 123:216–224

Ghosh M, Roy K, Roy S (2013) Immunomodulatory effects of antileishmanial drugs. J Antimicrob Chemother 68:2834–2838

Gomes LI, Gonzaga FM, de Morais-Teixeira E, de Souza-Lima BS, Freire VV, Rabello A (2012) Validation of quantitative real-time PCR for the in vitro assessment of antileishmanial drug activity. Exp Parasitol 131:175–179

Guevara P, Pinto-Santini D, Rojas A, Crisante G, Anez N, Ramirez JL (2001) Green fluorescent protein-tagged Leishmania in phlebotomine sand flies. J Med Entomol 38:39–43

Ha DS, Schwarz JK, Turco SJ, Beverley SM (1996) Use of the green fluorescent protein as a marker in transfected Leishmania. Mol Biochem Parasitol 77:57–64

Hailu A, Van der poll to, Berhe N, Kager PA (2004) Elevated plasma levels of interferon (IFN)-γ, IFN-γ inducing cytokines, and IFN-γ inducible CXC chemokines in visceral leishmaniasis. Am J Trop Med Hyg 71:561–567

Haldar AK, Sen P, Roy S (2011) Use of antimony in the treatment of leishmaniasis: current status and future directions. Mol Biol Int 2011:1–23

Halim MA, Alfurayh O, Kalin ME, Dammas S, Al-Eisa A, Damanhouri G (1993) Successful treatment of visceral leishmaniasis with allopurinol plus ketoconazole in a renal transplant recipient after the occurrence of pancreatitis due to stibogluconate. Clin Infect Dis 16:397–399

Hamerlinck FF, Van Gool T, Faber WR, Kager PA (2000) Serum neopterin concentrations during treatment of leishmaniasis: useful as test of cure? FEMS Immunol Medl Microbiol 27:31–34

Hamill RJ (2013) Amphotericin B formulations: a comparative review of efficacy and toxicity. Drugs 73:919–934

Hamza T, Barnett JB, Li B (2010) Interleukin 12 a key immunoregulatory cytokine in infection applications. Int J Mol Sci 11:789–806

Handman E, Burgess AW (1979) Stimulation by granulocyte-macrophage colony-stimulating factor of *Leishmania tropica* killing by macrophages. J Immunol 122:1134–1137

Hasker E, Singh SP, Malaviya P, Picado A, Gidwani K, Singh RP et al (2012) Visceral leishmaniasis in rural Bihar, India. Emerg Infect Dis 18:1662–1664

Heinzel FP, Schoenhaut DS, Rerko RM, Rosser LE, Gately MK (1993) Recombinant interleukin 12 cures mice infected with *Leishmania major*. J Exp Med 177:1505–1509

Islam MZ, Itoh M, Mirza R, Ahmed I, Ekram AS, Sarder AH et al (2004) Direct agglutination test with urine samples for the diagnosis of visceral leishmaniasis. Am J Trop Med Hyg 70:78–82

Jha TK (1983) Evaluation of diamidine compound (pentamidine isethionate) in the treatment of resistant cases of kala-azar occurring in North Bihar. India Trans R Soc Trop Med Hyg 77:167–170

Jha TK, Sundar S, Thakur CP, Felton JM, Sabin AJ, Horton J (2005) A phase II dose-ranging study of sitamaquine for the treatment of visceral leishmaniasis in India. Am J Trop Med Hyg 73:1005–1011

Kamhawi S (2006) Phlebotomine sand flies and Leishmania parasites: friends or foes? Trends Parasitol 22:439–445

Kaur A, Kinhikar AG, Singh PP (2004) Bioimmunotherapy of rodent malaria: co-treatment with recombinant mouse granulocyte-macrophage colony-stimulating factor and an enkephalin fragment peptide Tyr-Gly-Gly. Acta Trop 91:27–41

Kima PE, Soong L (2013) Interferon gamma in leishmaniasis. Front Immunol 4:156–160

Kip AE, Balasegaram M, Beijnen JH, Schellens JH, de Vries PJ, Dorlo TP (2015) Systematic review of biomarkers to monitor therapeutic response in leishmaniasis. Antimicrob Agents Chemother 59:1–4

Kolaczinski JH, Reithinger R, Worku DT, Ocheng A, Kasimiro J, Kabatereine N et al (2008) Risk factors of visceral leishmaniasis in East Africa: a case-control study in Pokot territory of Kenya and Uganda. Int J Epidemiol 37:344–352

Kumar R, Engwerda C (2014) Vaccines to prevent leishmaniasis. Clin Transl Immunol 3:3–9

Lafuse WP, Story R, Mahylis J, Gupta G, Varikuti S, Steinkamp H et al (2013) *Leishmania donovani* infection induces anemia in hamsters by differentially altering erythropoiesis in bone marrow and spleen. PLoS One 8:59509

Lindoso JA, Cota GF, da Cruz AM, Goto H, Maia-Elkhoury AN, Romero GA et al (2014) Visceral leishmaniasis and HIV coinfection in Latin America. PLoS Negl Trop Dis 8:3136–3144

Loiseau PM, Cojean S, Schrevel J (2011) Sitamaquine as a putative antileishmanial drug candidate: from the mechanism of action to the risk of drug resistance. Parasite 18:115–119

Loria-Cervera EN, Andrade-Narvaez FJ (2014) Animal models for the study of leishmaniasis immunology. Rev Inst Med Trop Sao Paulo 56:1–11

Maarouf M, de Kouchkovsky Y, Brown S, Petit PX, Robert-Gero M (1997) In vivo interference of paromomycin with mitochondrial activity of Leishmania. Exp Cell Res 232:339–348

Maltezou HC (2009) Drug resistance in visceral leishmaniasis. Biomed Res Int 2010:1–8

Mastroianni A (2004) Liposomal amphotericin B and rHuGM-CSF for treatment of visceral leishmaniasis in AIDS. Infez Med 12:197–204

Matte C, Marquis JF, Blanchette J, Gros P, Faure R, Posner BI et al (2000) Peroxovanadium-mediated protection against murine leishmaniasis: role of the modulation of nitric oxide. Eur J Immunol 30:2555–2564

Mohapatra S (2014) Drug resistance in leishmaniasis: newer developments. Trop Parasitol 4:4–9

Moore EM, Lockwood DN (2010) Treatment of visceral leishmaniasis. J Global Infect Dis 2:151–158

Moreno J, Alvar J (2002) Canine leishmaniasis: epidemiological risk and the experimental model. Trends Parasitol 18:399–405

Munoz DL, Robledo SM, Kolli BK, Dutta S, Chang KP, Muskus C (2009) *Leishmania (Viannia) panamensis*: an in vitro assay using the expression of GFP for screening of antileishmanial drug. Exp Parasitol 122:134–139

Murray HW (2000) Suppression of posttreatment recurrence of experimental visceral leishmaniasis in T-cell-deficient mice by oral miltefosine. Antimicrob Agents Chemother 44:3235–3236

Murray HW, Hariprashad J (1995) Interleukin 12 is effective treatment for an established systemic intracellular infection: experimental visceral leishmaniasis. J Exp Med 181:387–391

Murray HW, Hariprashad JU, Fichtl RE (1993) Treatment of experimental visceral leishmaniasis in a T-cell-deficient host: response to amphotericin B and pentamidine. Antimicrob Agents Chemother 37:1504–1505

Murray HW, Montelibano C, Peterson R, Sypek JP (2000) Interleukin-12 regulates the response to chemotherapy in experimental visceral leishmaniasis. J Infect Dis 182:1497–1502

Nacher M, Carme B, Sainte Marie D, Couppie P, Clyti E, Guibert P et al (2001) Influence of clinical presentation on the efficacy of a short course of pentamidine in the treatment of cutaneous leishmaniasis in French Guiana. Ann Trop Med Parasitol 95:331–336

Nieto A, Dominguez-Bernal G, Orden JA, De La Fuente R, Madrid-Elena N, Carrion J (2011) Mechanisms of resistance and susceptibility to experimental visceral leishmaniosis: BALB/c mouse versus Syrian hamster model. Vet Res 42:39–51

Okwor I, Uzonna JE (2013) The immunology of Leishmania/HIV co-infection. Immunol Res 56:163–171

Olobo JO, Gicheru MM, Anjili CO (2001) The African Green Monkey model for cutaneous and visceral leishmaniasis. Trends Parasitol 17:588–592

Ota H, Takashima Y, Matsumoto Y, Hayashi Y, Matsumoto Y (2008) Pretreatment of macrophages with the combination of IFN-γ and IL-12 induces resistance to *Leishmania major* at the early phase of infection. J Vet Med Sci 70:589–593

Patel TA, Lockwood DN (2009) Pentamidine as secondary prophylaxis for visceral leishmaniasis in the immunocompromised host: report of four cases. Tropical Med Int Health 14:1064–1070

Perez LE, Chandrasekar B, Saldarriaga OA, Zhao W, Arteaga LT, Travi BL et al (2006) Reduced nitric oxide synthase 2 (NOS2) promoter activity in the Syrian hamster renders the animal functionally deficient in NOS2 activity and unable to control an intracellular pathogen. J Immunol 176:5519–5528

Perez-Victoria JM, Bavchvarov BI, Torrecillas IR, Martinez-Garcia M, Lopez-Martin C, Campillo M et al (2011) Sitamaquine overcomes ABC-mediated resistance to miltefosine and antimony in Leishmania. Antimicrob Agents Chemother 55:3838–3344

Porrozzi R, Pereira MS, Teva A, Volpini AC, Pinto MA, Marchevsky RS et al (2006) Leishmania infantum-induced primary and challenge infections in rhesus monkeys (*Macaca mulatta*): a primate model for visceral leishmaniasis. Trans R Soc Trop Med Hyg 100:926–937

Rosenthal E, Marty P, del Giudice P, Pradier C, Ceppi C, Gastaut JA et al (2000) HIV and Leishmania coinfection: a review of 91 cases with focus on atypical locations of Leishmania. Clin Infect Dis 31:1093–1095

Rybniker J, Goede V, Mertens J, Ortmann M, Kulas W, Kochanek M et al (2010) Treatment of visceral leishmaniasis with intravenous pentamidine and oral fluconazole in an HIV-positive patient with chronic renal failure—a case report and brief review of the literature. Int J Infect Dis 14:522–525

Sah SP, Sharma SK, Rani S (2002) Kala azar associated with malaria. Arch Pathol Lab Med 126:382–383

Saha B, Saini A, Germond R, Perrin PJ, Harlan DM, Davis TA (1999) Susceptibility or resistance to Leishmania infection is dictated by the macrophages evolved under the influence of IL-3 or GM-CSF. Eur J Immunol 29:2319–2329

Saraiva EM, de Figueiredo Barbosa A, Santos FN, Borja-Cabrera GP, Nico D, Souza LO, de Oliveira Mendes-Aguiar C et al (2006) The FML-vaccine (Leishmune®) against canine visceral leishmaniasis: a transmission blocking vaccine. Vaccine 24:2423–2431

Schriefer A, Barral A, Carvalho EM, Barrel-Nettom (1995) Serum soluble markers in the evaluation of treatment in human visceral leishmaniasis. Clin Exp Immunol 102:535–540

Seifert K, Croft SL (2006) In vitro and in vivo interactions between miltefosine and other antileishmanial drugs. Antimicrob Agents Chemother 50:73–79

Sereno D, Roy G, Lemesre JL, Papadopoulou B, Ouellette M (2001) DNA transformation of Leishmania infantum axenic amastigotes and their use in drug screening. Antimicrob Agents Chemother 45:1168–1173

Singal P, Singh PP (2005) *Leishmania donovani* amastigote component-induced colony-stimulating factor production by macrophages: modulation by morphine. Microbes Infect 7:148–156

Singh N, Dube A (2004) Fluorescent Leishmania: application to anti-leishmanial drug testing. Am J Trop Med Hyg 71:400–402

Singh N, Kumar R, Gautam S, Singh OP, Gidwani K, Rai M et al (2014) Leishmania specific CD4 T cells release IFN-γ that limits parasite replication in patients with visceral leishmaniasis. Int J Infect Dis 21:158–166

Stauber LA, Franchino EM, Grun J (1958) An eight-day method for screening compounds against *Leishmania donovani* in the golden hamster. J Eukaryot Microbiol 5:269–273

Suman Gupta, Nishi (2011) Visceral leishmaniasis: experimental models for drug discovery. Indian J Med Res 133:27–39

Sundar S, Chatterjee M (2006) Visceral leishmaniasis-current therapeutic modalities. Indian J Med Res 123:345–352

Sundar S, Olliaro PL (2007) Miltefosine in the treatment of leishmaniasis: clinical evidence for informed clinical risk management. Ther Clin Risk Manag 3:733–740

Sundar S, Rai M (2002) Laboratory diagnosis of visceral leishmaniasis. Clin Diagn Lab Immunol 9:951–958

Sundar S, Rosenkaimer F, Lesser ML, Murray HW (1995) Immunochemotherapy for a systemic intracellular infection: accelerated response using interferon-γ in visceral leishmaniasis. J Infect Dis 171:992–996

Sundar S, Rai M, Chakravarty J, Agarwal D, Agrawal N, Vaillant M et al (2008) New treatment approach in Indian visceral leishmaniasis: single-dose liposomal amphotericin B followed by short-course oral miltefosine. Clin Infect Dis 47:1000–1006

Sundar S, Agrawal N, Arora R, Agarwal D, Rai M, Chakravarty J (2009) Short-course paromomycin treatment of visceral leishmaniasis in India: 14-day vs 21-day treatment. Clin Infect Dis 49:914–918

Tavora LG, Nogueira MB, Gomes ST (2015) Visceral leishmaniasis/HIV co-infection in Northeast Brazil: evaluation of outcome. Braz J Infect Dis 19:651–656

Tewary P, Saxena S, Madhubala R (2006) Co-administration of IL-12 DNA with rORFF antigen confers long-term protective immunity against experimental visceral leishmaniaisis. Vaccine 24:2409–2416

Wadhone P, Maiti M, Agarwal R, Kamat V, Martin S, Saha B (2009) Miltefosine promotes IFN-γ-dominated anti-leishmanial immune response. J Immunol 182:7146–7154

Wasunna MK, Rashid JR, Mbui J, Kirigi G, Kinoti D, Lodenyo H et al (2005) A phase II dose-increasing study of sitamaquine for the treatment of visceral leishmaniasis in Kenya. Am J Trop Med Hyg 73:871–876

Weiser WY, Van Niel AN, Clark SC, David JR, Remold HG (1987) Recombinant human granulocyte/macrophage colony-stimulating factor activates intracellular killing of *Leishmania donovani* by human monocyte-derived macrophages. J Exp Med 166:1436–1446

WHO (2015) Media centre Leishmaniasis. Leishmaniasis Fact sheet N°375, 1–5.

Wolday D, Akuffo H, Fessahaye G, Valantine A, Britton S (1998) Live and killed human immunodeficiency virus type-1 increases the intracellular growth of *Leishmania donovani* in monocyte-derived cells. Scand J Infect Dis 30:29–34

Yared S, Deribe K, Gebreselassie A, Lemma W, Akililu E, Kirstein OD et al (2014) Risk factors of visceral leishmaniasis: a case control study in north-western Ethiopia. Parasit Vectors 7:470–480

Zvulunov A, Klaus S, Vardy D (2002) Fluconazole for the treatment of cutaneous leishmaniasis. N Engl J Med 347:370–371

Chapter 6
Human Trichomoniasis

Sumeeta Khurana and Shreya Singh

Abstract Human trichomoniasis is a sexually transmitted disease of global concern with millions of cases reported worldwide. The causative agent is *Trichomonas vaginalis*, a flagellated protozoan, which primarily resides in the genitourinary tract of affected women and men. Apart from causing discomfort due to malodourous discharge, vaginal irritation, and dysuria, it can also cause complications such as infections of the adjoining glands and endometrium, premature labor in pregnant women, and even cervical dysplasia. The diagnosis can be made on clinical examination, wet mount microscopy, stained smear examination, and antigen detection using commercial kits. Rapid nucleic acid amplification-based platforms are also available which demonstrate high sensitivity and specificity for detection of *T. vaginalis*. The management primarily consists of administration of 2 g of a single dose of oral metronidazole or tinidazole, with concomitant treatment of the sexual partner. Prevention of infection in high-risk individuals can be ensured by using condoms, microbicidal vaginal suppositories, and hydrogels containing antimicrobial peptides. Vaccination has shown promise in animal trichomoniasis; however, identification of candidates for human use and subsequent development of vaccines for humans are still in the pipeline.

Keywords Hydrogels · Microbicidal vaginal suppositories · Sexually transmitted disease · Trichomoniasis · *Trichomonas vaginalis*

6.1 Introduction

Human trichomoniasis is perhaps the most common sexually transmitted infection (STI) after viral STI's in the world. Even though various serious health-related consequences can result from this infection, it still remains underrecognized, underreported, and largely ignored. This disease is caused by *Trichomonas vaginalis*, an anaerobic, flagellated parasite, and its trophozoite stage commonly resides in the

S. Khurana (✉) · S. Singh
Department of Medical Parasitology, Postgraduate Institute of Medical Education and Research, Chandigarh, India

P. P. Singh (ed.), *Infectious Diseases and Your Health*,
https://doi.org/10.1007/978-981-13-1577-0_6

genital and urinary tract of men and women. The trophozoite multiplies in the host and is transmitted via unprotected sexual intercourse. Although commonly asymptomatic, particularly in men, unusual genital discharge, burning or pain during micturition, and genital irritation are the most common clinical symptoms. Trichomoniasis has been linked to the facilitation of the HIV pandemic and can also result in pelvic inflammatory disease and poor outcomes in pregnant women if not recognized as a public health concern by clinicians and healthcare authorities. Global estimates suggest that there are an estimated 143 million cases of trichomoniasis, and approximately 90% of these infections are prevalent in patients from resource-limited settings (WHO 2008; Newman et al. 2015).

6.2 Epidemiology

The true epidemiology of trichomoniasis is largely unknown due to lack of surveillance programs. However, there is considerable variability depending on patient population, the region of study, and diagnostic test used. While studies in women from the United States (USA) have described prevalence rates of 2–3% in the 14–49 age group, studies from African patients show a particularly higher prevalence rate (Kissinger 2015). Screening studies in women attending antenatal clinics have been used to indicate the trends of prevalence in the general population, and using this approach, prevalence rates ranging from 3.2% to 52% were noted in resource-limited settings, while they were much lower at 7.6–12.6% in the developed countries like the United States (Johnston and Mabey, 2008). The region-wise global distribution of trichomoniasis is depicted in Fig. 6.1.

A prevalence of trichomoniasis ranging from 3.6% to 31% has been documented in Indian studies, the details of which are described in Table 6.1. It has been noted that the annual years of healthy life lost per 100,000 people due to trichomoniasis in Indian women has risen by 4% since 1990 at rate of 0.2% healthy life lost per year (Global Disease Burden 2017).

Risk factors associated with the development of trichomoniasis include commercial sex work, use of intravenous drugs, older age, African descent, and bacterial vaginosis. Among Indian women, it has been noted that those with infection are more likely to have lower levels of education, be married to an uneducated partner, belong to poor socioeconomic status, and report having sexual relations with more than one partner (Madhivanan et al. 2009).

6.3 Etiological Agent

The causative agent of trichomoniasis in humans is *Trichomonas vaginalis*, a flagellated protozoan. It has a primarily anaerobic lifestyle and is present extracellular to the genital and urinary tract epithelium (Kissinger 2015). Typically it exists in a

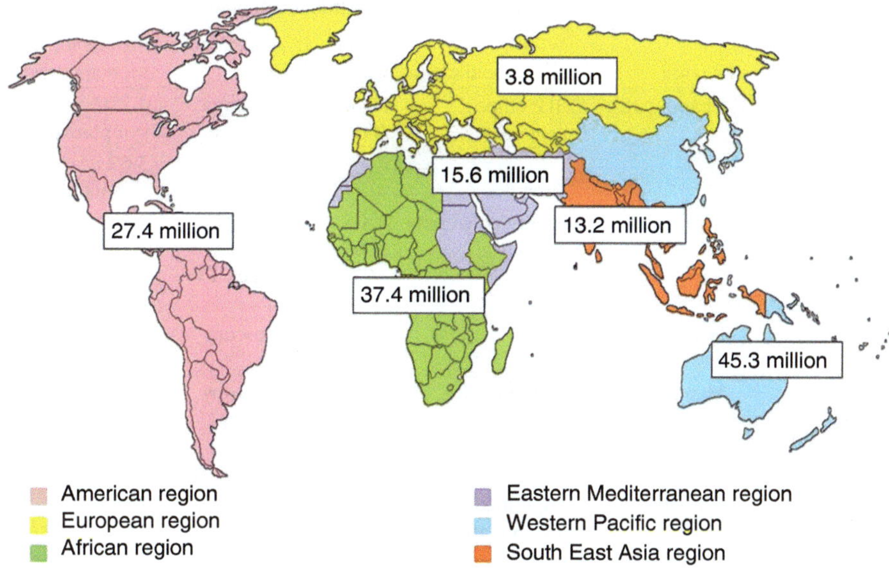

Fig. 6.1 WHO region-wise global estimates of trichomoniasis as per systematic review and global analysis in 2015 (Newman et al. 2015)

trophozoite form which is pyriform in shape but can occasionally be amoeboid. It is approximately 10–20 μm in length and 2–14 μm in width with four flagella projecting from the anterior end and one flagellum extending posteriorly toward the middle of the trophozoite to form an undulating membrane. An axostyle can be seen extending from the posterior aspect of the trophozoite. The illustrated diagram and scanning electron microscopy image of *T. vaginalis* trophozoite are shown in Fig. 6.2a, b, respectively.

Apart from trophozoites, pseudocyst forms of *T. vaginalis* have also been found in animals and could have relevance in humans. The flagella in the pseudocyst form are located inside endocytic vacuoles and remain beating, the axostyle is present in a curved shape, and a distinct mitotic process has also been described (Pereira-Neves et al. 2003). The pseudocyst is round, shows no motility, and does not have a true cyst wall on light microscopy (Afzan and Suresh 2012). Cyst forms of T. vaginalis have not been demonstrated to date.

6.4 Pathogenesis

T. vaginalis is an obligate parasite which infects the squamous epithelium of the genital tract. It resides and replicates in the lower female genital tract and the male urethra and prostate. It is a predatory parasite which can phagocytose vaginal epithelial cells, bacteria, as well as erythrocytes for nutrition. Since no cyst form is

Table 6.1 Studies on prevalence of trichomoniasis among Indian population

Study	Region	Patient population	Study year	Diagnostic test used	Prevalence of trichomoniasis
Malla et al. (2008)	Chandigarh	OBG[a] clinic outpatients	2004–2006	Microscopy and culture	4.28%: symptomatic
					3.6%: asymptomatic
Madhivanan et al. (2009)	Mysore	Sexually active, not pregnant	2005–2006	Microscopy, culture (InPouch TV kit)	8.5%
Das et al. (2011)	Delhi	CSW[b]	2008–2009	Microscopy	31.1%
Fule et al. (2012)	Maharashtra	Symptomatic, reproductive age, OBG clinic outpatients	2010–2011	Microscopy	12.06%
Arora et al. (2014)	Gurgaon	Symptomatic OBG clinic outpatients	2007–2013	Microscopy, Pap stain	15.7%
Deivam et al. (2017)	Tiruchirapalli	Sexually active, reproductive age group	2011–2013	Microscopy	8.1%
Muthusamy and Elangovan (2017)	Chennai	High-risk group	2012–2013	Microscopy, culture (CPLM[c] medium)	5%: high-risk group
		Symptomatic low-risk group			1.67%: low-risk group

[a]Obstetrics and gynecology
[b]Commercial sex worker
[c]Cysteine-peptone-liver-maltose

present, the parasite cannot survive in the external environment. However, in a moist environment outside the human body, reports of surviving up to 3 h have been documented (Burch et al. 1959). Humans are its only known hosts, and transmission among them is primarily by sexual intercourse. Although rare, there is evidence of nonsexual transmission, with reports of transmission via fomites and possibly water (Crucitti et al. 2011).

6.5 Clinical Features

The incubation time for trichomoniasis is generally between 4 and 28 days, although most of the women (85%) and men (77%) with infection are asymptomatic. Some studies even suggest that asymptomatic women can harbor *T. vaginalis* in their genital tract for as many as 3–5 years. This could be due to the type of infecting strain, variations in genitourinary anatomy, or merely lack of imperfect tools for screening

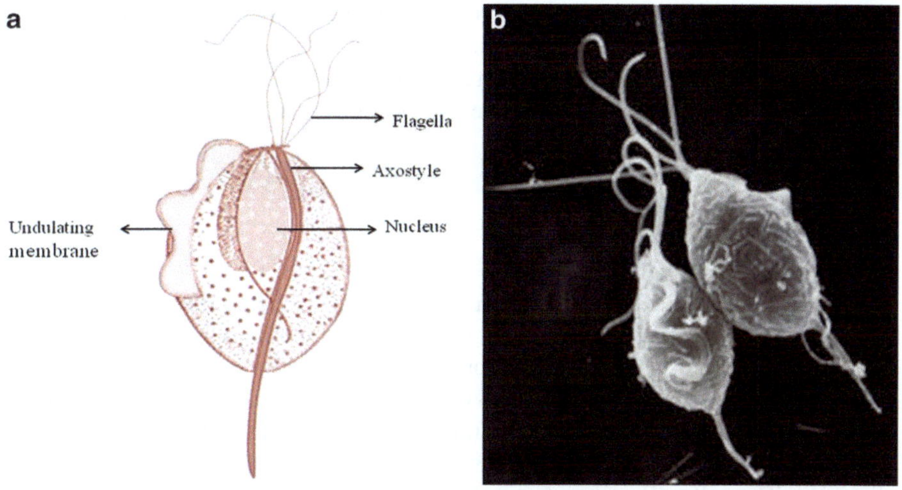

Fig. 6.2 (**a**) Illustrated diagram of T. vaginalis; (**b**) scanning electron microscopy image of T. vaginalis. (Source : Adapted from Korean J Parasitol Vol. 51, No. 2: 243-246, April 2013 BRIEF COMMUNICATION http://dx.doi.org/10.3347/kjp.2013.51.2.243)

(Swygard et al. 2004). Among the asymptomatic women, nearly a third of them become symptomatic after 6 months.

The most commonly infected sites in women include the vagina, the urethra, and lastly the endocervix. The chief presenting complaint is most commonly vaginal discharge, noted in >50% cases, followed by itching and dysuria. Irritation of the vulvar region and abdominal pain may also be present. Per speculum examination reveals vaginal discharge which is characteristically frothy green and malodorous (Swygard et al. 2004). The vaginal pH is often markedly increased, often to more than five. Cervical pathology can be observed in the form of "colpitis macularis" also called "strawberry cervix," which can be seen in 5% of women. When a colposcopic examination is carried out, it may be observed in as high as 50% of the infected women. It results from microscopic, punctate hemorrhages over the cervix causing the cervix to appear erythematous, edematous, and friable. Cervical mucopurulent discharge and other complications such as infections of the adnexa, Skene and Bartholin glands, and even the endometrium can occur.

There are reports of the pseudocyst forms of this agent being described from patients suffering from cervical neoplasia and may potentially play a role in exacerbating cervical cancer (Afzan and Suresh 2012). However, it is too early to come to such a conclusion with the currently available data. It has been noted that *T. vaginalis* infection has a significant association with the risk of developing pelvic inflammatory disease particularly in women with HIV infection (Moodley et al. 2002). Association between preterm labor and *T. vaginalis* infection has also been demonstrated in pregnant women (Hosney et al. 2017).

Information regarding *trichomoniasis* in men is limited, and they may often be asymptomatic. Common symptoms include urethral discharge (often less profuse and purulent than women) and dysuria. Seroepidemiologic evidence indicating an association of *T. vaginalis* with prostate cancer has emerged suggesting that infection may increase the risk of more aggressive cancer. However, most recent studies have found no associations between the two (Marous et al. 2017; Shui et al. 2016).

6.6 Diagnosis

The appropriate clinical specimens for diagnosis of trichomoniasis include urine, vaginal fluid, or endocervical swabs in women and in men; urethral discharge, urethral scrapings, urine, prostate secretions or semen may be used. Testing multiple sites is advisable especially in men (Swygard et al. 2004). Vaginal secretions can be obtained from the fornices and the lateral vaginal wall using a plastic loop or swab. They may be collected by the patients themselves or collected by the concerned physician. The self-collection of samples tends to have better acceptability among patients and has a similar performance when compared to clinician acquired samples.

1. Direct microscopic examination: For the wet mount preparation, collected secretions are suspended in normal saline and subjected to light microscopic examination under high-power field (40×). Diagnosis relies on visual detection of viable trophozoites of *T. vaginalis* which can be identified based on their appearance and characteristic motility. The size of the trichomonas trophozoite is comparable to a lymphocyte (10–20 μm), and it may, therefore, be difficult to identify in case the motility is lost. Motility depends on the ambient temperature and moisture in the sample, and since it can be lost quickly, the specimen must be examined within 10–20 min of sample collection. Although it is quick and inexpensive, the sensitivity of direct microscopy varies from 38% to 82% and depends on the inoculum size (should be $>10^4$ organisms/mL), the time interval between sample collection and examination, and the experience of the microbiologist (Garber 2005; Schwebke and Burgess 2004).
2. Isolation: Culture has been traditionally considered as the gold standard for diagnosing *T. vaginalis* infection. It can detect as low as 10^2 organisms/mL in the sample, and results are also easy to interpret (Garber 2005). Several media are available for culture, such as Diamond's TYI liquid broth media, cysteine-peptone-liver-maltose (CPLM), Feinberg and Whittington's medium, and self-contained culture systems (InPouch TV kit). Culture techniques have 100% specificity and a higher sensitivity (75–85%) compared to microscopy as a much lower concentration of parasite is needed for culture positivity (Pattullo et al. 2009). The drawbacks are higher cost and long turnaround time for diagnosis (3–7 days). Hence, screening for trichomoniasis using the only culture may not be a convenient approach. Cultivation of *T. vaginalis* on cell cultures has an even higher sensitivity and enables diagnosis form samples containing as few as three parasites/mL (Garber 2005). However, this technique is expensive and impractical and is highly prone to contamination by vaginal flora.

3. Nucleic acid amplification tests (NAAT): Molecular tests have become common for diagnosing infection with *T. vaginalis* and can detect as low as one organism. Polymerase chain reaction has been performed most notably on vaginal fluids and urine. Urine-based PCR detection of *T. vaginalis* was performed with sensitivity and specificity of 64–90.8% and 93.4–100%, respectively, in women (Lawing et al. 2000) and 92.7% and 88.6%, respectively, in men (Kaydos-Daniels et al. 2003). The PCR amplification of 18S rRNA and *pfoB* gene of *T. vaginalis* has also been shown to have a specificity of 95.1% and 94.8%, respectively, in symptomatic subjects and 90.2% and 88.5% in asymptomatic patients, while a sensitivity of 100% has been noted in all cases (Sonkar et al. 2016). Dot-blot hybridization has also been used, employing a 2.3 kb *T. vaginalis* DNA fragment as a probe. The APTIMA® *Trichomonas vaginalis* assay is an amplification-based assay for *T. vaginalis* cleared by the USFDA which utilizes target capture followed by transcription-mediated amplification and chemiluminescent probe hybridization to detect *T. vaginalis* ribosomal RNA. Endocervical or vaginal swabs and urine can be tested, and a sensitivity and specificity of 95% and 98%, respectively, have been documented (Chapin and Andrea 2011). GeneXpert platform for the detection of *T. vaginalis*, the assay has also been evaluated and found to be 95% sensitive and 95–100% specific for diagnosing trichomoniasis when compared to NAAT (Badman et al. 2016). Additional advantages of GeneXpert include small platform requirement and rapid (<1 h) and direct detection from self-collected vaginal swabs and urine (Gaydos et al. 2017).
4. Stained smears: Examination of Papanicolaou (Pap) smear has shown a low sensitivity (61%) for detection of *T. vaginalis*. Since *T. vaginalis* resides primarily in the vagina, ectocervical smears have better utility compared to endocervical smears. In a meta-analysis of the performance of Pap smear compared to wet mount preparation, it has been seen that in high prevalence populations, a positive Pap smear had a positive predictive value (PPV) of 83%, whereas it decreased in lower prevalence populations requiring a confirmatory culture for *T. vaginalis* (Wiese et al. 2000). The use of acridine orange and periodic acid-Schiff have been shown to be valuable by some investigators.
5. Antigen detection tests: A rapid antigen detection-based test by Sekisui Diagnostics called the OSOM *Trichomonas* rapid test can be applied as a point of care test. It is based on the use of immunochromatographic capillary flow dipstick technology. The results are available in nearly 10 min. A high sensitivity and specificity of 82–95% and 97–100%, respectively, have also been noted (Meites et al. 2015). The details of rapid tests for diagnosis of *T. vaginalis* are depicted in Table 6.2.
6. Indirect evidence: These include antibody-based methods which include complement fixation, gel diffusion techniques, hemagglutination, and ELISA to detect anti-trichomonas antibodies (Garber 2005). These, however, cannot be used to differentiate recent and remote infections as they are certainly not specific and could even reflect host interaction with nonpathogenic or commensal trichomonas and have thus been abandoned.

Table 6.2 Rapid tests for point of care diagnosis of trichomoniasis

Test	Principle	Sample	Sensitivity	Specificity
OSOM *Trichomonas* test (Sekisui diagnostics)	Lateral flow antigen detection assay	VS[a]	82–90%	97–100%
GeneXpert (Cepheid)	Real time PCR	VS, endocervical swabs, urine	95%	95.7–100%
Affirm VPIII microbial identification test (Dickinson 2017)	Nucleic acid hybridization	VS	46–90%	99–100%
Xenostrip-Tv (Xenotope diagnostics) (Pillay et al. 2004)	Dipstick assay	VS	66.7%	100%

[a]*VS* vaginal swab

7. Others: Presence of pus cells in the vaginal fluid and an elevated pH (>4.5) can indicate infection. The whiff test which is carried out by adding potassium hydroxide to the vaginal fluid for olfactory detection of amines gives variable results.

6.7 Treatment

In a resource-challenged setting, the screening for cases of trichomoniasis may be difficult and therefore the WHO promotes adopting the syndromic approaches for managing STIs. However, some experts suggest that this approach of managing trichomoniasis has minimal impact on the actual disease prevalence in endemic regions and may, in fact, lead to overtreatment of many cases (Bowden and Garnett 2000). A more fulfilling approach may thus be screening for trichomoniasis followed by treatment of the positive cases.

The treatment of *T. vaginalis* infection is essential as it reduces the clinical symptoms and prevents further transmission. At present, the only class of antimicrobial agents with known activity against *T. vaginalis* is nitroimidazoles such as metronidazole and tinidazole. According to the Centers for Disease Control and Prevention (2015), the recommended treatment regimen for trichomoniasis is 2 g of oral metronidazole or tinidazole provided as a single dose. Cure rates of approximately 84–98% and 92–100% have been documented in trials using the recommended metronidazole and tinidazole regimens, respectively. Generally, tinidazole is more expensive than metronidazole, but it reaches higher concentrations in the serum and genitourinary tract and also has a longer half-life and fewer side effects. The comparison of single-dose (2 g) metronidazole versus tinidazole suggests that tinidazole has equal or rather superior activity in achieving clinical relief and parasitological cure (O-Prasertsawat and Jetsawangsri 1992; Anjaeyulu et al. 1977). A 500 mg twice daily administration of metronidazole for 7 days has been found to be more effective than the traditional 2 g single dose in treating trichomoniasis in women with concomitant HIV infection (Kissinger et al. 2010). Gel formulations of metronidazole are available for topical application, but therapeutic levels of metronidazole are seldom reached in the urethra and perivaginal glands, and therefore the topical application of gels is not commonly used for treatment. However, high dose of intravaginal metronidazole (750 mg) with miconazole combination can be given as a vaginal suppository twice a day for 7 days and has been shown to offer well-tolerated treatment avoiding the systemic adverse effects of nitroimidazoles (Schwebke et al. 2013).

Another patient group of concern is pregnant and symptomatic women, who regardless of the stage of pregnancy must be tested and treated. The transmission of trichomoniasis in the perinatal period is uncommon. It is a yet to be proven if treatment of trichomoniasis reduces the risk of preterm labor in pregnant women as many studies with conflicting results are available (Stringer et al. 2010; Mann et al. 2009). The patient should be counseled by the treating physician regarding the potential benefits of treatment. Metronidazole does cross the placenta, but there is no evidence of mutagenicity or teratogenic effects in infants (Meites et al. 2015). The treatment is similar as in nonpregnant women (2 g metronidazole single dose). However, during breast feeding, 500 mg three times daily for 7 days is considered more compatible since it produces lower drug levels in breast milk.

Sexual intercourse must be avoided by all patients until they and their sexual partners are treated and testing for other STDs including HIV must be carried out concomitantly. Abstinence from alcohol, at least for 24–72 h after completion of treatment, is advised to prevent severe side drug-related side effects known as "disulfiram reaction" after metronidazole administration.

Following the treatment, retesting is recommended within 3 months for all patients who are sexually active (CDC 2017).Recurrent or persistent infection can occur most commonly due to reinfection from an untreated sexual partner or occasionally due to antimicrobial-resistant *T. vaginalis.*

6.8 Drug Resistance and Antimicrobial Susceptibility Testing

Because most organisms are susceptible to the drug of choice, i.e., metronidazole, antimicrobial susceptibility testing (AST) is not performed routinely. However, there are rising reports of women having clinically resistant *T. vaginalis* with studies showing clinical resistance to metronidazole in 4–10% cases and tinidazole resistance in 1% (Schwebke and Barrientes 2006; Cudmore and Garber 2010). Some studies suggest that the presence of *Mycoplasma hominis* symbionts might be associated with the metronidazole resistance of *T. vaginalis* although concrete evidence is still warranted (Wang and Xie 2012). These reports of resistance are concerning since very few alternatives to the standard treatment are available. If needed, AST can be performed using micro-broth dilution methods or shell vial cultures to determine the minimal inhibitory drug concentration. There are, however, no standardized protocols or proficiency testing available for the same. In drug-resistant cases, a higher dose (2–3 g) of tinidazole can be given for 14 days, often with intravaginal tinidazole or intravaginal paromomycin (CDC 2017; Tayal et al. 2010).

6.9 Prevention of *T. vaginalis* Infection

The most reliable protection against sexually transmitted infections such as trichomoniasis is barrier methods, particularly condoms. Other preventive strategies include male circumcision, local application of intravaginal microbicide formulations, and vaccines (Bouchemal et al. 2017). Intravaginal microbicides can be self-administered by women before sexual intercourse, and hydrogels of hydroxyethyl cellulose with antimicrobial peptides or metronidazole have been investigated and found effective in mice models (Bouchemal et al. 2017).

The strategies to combat STIs such as trichomoniasis focus on the following main directions: firstly, knowing the extent of the epidemic by monitoring case counts and infection distribution by region and in time; secondly, by providing a good coverage of quality health services; thirdly, making available adequate funds to minimize the financial constraints of patients requiring services; and finally, innovation and development of technology for rapid diagnosis, effective treatment, and prophylactic strategies to combat this disease (Klausner and Broutet 2017).

6.10 Vaccine

The development of a vaccine against *T. vaginalis* could reduce the medical and societal costs associated with trichomoniasis (Cudmore and Garber 2010). The use of *T. foetus*, a natural pathogen of cattle, in vaccines have shown promise in reducing the duration of bovine infection, and similarly, efforts are ongoing to create a vaccine for human trichomoniasis (Smith and Garber 2014). Establishing good animal models, studies on the *T. vaginalis* immunity and details of cross-isolate protection are all warranted to accelerate this process. Determining the appropriate components of a vaccine is problematic and may be elucidated following more genomic and proteomic studies on *T. vaginalis*. These can contribute valuable information to the identification of unique proteins of *T. vaginalis* that can in future be potential vaccine targets.

References

Afzan MY, Suresh K (2012) Pseudocyst forms of Trichomonas vaginalis from cervical neoplasia. Available from: http://repository.um.edu.my/29663/1/AfzanandSuresh.pdf

Anjaeyulu R, Gupte SA, Desai DB (1977) Single-dose treatment of trichomonal vaginitis: a comparison of tinidazole and metronidazole. J Int Med Res 5:438–441 Available from: http://www.ncbi.nlm.nih.gov/pubmed/590601

Arora BB, Maheshwari M, Devgan N, Arora DR (2014) Prevalence of trichomoniasis, vaginal candidiasis, genital herpes, chlamydiasis, and actinomycosis among urban and rural women of Haryana. Indian J Sex Transm Dis 29:1–5 Available from: http://www.hindawi.com/journals/jstd/2014/963812/

Badman SG, Causer LM, Guy R, Tabrizi SN, Francis F, Donovan B et al (2016) A preliminary evaluation of a new GeneXpert (Gx) molecular point-of-care test for the detection of *Trichomonas vaginalis*. Sex Transm Infect 92:350–352 Available from: http://sti.bmj.com/lookup/doi/10.1136/sextrans-2015-052384

Bouchemal K, Bories C, Loiseau PM (2017) Strategies for prevention and treatment of *Trichomonas vaginalis* infections. Clin Microbiol Rev 30:811–825 Available from: http://www.ncbi.nlm.nih.gov/pubmed/28539504

Bowden FJ, Garnett GP (2000) *Trichomonas vaginalis* epidemiology: parameterising and analysing a model of treatment interventions. Sex Transm Infect 76:248–256 Available from: http://www.ncbi.nlm.nih.gov/pubmed/11026878

Burch TA, Rees CW, Reardon L (1959) Diagnosis of *Trichomonas vaginalis* vaginitis. Am J Obstet Gynaecol 77:309–313. https://doi.org/10.1016/0002-9378(59)90232-7

Centres for Disease Control and Prevention (CDC). Trichomoniasis-2015 (2017) STD treatment guidelines [internet]. [Cited 2017, July 12]. Available from: https://www.cdc.gov/std/tg2015/trichomoniasis.htm

Chapin K, Andrea S (2011) APTIMA *Trichomonas vaginalis*, a transcription-mediated amplification assay for detection of *Trichomonas vaginalis* in urogenital specimens. Expert Rev Mol Diagn 11:679–688 Available from: http://www.tandfonline.com/doi/full/10.1586/erm.11.53

Crucitti T, Jespers V, Mulenga C, Khondowe S, Vandepitte J, Buvé A (2011) Non-sexual transmission of Trichomonas vaginalis in adolescent girls attending school in Ndola, Zambia. PLoS One 6:e16310 Available from: http://www.ncbi.nlm.nih.gov/pubmed/21305023

Cudmore SL, Garber GE (2010) Prevention or treatment: the benefits of *Trichomonas vaginalis* vaccine. J Infect Public Health 3:47–53 Available from: http://www.ncbi.nlm.nih.gov/pubmed/20701891

Das A, Prabhakar P, Narayanan P, Neilsen G, Wi T, Kumta S et al (2011) Prevalence and assessment of clinical management of sexually transmitted infections among female sex workers in two cities of India. Infect Dis Obstet Gynaecol 2011:494769 Available from: http://www.ncbi.nlm.nih.gov/pubmed/21747642

Deivam S, Rajalakshmi R, Priyadharshini S, Seethalaksmi R, Balasubramanian N, Brinda T et al (2017) Prevalence of Trichomonas vaginalis infection among patients that presented to Rural Tertiary Care Hospital in Tiruchirapalli, India in 2011 and 2013, vol 2, pp 255–260. Available from: http://www.pharmahealthsciences.net/pdfs/urgent/10_MS_1466.pdf

Dickinson B (2017) BD affirm™ VPIII microbial identification system [internet]. [Cited Jul 13]. Available from: http://moleculardiagnostics.bd.com/product/affirm/

Fule R, Fule S, Tankhiwale N (2012) Clinical and laboratory evidence of *Trichomonas vaginalis* infection among women of reproductive age in rural area. Indian J Med Microbiol 30:314 Available from: http://www.ijmm.org/text.asp?2012/30/3/314/99493

Garber GE (2005) The laboratory diagnosis of *Trichomonas vaginalis*. Can J Infect Dis Med Microbiol 16:35–38 Available from: http://www.ncbi.nlm.nih.gov/pubmed/18159526

Gaydos CA, Klausner JD, Pai NP, Kelly H, Coltart C, Peeling RW (2017) Rapid and point-of-care tests for the diagnosis of *Trichomonas vaginalis* in women and men. Sex Transm Infect 93:S31–S35 [Internet]. Jul 6 [cited 2017 Jul 21];sextrans-2016-053063. Available from: http://www.ncbi.nlm.nih.gov/pubmed/28684611

Global-Disease-Burden (2017) Trichomoniasis in India – statistics, demographics affected [internet]. [Cited 2017 Jul 13]. Available from: http://global-disease-burden.healthgrove.com/l/25879/Trichomoniasis-in-India

Hosny AE-DMS, El-Khayat W, Kashef MT, Fakhry MN (2017) Association between preterm labour and genitourinary tract infections caused by *Trichomonas vaginalis, Mycoplasma hominis*, gram-negative bacilli, and coryneforms. J Chin Med Assoc 80:575–581 Available from: http://www.ncbi.nlm.nih.gov/pubmed/28094234

Johnston VJ, Mabey DC (2008) Global epidemiology and control of *Trichomonas vaginalis*. Curr Opin Infect Dis 21:56–64 Available from: http://www.ncbi.nlm.nih.gov/pubmed/18192787

Kaydos-Daniels SC, Miller WC, Hoffman I, Banda T, Dzinyemba W, Martinson F et al (2003) Validation of a urine-based PCR-enzyme-linked immunosorbent assay for use in clinical research settings to detect *Trichomonas vaginalis* in men. J Clin Microbiol 41:318–323 Available from: http://www.ncbi.nlm.nih.gov/pubmed/12517867

Kissinger P (2015) *Trichomonas vaginalis*: a review of epidemiologic, clinical and treatment issues. BMC Infect Dis 15:307 Available from: http://www.ncbi.nlm.nih.gov/pubmed/26242185

Kissinger P, Mena L, Levison J, Clark RA, Gatski M, Henderson H et al (2010) A randomized treatment trial: single versus 7-day dose of metronidazole for the treatment of *Trichomonas vaginalis* among HIV-infected women. J Acquir Immune Defic Syndr 55:565–571 Available from: http://www.ncbi.nlm.nih.gov/pubmed/21423852

Klausner JD, Broutet N (2017) Health systems and the new strategy against sexually transmitted infections. Lancet Infect Dis 3099:9–10 Available from: http://linkinghub.elsevier.com/retrieve/pii/S1473309917303614

Lawing LF, Hedges SR, Schwebke JR (2000) Detection of trichomonosis in vaginal and urine specimens from women by culture and PCR. J Clin Microbiol 38:3585–3588 Available from: http://www.ncbi.nlm.nih.gov/pubmed/11015368

Madhivanan P, Bartman MT, Pasutti L, Krupp K, Arun A, Reingold AL et al (2009) Prevalence of *Trichomonas vaginalis* infection among young reproductive age women in India: implications for treatment and prevention. Sex Health 6:339–344 Available from: http://www.ncbi.nlm.nih.gov/pubmed/19917204

Malla N, Kaur S, Khurana S, Bagga R, Wanchu A (2008) Trichomoniasis among women in North India: a hospital based study. Indian J Sex Transm Dis AIDS 29:76 Available from: http://www.ijstd.org/text.asp?2008/29/2/76/48729

Mann JR, McDermott S, Zhou L, Barnes TL, Hardin J (2009) Treatment of trichomoniasis in pregnancy and preterm birth: an observational study. J Women's Health 18:493–497 Available from: http://www.ncbi.nlm.nih.gov/pubmed/19361316

Marous M, Huang W-Y, Rabkin CS, Hayes RB, Alderete JF, Rosner B et al (2017) *Trichomonas vaginalis* infection and risk of prostate cancer: associations by disease aggressiveness and race/ethnicity in the PLCO trial. Cancer Causes Control 28:889–898 Available from: http://www.ncbi.nlm.nih.gov/pubmed/28669054

Meites E, Gaydos CA, Hobbs MM, Kissinger P, Nyirjesy P, Schwebke JR et al (2015) A review of evidence-based care of symptomatic trichomoniasis and asymptomatic *Trichomonas vaginalis* infections. Clin Infect Dis 61(Suppl. 8):S837–S848 Available from: http://www.ncbi.nlm.nih.gov/pubmed/26602621

Moodley P, Wilkinson D, Connolly C, Moodley J, Sturm AW (2002) *Trichomonas vaginalis* is associated with pelvic inflammatory disease in women infected with human immunodeficiency virus. Clin Infect Dis 34:519–522 Available from: http://www.ncbi.nlm.nih.gov/pubmed/11797180

Muthusamy S, Elangovan S (2017) A study on the prevalence of genital trichomoniasis among female outpatients attending sexually transmitted infection clinic in a tertiary care hospital. J Lab Phys 9:16–19 Available from: http://www.ncbi.nlm.nih.gov/pubmed/28042211

Newman L, Rowley J, Vander Hoorn S, Wijesooriya NS, Unemo M, Low N et al. 2015. Global estimates of the prevalence and incidence of four curable sexually transmitted infections in 2012 based on systematic review and global reporting (Meng Z (ed)). PLoS One 10:e0143304. Available from: http://dx.plos.org/10.1371/journal.pone.0143304

O-Prasertsawat P, Jetsawangsri T (1992) Split-dose metronidazole or single-dose tinidazole for the treatment of vaginal trichomoniasis. Sex Transm Dis 19:295–297 Available from: http://www.ncbi.nlm.nih.gov/pubmed/1411848

Pattullo L, Griffeth S, Ding L, Mortensen J, Reed J, Kahn J et al (2009) Stepwise diagnosis of *Trichomonas vaginalis* infection in adolescent women. J Clin Microbiol 47:59–63 Available from: http://www.ncbi.nlm.nih.gov/pubmed/18987174

Pereira-Neves A, Ribeiro KC, Benchimol M (2003) Pseudocysts in trichomonads – new insights. Protist 154:313–329 Available from: http://www.ncbi.nlm.nih.gov/pubmed/14658492

Pillay A, Lewis J, Ballard RC (2004) Evaluation of Xenostrip-Tv, a rapid diagnostic test for *Trichomonas vaginalis* infection. J Clin Microbiol 42:3853–3856 Available from: http://www.ncbi.nlm.nih.gov/pubmed/15297548

Schwebke JR, Barrientes FJ (2006) Prevalence of *Trichomonas vaginalis* isolates with resistance to metronidazole and tinidazole. Antimicrob Agents Chemother 50:4209–4210 Available from: http://www.ncbi.nlm.nih.gov/pubmed/17000740

Schwebke JR, Burgess D (2004) Trichomoniasis. Clin Microbiol Rev 17:794–803 Available from: http://cmr.asm.org/content/17/4/794.short

Schwebke JR, Lensing SY, Sobel J (2013) Intravaginal metronidazole/miconazole for the treatment of vaginal trichomoniasis. Sex Transm Dis 40:710–714 Available from: http://www.ncbi.nlm.nih.gov/pubmed/23949586

Shui IM, Kolb S, Hanson C, Sutcliffe S, Rider JR, Stanford JL (2016) *Trichomonas vaginalis* infection and risk of advanced prostate cancer. Prostate 76:620–623 Available from: https://insights.ovid.com/pubmed?pmid=26818005

Smith J, Garber GE (2014) Current status and prospects for development of a vaccine against *Trichomonas vaginalis* infections. Vaccine 32:1588–1594 Available from: http://www.ncbi.nlm.nih.gov/pubmed/23916988

Sonkar S, Yadav S, Malla N, Dhanda R, Khurana S, Bagga R et al (2016) Evaluation of DNA based techniques for the diagnosis of human vaginal trichomoniasis in North Indian population. Br Microbiol Res J 17:1–12 Available from: http://www.sciencedomain.org/abstract/17070

Stringer E, Read JS, Hoffman I, Valentine M, Aboud S, Goldenberg RL (2010) Treatment of trichomoniasis in pregnancy in sub-Saharan Africa does not appear to be associated with low birth weight or preterm birth. S Afr Med J 100:58–64 Available from: https://pdfs.semanticscholar.org/4c44/f28a3a94532bca269390b3aa5a9c6ea4450e.pdf

Swygard H, Seña AC, Hobbs MM, Cohen MS (2004) Trichomoniasis: clinical manifestations. Diagn Manag Sex Transm Infect 80:91–95 Available from: http://sti.bmj.com/content/sextrans/80/2/91.full.pdf

Tayal SC, Ochogwu SA, Bunce H (2010) Paromomycin treatment of recalcitrant *Trichomonas vaginalis*. Int J STD AIDS 21:217–218 Available from: http://www.ncbi.nlm.nih.gov/pubmed/20215633

Wang P-J, Xie C-B (2012) *Mycoplasma hominis* symbiosis and *Trichomonas vaginalis* metronidazole resistance. Zhongguo Ji Sheng Chong Xue Yu Ji Sheng Chong Bing Za Zhi 30:210–213 Available from: http://www.ncbi.nlm.nih.gov/pubmed/23072138

Wiese W, Patel SR, Patel SC, Ohl CA, Estrada CA (2000) A meta-analysis of the Papanicolaou smear and wet mount for the diagnosis of vaginal trichomoniasis. Am J Med 108:301–308 Available from: http://linkinghub.elsevier.com/retrieve/pii/S0002934399004660

World Health Organization (2008) Global incidence and prevalence of selected curable sexually transmitted infections – 2008. WHO. [Internet]. [cited 2017 Jul 10]. Available from: http://www.who.int/reproductivehealth/publications/rtis/stisestimates/en/

Pillay A, Lewis J, Ballard RC (2004) Evaluation of Xenostrip-Tv, a rapid diagnostic test for *Trichomonas vaginalis* infection. J Clin Microbiol 42:3853–3856 Available from: http://www.ncbi.nlm.nih.gov/pubmed/15297548

Schwebke JR, Barrientes FJ (2006) Prevalence of *Trichomonas vaginalis* isolates with resistance to metronidazole and tinidazole. Antimicrob Agents Chemother 50:4209–4210 Available from: http://www.ncbi.nlm.nih.gov/pubmed/17000740

Schwebke JR, Burgess D (2004) Trichomoniasis. Clin Microbiol Rev 17:794–803 Available from: http://www.ncbi.nlm.nih.gov/pubmed/15489349

Schwebke JR, Lensing SY, Sobel J (2013) Intravaginal metronidazole/miconazole for the treatment of vaginal trichomoniasis. Sex Transm Dis 40:710–714 Available from: http://www.ncbi.nlm.nih.gov/pubmed/23949585

Sood [illegible]

Smith [illegible]

Sparks [illegible]

[illegible]

Swygard [illegible]

Tsai [illegible]

[illegible]

[illegible]

[illegible]

Chapter 7
Taeniasis and Neurocysticercosis: Emerging Public Health Problems

Kashi Nath Prasad and Satyendra Kumar Singh

Abstract *Taenia*, one of the earliest recognized helminths, is a comparatively large genus with 42 known valid species. Three most commonly recognized species of human importance are *T. solium* and *T. asiatica* (pork tapeworms) and *T. saginata* (beef tapeworm). Adult tapeworms reside in human intestine, and the disease is called taeniasis. The larva (cysticercus) of only *T. solium* can infect internal organs of human, known as cysticercosis; when the larva infects the central nervous system (CNS) of the host, it is known as neurocysticercosis (NCC). NCC is the most severe form of the disease with considerable morbidity and mortality. It is considered as the most common cause of community-acquired epileptic seizure disorders. The life cycle of *T. solium* involves two hosts: humans are the only definitive host and accidental intermediate host, while pig is the natural intermediate host. The eggs passed through faeces by *T. solium* carriers contaminate the environment. Both human and pig can get infection (cysticercosis/NCC) through ingestion of eggs, while human acquires taeniasis through consumption of cysticercotic pork. *T. solium* infection is highly endemic in Africa, Asia and Latin America. Now developed world is also facing this problem due to human migration from *Taenia* endemic areas. Therapeutic measures for NCC-related active epilepsy include anti-epileptic drug(s) with or without steroids: surgery/placement of shunt is indicated for patients with raised intracranial pressure. Antiparasitic drug to kill the brain cysticerci remains controversial. Treatment of choice for taeniasis is niclosamide; alternatives are praziquantel and albendazole. The following measures such as antiparasitic therapy to eliminate *Taenia* carriers in endemic populations, health education, toilet facilities and handwash with soap, control on sale of measly pork, restriction on pig roaming and pig vaccination, etc. may help to control the disease.

Keywords Albendazole · Cysticercosis · Neurocysticercosis · Praziquantel · Taeniasis · Vaccination

---

K. N. Prasad (✉) · S. K. Singh
Department of Microbiology, Sanjay Gandhi Postgraduate Institute of Medical Sciences, Lucknow, India

P. P. Singh (ed.), *Infectious Diseases and Your Health*,
https://doi.org/10.1007/978-981-13-1577-0_7

7.1 Introduction

Taenia species are among the earliest recognized helminths. Life history and ecology of *Taenia* species are well studied and drawn considerable attention among the eucestodes. *Taenia* is a comparatively large genus with 42 known valid species and 3 subspecies (Hoberg 2006). Morphologically, they are ribbon-like structure with series of well-defined segments known as proglottids and therefore the name *Taenia* (Greek ταίνια meaning ribbon/stripe). The most common species infecting humans are the pork tapeworm *T. solium* and *T. asiatica* and the beef tapeworm *T. saginata.* Infection with adult tapeworms of *T. solium*, *T. asiatica* or *T. saginata* leads to taeniasis in humans. Among all the species of *Taenia*, the larval form (cysticercus) of only *T. solium* can infect internal organs of human, and the disease is called cysticercosis. When the larva of *T. solium* is lodged in the brain, it is known as neurocysticercosis (NCC). NCC is considered as the severe form of the disease with considerable morbidity and mortality. The life cycle of *T. solium* involves two hosts; viz. human (definitive host) and pig (intermediate host). Human harbours the adult tapeworm, which produces thousands of eggs for years. These eggs are excreted through faeces and disseminated in the environment. When free-roaming pigs ingest egg-loaded human faeces, the hexacanth embryos are liberated from the eggs, which penetrate the mucosal layer of the intestinal wall and enter into the blood circulation and subsequently develop into cysticerci (larval forms) in different internal organs, especially in the muscles of pigs. Human can also develop cysticercosis/NCC through consumption of food (raw vegetables and salads, etc.) and water contaminated by *T. solium* eggs. When human eats undercooked cysticercotic pork, the larva (cysticercus) develops into an adult worm in human intestine, and the disease is called taeniasis. In brief, both human and swine can get cysticercosis/NCC by ingestion of *T. solium* eggs excreted by the adult worm carriers. Clinical features of NCC may vary from mild form with little or no symptoms to life-threatening medical emergency. *T. solium* infection is cosmopolitan but highly endemic in African, Asian and Latin American countries where open-field defaecation practice is common and pigs roam freely with easy access to human faeces, and humans consume undercooked pig meat and hygienic standards are poor.

7.2 Morphology/Structure

The body of a classical adult eucestode is divided into three distinct parts, namely, scolex, neck and strobila. The scolex, responsible for attachment to the host tissue, is present at the anterior end. The size and morphological features of scolex are used to identify the worms. The neck is present immediately posterior to the scolex. It is the narrowest unsegmented and poorly differentiated region. New segments or proglottids differentiate from the neck region, and they push the older segments gradually posterior forming a chain of proglottids called strobila, which is the last part of

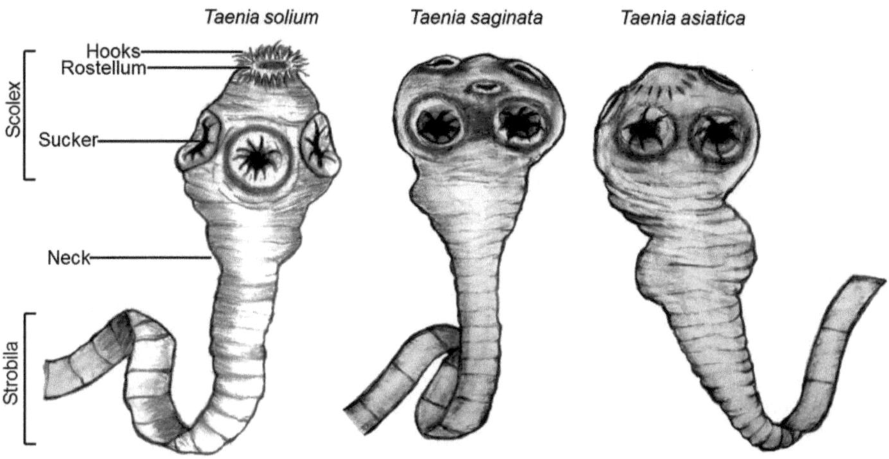

Fig. 7.1 Diagrammatic representation of the comparative morphological features of adult tapeworms, *Taenia solium*, *T. saginata* and *T. asiatica*

the adult parasite. As each proglottid moves posteriorly, its reproductive system starts maturing progressively. Therefore, the posterior most proglottids have the fully developed reproductive systems. The progressive maturity of the reproductive systems divides strobila into immature, mature and gravid proglottids (filled with egg) with often atrophied reproductive organs.

Three species of the *Taenia* genus, *T. solium*, *T. saginata* and *T. asiatica*, can cause taeniasis (intestinal infection) in humans. However, only the first two species have been considered for centuries except *T. asiatica* that remained undiscovered till recently. These tapeworms are flat and opaque white or yellowish with long segments measuring 1–12 m in the adult stage. An adult tapeworm has three segments: scolex, narrow neck and large strobila with hundreds of proglottids. The scolex is the organ for attachment with four suckers and a rostellum. Rostellum may be armed with hooks as in *T. solium*, or unarmed and sunken (*T. saginata*), or with rudimentary hooklets (*T. asiatica*) (Fig. 7.1) (Eom and Rim 1993; Flisser 1994). The scolex is a muscular structure with mesenchymal tissues that lodges the main part of the nervous system of the parasite and nephridial canals. Proximal segments are immature, followed by mature and gravid proglottids filled with eggs. Mature proglottids are hermaphrodites and contain thousands of testes. The female sexual organs have one lobulated ovary, connected to an oviduct. Gravid proglottids are full of $50–60 \times 10^3$ fertile eggs. The egg-containing uterus develops 7–32 lateral branches; this feature allows identification of species. *T. solium* has 7–16 and *T. saginata* has 14–32 branches. The eggs are spherical and their size varies from 31 to 43 μm in diameter. When eggs are excreted, most of them are fully embryonated and infective, while others may be at different stages of maturation. The main features of tapeworms and eggs are shown in Table 7.1 and Fig. 7.2 (Eom and Rim 1993; Flisser 2013; Galan-Puchades and Fuentes 2013).

Table 7.1 Morphological differences among *Taenia solium*, *T. saginata* and *T. asiatica*

Morphological features	*T. solium*	*T. saginata*	*T. asiatica*
Scolex			
Shape	Globular	Quadrilateral	Quadrilateral
Rostellum	+	–	+
Number of hooks	22–32	–	–
Diameter (mm)	0.6–1.0	1.5–2.0	0.8
Mature proglottid			
Testis (number)	375–575	800–1200	324–1216
Ovary (number of lobes)	3	2	2
Vaginal sphincter	–	+	+
Gravid proglottid			
Uterine branches (each side number)	7–16	14–32	11–31
Pattern of uterine branching	Dendritic	Dichotomous	Dichotomous
Posterior protuberance	–	+	+
Size (length × width; mm)	3.1–10 × 3.8–8.7	10–20 × 6.5–9.5, longer than wide	4–22 × 3–12
Adult tapeworm			
Length (m)	1–5	4–12	1–8
Number of proglottids	700–1000	1000–1500	200–1200
Cysticercus			
Size (mm)	5–8 × 3–6	6–10 × 4–6	2 × 2
Hooks in scolex	+	–	Rudimentary

+ Present; – absent

After 15–48 h of ingestion of eggs, hexacanth embryos (oncopsheres) are liberated in the intestine of the host that pierce the intestine and enter into the blood circulation. When the embryos are lodged in the internal organs, they develop into the larval forms called cysticercus, which is a fluid-filled bladder-like structure with invaginated whitish scolex (de Aluja et al. 1998). When definitive host (human) consumes the viable cysticercus present in the muscles of intermediate host (pork/beef), bladder wall pore widens that allows the scolex and neck to evaginate, and the adult tapeworm develops (de Queiroz and Alkire 1998). The adult tapeworm resides in the small intestine of humans, the only definitive hosts. Stages of taeniid development are as follows: embryo inside the egg, circulating oncosphere or larva, post-oncosphere that is transformed into the cysticercus or metacestode in tissues (the post-larval or pre-adult stage) and the adult worm. Eggs and embryos are microscopic (Fig. 7.3), while cysticerci (Fig. 7.4) and adult worms are macroscopic (Fig. 7.1).

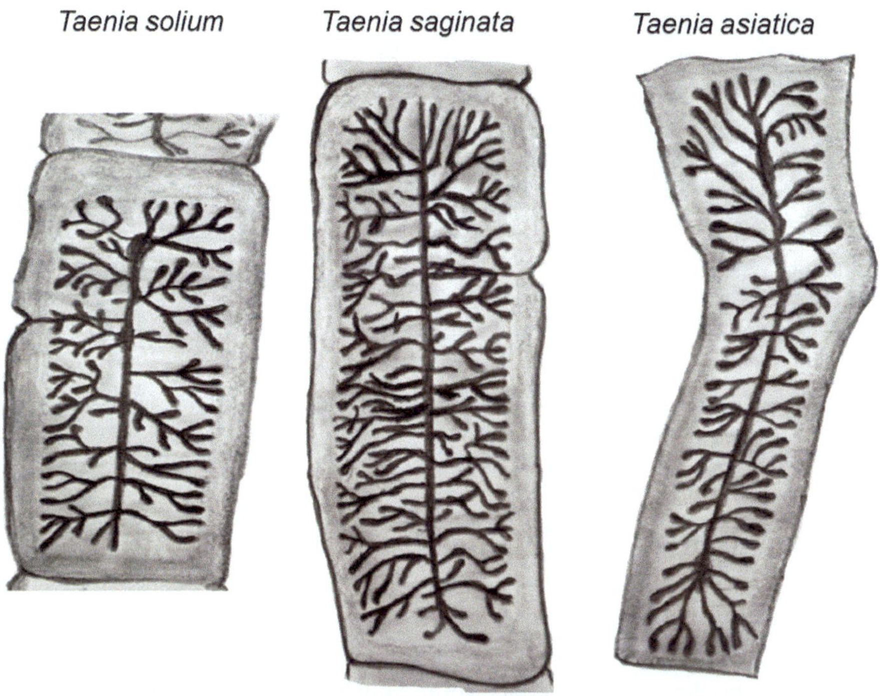

Fig. 7.2 Differences in uterine branching pattern of *Taenia solium*, *T. saginata* and *T. asiatica* proglottids

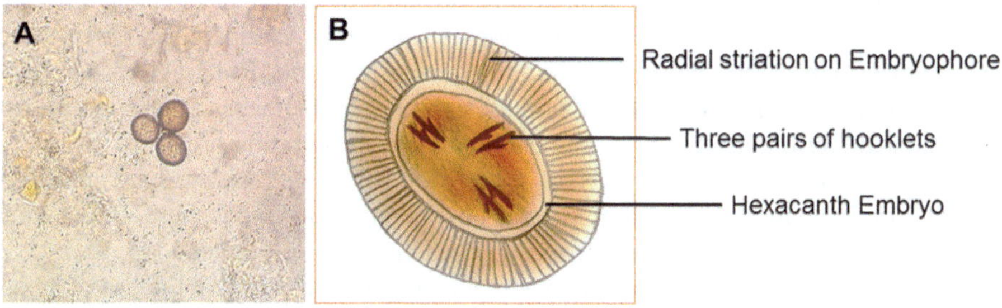

Fig. 7.3 Egg of *Taenia*: (**a**) as seen under microscope, (**b**) schematic diagram

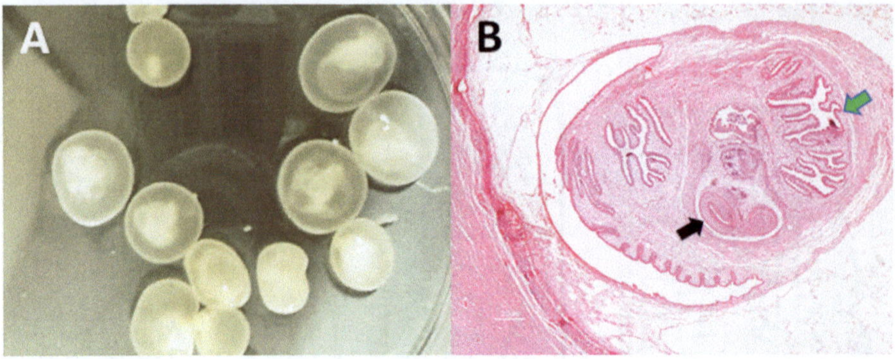

Fig. 7.4 (**a**) Cysticerci of *Taenia solium* obtained from pig muscles, (**b**) histologic section of *T. solium* cysticercus showing canalicular system (green arrow) and suckers (black arrows)

7.3 Biology/Life Cycle

7.3.1 T. solium

It is commonly known as pork tapeworm. Humans are the only definite host and harbour the adult tapeworm in their intestine. Both swine and human act as the intermediate hosts harbouring larvae in different internal organs (Fig. 7.5). The adult tapeworm residing in the small intestine of human produces thousands of eggs daily which are disseminated into the environment through faeces. The intermediate host (pig) ingests these eggs, which develop into cysticerci in different internal organs like muscles and brain. When human consumes cysticercotic pork, these cysts develop into adult worms residing the human intestine. The gravid proglottids start to separate from the distal end and are excreted in the faeces after 2 months of intestinal infection. Every day around four to five proglottids break off from the adult worm, and each segment has 50–60 × 10^3 fertile eggs. The eggs are spherical (31–43 μm in diameter) with thick striated cover containing oncosphere. Both swine and man can ingest the eggs, and these eggs reach digestive tract and lose their coat due to action of the gastric acid and pancreatic enzymes resulting in release of hexacanth embryos or oncospheres. Oncospheres pierce the intestinal wall using their hooklets to reach the blood circulation and further reach the different organs such as subcutaneous tissue, skeletal muscles, central nervous system (CNS) and eyes. Now, the oncospheres drop their hooklets, take a vesicular shape and develop into cysticerci by gradual invagination of scolex, and it takes approximately 2 months to develop (Escobar and Nieto 1972). Its life cycle is completed when human consumes the undercooked measly pork. After reaching the small intestine, the scolex evaginates from the cysticerci and attaches itself to the mucosal wall and gradually evolves into the adult tapeworm. However, humans may also be infected with eggs

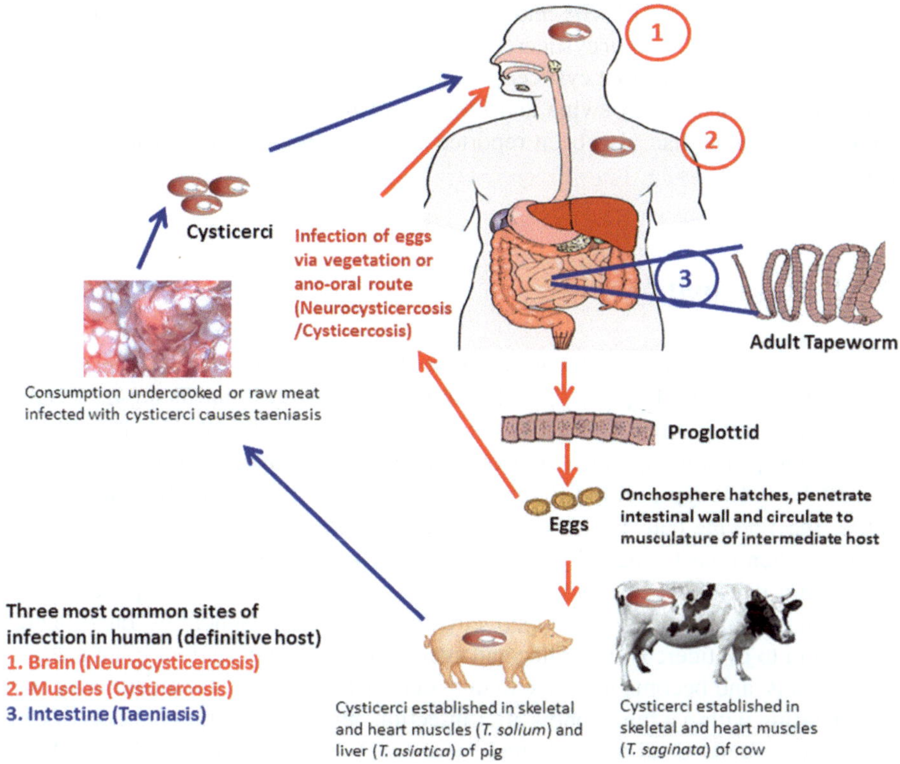

Fig. 7.5 Life cycles of three human *Taenia* tapeworms

in one of the following ways: (1) by hetero-infection through consumption of food/water contaminated with eggs of *Taenia*, (2) by exogenous autoinfection caused by ano-oral ingestion in *Taenia* carriers and (3) by endogenous autoinfection, in which the eggs present in the small intestine reach to the stomach due to reverse peristalsis. Hetero-infection is the most common route, while the last two modes of infection are very uncommon. The eggs of *Taenia* are sticky, and they may remain stuck to nail and nail beds of *Taenia* carriers if the hands are not properly washed with soap after defaecation. While handling such carriers can contaminate the food and disseminate the disease (cysticercosis/NCC).

In human, the parasite has a predilection to the brain and spinal cord, and when it gets lodged in these organs, it results in NCC. The larval stage also infects other tissues, especially those with pulsatile or contractile property and enriched in cholinergic innervations like skeletal muscle, diaphragm, heart, tissue and cavities subjected to contractile, rhythmic pulsatile movement by the internal organs like peritoneum, pleura and subcutaneous tissue (Del Brutto and Sotelo 1988).

Other animals have also been reported to harbour cysticerci of *T. solium*. An Indonesian study revealed that sera from 7 of 64 dogs were highly positive for

T. solium infection by ELISA and immunoblot. Subsequent examination of two such dogs showed *T. solium* cysticerci in their brain and heart muscle. Mitochondrial DNA analyses revealed that cysticerci from these dogs were similar with cysticerci obtained previously from swine and biopsies of local people (Ito et al. 2002). *T. solium* cysticercus had also been reported from the brain of a cat (Schwan et al. 2002).

7.3.2 T. saginata

Humans get infection through consumption of raw/undercooked beef infected with larvae of *T. saginata* (Fig. 7.5). The developmental stages of both *T. saginata* and *T. solium* are similar. After 3 months, *T. saginata* becomes sexually mature and produce gravid proglottids, which are expelled in faeces. Gravid proglottids may contain 50–80 × 10^3 eggs, with different maturation stages. Eggs are only infective to cattle, the intermediate host; hence *T. saginata* does not cause cysticercosis in human. When ingested eggs by the intermediate host (cattle) come in contact with gastric and intestinal juices, the active embryos are liberated, which penetrate the intestinal wall and reach the blood circulatory system. Generally, oncospheres develop in to cysticercus in cardiac and skeletal muscles and seldom in fat and visceral organs, and become infective to human by 10 week time. Cysticerci may start degenerating within a few months after infection, and a substantial numbers of cysticerci are usually dead and calcified by about 9 months (Pawlowski and Murell 2000).

7.3.3 T. asiatica

T. asiatica was first reported in Taiwan and later on in many other Asian countries such as Indonesia, Korea, the Philippines, and Thailand (Eom et al. 1992). It has recently been reported from India (Singh et al. 2016). Morphologically, *T. asiatica* is related to *T. saginata*; it has scolex without hooklets, a large number of dichotomous pattern of uterine branches in gravid proglottids and a posterior protuberance (Table 7.1). *T. asiatica* cysticerci develop in visceral organs of pigs such as the liver, omentum, lungs and serosa; however, *T. saginata* cysticerci infect only muscles of cattle. In general, *T. asiatica* life cycle is similar to *T. solium*; specifically the adult stage develops in humans (Fig. 7.5; Eom et al. 1992). Taeniasis due to *T. asiatica* occurs in human after consumption of raw/undercooked meat or viscera of pigs infected with cysticerci. Cysticercosis in human due to *T. asiatica* has not been reported till date.

7.4 Epidemiology

The incidence of *T. solium* infection significantly varies according to sanitation, eating habits and pig husbandry practices in a region. However, it is hard to estimate the exact prevalence of *T. solium*-associated taeniasis, because eggs of different *Taenia* species are indistinguishable by microscopic method.

Cysticercosis is an under-reported disease because of involvement of different internal organs. However, NCC is considered the severe form of the disease, and epilepsy is the common clinical manifestation of parenchymal NCC. Extra-parenchymal location of the cysticerci in the brain leads to raised intracranial pressure and hydrocephalus. The World Health Organization estimates approximately 50,000 deaths every year due to NCC. Overall, NCC is identified as a cause of active epilepsy in 26.3–53.8% of seizure disorders in the developing world (Del Brutto et al. 2005; Montano et al. 2005; Prasad et al. 2009a, b). It is prevalent in Asian countries such as China, Cambodia, India, Indonesia, Mongolia, Myanmar, Nepal, Thailand and Vietnam. *T. solium* infection has been eradicated from Japan and South Korea. Seroprevalence studies in Asian countries (Vietnam, China, Korea and Indonesia) indicate high rates of exposure to the parasite ranging from 0.02% to 12.6% (Kong et al. 1993; Margono et al. 2001; Rajshekhar et al. 2003).

Reports have demonstrated a potentially high risk of *T. solium* infection for inhabitants of many Latin American countries with variable incidence rates suggesting an active transmission involving pig and human cycle in the region (Fleury et al. 2006; Garcia-Noval et al. 1996). Studies from Guatemala, Honduras, Peru and Mexico demonstrated NCC infection in rural populations between 9% and 18% (Flisser et al. 2003). A study from Mexico had shown that up to 50% of patients with evidence of NCC were neurologically and systematically asymptomatic (Chavarria et al. 2003). Swine cysticercosis is also often reported at meat inspection in the slaughterhouses of Latin America.

T. solium is an emerging and expanding zoonosis in African countries like Cameroon, Mozambique, Nigeria, South Africa, Tanzania, Zimbabwe, etc. Data from African countries show low incidence of human infection and a high prevalence in pig populations. This discrepancy may be due to lack of suitable surveillance, monitoring and reporting systems. Moreover, a high prevalence of pig cysticercosis (20–40%) is reported from East and Southern Africa (Phiri et al. 2002).

NCC is now becoming prevalent in developed countries because of increased human travel from endemic areas and migration of tapeworm carriers (Burneo et al. 2009; Schantz et al. 1992). Individuals who have never gone outside the USA as well as those who travel to disease-endemic regions are at risk acquiring infection. Hospital-based data analysis showed that up to 2% of admissions in neurosciences in southern California were due to NCC (McCormick 1985). The USA reports more than 1000 cases of NCC per year (Shandera et al. 1994).

Only few population-based data are available that provide evidence about the community burden, risk factors, and geographical distribution of cysticercosis in India. Studies from India reported that 28.4–48.3% cases of active epilepsy are due to NCC by neuroimaging techniques such as CT and MRI (Prasad et al. 2009a, b; Rajshekhar et al. 2006). In Indian subcontinent, single cyst infection is more frequently reported, and its prevalence ranges from 47.7% to 53.4% of NCC cases (Prabhakaran et al. 2007).

7.5 Clinical Features

7.5.1 Clinical Manifestations of Taeniasis

Most people having taeniasis are either asymptomatic or have lowtomoderate symptoms. The most frequent complaint is passage of proglottids with slight discomfort. Other common symptoms are abdominal pain often colicky in nature (frequent in children), loss of appetite, nausea, constipation/diarrhoea, dizziness, headache, weakness, hyper-excitability, etc. These clinical symptoms may appear when the tapeworms become fully mature in human intestine, usually 6–8 weeks after ingestion of measly meat of swine (Ooi et al. 2013). Stomach pain and nausea are more frequent in the morning which is reduced after having small amounts of food. These symptoms are most common in children than adults and often lead to increased or decreased appetite. Infants may have diarrhoea, fever, irritability, vomiting and weight loss. The most severe complication of taeniasis is appendicitis.

7.5.2 Clinical Manifestations of NCC

The clinical manifestations of NCC vary with location, stage and number of *T. solium* cysticerci in the brain and immune response of the host to the parasite. It can affect the parenchyma, subarachnoid space or intraventricular systems within the CNS and rarely ocular and spinal region.

7.5.3 Parenchymal NCC

In parenchymal NCC, the most common clinical manifestation is seizure. It accounts up to 80% of patients (Ndimubanzi et al. 2010). Generalized tonic-clonic or simple partial seizure with motor symptomatology is mostly associated with NCC. However, some patients may have myoclonic, truncal or complex partial seizures (Sotelo et al. 1985). Presence of multiple degenerating parenchymal cysts with massive pericystic oedema may lead to cysticercotic encephalitis. The other most common clinical feature is headache (37% of NCC cases) followed by altered mental state (28% cases), neuro-focal deficits (16% cases), signs of increased intracranial pressure (12% cases) and cognitive decline (5% cases) (Carabin et al. 2011).

7.5.4 Subarachnoid NCC

Hydrocephalus is the most common clinical presentation of subarachnoid NCC due to increased intracranial pressure followed by stroke in about 12% of cases (Marquez and Arauz 2012). A variety of stroke syndromes have been described in patients with NCC such as cerebral infarctions, intracranial haemorrhages and transient ischemic attacks. Other manifestations are acute meningitis and cranial nerve involvement.

7.5.5 Ventricular NCC

The main clinical manifestation of intraventricular NCC is obstruction of cerebrospinal fluid (CSF) flow that leads to increased intracranial pressure leading to acute hydrocephalus. It usually happens when freely moving cysticerci reach the third ventricle from lateral ventricles or move up to cerebral aqueduct from fourth ventricle leading to blockage of CSF flow. It is an acute emergency and may need urgent shunt placement. Cysticerci present in the lateral ventricles may compress nearby tissues and are generally associated with focal neurological symptoms. Cysticerci present in the fourth ventricle may cause brainstem dysfunction because of compression of fourth ventricle floor (Madrazo et al. 1983; Sinha and Sharma 2012).

7.5.6 Spinal Cord NCC

In NCC, spinal cord involvement is very rare (1–5%); however, introduction of MRI has significantly improved the diagnosis. Generally, cysts are single or clusters of multiple cysts and present in the spinal cord parenchyma or spinal cord subarachnoid space, and the clinical symptoms are radicular pain and motor deficits of subacute onset with progressive course (Bandres et al. 1992; Park et al. 2011).

7.5.7 Ophthalmic Cysticercosis

Intraocular cysticerci (ophthalmic cysticercosis) may be found in the anterior chamber, the lens, the vitreous and the sub-retinal space. Although it is often asymptomatic, inflamed degenerating cysticerci cause progressive decrease in vision with symptoms like proptosis, diplopia, chorioretinitis, retinal detachment or vasculitis (Kruger-Leite et al. 1985; Madigubba et al. 2007).

7.5.8 Systemic Cysticercosis

Cysticerci can develop in almost any body site but tend to have a predilection for muscle or subcutaneous tissues. Cysticerci at these sites are usually asymptomatic, but the patient may notice subcutaneous, pealike or walnut-sized nodules. Subcutaneous nodules are more common in patients from Asia and Africa than from Latin America. In cases of major muscle involvement, acute myopathy can develop (Sawhney et al. 1976).

7.6 Pathogenesis of NCC

T. solium cysticercus is a structurally complex helminthic larva, which expresses diverse sets of antigens mounting variable immune response that leads to various clinical manifestations of the disease. Histological investigations on human and pig brain tissues have shown a very low or no inflammatory response around viable cysticerci. Viable cysticerci escape the host's immune response by blocking/inhibiting the complement system. The viable cysticerci produce excretory/secretory molecules such as paramyosin which inhibits Clq (Laclette et al. 1987); taeniaestatin identified in *T. taeniaeformis* inhibits both classical and alternate pathways of complement system. Taeniaestatin also decreases IL-1 and IL-2 production and lymphocyte proliferation (White et al. 1992). Cysticerci cell wall is rich in sulphated polysaccharides, which are to a large extent immunologically inert. Moreover, sulphated polysaccharides activate and consume complement and evade its response. Cysticercal cysteine proteinase degrades the host immunoglobulins, and its prostaglandin E2 (PGE2) induces Th2 response and suppresses the inflammation (Garcia et al. 2014; Terrazas 2008). However, when the cysticerci start to die/degenerate, granuloma is formed with inflammation around the dying cysts in both human and pig (Singh et al. 2013, 2015a, b). Immune cells such as eosinophils, lymphocytes, macrophages, and plasma cells infiltrate around the dying cysticerci to form granuloma. These cells secrete cytokines, chemokines and other inflammatory mediators that cause the various symptoms in patients. The infiltrating cells produce high level of Th1 (IFN-γ, TGF-β and IL-18) and lower level of Th2 cytokines (IL-4, IL-13 and IL-10) in the brain tissue surrounding dying cyst (Restrepo et al. 2001). Anti-helminthic treatment also induces Th1 (IFN-γ) and pleiotropic (IL-6) cytokine response with leukocyte infiltration around dead cysticerci in swine (Singh et al. 2015a, b).

The host immune response against cysticerci can be of two types, i.e. humoral and cellular. In humoral immune response, a number of immunoglobulin classes as specific antibodies against the parasite are produced. The most common immunoglobulin is IgG detected in patient's CSF, serum and saliva which is suggestive of long duration of infection (Grogl et al. 1985; Zini et al. 1990). The immune response against *T. solium* cysticerci display both Th1 and Th2 type; however, the basic mechanisms are not clear. Probably, the parasite is destroyed by eosinophils, which is supposed to be mediated by Th2 cytokines (Ostrosky-Zeichner and Estanol 1999). The inflammatory response that kills the parasite and leads to resolution of fibrosis is mediated by Th1 cytokines. Peripheral immune response is associated with reduced lymphocytic proliferation, inhibition of granulocyte aggregation and induction of Th2 response (IL-4, IL-5 and IL-13) (White et al. 1997). In chronic NCC, pro-inflammatory cytokines, up-regulated cellular adhesion molecules such as ICAM-1 and activation of MMP-9 and MMP-2 contribute to blood-brain barrier disruption resulting in seizure disorder (Alvarez et al. 2002; Prasad et al. 2009a, b; Verma et al. 2011). It has been reported that individuals with glutathione S-transferase (GST)-M1 and (GST)-T1 deletions (null genotypes) are low producers of GST enzymes and such individuals are at higher risk to develop seizures in NCC. GST enzymes are essential for the protection of cells from damage caused by reactive oxygen species (ROS) generated during inflammation. Hence, higher GST activity may maintain asymptomatic condition, possibly by neutralizing the ROS and free radicals (Singh et al. 2017).

7.7 Diagnosis

7.7.1 Diagnosis of Taeniasis

Intestinal *Taenia* infections in humans are diagnosed by the detection of gravid segments or eggs in faecal samples. However, there are various tools for diagnosis of taeniasis with variable sensitivity and specificity.

7.7.2 Microscopy

Traditional microscopy to detect *Taenia* eggs in the stool has poor sensitivity but a very high specificity. However, concentration methods, preferably using sedimentation, increase the sensitivity. *Taenia* egg appears as a thick radiate cover and hooks under the microscope. However, eggs of different *Taenia* species (*T. solium*, *T. asiatica* and *T. saginata*) cannot be differentiated by microscopic examination (Garcia and Del Brutto 2003). Detection of gravid segment with its uterine branching is a reliable test to differentiate different *Taenia* species.

7.7.3 Antigen Detection in Stool

Copro-antigen detection in stool by capture ELISA has enhanced the sensitivity two to three times for diagnosis of taeniasis. To detect tapeworm antigen in stool, generally polyclonal antibody is used against the adult tapeworm in copro-antigen ELISA (Bustos et al. 2012).

7.7.4 Molecular Method

The differentiation of the different species of *Taenia* is possible by mitochondrial DNA (*cox1*, *cob* and NADH dehydrogenase 1 gene)-based PCR, and the results are consistent if tapeworm material is available (Gonzalez et al. 2000, 2002; Yamasaki et al. 2004). However, the sensitivity of direct PCR assays in stool samples is yet to be defined.

7.7.5 Diagnosis of NCC

The clinical diagnosis of NCC is hard due to its polymorphic/nonspecific disease symptoms. However, a precise diagnosis can be done if clinical and epidemiologic data is interpreted together with the neuroimaging findings and the results of specific immunological tests.

7.7.6 Immunodiagnostic Techniques

In immunodiagnosis, specific antibodies and circulating cysts antigens are detected in serum or CSF. Various techniques such as complement fixation test, ELISA, indirect haemagglutination test, latex agglutination, radioimmunoassay and enzyme-linked immune-electrotransfer blot (EITB) have been used to detect antibodies. The serum-based ELISA to detect antibody is not reliable to diagnose NCC; however, it is very much in use due to simplicity of the technique. EITB is the most specific test so far. This immune-blot uses fractions of glycoproteins (50, 39–42, 24, 21, 18, 14 and 13 kDa) from crude extracts purified using a lentil-lectin column (Tsang et al. 1989).

7.7.7 Neuroimaging Techniques

Neuroimaging techniques such as computed tomography (CT) and magnetic resonance imaging (MRI) have significantly enhanced the diagnostic precision of NCC. CT is found to be more sensitive for the detection of calcified cysts; however, MRI is the most accurate as it gives information on the number and topography of lesions, stage of the cysts and the intensity of inflammatory response.

MRI has particular advantages such as it is non-invasive, it does not use any ionizing radiation and it has a high soft-tissue resolution. It may also provide both morphological and functional information. MRI is found to be accurate neuroimaging tool for the detection of dying and viable cysts, whereas CT is good for calcified lesions, but it cannot differentiate the stage of the parasite (Garcia and Del Brutto 2003). Moreover, a study has shown that MRI with gradient echo sequence phase imaging is equally good as CT for the detection of the scolex in lesions and also the calcified stage (Gupta et al. 2001).

7.8 Treatment

7.8.1 Anti-helminthic Therapy

Although treatment for NCC with anticysticercal drugs such as albendazole, mebendazole, praziquantel, metriphonate and flubendazole has been used, there has been a debate over safety and usefulness of such drugs due to lack of clinical trials to define doses and duration of therapy. Cysts location in the brain is a major factor that decides the success of anti-helminthic treatment. When the parasite infects the brain parenchyma, treatment with these drugs show radiological and clinical improvement in majority of patients with only a low adverse effect in some cases (Carpio et al. 2008; Del Brutto et al. 2006). However, when the parasite infects the subarachnoid basal cisterns, the prognosis is uncertain, and in such cases, either albendazole or praziquantel is preferred (Sotelo et al. 1988). However, controversies still persist about the use antiparasitic drugs in treatment of NCC, especially on seizure occurrence and other neurological disorders that result from inflammatory response triggered by antiparasitic treatment. Albendazole has been found to be more effective than praziquantel in clinical trials (Takayanagui and Jardim 1992). Another benefit of using albendazole is that it can also kill subarachnoid and ventricular cysts and has relatively less interaction with steroids and anti-epileptic drugs. However, increased dose of praziquantel (up to 30 mg/kg/day) or more prolonged or even repeated use has been recommended for treatment of patients with large subarachnoid cyst (Fleury et al. 2011). Further, surgery is also recommended by some clinicians as the best option for this form of infection. Albendazole is effective on 75–90% of cysts in the parenchyma when 15 mg/kg/day of albendazole is administered; however, the duration of treatment in such patients varies from 8 to 30 days (Castro et al. 2009). Further, antiparasitic therapy is recommended only in patients with low number of parenchymal cyst since the dying parasites release antigens to the surrounding tissues that initiate a strong inflammatory immune response, thus aggravating the symptoms. In order to reduce the risk, concomitant corticosteroids should be given, especially if there are large numbers of cysts. However, antiparasitic therapy is not recommended in patients with cysticercotic encephalitis. Recommendations for the treatment of NCC are given in Table 7.2 (Garcia et al. 2002).

Table 7.2 Guidelines for the use of antiparasitic treatment of neurocysticercosis (NCC)

Parenchymal NCC		
Viable cysts	1–5 cysts	Antiparasitic + steroid treatment
	More than 5 cysts	Antiparasitic + steroid treatment
	More than 100 cysts	Antiparasitic treatment + high-dose steroid Alternatively, chronic steroid management; no antiparasitic treatment
Enhancing lesion (dying cysts)	Mild or moderate	No antiparasitic treatment; neuroimaging follow-up. Alternatively, antiparasitic + steroids treatment
	Heavy (cysticercotic encephalitis)	No antiparasitic treatment + high-dose steroids and osmotic diuretics
Calcified cysticerci	Any number	No antiparasitic treatment
Extra-parenchymal NCC		
Ventricular cysticercosis		Neuroendoscopic removal, if not available:
		CSF diversion followed by antiparasitic + steroids treatment or open surgery
Subarachnoid cysts (racemose cysticercosis and chronic meningitis)		Antiparasitic + steroids treatment, ventricular shunt if there is hydrocephalus
Hydrocephalus with no visible cysts on neuroimaging		Ventricular shunt; no antiparasitic treatment
Other locations		
Spinal cysticercosis		Primarily surgical; anecdotal reports of successful use of albendazole with steroids
Ophthalmic cysticercosis		Surgery

7.8.2 Anti-inflammatory Therapy

Corticosteroids are commonly used to reduce the inflammatory response that leads to various neurological symptoms due to degenerating cysticerci. Drug of choice for this therapy is dexamethasone; dose varies between 4.5 and 12 mg/day. It is also used in chronic cysticercosis arachnoiditis or encephalitis, where up to 32 mg/day of dexamethasone may be given to decrease the brain oedema (Del Brutto et al. 1993). Another drug, prednisone at 1 mg/kg/day, may replace dexamethasone when long-term corticosteroid is given (Suastegui et al. 1996). To reduce the raised acute intracranial pressure secondary to NCC, mannitol (2 g/kg/day) is also recommended as an osmotic agent.

7.9 Anti-epileptic Drug Therapy

Other drugs generally used to treat symptomatic NCC patients are anti-epileptic drugs (AEDs). Carbamazepine, phenytoin and phenobarbitone are the first-line AEDs more often used for the control of seizures due to NCC. In recurrent severe headache associated with seizures, valproic acid/topiramate is preferred. Addition of praziquantel and albendazole with AED may improve seizure control. One study showed that 83% patients were seizure-free when treated with combination of anti-epileptic and anticysticercal drugs, whereas only 26% patients were seizure-free when treated with AEDs alone over a period of 28 months. However, despite better seizure control, albendazole and praziquantel may not be a definitive therapy for NCC-related active epilepsy, and AEDs need to be continued (Del Brutto et al. 1992). There is no definite time frame how long the anti-epileptic treatment should be given. Generally, AED is prescribed until a 2-year seizure-free period followed by gradual withdrawal.

7.10 Prevention

Cysticercosis/NCC is recognized as an eradicable disease. Eradication of swine cysticercosis through better animal husbandry and pig meat inspection are the important parameters adopted by the developed countries. These measures helped in breaking the transmission cycle of *T. solium* infection in the USA and Western Europe (Ferreira et al. 1997). Tapeworm carriers are the main target for anti-helminthic therapy for effective control of *T. solium* infection, since a small number of such carriers are likely to infect large numbers of healthy individuals. In developing countries the control measures should focus on mass administration of anti-helminthic drugs in endemic regions to cure tapeworm carriers and public awareness programme through health education. Other measures include toilet facilities to discourage open-field defaecation, handwash with soap after defaecation, restriction on sale of measly pork, restriction on pig roaming and pig vaccination, etc. Mass treatment of pigs in highly endemic regions with albendazole for 1 week followed by at least 2 weeks window period before slaughter for human consumption can reduce the incidence of *T. solium* carriers.

7.11 Treatment of Taeniasis

The only source of NCC/cysticercosis both for humans and pigs is the adult *T. solium* carrier individuals; a single tapeworm carrier can infect large number of hosts. Therefore, tapeworm carriers are the appealing target for control of the disvease. Tapeworm carriers can be cured either by individualized or mass treatment of the population in the endemic area with a single oral dose of either niclosamide (2 g in adults) or praziquantel (5–10 mg/kg; Garcia et al. 2007). Albendazole 15 mg/kg is a good alternative to praziquantel especially in the developing countries.

7.12 Vaccination of Swine

Vaccination/immunization of porcine population in endemic region may be good approach to interrupt the *T. solium* life cycle thus preventing taeniasis and NCC. Many research groups had evaluated several parasite antigens derived from different developmental stages of the related cestodes (*T. crassiceps*, *T. solium*, *T. saginata* and *T. ovis*) or of synthetic origin with variable results. Three different protective antigens (TSOL18, TSOL45-1A and TSOL16) were identified from *T. solium* oncospheres and evaluated as vaccines (Lightowlers 2004). These antigens were found to induce almost complete protection in vaccinated pigs challenged with *T. solium* in experiments. Studies showed that the two doses of TSOL18 vaccine had given almost complete protection from infection in swine population (Flisser et al. 2004; Gonzalez et al. 2005).

7.13 Treatment of Infected Swine

Treatment of cysticercotic swine can reduce the taeniasis burden and help in *T. solium* elimination. Praziquantel and albendazole are proven to be highly effective in treatment of swine cysticercosis. However, oxfendazole at a single oral dose of 30 mg/kg demonstrated 100% effective to kill muscle cysts with no major side effects (Pondja et al. 2012). In naturally infected swine, treatment with albendazole (15 mg/kg) for 2 weeks was found 94% and 100% effective to kill the cysticerci in the brain and muscles, respectively (Singh et al. 2015a, b).

7.14 Conclusions

T. solium infection (taeniasis and cysticercosis/NCC) is a serious public health issue worldwide, particularly in developing nations including India. Swine cysticercosis is an economic loss to the pig farmers. Pig health and management would produce benefits to pig farmers. Avoidance of open defaecation and practice of using latrines may prevent roaming pigs or piglets from infection. Proper porcine meat inspection in slaughterhouses and other measures such as health education, availability of health-care services, elimination of *T. solium* carriers by mass treatment and control on sale of cysticerci infected pork may help in decreasing the infection rate in the endemic areas.

References

Alvarez JI, Colegial CH, Castano CA, Trujillo J, Teale JM, Restrepo BI (2002) The human nervous tissue in proximity to granulomatous lesions induced by Taenia solium metacestodes displays an active response. J Neuroimmunol 127:139–144

Bandres JC, White AC Jr, Samo T, Murphy EC, Harris RL (1992) Extraparenchymal neurocysticercosis: report of five cases and review of management. Clin Infect Dis 15:799–811

Burneo JG, Plener I, Garcia HH (2009) Neurocysticercosis in a patient in Canada. CMAJ 180:639–642

Bustos JA, Rodriguez S, Jimenez JA, Moyano LM, Castillo Y, Ayvar V, Allan JC, Craig PS, Gonzalez AE, Gilman RH, Tsang VC, Garcia HH (2012) Detection of Taenia solium taeniasis coproantigen is an early indicator of treatment failure for taeniasis. Clin Vaccine Immunol 19:570–573

Carabin H, Ndimubanzi PC, Budke CM, Nguyen H, Qian Y, Cowan LD, Stoner JA, Rainwater E, Dickey M (2011) Clinical manifestations associated with neurocysticercosis: a systematic review. PLoS Negl Trop Dis 5:e1152

Carpio A, Kelvin EA, Bagiella E, Leslie D, Leon P, Andrews H, Hauser WA (2008) Effects of albendazole treatment on neurocysticercosis: a randomised controlled trial. J Neurol Neurosurg Psychiatry 79:1050–1055

Castro N, Marquez-Caraveo C, Brundage RC, Gonzalez-Esquivel D, Suarez AM, Gongora F, Jara A, Urizar J, Lanao JM, Jung H (2009) Population pharmacokinetics of albendazole in patients with neurocysticercosis. Int J Clin Pharmacol Ther 47:679–685

Chavarria A, Roger B, Fragoso G, Tapia G, Fleury A, Dumas M, Dessein A, Larralde C, Sciutto E (2003) TH2 profile in asymptomatic Taenia solium human neurocysticercosis. Microbes Infect 5:1109–1115

de Aluja AS, Martinez MJ, Villalobos AN (1998) Taenia solium cysticercosis in young pigs: age at first infection and histological characteristics. Vet Parasitol 76:71–79

de Queiroz A, Alkire NL (1998) The phylogenetic placement of Taenia cestodes that parasitize humans. J Parasitol 84:379–383

Del Brutto OH, Sotelo J (1988) Neurocysticercosis: an update. Rev Infect Dis 10:1075–1087

Del Brutto OH, Santibanez R, Noboa CA, Aguirre R, Diaz E, Alarcon TA (1992) Epilepsy due to neurocysticercosis: analysis of 203 patients. Neurology 42:389–392

Del Brutto OH, Sotelo J, Roman GC (1993) Therapy for neurocysticercosis: a reappraisal. Clin Infect Dis 17:730–735

Del Brutto OH, Santibanez R, Idrovo L, Rodriguez S, Diaz-Calderon E, Navas C, Gilman RH, Cuesta F, Mosquera A, Gonzalez AE, Tsang VC, Garcia HH (2005) Epilepsy and neurocysticercosis in Atahualpa: a door-to-door survey in rural coastal Ecuador. Epilepsia 46:583–587

Del Brutto OH, Roos KL, Coffey CS, Garcia HH (2006) Meta-analysis: cysticidal drugs for neurocysticercosis: albendazole and praziquantel. Ann Intern Med 145:43–51

Eom KS, Rim HJ (1993) Morphologic descriptions of Taenia asiatica sp. n. Kor J Parasitol 31:1–6

Eom KS, Rim HJ, Geerts S (1992) Experimental infection of pigs and cattle with eggs of Asian Taenia saginata with special reference to its extrahepatic viscerotropism. Kisaengchunghak Chapchi 30:269–275

Escobar A, Nieto D (1972) Parasitic diseases. In: Minckler J (ed) Pathology of the nervous system. McGraw-Hill, New York, pp 2503–2521

Ferreira AP, Vaz AJ, Nakamura PM, Sasaki AT, Ferreira AW, Livramento JA (1997) Hemagglutination test for the diagnosis of human neurocysticercosis: development of a stable reagent using homologous and heterologous antigens. Rev Inst Med Trop Sao Paulo 39:29–33

Fleury A, Morales J, Bobes RJ, Dumas M, Yanez O, Pina J, Carrillo-Mezo R, Martinez JJ, Fragoso G, Dessein A, Larralde C, Sciutto E (2006) An epidemiological study of familial neurocysticercosis in an endemic Mexican community. Trans R Soc Trop Med Hyg 100:551–558

Fleury A, Carrillo-Mezo R, Flisser A, Sciutto E, Corona T (2011) Subarachnoid basal neurocysticercosis: a focus on the most severe form of the disease. Expert Rev Anti-Infect Ther 9:123–133

Flisser A (1994) Taeniasis and cysticercosis due to Taenia solium. Prog Clin Parasitol 4:77–116

Flisser A (2013) State of the art of Taenia solium as compared to Taenia asiatica. Kor J Parasitol 51:43–49

Flisser A, Sarti E, Lightowlers M, Schantz P (2003) Neurocysticercosis: regional status, epidemiology, impact and control measures in the Americas. Acta Trop 87:43–51

Flisser A, Gauci CG, Zoli A, Martinez-Ocana J, Garza-Rodriguez A, Dominguez-Alpizar JL, Maravilla P, Rodriguez-Canul R, Avila G, Aguilar-Vega L, Kyngdon C, Geerts S, Lightowlers MW (2004) Induction of protection against porcine cysticercosis by vaccination with recombinant oncosphere antigens. Infect Immun 72:5292–5297

Galan-Puchades MT, Fuentes MV (2013) Taenia asiatica: the most neglected human Taenia and the possibility of cysticercosis. Kor J Parasitol 51:51–54

Garcia HH, Del Brutto OH (2003) Imaging findings in neurocysticercosis. Acta Trop 87:71–78

Garcia HH, Evans CA, Nash TE, Takayanagui OM, White AC Jr, Botero D, Rajshekhar V, Tsang VC, Schantz PM, Allan JC, Flisser A, Correa D, Sarti E, Friedland JS, Martinez SM, Gonzalez AE, Gilman RH, Del Brutto OH (2002) Current consensus guidelines for treatment of neurocysticercosis. Clin Microbiol Rev 15:747–756

Garcia HH, Gonzalez AE, Del Brutto OH, Tsang VC, Llanos-Zavalaga F, Gonzalvez G, Romero J, Gilman RH (2007) Strategies for the elimination of taeniasis/cysticercosis. J Neurol Sci 262:153–157

Garcia HH, Rodriguez S, Friedland JS (2014) Immunology of Taenia solium taeniasis and human cysticercosis. Parasite Immunol 36:388–396

Garcia-Noval J, Allan JC, Fletes C, Moreno E, DeMata F, Torres-Alvarez R, Soto de Alfaro H, Yurrita P, Higueros-Morales H, Mencos F, Craig PS (1996) Epidemiology of Taenia solium taeniasis and cysticercosis in two rural Guatemalan communities. Am J Trop Med Hyg 55:282–289

Gonzalez LM, Montero E, Harrison LJ, Parkhouse RM, Garate T (2000) Differential diagnosis of Taenia saginata and Taenia solium infection by PCR. J Clin Microbiol 38:737–744

Gonzalez LM, Montero E, Puente S, Lopez-Velez R, Hernandez M, Sciutto E, Harrison LJ, Parkhouse RM, Garate T (2002) PCR tools for the differential diagnosis of Taenia saginata and Taenia solium taeniasis/cysticercosis from different geographical locations. Diagn Microbiol Infect Dis 42:243–249

Gonzalez AE, Gauci CG, Barber D, Gilman RH, Tsang VC, Garcia HH, Verastegui M, Lightowlers MW (2005) Vaccination of pigs to control human neurocysticercosis. Am J Trop Med Hyg 72:837–839

Grogl M, Estrada JJ, MacDonald G, Kuhn RE (1985) Antigen-antibody analyses in neurocysticercosis. J Parasitol 71:433–442

Gupta RK, Rao SB, Jain R, Pal L, Kumar R, Venkatesh SK, Rathore RK (2001) Differentiation of calcification from chronic hemorrhage with corrected gradient echo phase imaging. J Comput Assist Tomogr 25:698–704

Hoberg EP (2006) Phylogeny of Taenia: species definitions and origins of human parasites. Parasitol Int 55(Suppl):S23–S30

Ito A, Putra MI, Subahar R, Sato MO, Okamoto M, Sako Y, Nakao M, Yamasaki H, Nakaya K, Craig PS, Margono SS (2002) Dogs as alternative intermediate hosts of Taenia solium in Papua (Irian Jaya), Indonesia confirmed by highly specific ELISA and immunoblot using native and recombinant antigens and mitochondrial DNA analysis. J Helminthol 76:311–314

Kong Y, Cho SY, Cho MS, Kwon OS, Kang WS (1993) Seroepidemiological observation of Taenia solium cysticercosis in epileptic patients in Korea. J Kor Med Sci 8:145–152

Kruger-Leite E, Jalkh AE, Quiroz H, Schepens CL (1985) Intraocular cysticercosis. Am J Ophthalmol 99:252–257

Laclette JP, Merchant MT, Willms K (1987) Histological and ultrastructural localization of antigen B in the metacestode of Taenia solium. J Parasitol 73:121–129

Lightowlers MW (2004) Vaccination for the prevention of cysticercosis. Dev Biol (Basel) 119:361–368

Madigubba S, Vishwanath K, Reddy G, Vemuganti GK (2007) Changing trends in ocular cysticercosis over two decades: an analysis of 118 surgically excised cysts. Indian J Med Microbiol 25:214–219

Madrazo I, Garcia-Renteria JA, Sandoval M, Lopez Vega FJ (1983) Intraventricular cysticercosis. Neurosurgery 12:148–152

Margono SS, Subahar R, Hamid A, Wandra T, Sudewi SS, Sutisna P, Ito A (2001) Cysticercosis in Indonesia: epidemiological aspects. Southeast Asian J Trop Med Public Health 32(Suppl 2):79–84

Marquez JM, Arauz A (2012) Cerebrovascular complications of neurocysticercosis. Neurologist 18:17–22

McCormick GF (1985) Cysticercosis – review of 230 patients. Bull Clin Neurosci 50:76–101

Montano SM, Villaran MV, Ylquimiche L, Figueroa JJ, Rodriguez S, Bautista CT, Gonzalez AE, Tsang VC, Gilman RH, Garcia HH (2005) Neurocysticercosis: association between seizures, serology, and brain CT in rural Peru. Neurology 65:229–233

Ndimubanzi PC, Carabin H, Budke CM, Nguyen H, Qian YJ, Rainwater E, Dickey M, Reynolds S, Stoner JA (2010) A systematic review of the frequency of neurocyticercosis with a focus on people with epilepsy. PLoS Negl Trop Dis 4:e870

Ooi HK, Ho CM, Chung WC (2013) Historical overview of Taenia asiatica in Taiwan. Kor J Parasitol 51:31–36

Ostrosky-Zeichner L, Estanol B (1999) Immunopathogenesis of neurocysticercosis: is damage mediated by the host immune response? Int J Parasitol 29:649–650

Park YS, Lee JK, Kim JH, Park KC (2011) Cysticercosis of lumbar spine, mimicking spinal subarachnoid tumor. Spine J 11:e1–e5

Pawlowski Z, Murell KD (2000) Taeniasis and cysticercosis. In: Hui Y, Murrell KD, Nip WK, Stanfield P, Satter SA (eds) Foodborne diseases handbook. Marcel Dekker, New York, pp 217–227

Phiri IK, Dorny P, Gabriel S, Willingham AL 3rd, Speybroeck N, Vercruysse J (2002) The prevalence of porcine cysticercosis in Eastern and Southern provinces of Zambia. Vet Parasitol 108:31–39

Pondja A, Neves L, Mlangwa J, Afonso S, Fafetine J, Willingham AL 3rd, Thamsborg SM, Johansen MV (2012) Use of oxfendazole to control porcine cysticercosis in a high-endemic area of Mozambique. PLoS Negl Trop Dis 6:e1651

Prabhakaran V, Rajshekhar V, Murrell KD, Oommen A (2007) Conformation-sensitive immunoassays improve the serodiagnosis of solitary cysticercus granuloma in Indian patients. Trans R Soc Trop Med Hyg 101:570–577

Prasad A, Prasad KN, Gupta RK, Pradhan S (2009a) Increased expression of ICAM-1 among symptomatic neurocysticercosis. J Neuroimmunol 206:118–120

Prasad KN, Prasad A, Gupta RK, Nath K, Pradhan S, Tripathi M, Pandey CM (2009b) Neurocysticercosis in patients with active epilepsy from the pig farming community of Lucknow district, north India. Trans R Soc Trop Med Hyg 103:144–150

Rajshekhar V, Joshi DD, Doanh NQ, van De N, Xiaonong Z (2003) Taenia solium taeniosis/cysticercosis in Asia: epidemiology, impact and issues. Acta Trop 87:53–60

Rajshekhar V, Raghava MV, Prabhakaran V, Oommen A, Muliyil J (2006) Active epilepsy as an index of burden of neurocysticercosis in Vellore district, India. Neurology 67:2135–2139

Restrepo BI, Alvarez JI, Castano JA, Arias LF, Restrepo M, Trujillo J, Colegial CH, Teale JM (2001) Brain granulomas in neurocysticercosis patients are associated with a Th1 and Th2 profile. Infect Immun 69:4554–4560

Sawhney BB, Chopra JS, Banerji AK, Wahi PL (1976) Pseudohypertrophic myopathy in cysticerosis. Neurology 26:270–272

Schantz PM, Moore AC, Munoz JL, Hartman BJ, Schaefer JA, Aron AM, Persaud D, Sarti E, Wilson M, Flisser A (1992) Neurocysticercosis in an Orthodox Jewish community in New York City. N Engl J Med 327:692–695

Schwan EV, de Scally MP, van Rensburg CL, Durand DT (2002) Cerebral cysticercosis in a cat. J S Afr Vet Assoc 73:219–221

Shandera WX, White AC Jr, Chen JC, Diaz P, Armstrong R (1994) Neurocysticercosis in Houston, Texas. A report of 112 cases. Medicine (Baltimore) 73:37–52

Singh AK, Prasad KN, Prasad A, Tripathi M, Gupta RK, Husain N (2013) Immune responses to viable and degenerative metacestodes of Taenia solium in naturally infected swine. Int J Parasitol 43:1101–1107

Singh AK, Singh SK, Singh A, Gupta KK, Khatoon J, Prasad A, Rai RP, Gupta RK, Tripathi M, Husain N, Prasad KN (2015a) Immune response to Taenia solium cysticerci after anti-parasitic therapy. Int J Parasitol 45:749–759

Singh SK, Singh AK, Prasad KN, Singh A, Rai RP, Tripathi M, Gupta RK, Husain N (2015b) Expression of adhesion molecules, chemokines and matrix metallo- proteinases (MMPs) in viable and degenerating stage of Taenia solium metacestode in swine neurocysticercosis. Vet Parasitol 214:59–66

Singh SK, Prasad KN, Singh AK, Gupta KK, Chauhan RS, Singh A, Rai RP, Pati BK (2016) Identification of species and genetic variation in Taenia isolates from human and swine of North India. Parasitol Res 115:3689–3693

Singh A, Prasad KN, Singh AK, Singh SK, Gupta KK, Paliwal VK, Pandey CM, Gupta RK (2017) Human glutathione s-transferase enzyme gene polymorphisms and their association with neurocysticercosis. Mol Neurobiol 54:2843–2851

Sinha S, Sharma BS (2012) Intraventricular neurocysticercosis: a review of current status and management issues. Br J Neurosurg 26:305–309

Sotelo J, Guerrero V, Rubio F (1985) Neurocysticercosis: a new classification based on active and inactive forms. A study of 753 cases. Arch Intern Med 145:442–445

Sotelo J, Escobedo F, Penagos P (1988) Albendazole vs praziquantel for therapy for neurocysticercosis. A controlled trial. Arch Neurol 45:532–534

Suastegui Roman RA, Soto-Hernandez JL, Sotelo J (1996) Effects of prednisone on ventriculo-peritoneal shunt function in hydrocephalus secondary to cysticercosis: a preliminary study. J Neurosurg 84:629–633

Takayanagui OM, Jardim E (1992) Therapy for neurocysticercosis. Comparison between albendazole and praziquantel. Arch Neurol 49:290–294

Terrazas LI (2008) The complex role of pro- and anti-inflammatory cytokines in cysticercosis: immunological lessons from experimental and natural hosts. Curr Top Med Chem 8:383–392

Tsang VC, Brand JA, Boyer AE (1989) An enzyme-linked immunoelectrotransfer blot assay and glycoprotein antigens for diagnosing human cysticercosis (Taenia solium). J Infect Dis 159:50–59

Verma A, Prasad KN, Nyati KK, Singh SK, Singh AK, Paliwal VK, Gupta RK (2011) Association of MMP-2 and MMP-9 with clinical outcome of neurocysticercosis. Parasitology 138:1423–1428

White AC Jr, Tato P, Molinari JL (1992) Host-parasite interactions in Taenia solium cysticercosis. Infect Agents Dis 1:185–193

White AC Jr, Robinson P, Kuhn R (1997) Taenia solium cysticercosis: host-parasite interactions and the immune response. Chem Immunol 66:209–230

Yamasaki H, Allan JC, Sato MO, Nakao M, Sako Y, Nakaya K, Qiu D, Mamuti W, Craig PS, Ito A (2004) DNA differential diagnosis of taeniasis and cysticercosis by multiplex PCR. J Clin Microbiol 42:548–553

Zini D, Farrell VJ, Wadee AA (1990) The relationship of antibody levels to the clinical spectrum of human neurocysticercosis. J Neurol Neurosurg Psychiatry 53:656–661

Chapter 8
Helminth Parasites: The Cause of Distress and Diseases

Qudsia Tahseen

Abstract Helminth infections are one of the most common infections of mankind. The nematode and trematode parasites cause almost ten million disability-adjusted life years (DALYs) worldwide. The worm burden and the resulting morbidities are largely restricted to the children and those mainly belonging to poor countries. Although preventive chemotherapy is the most common method of treatment, it is associated with high-intensity recurrent infections. Therefore, the preventive strategies such as water, sanitation and health education along with chemotherapy are required to completely eliminate the soil-transmitted helminth (STH) infections. The frequent use of anthelmintic drugs because of reinfections results in their diminished efficacy leading to drug resistance. Therefore, mass drug administration is not an effective measure to combat STH infections. In such cases, vaccine too proves ineffective because these parasites often evade or modify the immune responses. Therefore, the approach now is to control the disease by reducing the worm burden below threshold level.

Keywords Disease burden · Epidemiology · Helminths · Neglected tropical diseases · Soil-transmitted helminths · Vaccination

8.1 Introduction

Water is a vital resource to life on the planet Earth. Early civilizations sprang fast mainly in the fertile flood plains along the banks of the rivers (Keddy 2000). According to US Geological Survey, freshwater habitats including lakes, rivers, swamps, etc. constitute about 0.3% of the total water of the world. Due to an increasing demand and pressures, these habitats have been overexploited and degraded. This was largely because of the lack of understanding of their nature and dynamics. Water quality is nowadays a major issue with drinking water sources in developing world getting scarcer as well as contaminated, thus impacting the health, economic

Q. Tahseen (✉)
Department of Zoology, Aligarh Muslim University, Aligarh, India

P. P. Singh (ed.), *Infectious Diseases and Your Health*,
https://doi.org/10.1007/978-981-13-1577-0_8

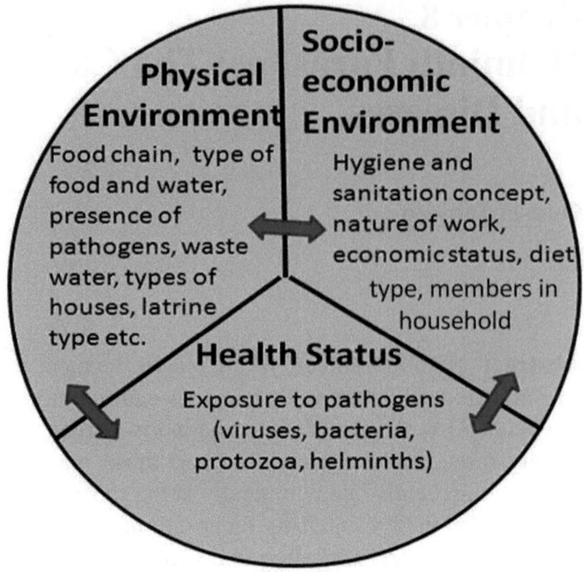

Fig. 8.1 Relationship of health status with the physical and socio-economic environments

and social development. The condition has considerably improved from the 1990s with greater proportion of the world population having access to water sources; however, quite often such waters may be contaminated and not be necessarily safe.

There may be various causes for this deteriorating water quality which deprives people of sustainable access to safe water. The contamination of water supply systems in nature or through xenobiotic pollutants is a matter of concern. Although the chemical pollutants like arsenic and fluoride cause health hazards to millions of people, the major contaminants happen to be the pathogens and parasites that enter aquatic bodies through domestic wastes and industrial and agricultural effluents. In that context, the faecal contamination of drinking water system is the root cause of morbidity and mortality of millions of children every year. The wastewater (Raschid-Sally and Jayakody 2008) including black water (having domestic and toilet wastes, urine, faecal slurry, etc.) and grey water (kitchen and bathing wastewater) has long been used as a source of nutrients in crop fields (Fig. 8.14a) in countries like India, China, Mexico, Vietnam, Morocco, etc. According to a report, an area of 4–6 million hectares utilizes wastewater or polluted water for irrigation (Jiménez and Asano 2008; Keraita et al. 2008). Earlier, in many European and North American cities, there was a practice of disposing wastewater in agricultural fields to prevent pollution of water bodies (Asano et al. 2007). The situation remains the same, by and large, in most developing countries in the absence of conventional treatment plants. Thus water-borne diseases due to faecal contamination are very common in developing countries where excreted pathogens such as viruses, bacteria, protozoan cysts and helminth eggs enter the human body through contaminated water and food (Table 8.1).

Table 8.1 Infection and dissemination route of different helminth parasites

Category	Type of helminth parasite	Source
Water borne/ water washed	*Ascaris lumbricoides*, *Enterobius vermicularis*, *Strongyloides stercoralis*, *Ancylostoma duodenale*, *Necator americanus*	Hands, drinking water contaminated with excreta; agricultural produce contaminated with night soil or irrigated with wastewater
Water based	*Dracunculus medinensis*	Water bodies holding population of *Cyclops*, used by cattle and cattle owners
Water related (vector)	*Onchocerca volvulus*, *Wuchereria bancrofti*	Stagnant water bodies serving as breeding grounds of vector fire flies or mosquitoes

Modified from Table 2.2 Chapter 2. UNICEF Handbook on Water Quality (2008)

Children are the worst sufferers of such ailments due to weak immunity and low resistance. As per UN estimates, about 33% of the preschool children of the world living in developing countries are underweight and stunted (ACC/SCN 2000) with greater death risks during childhood. Such children fail to demonstrate full abilities in education and physical activities (Martorell and Scrimshaw 1995). The health of a man in general and a child in particular is influenced mainly by the diet and the environment. The malnutrition-infection complex is a vicious cycle leading to ill health. Malnutrition leads to weaker immunity, thus making a child susceptible to diseases; likewise the infection reduces normal appetite and food intake, leading to stunted growth and impaired immune system. Malnutrition and parasitic infection also have a closer link with poverty (Fig. 8.1) that is further related to poor infrastructure and housing facility, low standards of education, poor health care and improper sanitation (Crompton and Nesheim 1982). Younger children of 6 months to 2 years tend to have more seriously affected health status due to these infections as the immune system is challenged to respond to new infections.

8.2 The Neglected Tropical Diseases

The name seems to have been coined somewhere in 2001 while reporting a case study from India which emphasized on the need for development of local research infrastructure to tackle many neglected infectious diseases prevalent in tropical region (Kettler and Modi 2001). The term was then picked up by researchers as there was a conspicuous association of these diseases with poverty or indirectly the substandard conditions and poor hygiene. The apathy and negligence were also reflected in the dearth of money allocated for research and development by the government and the insignificant investment by the industry. A comprehensive study by Trouiller et al. (2002) revealed that no new chemical formulations were introduced during the 25-year period (1975–2000) for these neglected diseases. Molyneux et al. (2005) discussed in detail the strategies to regulate these diseases and also suggested how chemotherapeutics could be integrated with other strategies.

Neglected tropical diseases represent a group of mainly chronic and incapacitating diseases that largely affect the poor class that resides in remote villages or in slum areas of cities in tropical and subtropical countries. The number of diseases considered under this category in 2005–2006 was 15, of which 13 diseases were distinct in causing greater annual mortality and global burden (Hotez et al. 2006, 2012; Weiss 2008; WHO 2012; Litt et al. 2012). These diseases also included helminth infections caused by cysticercus, trematodes/schistosomes and nematodes particularly *Dracunculus*, *Wuchereria*, *Onchocerca*, *Ascaris*, hookworm, *Trichuris*, etc. The list has further expanded to include more than 40 diseases caused by helminths, protozoan, bacterial, fungi, viruses and some invertebrate ectoparasites (Utzinger et al. 2012).

Soil-transmitted helminth (STH) infections are among the most prevalent yet neglected infections worldwide that affect the poor and deprived classes and hence often considered as the most common parasitic infection of mankind. The source of helminth infection can largely be the contaminated water as most of such parasites are released through faeces and complete their life cycle both in soil and human body. That's the reason they are referred to as soil-transmitted helminths (STH). Nevertheless, their routes of transmission and direct or indirect association with water can be elucidated hereunder along with other nematode parasites which are vector borne (*Dracunculus medinensis*, *Wuchereria bancrofti* and *Onchocerca volvulus*) or which may accidentally parasitize humans (*Toxocara cati/canis*) if proper sanitation and hygiene are not maintained.

8.3 Transmission Routes

The five Fs are considered to be the mode of transmission of these parasites, *viz*, faeces, fingers, flies, fields and feed. The infections can happen through hands or fingers contaminated with eggs or vegetables and fruits soiled with eggs due to improper peeling, washing or cooking. Children having nail-biting or thumb-sucking habits may ingest the eggs readily after scratching the perianal area. In some cases the eggs deposited in soil through faeces may not reach the host instantly and undergo development to form larvae that infect the host either through an insect vehicle or by penetrating the skin to enter the blood circulation.

8.4 Main Pathogens

The soil-transmitted helminths represent mainly the nematode parasites, *viz*. roundworm (*Ascaris lumbricoides*), pinworm (*Enterobius vermicularis*), hookworms (*Ancylostoma duodenale* and *Necator americanus*) and whipworm (*Trichuris trichiura*). Other parasites such as guinea worm (*Dracunculus medinensis*), filarial worm (*Wuchereria bancrofti*), *Onchocerca volvulus*, threadworm (*Strongyloides stercoralis*) and *Toxocara canis/cati* are also included in the list very often. The term helminth is a misnomer as it has been used since the early times when nematodes were being placed in *Aschelminthes/Nemathelminthes* group parallel to the *Platyhelminthes*. However, with the elevation of phylum *Nematoda*, all the above species have been placed under it. Therefore, instead of helminths, they should be called nematodes. Some of the commonly occurring infections of these nematodes are discussed hereunder, which spread mainly due to lack of hygiene or sanitation resulting in their direct or indirect transmission through the vectors bred under such conditions.

8.5 Ascariasis

Causative organism: An estimated 1.5 billion persons are infected with the largest nematode parasite *Ascaris lumbricoides* (WHO 2017) often called giant roundworm (Fig. 8.2) that has been found to infect urban, suburban and agricultural populations.

Distribution *Ascaris lumbricoides* has been reported to be cosmopolitan with records from more than 150 countries across the globe having tropical subtropical and temperate climates (Fig. 8.5a). Thus it is prevalent in all regions including Nearctic, Palaearctic, Oriental, Ethiopian, Neotropical and Australian ones. The eggs can resist adverse conditions and remain viable in the soil for up to 10 years (Chong 2003; Dora-Laskey et al. 2009; Khuroo 1996).

Morphology The females are 20–49 cm × 3–6 mm in size and can grow up to 49 cm in length, whereas males measure 12–31 cm × 2–4 mm. Individuals are pale in colour, often with a pinkish hue. Besides the location of genital opening, the sexes can be differentiated on the basis of sizes and genital papillae. Females can lay up to 200,000 eggs per day that may reach soil due to open defecation or through night used in agricultural fields. Fertilized eggs are ovoid and smaller 30–40 × 50–60 μm in dimension with golden brown mammillated shell (Fig. 8.2). Unfertilized eggs are relatively larger (up to 90 μm long) with a thinner outer coat (Chong 2003; Khuroo 1996).

Life Cycle The deposited (fertilized) eggs undergo cleavage leading to development of the embryos and, thereafter, into a full-fledged larvae within a period of 2–4 months. The embryonating eggs, when ingested by host through contaminated food and water, hatch into rhabditiform larvae which penetrate mucous membranes of the stomach, enter the bloodstream and reach the pulmonary capillaries to finally burrow into alveoli. The larvae then pass into the trachea by the movement of microvilli, reach the throat and are coughed to be swallowed a second time to enter the gut, thus reaching the intestine. After moulting several times, the larvae finally transform into adults which mate to initiate the process of egg laying. The time taken from infection to the passing of eggs in the stool is about 60–70 days.

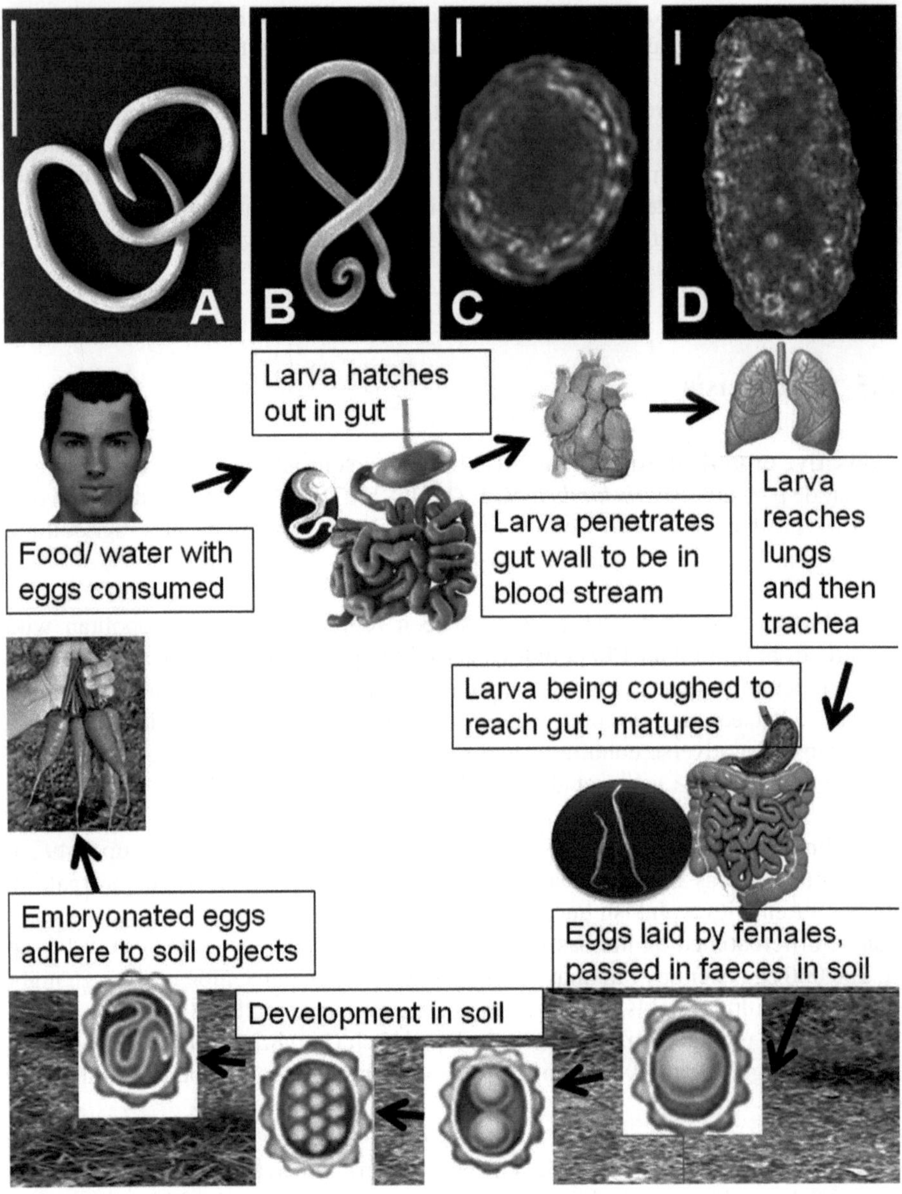

Fig. 8.2 *Ascaris lumbricoides*. (**a**) Female; (**b**) male; (**c**) fertilized egg; (**d**) unfertilized egg. Scale bar: (**a, b**) 5 cm; (**c, d**) 10 μm. Life cycle of *A. lumbricoides*

Disease Symptoms and Associated Morbidities The WHO definition of heavy infection estimates >50,000 eggs/g of faeces. The people having direct contact with wastewater showed high incidence of *A. lumbricoides* infections; however, the effect was more obvious in children than in adults. The acute symptoms of *Ascaris* infections in children are often reflected in the form of intestinal obstruction and hepatobiliary ascariasis. In heavy infections worms may be expelled through the mouth or nose. Heavy worm burden may lead to abdominal cramps, malabsorption and partial or complete intestinal blockage (Villamizar et al. 1996) in children. Coughing, wheezing, rales, nausea, vomiting and pulmonary eosinophilia occur when larvae migrate to the lungs (Loeffler 1956). In some cases, the invasive larvae may cause eosinophilic inflammatory response in the liver and pneumonitis known as Loeffler's syndrome in the lung. Occasionally, patients may experience asthma-like symptoms due to the release of volatile allergens by the parasite (Coles 1985; Kennedy 1992). The chronic infections can lead to physiological abnormalities resulting in malabsorption of nutrients, nutritional deficiency and vitamin A deficiency besides physical and cognitive impairments. These consequences have led to launch of school-based periodic anthelmintic drug deworming programs (Hotez et al. 2003).

Diagnosis The stool samples may be analysed for parasites and eggs. After an indication of the infection, the imaging tests such as X-rays, CT scans, ultrasounds, MRI scans and endoscopy may be some other ways of locating the worms and their impacts on the affected tissue.

Treatment and Prevention The drugs commonly used for treatment of ascariasis are albendazole (Albenza), ivermectin (Stromectol) and mebendazole (Vermox). Cleanliness is the best preventive measure. The vegetables/food and water contaminated/soiled with eggs should not be consumed. Hands must be washed thoroughly after using the lavatory or before eating.

8.6 Enterobiasis

Causative Organism Pinworm (*Enterobius vermicularis*) is a common intestinal parasite (Fig. 8.3) that primarily affects children, while parents usually get infected by transmission through their children.

Distribution *Enterobius vermicularis* is found worldwide infecting humans and is not specific to any biogeographic region (Garcia and Bruckner 1997) but well adjusted to temperate, tropical, subtropical or polar climates, and there is even no special preference to rural, urban or agricultural conditions (Fig. 8.5b). The parasite develops and reproduces in human intestine, and the eggs passed out in faeces may reach a new host to complete the cycle.

Morphology Pinworms are small nematodes that live parasitically in human appendix, caecum, colon or rectum. Adult nematodes are quite small and delicate with females attaining 8–13 mm length and 0.5 mm width and having a long, conical tail with acute tip (Fig. 8.3). The males are very small ranging between 2 and

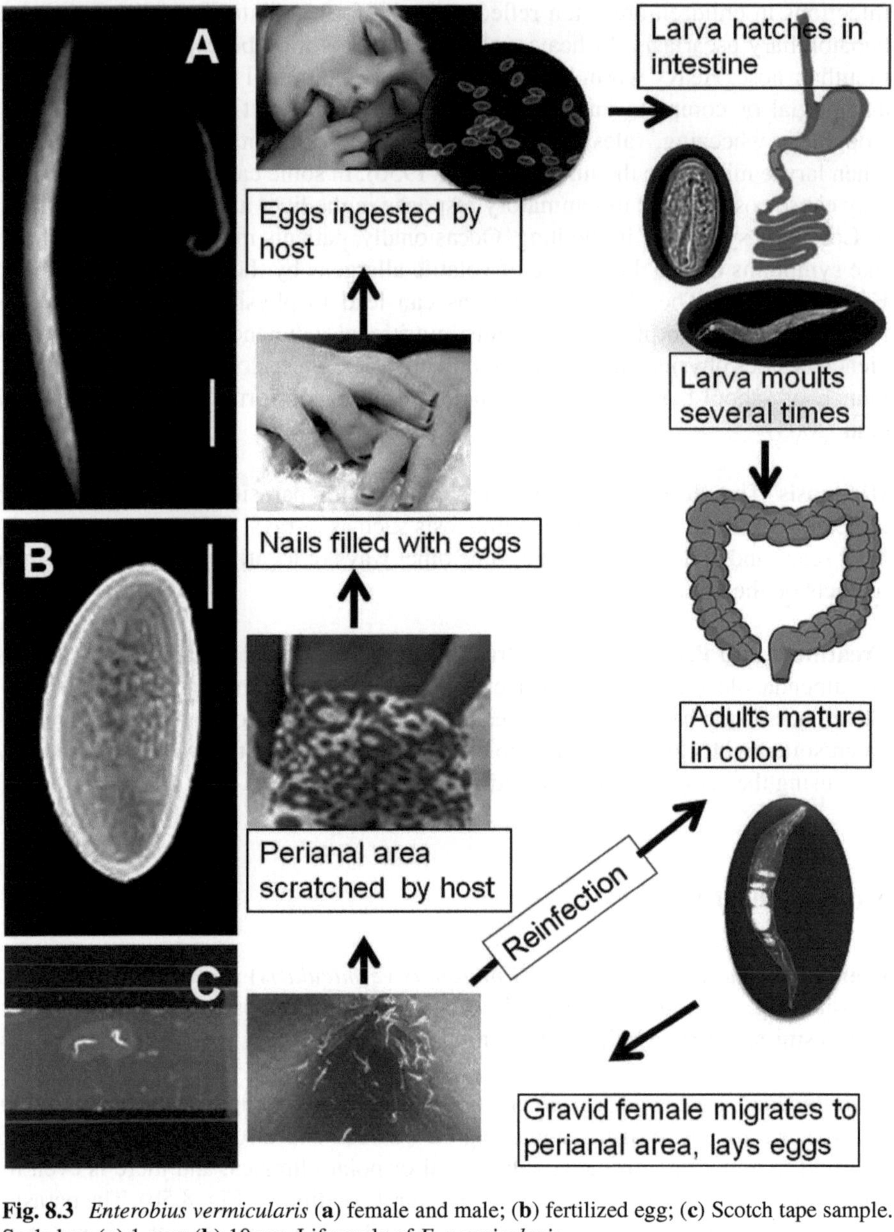

Fig. 8.3 *Enterobius vermicularis* (**a**) female and male; (**b**) fertilized egg; (**c**) Scotch tape sample. Scale bar: (**a**) 1 mm; (**b**) 10 um. Life cycle of *E. vermicularis*

5 mm in length and 0.2 mm in width with curved posterior end. The eggs are oval to asymmetrical in shape and appear translucent due to thick and sticky shells. The laterally compressed eggs with one side flattened measure 50–60 × 20–30 μm in dimension (Cook and Alimuddin 2009). The eggs may remain viable for longer time at low temperature and high humidity.

Life Cycle Pinworms can only infect humans. After being ingested, the embryonating eggs either hatch in the stomach or in the anterior part of the intestine. The larvae reach the posterior part of the small intestine, caecum and appendix and undergo two further moults. The adult worms remain in the ascending colon, caecum or appendix and copulate (Bogitsh and Cheng 1998; Chitwood and Chitwood 1950; Despommier et al. 1994; Donowitz 1999; Garcia and Bruckner 1997). They cling to the intestinal tissue producing microscopic ulcerations and haemorrhages. Typically at night, the gravid females with uteri filled with fertilized eggs move from the colon to perianal area and deposit eggs causing severe perianal itching to the host. A single female can lay up to 11,000 eggs. The scratching of the perianal skin may also kill the females but at the same time is the source of transfer of eggs to the nails and fingers of the host (Fig. 8.3). Occasionally, it is also possible that the larvae hatch and move back into the rectum.

Disease Symptoms and Associated Morbidities Often the infections are asymptomatic. However, severe infections may lead to intense itchiness around the anus and vagina caused by the movement of gravid female pinworms. This itchiness can disrupt sleep, cause irritability and may lead to bacterial infection after the area being scratched. In rare instances, the nematodes can spread to the vaginal area causing urinary tract infections or invade the appendix causing appendicitis-like symptoms (vomiting, abdominal pain and reduced appetite) without causing any inflammation of the appendix (Dundas et al. 1999; Hong et al. 2002; Smolyakov et al. 2003; Burkhart and Burkhart 2005) or occasionally move to some unusual sites (McDonald and Hourihane 1972; Gargano et al. 2003) like kidneys.

Diagnosis The common and cost-effective diagnostic method of pinworm infection is the "Scotch tape" test (where a transparent adhesive tape is used to sample the anal area early in the night or early morning before defecation). The tape is treated with lactophenol+ cotton blue before examining under microscope for pinworm eggs. This test is the most successful with 90% sensitivity if performed consecutively for 3 days. Anal swabs and stool examination are other methods of detection of the parasites. Invasive techniques are used for extra-intestinal infections, viz. colonoscopy and laparotomy (Petro et al. 2005).

Treatment and Prevention Mebendazole is an effective drug for pinworm infections as it blocks glucose uptake in adult worms, thus resulting in decreased ATP production, energy depletion, immobilization and death. Another drug, pyrantel pamoate inhibits cholinesterases, thus causing spastic paralysis of the adult worms. However, there may be contraindications including GI distress, neurotoxicity and temporary increases in liver enzymes. Other effective drugs suggested for pinworm infections include piperazine citrate, pyrvinium pamoate, oxantel, oxantel-pyrantel, fenbendazole and nitazoxanide. The treatment of the entire household is strongly recommended in all cases of pinworm infection even if they are asymptomatic.

Personal hygiene is the best strategy to prevent the infection. Cleanliness should be maintained in household tasks related with consumption of food or drinking. Hand must be washed thoroughly after any likely act of contamination and also before eating. The bed linen of the infected person should be changed twice a week for 3 weeks, and the infected clothes must be washed daily for 2 weeks because live eggs continue to pass in faeces for up to a week even after the treatment.

8.7 Trichuriasis

Causative Organism The whipworm (*Trichuris trichiura*) is a cosmopolitan intestinal parasite. Called as whipworms, this species has been known to parasitize humans and monkeys. It is unique in lacking tissue migration phase and also lacks specific symptoms of infection.

Distribution *T. trichiura* is found throughout the world within temperate and tropical environments with one fourth of the world's population infected with the parasite (Fig. 8.4). Two conditions promote spreading of *Trichuris trichiura*, e.g. poor hygiene and a warm, moist climate that are suitable for the worm's propagation (Smyth 1994; Roberts and Janovy 2000); however, there is no special affiliation to rural or urban conditions (Fig. 8.5c). Largely, it is considered a problem in tropical Asia and to a lesser extent in Africa and South America.

Morphology *T. trichiura* has a whip-like, pinkish grey colour body. Females measure 35–50 mm in length with a slender anterior and a thicker posterior end (Fig. 8.4). The finely attenuated anterior part shows the stichosome along with elongated capillary-like pharynx (Cooper 1995). The males are smaller, 30–45 mm in length with coiled posterior end.

Life Cycle The fertilized females lay several thousands of eggs of pale or brown colour. The eggs vary from barrel-shaped to lemon-, baseball-shaped or spindle-shaped with polar plugs at both ends. The eggs (49–65 × 20–29 μm) are passed in faeces of infected person and may be deposited in soil in cases of open defaecation. In moist soil, the eggs embryonate within 2–3 weeks. However, the eggs are more vulnerable to desiccation, heat and cold than *Ascaris* eggs and lose viability at 37 °C within 15 min. Infection occurs by ingestion of embryonated eggs through contaminated food and water. The larva hatches in the small intestine, embeds its thin, thread-like anterior part and feeds on tissue secretions of the intestinal mucosa. It stays there for 3–10 days before reaching the colon where it again penetrates the

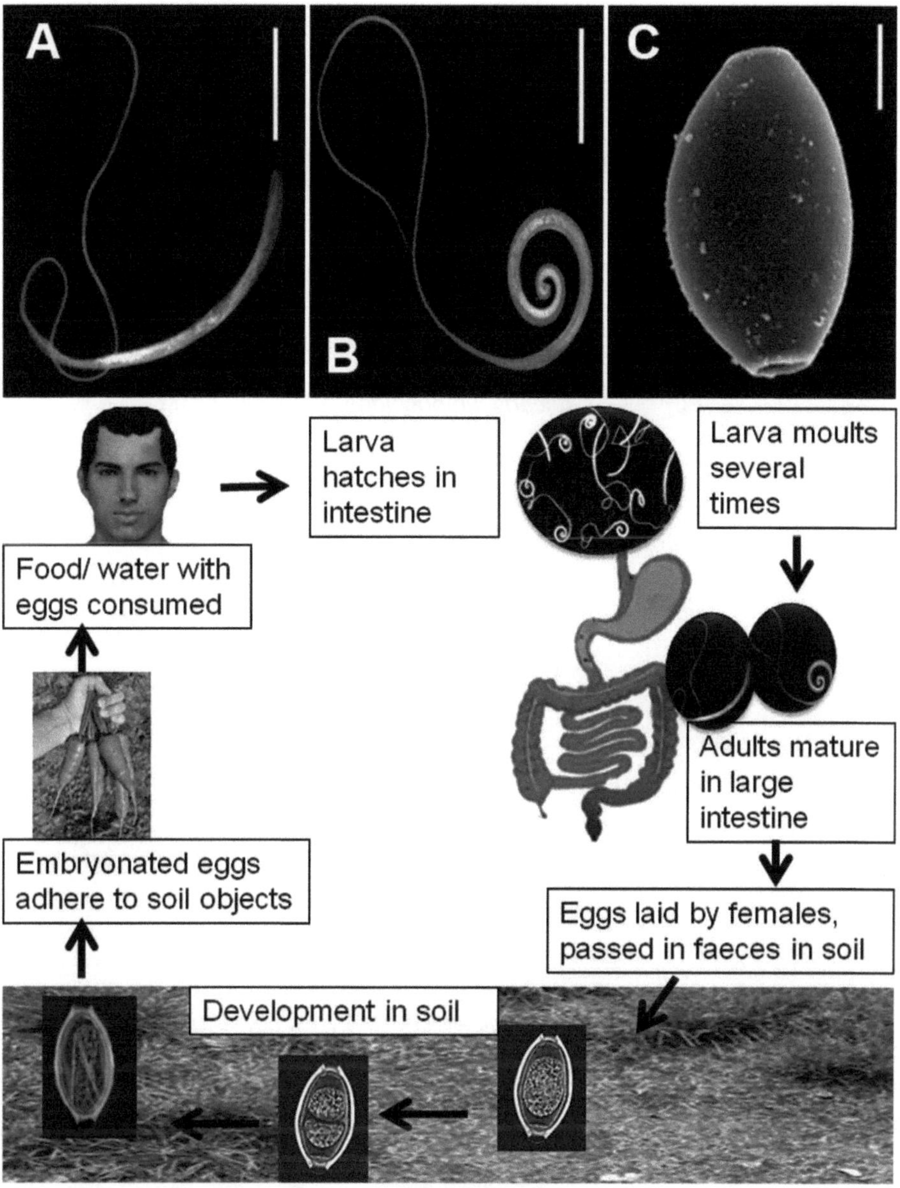

Fig. 8.4 *Trichuris trichiura* (**a**) female; (**b**) male; (**c**) fertilized egg. Scale bar: (**a, b**) 5 mm; (**c**) 10 um. Life cycle of *T. trichiura*

mucosa and transforms into adult by moulting several times. The anterior end remains buried; the posterior end is hanging loose for mating. The time taken from the infection to egg laying may vary from 30 to 90 days, and 3000–10,000 eggs may be produced per day by the females. The lifespan of females may vary from 5 to 6 years.

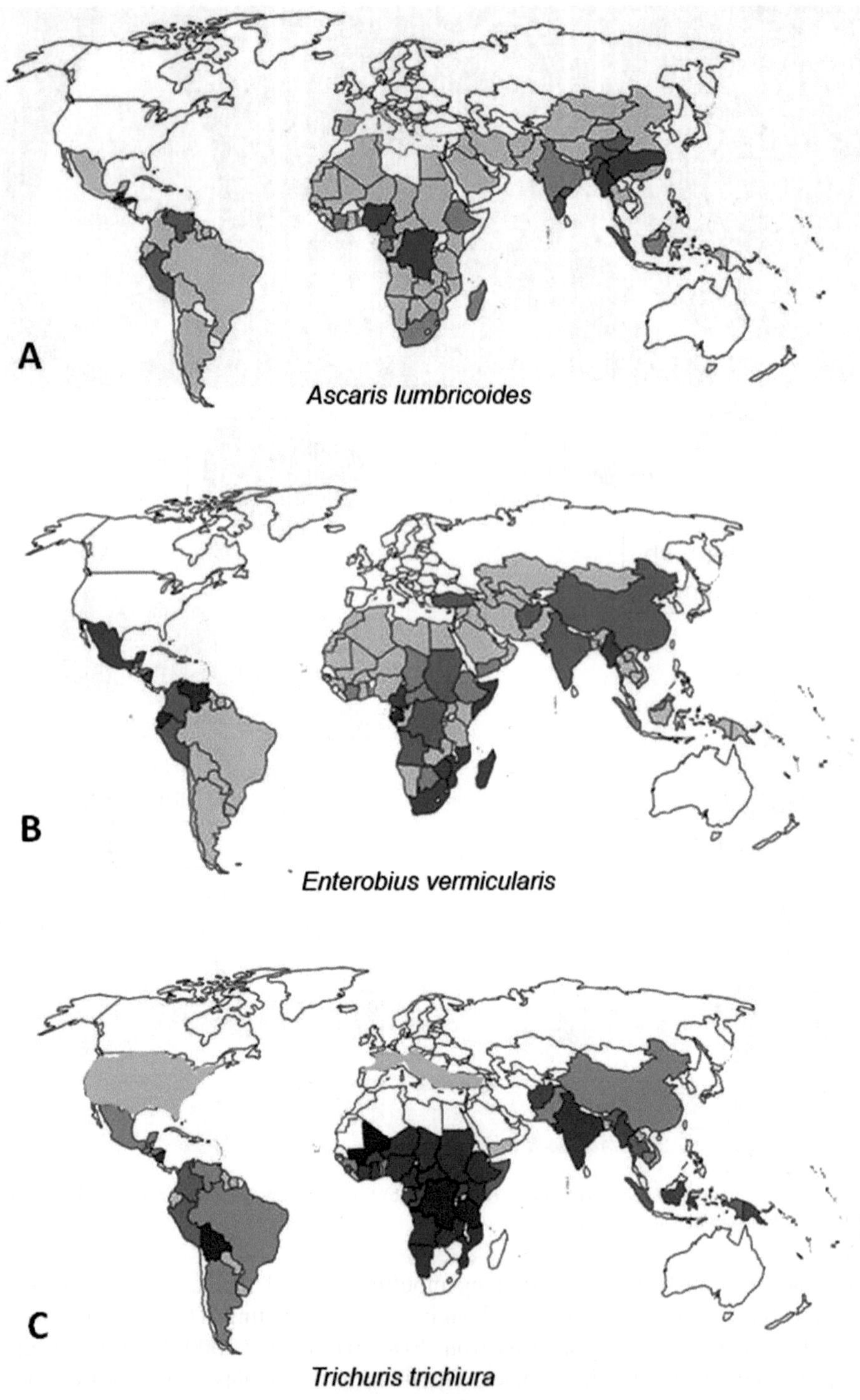

Fig. 8.5 Global distribution of *Ascaris lumbricoides*, *Enterobius vermicularis* and *Trichuris trichiura* as per data published by the World Health Organization

Disease Symptoms and Associated Morbidities *T. trichiura* remains asymptomatic in light infections in most instances. However, the heavy infections show conspicuous symptoms and are common in malnourished children. Early symptoms include diarrhoea, dysentery, tenesmus and abdominal pain, while the chronic symptoms reflect greater impact of parasitic burden resulting in severe diarrhoea or dysentery with blood and mucus that may continue for 6 months to 3 years. The symptoms resemble those of inflammatory bowel disease and can cause headache, insomnia, irritability, vomiting and abdominal distention besides retarding the cognitive ability and growth. Additional symptoms, viz. urticaria, rhinitis and eosinophilia, can also occur. Vitamin A deficiency is also reported in whipworm infection (Albonico et al. 2008).

Diagnosis Microscopic identification of eggs in faeces forms the primary diagnosis. In light infections, the eggs may be rare; hence a concentration procedure should be adopted. Other effective methods of detection are Kato-Katz technique and rectal proctoscopy.

Treatment and Prevention Albendazole as used in other worm infections is effective in decreasing whipworm ATP production, thus leading to their death. Drugs, like mebendazole, kill the parasite by blocking its glucose uptake.

Improved hygiene and sanitation are the most effective ways to prevent the spread of whipworm infections. Besides, cooking fresh produce also reduces the risk of infection. Thumb-sucking or nail-biting habits of children should be discouraged.

8.8 Hookworm Infections (Ancylostomiasis)

Causative Organism The hookworm species infecting humans are *Ancylostoma duodenale* and *Necator americanus* (Fig. 8.6). After ascariasis, hookworm infections are the most common chronic infections in the world, affecting around three quarters of a billion people in the tropics and subtropics (de Silva et al. 2003). These worms cannot be transmitted from person to person through eggs but require infective larvae that survive in soil for longer duration under suitable climatic conditions. The hookworm species parasitic on other animals can enter human body by penetrating the skin and cause cutaneous larva migrans.

Distribution Human hookworms are reported to occur in tropical and subtropical regions of the world, particularly between 30° north and south of the equator. *Ancylostoma duodenale* is predominantly found (Fig. 8.8a) in the Mediterranean region, Southeast Asia and Southern America (Beigal et al. 2000; Changhua et al. 1999; Roberts and Janovy 2000). *Necator americanus* is found (Fig. 8.8b) to exist both in tropical and temperate climates, viz. in Africa, Asia and Europe, but is pre-

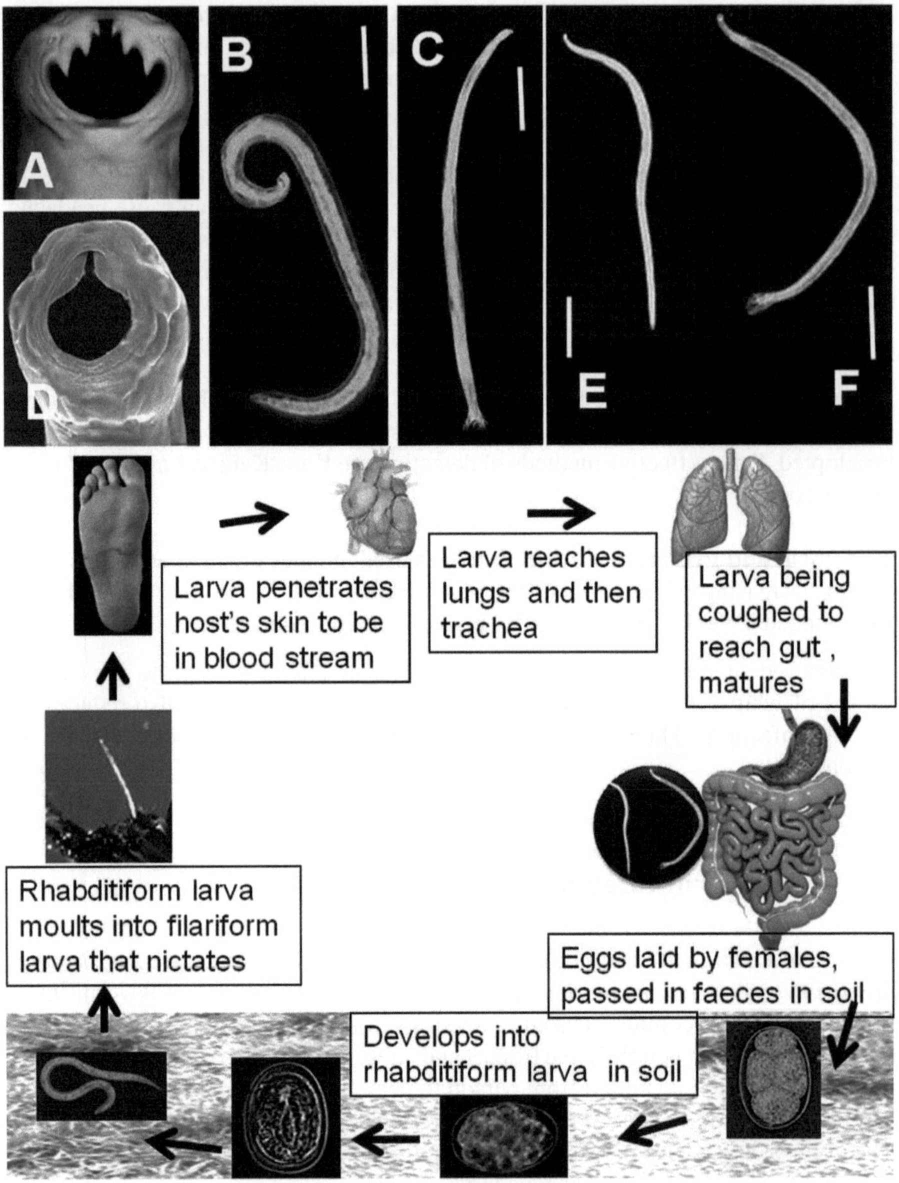

Fig. 8.6 Hookworms. *Ancylostoma duodenale* (**a**) anterior end; (**b**) female; (**c**) male. *Necator americanus* (**d**) anterior end; (**e**) female; (**f**) male. Scale bar: 1 mm. Life cycle of hookworm

dominately found in the Americas and in Australia. It is mostly prevalent in southern and southwestern states of the United States. In the rest of the world, *N. americanus* is found in tropical climates (Behnke et al. 2000; Romstad et al. 1998).

Morphology *A. duodenale* possesses an S-shaped, light pink colour body with buccal cavity bearing two pairs of large onchia or teeth, post-equatorial vulva in female and free spicules and closed-circular copulatory bursa in male. The females *Ancylostoma duodenale* measure about 8–13 × 0.6 mm, while the males measure about 8–11 × 0.5 mm. The eggs laid by females measure about 55–68 × 35–40 μm and are oval and colourless with broadly rounded extremities (Ashton et al. 1999; Carson-DeWitt 1999; Roberts and Janovy 2000). *N. americanus* is smaller than *A. duodenale* possessing one pair of cutting plates on the ventral surface and small teeth inside the dorsal surface of buccal cavity, preequatorial vulva in female and fused spicules and closed-oval copulatory bursa in male (Brusca and Brusca 2003; Roberts and Janovy 2000). The males of *Necator americanus* measure 7–9 mm in length, while the females are slightly larger measuring 9–11 mm. Eggs range in size from 65–75 × 36–40 μm and resemble those of *Ancylostoma duodenale* in appearance.

Life Cycle In *Ancylostoma duodenale*, the eggs passed out in faeces hatch into rhabditiform larvae in 1–2 days under favourable environment with appropriate soil type, moisture, warmth and shade. These larvae transform into infective (filariform) larvae in 7–10 days which can penetrate human skin, usually of the foot resulting in itchy skin or ground itch. The larvae stand on their tail and wave to make contact with the host. This characteristic nictation is to establish contact with the host. The infective larvae can remain in soil for 3–4 weeks in suitable environment. On entry, the larvae make their way to the lungs through blood circulation. On reaching the wind pipe, the larvae are coughed up and swallowed into the digestive system. They later attach to the wall of the small intestine by the onchia and suck blood to further moult and to finally develop into adults in about 6–7 weeks. The adult females and males mate, and females lay a large number of eggs to be passed in faeces.

The life cycle of *Necator americanus* begins with unembryonated eggs being passed in the faeces of host and involves similar phases and stages as found in *Ancylostoma duodenale*. The migration and maturation of *Necator americanus* larvae take approximately 6–8 weeks, while the lifespan of an adult worm is typically around 2–5 years.

Disease Symptoms and Associated Morbidities There may an exception from the normal route of infection in *A. duodenale* infective larvae which can become dormant in the intestine or muscles. Besides, infection also occurs occasionally by the oral and transmammary route, whereas *Necator americanus* always requires a transpulmonary migration phase. Anaemia due to blood loss is the most common symptom of hookworm infection that can be accompanied by gastrointestinal/metabolic disorders and cardiac complications. In addition, local skin irritation ("ground itch") can occur during penetration by the larvae.

Because of the anchorage of worms to the intestinal wall to draw blood, the infected individuals can also experience bloody diarrhoea and abdominal pain. Usually the light infections are asymptomatic, but heavier chronic infections can cause malaise, weakness and impaired cognitive and physical development in the host. Infants are more vulnerable than adults, and infections can lead to permanent growth deficiencies. In 95% cases, anaemia gets serious during pregnancy due to inadequate food intake (Porter and Kaplan 1999).

Diagnosis Hookworm infections can be diagnosed by examining eggs in the stool sample, and the condition of resulting anaemia can be determined by blood tests.

Treatment and Prevention The anthelmintic drugs used commonly for treatment are albendazole and mebendazole; however, these drugs cause adverse effects on the foetus if taken by pregnant women. The patients are advised iron supplements to counter anaemia.

Use of sanitary toilet, preventing the direct contact of the skin to the soil, and treatment of the pets to prevent them from spreading infective juveniles may prove effective in preventing the infections. In infection-prone areas, the periodical treatment of people likely to be exposed to hookworms is a good preventive strategy.

8.9 Strongyloidiasis

Causative Organism The species of threadworm parasitizing humans, is *S. stercoralis* (Fig. 8.7). Another species, *S. fuelleborni* is a zoonotic parasite that infects other primates, but has been occasionally reported in humans in Africa and Southeast Asia.

Distribution *S. stercoralis* is a cosmopolitan parasite distributed in tropical and subtropical areas as well as temperate regions (Fig. 8.8c) infecting an estimated 30–100 million people worldwide. Although the parasite shows a high incidence in Africa, Asia, tropical America and the Pacific Islands (Cheng 1986; Roberts and Janovy 2000), it is also endemic in southern regions of Europe and Japan besides the United States. The common occurrence is due to the variety of hosts which it can parasitise. Its prevalence is equally there in rural as well as urban areas including large cities of the United States such as New York City, Chicago and Montreal.

Morphology *Strongyloides stercoralis* also infects humans percutaneously to reach the intestine. It is a 2–3 mm long intestinal worm. Usually males are rare and not parasitic, measuring about 0.9 mm. Small buccal capsule leads to a cylindrical pharynx. Eggs are ovoid and smooth-shelled, about 40–70 × 20–35 μm in dimension (Cheng 1986; Roberts and Janovy 2000).

Life Cycle The high prevalence of *Strongyloides stercoralis* is due to its heterogonic life cycle with the parasitic as well as free-living phases. During the latter phase, it lives and propagates in the soil without a host. Strongyloidiasis cannot be cured easily because the parasite can cause autoinfection in the same host repeatedly.

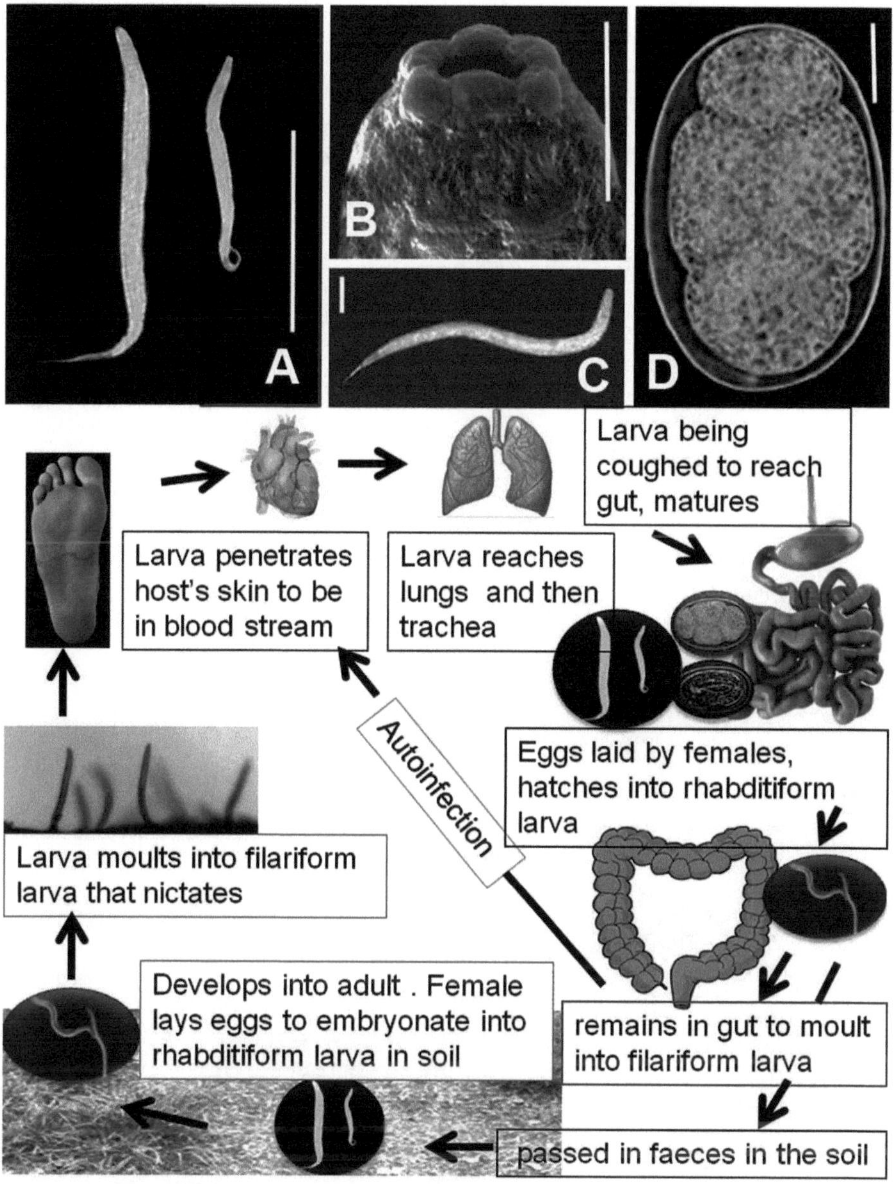

Fig. 8.7 *Strongyloides stercoralis* (**a**) female and male; (**b**) anterior end; (**c**) infective larva; (**d**) embryonating egg. Scale bar: (**a**) 1 mm; (**b**, **c**) 50 um; (**d**) 10 um. Life cycle of *S. stercoralis*

The eggs are deposited in soil through faeces where they hatch into first-stage larvae called rhabditiform larvae. The larvae moult twice into L3 filariform (infective) larvae. These larvae (L3) enter through the skin (usually of the foot) that touches worm-infested soil. The larvae enter the bloodstream and reach lungs and trachea to

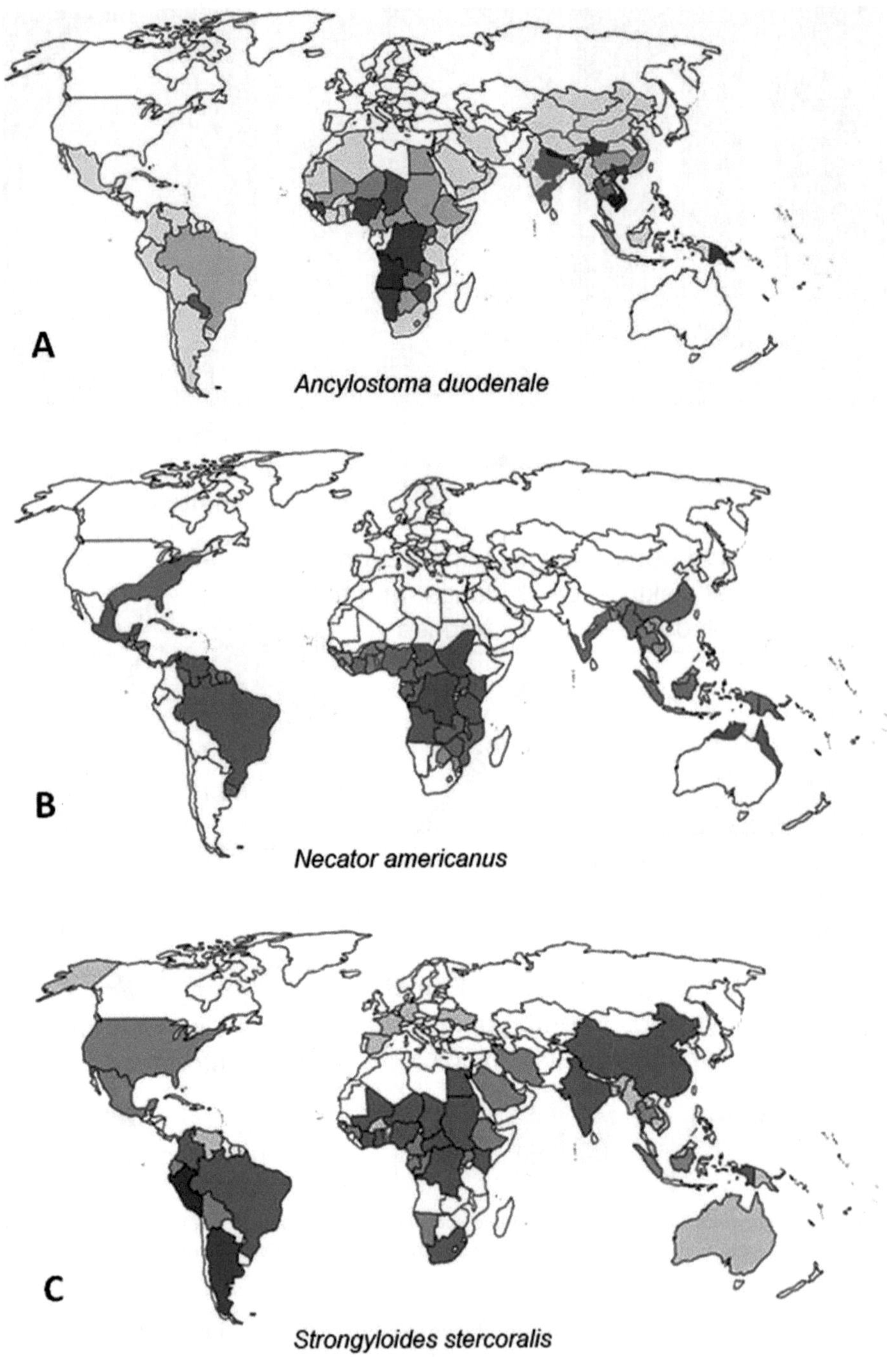

Fig. 8.8 Global distribution of hookworms (*Ancylostoma duodenale* and *Necator americanus*) and *Strongyloides stercoralis* as per data published by the World Health Organization

eventually get swallowed to lie in the small intestine. In the intestine, the larvae moult two times usually to develop into adult females. The females live burrowed into the intestinal wall and reproduce by parthenogenesis. In another instance, the eggs if not passed in faeces hatch into the rhabditiform larvae which transform into filariform larvae. The latter penetrate the intestinal wall to pass into the bloodstream (demonstrating the phenomenon of internal autoinfection) or penetrate the outside/ perianal skin (external autoinfection) to reach lungs via the bloodstream and are swallowed back to reach the small intestine to further develop into adults.

Disease Symptoms and Associated Morbidities The adaptability of the parasite is the reason for its cosmopolitan distribution, persistent occurrence and hyperinfection. *S. stercoralis* infections result in severe abdominal pain, bloody stool, cough and skin allergies that are often ignored (Becker et al. 2011). Acute infections result in Loffler's syndrome, whereas chronic infections may result in hyperinfection including fever, nausea and vomiting, anorexia, diarrhoea, abdominal pain, dyspnoea, wheezing, haemoptysis, cough and pulmonary infiltrates. Severe infections have been reported to lead to serious ailments including intestinal obstruction, pneumonia, meningitis and septicaemia (Grove 1989). The mortality rate of disseminated infections has been estimated to be as high as 87% (Siddiqui and Berk 2001). Patients using systemic corticoids and cytotoxic medication or having immune suppression (Kramer et al. 1990; Zygmunt 1990; Gill et al. 2004) due to achlorhydria, haematological malignancies or nephrosis may suffer from hyperinfection (Igra-Siegman et al. 1981; Morgan et al. 1986; Ghosh and Ghosh 2007; Marcos et al. 2008). The hyperinfection leads to the spread of larvae to various organs including the brain and occasionally is followed by bacterial septicaemia.

Diagnosis The diagnoses of *Strongyloides* by traditional stool examinations and the Kato–Katz technique do not give 100% success. Duodenal aspirate has been found to be more sensitive than stool sample; or the fluid from a bronchoalveolar lavage (BAL) can also indicate the presence of larvae.

Serological tests have demonstrated higher sensitivity particularly; the luciferase immune-precipitation system technique combined with a recombinant antigen (NIE) demonstrated almost 100% specificity. ELISA coproantigen detection and Gram staining of sputum sample are also excellent tools for diagnosing strongyloidiasis.

Treatment and Prevention Albendazole can be used if there is no hypersensitivity to benzimidazole compounds or condition of pregnancy. If recrudescence of larvae is observed, retreatment is indicated. In case of hyperinfection syndrome, ivermectin can be used. The patients with ileum obstruction or malabsorption can be given rectal administration.

The best preventive strategy for *Strongyloides* infection is to wear shoes and to avoid contact with soil, faecal matter or sewage. Proper sewage disposal and faecal management besides sanitation and hygiene are keys to prevention.

There are few parasites which are not soil-transmitted or water-borne but can be disseminated due to contaminated water harbouring the agents of transmission, viz. water fleas or mosquitoes.

8.10 Dracunculiasis

Causative Organism The parasite (*Dracunculus medinensis*) was first reported in the seventeenth century from Guinea coast of West Africa; hence the common name "Guinea worm" (Fig. 8.9) has been documented in literature. However, the worm showed a worldwide distribution with reports from throughout Europe, Asia and Northern Africa. The parasite is also called *the fiery serpent* due to causing a burning and painful sensation.

Distribution The parasite has a prehistoric presence as it was spotted in the 3000-year-old mummified remains. The parasite was earlier reported to be found in 13 sub-Saharan African countries which account for 93% of all cases worldwide. However, it was also reported from South Asia (Fig. 8.12a), whereas other species of the genus *Dracunculus* have been reported from Americas (Cairncross et al. 2002; Greenaway 2004; Hopkins et al. 2008). It is still prevalent in those areas of Africa where water contaminated with water fleas is used for drinking or for other domestic needs.

Morphology The adult female has 50–120 cm long and 1 mm wide body, while the male is about 12–29 mm long and 0.4 mm wide. Both male and female worms have a small, triangular mouth surrounded by a quadrangular, sclerotized plate, but no lips. The intestine becomes squashed and nonfunctional in gravid females. The infective larvae tend to range from 500 to 700 μm in length.

Life Cycle The primary or definitive host is man, while the secondary or intermediate host is water flea, *Cyclops* (a crustacean). The infection is caused by drinking water contaminated with water fleas (*Cyclops*) harbouring larvae. In human stomach, *Cyclops* is digested by the effect of the gastric juices, and *D. medinensis* larvae are liberated which undergo moulting. The infective larvae penetrate through the digestive tract wall, get lodged in the loose connective tissues and moult several times to mature into the adults within 10–12 weeks. The male dies after mating. In about 10–12 months, the gravid females tend to move to subcutaneous tissue of arms, legs, knee, ankle, etc. (the body parts normally coming in contact with water) and secrete a toxin that causes inflammation and forms sterile blister. The blister reaches about 5–7 cm in diameter on the skin surface and eventually breaks to form an ulcer with a minute pore at its base. The pore leads to subcutaneous tunnel where the female lies with the posterior end towards the blister. Contamination of the blister may lead to abscesses, cellulitis and necrosis. If the female does not migrate to the subcutaneous tissue, it dies, and no blister is formed. Contact with water stimu-

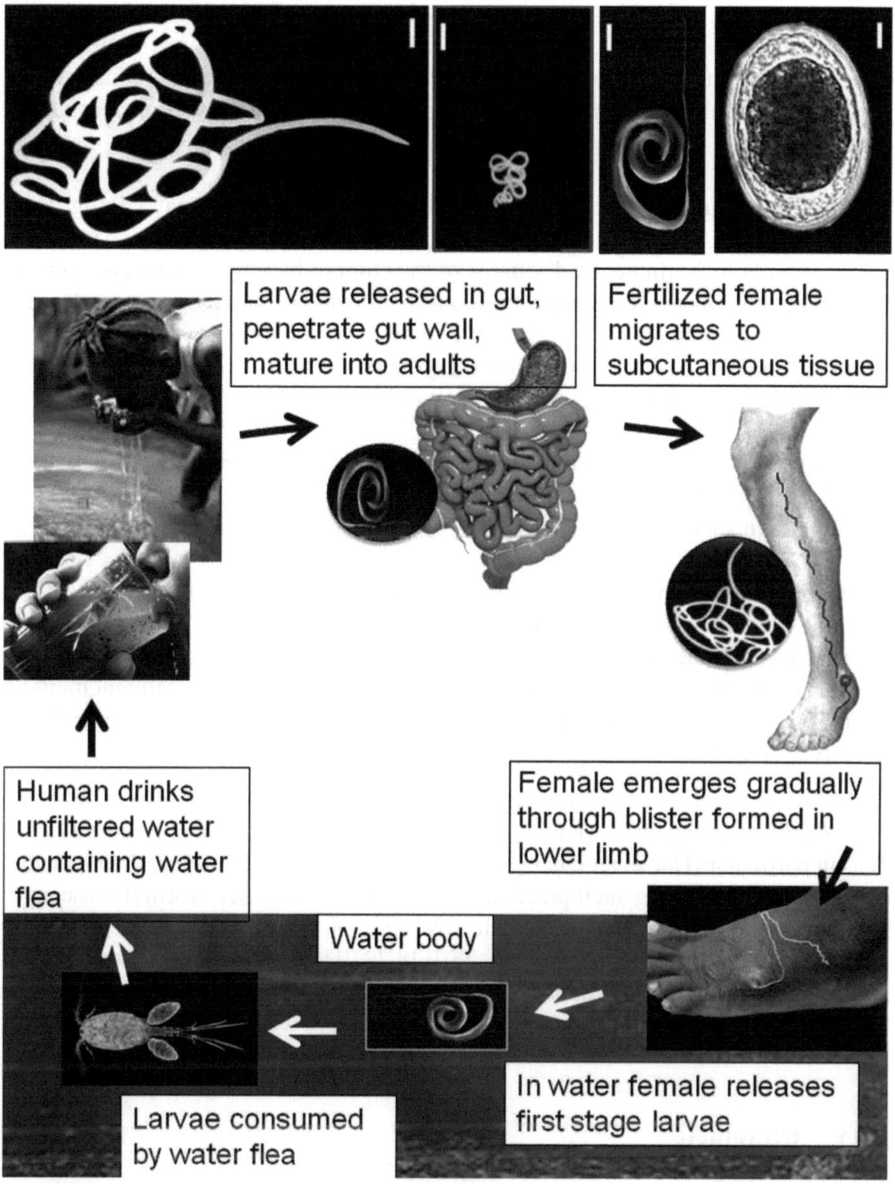

Fig. 8.9 *Dracunculus medinensis* (**a**) female; (**b**) male; (**c**) sheathed microfilaria; scale bar: (**a, b**) 1 cm; (**c**) 50 um; (**d**) 10 um. Life cycle of *D. medinensis*

lates the gravid female to discharge milky fluid containing large number of embryos. *D. medinensis* is ovoviviparous as it releases eggs containing fully developed larvae. The eggs are ingested by *Cyclops*, and within 1–6 h of ingestion, the larvae hatch, penetrate the gut wall and enter the coelomic cavity where they moult twice and transform into infective stage larvae by 21st day. The infective larvae do not develop further and remain in the coelomic cavity of the *Cyclops* in inactive, tightly coiled state up to 3 months or till the death of *Cyclops*. When the water containing infected *Cyclops* is taken in by man along with drinking water, the infection reaches the definitive host.

Disease Symptoms and Associated Morbidities The adult females lying in the subcutaneous tissues produce an irritant to form a blister on the skin. The blister bursts in order to facilitate the discharge of the embryo by female. However, opened blister may later have secondary bacterial infection and septic. During the release of embryos, the female also secretes "toxin" which causes allergic symptoms characterized by nausea, vomiting, diarrhoea, erythema (redness of the skin), giddiness, dyspnoea (difficult breathing) and eosinophilia.

Diagnosis The diagnosis is made from the erupted blister or the worm emerging from the blister. The worms come to the surface as soon as the blister area is wetted. The female in subcutaneous location can be seen by a reflection of light.

Treatment and Prevention The treatment prevalent in ancient days of slowly winding the protruding worm on a stick is still practised today. With extraction of worm a few centimetres per day, 2–3 months are required to completely remove the worm. The worm can also be removed surgically. However, such treatment methods do not kill worms or prevent any damage caused by them. The drug metronidazole effectively kills the worms.

There is a worldwide effort to eradicate the *Dracunculus medinensis*. The spread of *D. medinensis* can be prevented by ensuring clean drinking water free from water fleas in particular. The water can be filtered using cotton cloth or treated with chemical Abate. By following such practices, the numbers of Guinea worm disease cases have been reportedly reduced worldwide by 98%, i.e. from 3.5 million cases in 1986 to fewer than 65,000 cases reported in 2001. In 2010 there were 1797 cases reported, while in 2011 the number reduced to 1035 which further dropped down to 536. In 2013 there were 148 reports, and in 2015 only 20 cases were reported (WHO 2015).

8.11 Filariasis

Causative Organism *Wuchereria bancrofti* is exclusively a human parasite (Fig. 8.10), whereas another parasite *Brugia malayi* infects a number of wild and domestic animals and is restricted to Southeast Asia. Mosquitoes are vectors for both parasites.

Distribution *Wuchereria bancrofti* is predominantly found in tropical regions worldwide. The parasite has been reported from South and Central America, Africa,

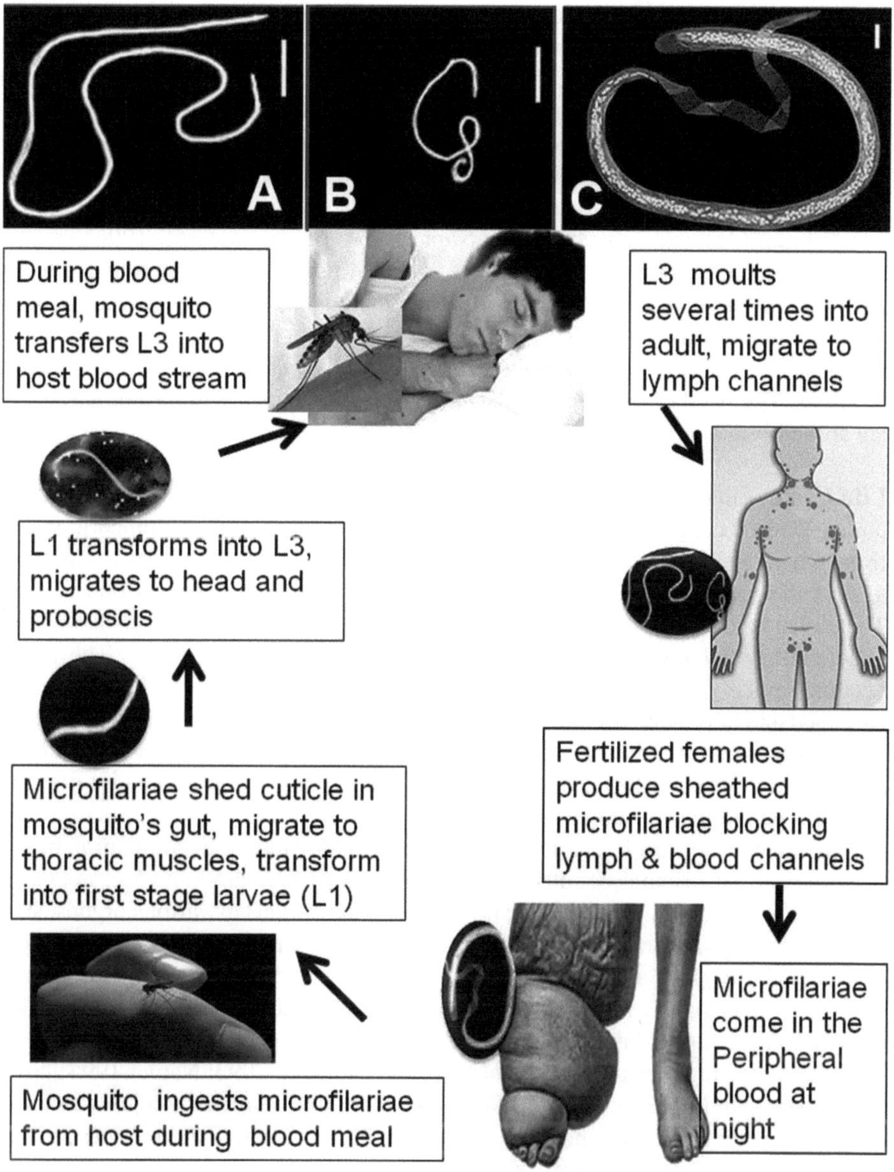

Fig. 8.10 *Wuchereria bancrofti* (**a**) female; (**b**) male; (**c**) sheathed microfilaria; scale bar: (**a**, **b**) 5 cm; (**c**) 10 um. Life cycle of *W. bancrofti*

the Southeast Asia, Southern China and the Pacific (Fig. 8.12b). The major reason of restricted geographical distribution of *W. bancrofti* is the specific climatic conditions required by mosquito and its intermediate host (de Almeida and Freedman 1999; Nematode and Neglected Genomics 2008; CDC 2009); thus areas with proper humidity and moderate temperatures with aquatic sites are preferred.

Morphology *W. bancrofti* and *B. malayi* are very similar in morphologies and the diseases they cause. Both complete the life cycle in man, the definitive host, and mosquitoes (*Culex* and *Aedes*), the intermediate host. Adult *B. malayi* are only half the size of *W. bancrofti*, but their microfilariae are relatively larger and only a bit smaller than those of *W. bancrofti*. The worms are translucent white with a smooth cuticle. The adult *W. bancrofti* females range 3–10 cm in length and 100–300 μm in diameter, whereas males are smaller and show an average length and width of 40 × 0.1 mm (Ash and Schacher 1971; Napier 1994; Baron and Cross 1996), respectively. The head is globular and sets off from the body. The female tail gradually tapers to the end in a rounded terminus, while male tail is ventrally curved. Usually adults remain in coiled state. The microfilariae emerging from egg show a surrounding sheath. They show striated cuticle and gradually tapering tail with a rounded terminus. The head of the infective larva is trapezoid-shaped with prominent papillae around oral aperture and three caudal papillae.

Life Cycle Filariform larvae or the third-stage infective larvae gain entry into the human body through a mosquito bite. They soon enter the main lymphatic trunks from where they find their way into the circulating blood. They move along or against the bloodstream and ultimately migrate to reside in deeper blood vessels. The larvae tend to accumulate at a particular region, especially inguinal, scrotal and abdominal lymphatic, undergo two moults and transform into adults. The larvae become sexually mature 5–18 months after infection. The male and female remain entwined and copulate. The females are ovoviviparous as the larvae hatch inside the body. These microfilariae fail to undergo any further development in human body due to requirement of lower ambient temperature. Thus, if not sucked by mosquitoes, the larvae die in human body in approximately 70 days. They pass through lymph nodes and reach the main lymphatic trunks into the circulating blood. Microfilariae of oriental countries, like India and China, show marked nocturnal periodicity, i.e. remain in the large and deeper blood vessels of various organs, such as lungs, kidneys, heart and large arteries, during daytime. The larvae show remarkable periodicity (Ash and Schacher 1971) and possess distinction skills of arteries and veins on the basis of oxygen concentration levels and reach the peripheral blood vessels during night (between 10 p.m. and 4 p.m.) where they are likely to be ingested by mosquitoes in blood meal (Cox and Chappell 1993; Napier 1994). Such behaviour is common in countries like India, where *Culex pipiens fatigans*, the principal intermediate host, is a nocturnal feeder. In Pacific islands where *Aedes polynesiensis* is the intermediate host of *Wuchereria bancrofti*, the microfilariae do not exhibit any periodicity and remain in the peripheral blood throughout because of no specific biting time of the vector. When a mosquito takes blood meal from a diseased person, microfilariae are sucked from peripheral blood to reach mosquito's stomach where their sheaths get digested. Then larvae penetrate the stomach wall and reach thoracic and wing muscles. The slender microfilariae change to a thick, short sausage form first-stage larvae with a short spiky tail. In 3–7 days, they grow and moult into second-stage larvae which moult into third-stage infective larvae with degenerated tail. The latter migrate to the mosquito's labium and get transferred to a new host during mosquito bite by coming out of the labium and penetrating through the wound made by the mosquito.

Disease Symptoms and Associated Morbidities The symptoms of filariasis include lymphadenitis because of the inflammatory reaction to the parasite that occupies lymphatic channels. As a result of inflammation, there is high-grade fever every 8–10 weeks. On the death of worms, a fibro-proliferative granuloma is formed that obstructs lymph channels, and lymphedema and elephantiasis result. The stretched skin may further be vulnerable to injuries and infections. Microfilariae may also cause eosinophilia and splenomegaly up to some extent. The complications such as hydrocele or chyluria may also develop.

The acute effects include filarial fever, tenderness of infected parts, eosinophilia, head ache, anxiety and mental depression. In heavy infections the lymphatic vessels and glands are blocked by living or dead adult worms causing an inflammatory reaction or lymphangitis, transient swelling and enlargement of lymphatic glands (lymphadenitis) and organs due to lymph channel obstruction (lymphedema) and hyperplasia of muscle fibres. The typical symptoms of elephantiasis include the enlargement of organs such as the scrotum, breasts or legs. Microfilariae and adults occasionally produce lesions in lymph node and granulomas in the spleen.

Diagnosis Blood test can be done by taking night samples from 8 pm to 4 am. The microfilariae can be examined through microscope. Microfilaria in chylous urine, lymph exudates and hydrocoele fluid can also be looked for. Antigen-detection (ELISA) to detect CFA (circulating filarial antigen) can also be tested. Lymph node biopsy, X-Ray examination (calcified worm), ultrasonography (dancing worm/filarial dance sign), etc. may also indicate the presence of infection.

Treatment and Prevention Antifilarial drug such as diethylcarbamazine (DEC) and ivermectin or a combination of both can be effective. However, a doctor's advice is necessary as DEC in certain cases may lead to encephalitis (inflammation and swelling of the brain). In case of severely blocked lymph nodes and enlarged limbs, surgery may be done to relieve the obstruction. Antihistamines and pain medication may help with inflammation and pain in swollen parts. The infected limbs should be cleaned to avoid bacterial and fungal infections. Exercising and massaging of an infected limb may also improve lymph flow. Mosquito bites can be avoided by using a mosquito net, mosquito repellent or wearing long-sleeved shirts and trousers.

8.12 Onchocerciasis

Causative Organism *Onchocerca volvulus* (Fig. 8.11) is the major cause of onchocerciasis (river blindness or blinding filariasis).

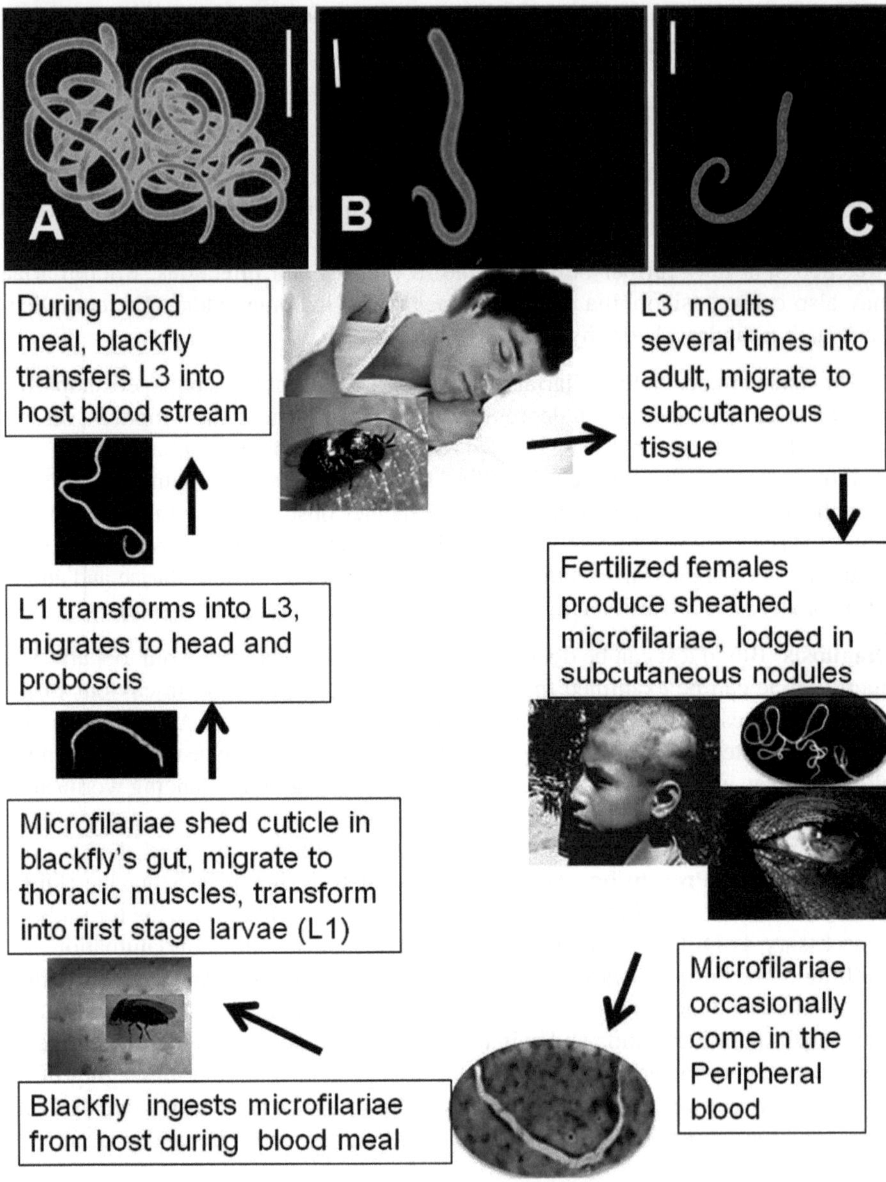

Fig. 8.11 *Onchocerca volvulus* (**a**) female; (**b**) male; (**c**) sheathed microfilaria; scale bar: (**a**, **b**) 1 cm; (**c**) 50 um. Life cycle of *O. volvulus*

Distribution *Onchocerciasis* is mainly confined to Ethiopian and Neotropical regions. *Onchocerca volvulus* is found mainly in eastern, central and western regions of Africa besides being its occurrence in Central and South America (Fig. 8.12c). The species is considered to be native to Africa which was introduced

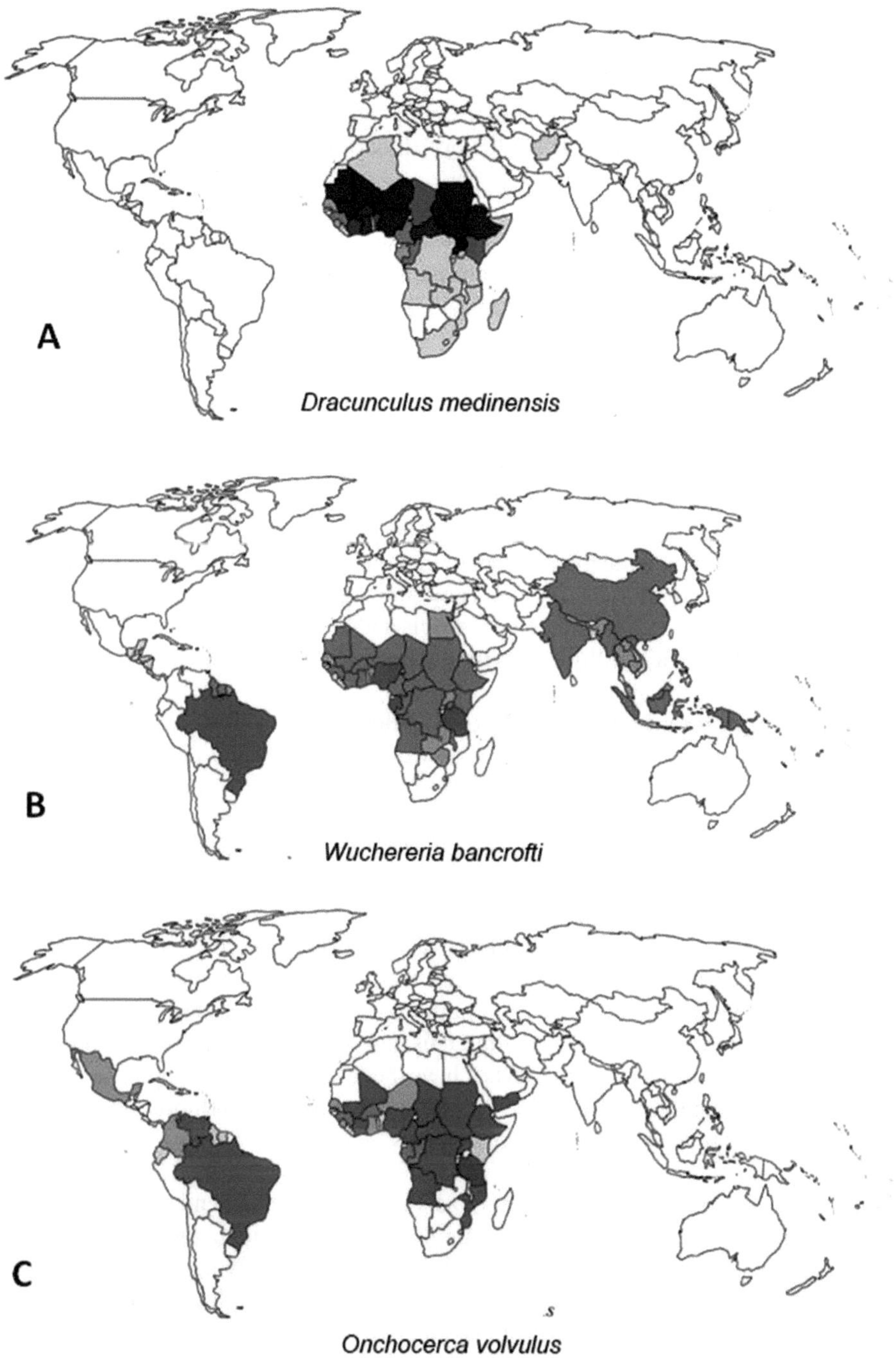

Fig. 8.12 Global distribution of *Dracunculus medinensis*, *Wuchereria bancrofti* and *Onchocerca volvulus* as per data published by the World Health Organization

to American continent by African slaves. Two pathotypes of this species were selected based on DNA sequences. One is typically found in the savanna regions, while the other is prevalent in rain forests as reported by Ogunrinade et al. (1999) and Roberts and Janovy (2000). The disease is usually prevalent in low areas with streams providing ideal habitats for the breeding of black flies.

Morphology Adult females measure 50 cm × 300 μm, while males are smaller measuring 19–42 cm × 130–210 μm. Lips or buccal capsule is absent and oral aperture is surrounded by two circlets of four papillae each. The microfilariae that measure 250–300 μm long are devoid of sheath but possess arcuate and acutely pointed tails.

Life Cycle Infective third-stage larvae gain entry into human skin by the bite of female black fly (*Simulium damnosum*). The larvae move to the subcutaneous layer and form nodules before maturing into adults in 6–12 months. The adults mate, and the eggs hatch inside the uterus of gravid females, and first-stage microfilariae are released in the skin. They are again taken up by the black fly while sucking the blood. The larvae reach the gut and then penetrate the gut wall of the fly to reach the flight muscles in the thoracic region. They moult twice to transform into infective stages in 6–8 days which migrate to proboscis and remain in the saliva. These larvae are again injected to a healthy host with the bite of fly.

Disease Symptoms and Associated Morbidities Acute symptoms include the formation of subcutaneous nodules and a pruritic hypopigmented or hyperpigmented papular dermatitis. Oedema and thickening of the skin with troublesome itching, dermal atrophy, epidermal atrophy, elastic fibre breakdown and fibrosis are some other complications. During 10–12 months of incubation, eosinophilia and urticaria are the common symptoms. Sometimes in persistent infections, many nodular and erythematous lesions develop in the skin or subcutaneous tissue. Ocular complications involve embedding of microfilaria in the cornea, choroid, iris and anterior chambers which leads to photophobia and lacrimation ultimately terminating into blindness with glaucoma and phthisis bulbi, commonly referred to as "river blindness". A degeneration in the crystalline lens and clogging of anterior chamber due to exudation of fibrin and leukocytes have been reported with optic nerve also affected occasionally. Obstructive lymphadenitis involving femoral and inguinal lymph nodes may lead to genital elephantiasis.

Diagnosis The diagnosis of onchocerciasis in usual cases depends on the presence of microfilariae in a skin-snip biopsy sample. This technique though highly specific (100%) requires expertise but demonstrates low sensitivity (20–50%) in initial stages of infection. The nodulectomy of patient with skin nodules can also help to identify parasites. Slit lamp eye exam is effective to visualize the microfilariae present in eyes. Ultrasonography may also be a help in detecting worms in nodules. Polymerase chain reaction (PCR) provides high sensitivity and specificity but not generally available due to the high cost involved.

Treatment and Prevention Ivermectin lowers the occurrence of blindness and nodule formation by killing the microfilariae but not the adults. Treatment with doxycycline is effective as it kills *Wolbachia*, an endosymbiotic rickettsia-like bacterium which seems crucial for survival of adults and for intrauterine development. It has been found to cause mortality of more than 60% of the adult female worms. There 1 week ivermectin treatment prior to doxycycline dose is considered effective. The older drugs suramin and diethylcarbamazine cause toxic and adverse effects and hence not advised.

There are no vaccines or medications available to prevent *Onchocerciasis*, and the best preventive strategy is to be protected from insect bites by using insect repellents, wearing long sleeves and long pants during the day when blackflies bite.

8.13 Toxocariasis (Visceral Larva Migrans)

Causative Organism These are primarily roundworm parasites of dogs (*Toxocara canis*) and cats (*T. catti*) but can infect humans (Fig. 8.13) having close proximity to dogs and cats under unhygienic conditions and cause damage to the visceral organs.

Distribution *Toxocara cati* are found in domestic or wild cats and are prevalent in those regions of the world where proper hygiene is not maintained with pets. However, the spread of *T. cati* mainly depends on the ingestion of eggs excreted in cat's faeces. Mild temperate climates are favourable for maintaining the eggs viable within faeces. *Toxocara cati* is predominantly parasitic on genus *Felis* (the cats) but can also be infect earthworms, cockroaches, birds, rodents, dogs and humans (Schierenberg 1997). However, since these animals serve as paratenic hosts, *T. cati* cannot develop into adults and keeps on migrating in host's body.

Morphology *Toxocara cati* females are about 10 cm long, while males are about 6 cm long (Uga et al. 2000). Eggs have rough and pitted surface, measure 75 × 65 μm in dimension, and can remain viable for years (Uga et al. 2000).

Life Cycle The life cycle of *Toxocara cati* is completed in the main host, viz. dog or cat. Man is the accidental host; hence the development does not take place in humans. Eggs from faeces of infected animals when swallowed by man hatch in the intestine, and the larvae by penetrating intestinal mucosa enter to the blood circulation to reach different organs. During the course of their journey, the larvae do not develop but cause inflammatory necrosis. The serious consequences include loss of sight due to obstruction caused by larva.

Disease Symptoms and Associated Morbidities Sometimes infections are asymptomatic. When migrate to the eye, the larvae can cause ocular toxocariasis involving inflammation of the eye, damage to the retina and loss of sight. Other body organs may also be invaded by larvae causing visceral toxocariasis which is manifested by fever, exhaustion, coughing, wheezing and pain in the abdomen.

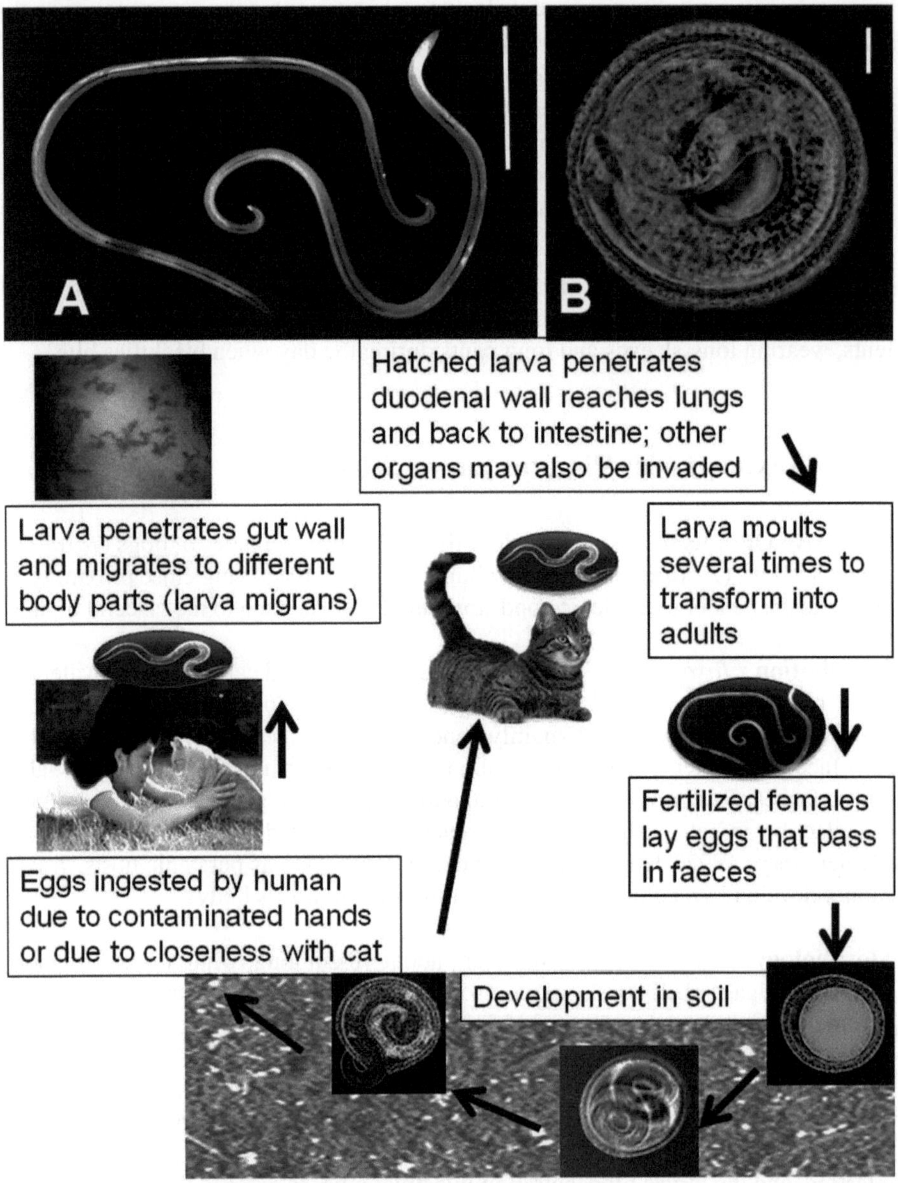

Fig. 8.13 *Toxocara cati* (**a**) female and male; (**b**) embryonating egg; scale bar: (**a**) 1 cm; (**b**) 10 um. Life cycle of *T. cati*

Diagnosis The diagnosis is difficult since adults are never present in humans, and stool examination of patient would not show eggs. The diagnosis in this case is based on clinical signs, history of exposure to feline pets and other symptoms like eosinophilia. Antibody testing is effective in cases of larva migrans.

Treatment and Prevention Visceral toxocariasis can be treated by the usual drugs, albendazole or mebendazole. Treatment of ocular toxocariasis requires ocular surgery. For ocular involvement with retinal detachment, laser treatment may be considered. The drugs, thiabendazole and corticosteroids, are used for specific organ treatment. Children are more vulnerable to infection by eggs in the soil, and the migrating larvae can infect the organs and tissues of the body. Various risk factors to the disease are pet ownership, poverty, male gender, asthmatic condition, etc. Another preventive strategy is regular deworming of pets like dogs and cats and washing hands after playing with pets or before handling food.

8.14 General Impact of Soil-Transmitted Helminthiasis and Its Epidemiology

Neglected tropical diseases (NTDs) primarily affect the world's poorest people in rural as well as urban areas particularly those estimated 2.7 billion people who live on less than $2 per day (Hotez et al. 2007). NTDs are usually represented by 14 diseases, viz. ascariasis, trichuriasis, leptospirosis, hookworm infection, schistosomiasis, lymphatic filariasis, trachoma, onchocerciasis, leishmaniasis, Chagas disease, leprosy, human African trypanosomiasis, dracunculiasis and Buruli ulcer (Hotez et al. 2007). Soil-transmitted helminthiasis (STH) has been placed in the category of neglected tropical disease as these infections are often not given priority by authorities because of their inconspicuous symptoms and indistinct morbidity and mortality. Evidences have proved that helminth infections result in reduced growth among children (Stephenson et al. 1989; Hotez 1989, 2000), reduced physical fitness (Hotez 1989, 2000) besides slowed mental development and impaired cognitive function including learning capabilities and other skills leading to lower educational achievement. According to a 2005 report by the WHO, approximately 0.807–1.221 billion people suffered from ascariasis and 604–795 million from trichuriasis, and 576–740 million had hookworm infections. A later report of WHO estimates 1.5 billion people to be infected with helminths in general and the soil-transmitted helminths in particular worldwide (WHO 2017) and another 4 billion are at risk. Greater prevalence of soil-transmitted helminth infections occur in tropical and subtropical regions, viz. impoverished rural areas of sub-Saharan Africa, Latin America, Southeast Asia and China. According to the WHO estimates, about 870 million children live in the most affected areas such as Africa, South Asia and South America (Lobo et al. 2011). India alone projects 25% of the total global cases with 220.6 million children requiring preventive chemotherapy (WHO 2015). Besides the conducive warm and moist climate of tropical and subtropical countries for the parasites, the filthy, unhygienic and water-logged conditions serve as breeding grounds of vectors. The presence of such pathogens poses threat to human health where infections can be acquired subject to conditions like the virulence of the pathogen, survival and fecundity, its incubation or latency period and the resistance of the human population towards infections (van Eijk et al. 2009). A number of features account for the high prevalence of these parasites including a ubiquitous distribution, the resistance of eggs towards a variety of environmental conditions, the reproductive potential of the parasite and the poor socioeconomic conditions. In absence of any protective immunity, transmission rates keep increasing if individuals are asymptomatically infected as they continue to shed eggs for years. The transmission of disease is also dependent on the host health, the immunity, nutritional status, age, sex, personal hygiene, etc. Usually the farm workers with prolonged wastewater contact, their families, crop handlers and closely located communities exposed to contaminated water, sludge or excreta (Stoltzfus et al. 1997) are the worst affected ones. Almost 20 countries have been identified for using untreated wastewaters in agricultural fields to enhance fertility of soil (Fig. 8.14a). However, crop samples have revealed (Durán–Álvarez and Jiménez–Cisneros 2014) as many as 100 helminth eggs/kg of soil. Of the samples contaminated with helminth eggs, the maximum representation (Figs. 8.14b and 8.15) was by *Ascaris* (75%) followed by hookworms (15%) and *Trichuris* (10%). India has a prevalence of about 10–15% STH infection among school-going children. The infection rate varies in different parts and communities of the country. In some regions of the world, 95% of the children have been reported to be infected (Stephenson et al. 2000). Epidemiologic studies conducted have indicated that school-aged children face greatest risk for acquiring persistent infections. The majority of people with ascariasis live in Asia (73%), Africa (12%) and South America (8%). Often children are found to be coinfected with two parasites, viz. *Ascaris* and *Trichuris*, in developing countries, and there is a statistically significant coexistence of the two (Howard et al. 2002). The poor, unprivileged classes of the developed/industrialized countries (Blumenthal and Schultz 1975; Jones 1983; Kappus et al. 1994), including the United States and the United Kingdom (Crompton 1989), also show prevalence of these infections.

8.15 Worm Burden/Disease and Genetics of Distribution

The severity of a disease is dependent on the intensity of infection or indirectly the worm burden, i.e. the number of parasites per individual. Often most intense infection are born by a minority of infected individuals in the population. Hence, the distribution of helminth parasites is found highly over dispersed with often 10–15% of people carrying 70% of the parasite (Anderson and May 1985) populations. Despite the near-ubiquitous nature of soil-transmitted helminths from the global perspective, it is difficult to estimate the true worm burden due to limited work carried out on the aspect, and information related to associated impacts are not being regularly updated (Utzinger and de Savigny 2006; Nagpal et al. 2013). Brooker (2010) reviewed the efforts to quantify morbidity and mortality from the common intestinal nematode parasites.

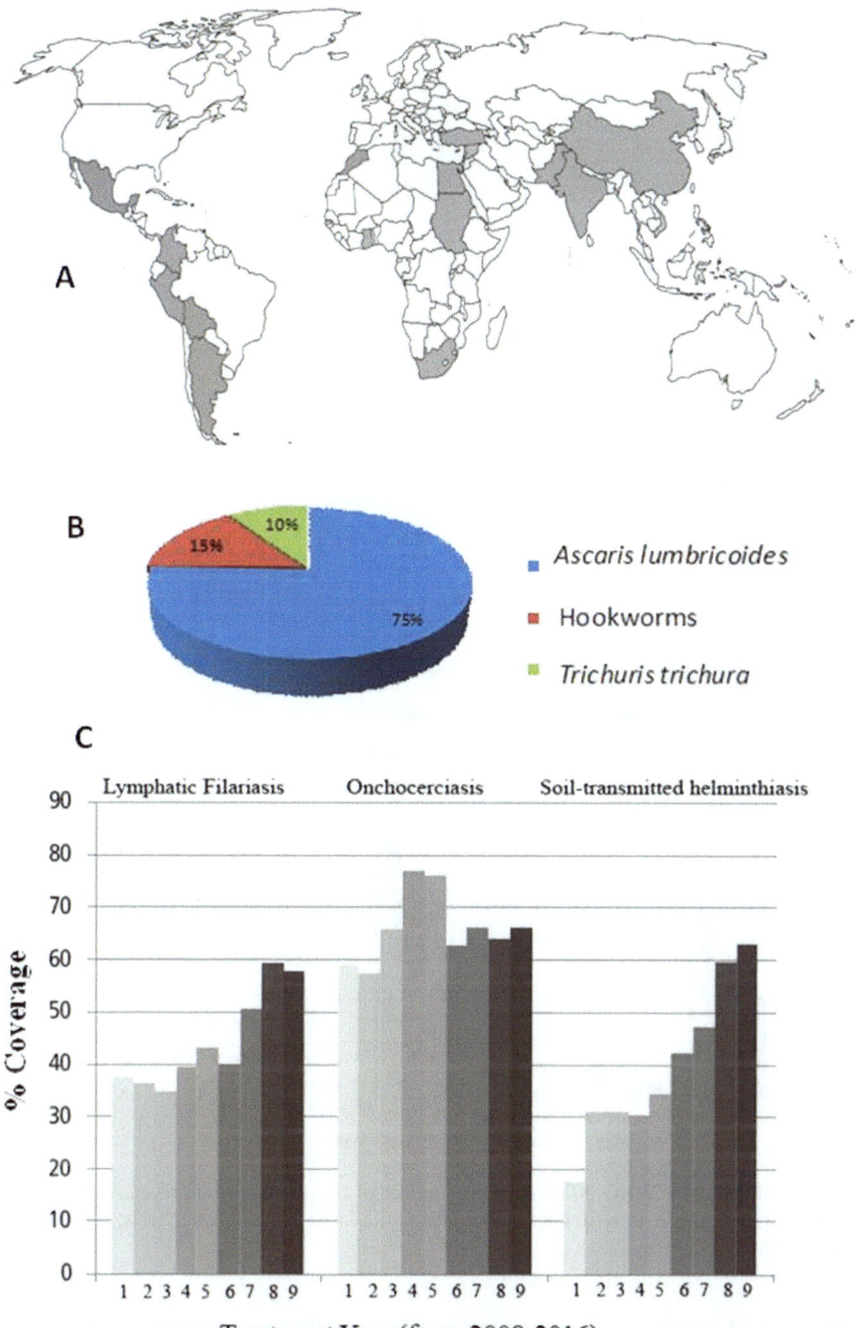

Fig. 8.14 (**a**) Countries using wastewater in agricultural fields are highlighted with dark shade; (**b**) Percent occurrence of helminth eggs in vegetables irrigated with wastewater; (**c**) worldwide percent coverage of the important helminth diseases during the past 8 years

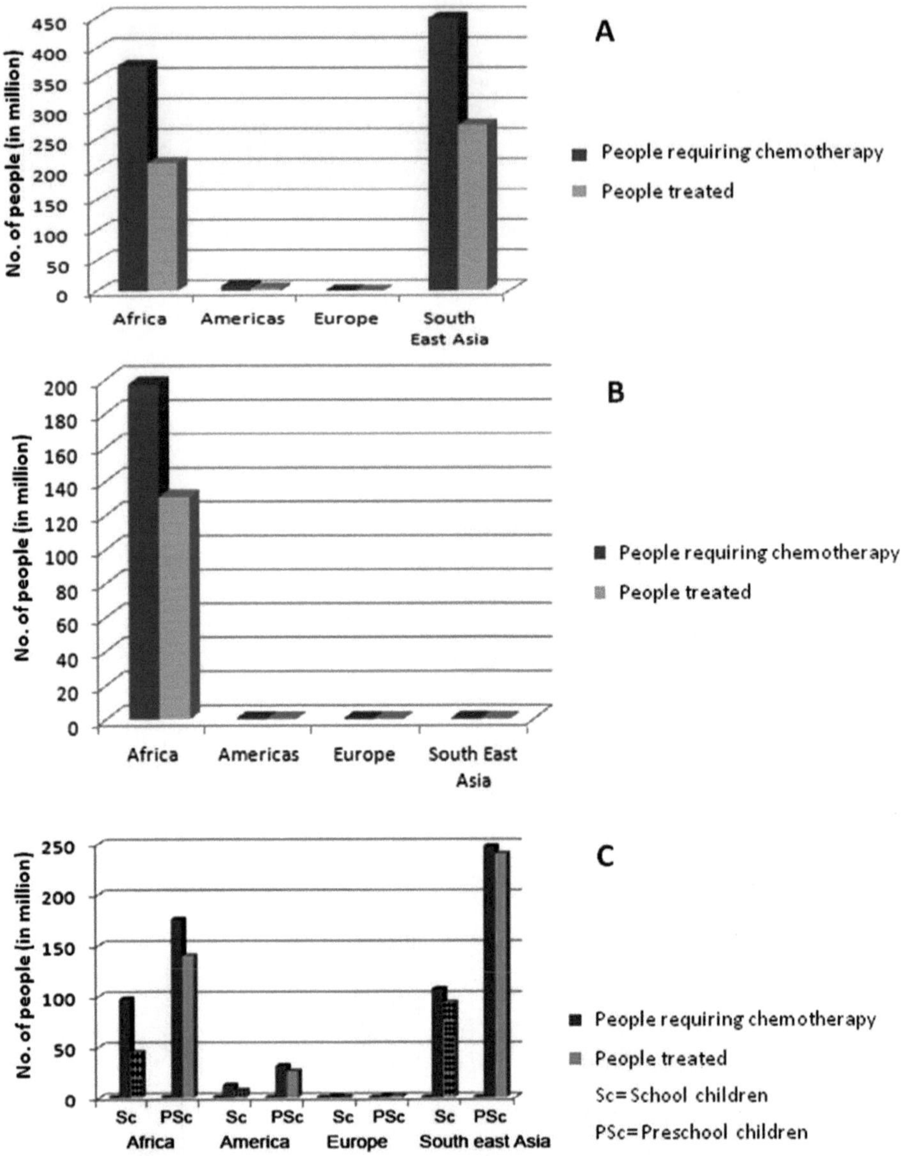

Fig. 8.15 Global data (WHO 2017) showing number of people requiring preventive chemotherapy (PC) against those treated for (**a**) lymphatic filariasis; (**b**) onchocerciasis; and (**c**) soil-transmitted helminthiasis

It is considered that less than ten worms are usually asymptomatic, although few but the infected individuals are also a greater source of parasite propagation (Bundy 1995). Genetic epidemiology of gastrointestinal helminth infection focuses more on phenotypic differences between individuals to be the key factor for the intensity of infection (worm burden) (May and Anderson 1979). The phenotypic variations of host are due to variation in both exposure and sensitivity to disease and familial history of infection (Crofton 1971a, b; Anderson and Gordon 1982). The likelihood of individuals to high or low intensity of infection is generally assessed by a positive correlation between the worm burden before treatment and then at reinfection after the treatment. The susceptibility to helminth infections varies in people of different ethnicities. A study conducted in Southern United States revealed a greater prevalence and intensity of hookworm infection in people of European ancestry than African ethnicity (Smillie and Augustine 1925; Keller et al. 1937). Household clustering of infection for helminth species is another feature reported by workers which includes *Ascaris* and *Trichuris* (Williams et al. 1974; Forrester et al. 1988; Anderson et al. 1993; Chan et al. 1994), hookworm (Behnke et al. 2000) and *Strongyloides stercoralis* (Conway et al. 1995). The genetic control of worm burden is likely to involve several genes and also exhibit ages in certain helminth infections, with a peak in the child age class but with a subsequent decline among adults (Bundy 1988, 1995).

The most reliable way (90% accurate) to obtain worm burden is to first treat a given population with chemotherapy and then to estimate the expelled worms (Bundy 1995). The studies of different geographic regions of the world indicated maximum worm burden for *Ascaris* and *Trichuris* infection among all helminths. However, the disease burden estimates over the past decade indicate greatest variation between ascariasis, hookworm disease and trichuriasis. A reason to this is the enormity of the cases that suffer with these diseases.

Multiparasitism denotes the simultaneous infection of host with two or more parasite species and is of common occurrence in individuals infected with soil-transmitted helminths (Steinmann et al. 2010). The data from different countries have shown the prevalence and significance of multiparasitism (Raso et al. 2004; Ohta and Waikagul 2007; Jardim-Botelho et al. 2008). Howard et al. (2002) reported strong association between *Ascaris* and *Trichuris* where infection with one increases the probability of infection with the other. Although coinfections may simply arise by chance, shared risk factors have been a major contribution (Utzinger and Keiser 2004; Pullan and Brooker 2008; Gilgen et al. 2001; Mwangi et al. 2006). Fewer studies, however, have reported such (Bethony et al. 2006; Utzinger and Keiser 2004; Pullan and Brooker 2008; Gilgen et al. 2001; Mwangi et al. 2006) cases of multiparasitism.

Jardim-Botelho et al. (2008) found that almost half (48%) of the tested children had infection of two parasites, while another 19% were infected with three, i.e. *A. lumbricoides*, hookworm and *Schistosoma mansoni*. In conditions of multiparasitism, both synergism and antagonism can occur due to the different immune responses evoked by the parasites. The concurrent infection of *Plasmodium* spp. with intestinal helminths results in persistent malarial infection (Brooker et al. 2006). Nevertheless, this process protects the host against severe inflammation (Helmby 2009; Mupfasoni et al. 2009; Supali et al. 2010). Likewise, the coinfection of HIV with intestinal helminth increases the vulnerability of host to HIV because of suppression of T-helper 1 (Th1) responses (Brooker et al. 2004; Lewthwaite et al. 2005; Secor 2012).

Jain et al. (2016) published a review on the state of helminth infections in India. They reported interesting facts using a series of parameters. Most studies revealed occurrence of multiparasitism (94.4%) in Indian populations. The workers (l. c.) also reported the occurrence of *Ascaris lumbricoides* (100%), *Ancylostoma duodenale* (66.6%) and *Trichuris trichiura* (77.7%) in the county between the study periods 2008–2015. The prevalence in the northern, southern and eastern states was reportedly high compared to the prevalence in western states of the country.

8.16 Special Adaptations of Soil-Transmitted Helminths

Soil-transmitted helminths are not vulnerable to host immunity which is a reason for their chronic intervention in the host physiology. The life stages of parasites move to different tissues carrying stage-specific antigens (Maizels et al. 1999) in order to weaken the immune response of host for their successful establishment. Adult and larvae of *Ascaris* have been reported (Coles 1985; Kennedy 1992) to release volatile allergens and cause asthma-like symptoms occasionally. However, a series of host responses are reported which include eosinophil-mediated larval killing, production of specific polyclonal antibodies, degranulation of mast cell, multiplication of goblet cells and excessive mucus release. Specificity was observed in several instances (Finkelman et al. 1997; Bradley and Jackson 2004), e.g., mast cells were effective in countering hookworms and *Ascaris* but not *Trichuris*.

In order to sustain on the host for a long time, STHs maintain a moderate transmission and do not cause acute pathogenicity and morbidity. Their secretions modulate host's immune responses or interrupt host physiological processes, for example, *Ancylostoma* secretes anticoagulant at its feeding site; *Ascaris* larvae secrete a pepsin inhibitor to resist digestive enzymes of stomach before reaching the small intestine; *T. trichiura* creates ion-conducting pores in lipid bilayers to penetrate through caecal epithelium using protein TT47 (Drake et al. 1994).

8.17 General Morbidities Associated with Helminth Infections

The symptoms of single-species helminth (nematode) infections range from abdominal pain, dysentery, anaemia, pruritus, allergies and impairment of cognitive abilities. Due to the chronic and meandering nature of most infections, the morbidity symptoms do not appear obviously different from normal condition; however, they have long-term negative impacts on the health, physical and cognitive skills and working efficiency of children in particular and humans in general (Fenwick and Figenschou 1972; Stephenson et al. 2000; Crompton and Nesheim 2002; Jardim-Botelho et al. 2008; Eppig et al. 2010). The individuals harbouring many parasite species experience increased morbidities compared to those with single-species infection (Brooker 2010). Lustigman et al. (2012) reported that about 300 million people worldwide suffer from helminth morbidities which also lead to 10,000–135,000 annual deaths. Lozano et al. (2012) reported an estimated 204,111 years of life lost (YLLs) globally due to *A. lumbricoides* infections. Stephenson et al. (2000) suggested anorexia to be the reason of malnourishment in the host. The loss of appetite and the resulting lowered dietary intake (Hadju et al. 1996) may lead to deficiencies of iron, iodine, folate, zinc, vitamins A, B_{12}, etc. The rate of absorption is also compromised in infections due to intestinal inflammation. Other related malfunctions are poor fat absorption due to mucosal damage, obstruction and cellular damage in the liver and reduced energy production (Stephenson et al. 2000) that retard immune functions and defences. The nutrients have their crucial role in body's health and maintenance: proteins required for antibody and interleukin formation (Malafaia et al. 2009), while lipids and carbohydrates are crucial in T-cell production and function (Gershwin et al. 2000). Any disruption in the immune response eventually makes the host susceptible to infections (Gershwin et al. 2000). Deficiency of vitamin A increases risk for diarrhoeal infections due to damage of epithelium (Scrimshaw and SanGiovanni 1997; Katona and Katona-Apte 2008); both vitamins C and E deficiency causes oxidative stress; vitamin E deficiency affects cell division and functioning of T cells (Carr and Frei 1999; Maggini et al. 2010). Zinc deficiency impedes antibody production and decreases the number of T and B cells (Katona and Katona-Apte 2008; Maggini et al. 2010), whereas iron deficiency makes immune system vulnerable against infections (Ayoya et al. 2006; Katona and Katona-Apte 2008). Anaemia is aggravated during pregnancy due to low nutritional status in poor families where food comprises of rice, cassava and maize that are poor sources of iron (Ayoya et al. 2006; Pasricha et al. 2008). Thus stunting with low height and weight and anaemia are common indicators for such infections.

Due to the ignorance and conflicting results, the gross morbidities of soil-transmitted helminth are anaemia, rectal prolapse and bile duct or intestinal obstruction (Lustigman et al. 2012). Salomon et al. (2012) estimated the collective disability weight of intestinal nematode infections to be 0.03 on a scale from 0 (perfect health) to 1 (death). Quantifying the real burden of soil-transmitted helminths is mandatory for the health-care services (Nagpal et al. 2013). Hence the concept of disability-adjusted life years (DALYs), for quantifying ill-health, was proposed by Murray et al. (1994). *A. lumbricoides*, *T. trichiura*, and hookworms were calculated to cause a loss of 4.2, 0.9 and 3.9 million DALYs, respectively (Murray and Lopez 1994; Murray et al. 1994). Bundy (1994) gave an estimate of 10.5, 6.4 and 22.1 million DALYs due to *A. lumbricoides*, *T. trichiura* and hookworm infections, with a collective burden of 39.0 million DALYs. Most of the DALYs were calculated for the school-aged population, as they have the highest prevalence and infection intensity of soil-transmitted helminths, and a younger person with a permanent disability would incur more years lost than an older person (Chan 1997). In another study, the morbidities of *A. lumbricoides*, *T. trichiura* and hookworm were calculated to have 1.3, 0.6 and 3.2 million DALYs, respectively, giving a collective burden of 5.2 mil-

Table 8.2 The global burden and associated loss due to different diseases

Disease	DALYs (in millions)	YLDs (in millions)
All neglected tropical diseases	26.06	18.22
Total intestinal nematode infection	5.19	4.98
Hookworm disease	3.23	3.23
Ascariasis	1.32	1.11
Trichuriasis	0.64	0.64
Lymphatic filariasis	2.78	2.77
Onchocerciasis	0.49	0.49

lion DALYs (Lozano et al. 2012). In most cases, *A. lumbricoides* infections lead to impaired growth and malnutrition in children (Crompton and Nesheim 2002). However, the exact differences of morbidities in *Ascaris* infection and the coinfection with other helminths could not be explicitly compared. Therefore, assigning individual disability weights to each helminth species in case of multiparasitism is not the accurate way of estimating DALYs. Therefore, another parameter, quality-adjusted life years (QALYs), was introduced where assessment involved health-related quality-of-life (QoL) questionnaires that were answered by infected communities. In 2010, 4.98 million YLDs were attributed of which 65% assigned to hookworm, 22% to *A. lumbricoides,* and 13% to *T. trichiura*. Much of the disability due to STH infection occurred in Asia compared to other regions of the world. Table 8.2 gives an account of DALYs and YLDs due to various helminth infections (PATH 2016).

8.18 Genetics of Pathology in Soil-Transmitted Helminth Infection

Any genetic control of worm burden is manifested in variation in severity of disease in people of different genders or ethnicities with some populations showing greater predisposition towards infections as observed in the case of *Trichuris* and *Ascaris* (Bundy 1986; Haswell-Elkins et al. 1987; Hlaing et al. 1987). The variation in pathological condition in different hosts (Roche and Layrisse 1966) is an indicator of the genetic basis of such selection. With similar intensities of infection, two populations may have different frequencies of an infection as observed (Nunesmaia et al. 1975; Bina et al. 1978; Prata 1992) in African Brazilian and Caucasian populations where the former showed much lower prevalence of hepatosplenic disorder. Studies have revealed that the severity of disease depends on the gender and pedigree type (Mohammed-Ali et al. 1999; Tavares-Neto and Prata 1989). Sudanese population was reportedly carrying hepatic fibrosis controlling gene SM2 (Dessein et al. 1999), whereas similar gene SM1 controlling infection intensity was found to be present in Brazilian population. Two quantitative trait loci on chromosomes 9 and 18 were reported to be associated with susceptibility to *T. trichiura* infection (Ellis et al.

2007a, b). Likewise, an increased parasite-specific IgE and eosinophilic responses were found to be responsible for resistance to reinfection or reducing worm burden (Quinnell et al. 1995; Pritchard et al. 1995; Faulkner et al. 2002). IgG4 host antibody responses also revealed the correlation of immunoglobulin levels with worm burdens (Haswell-Elkins et al. 1989; Palmer et al. 1996; Xue et al. 2000).

The variation in susceptibility towards parasite can be due to variations in sensitivities due to phenotypic or genetic variability of host and also due to variability in immune responses of individuals to parasitic antigens (Lammie et al. 1991; Ottesen 1992; King et al. 2001).

8.19 Impact of Treatment

Most anthelmintic medicines inhibit microtubule polymerization by binding to tubulin protein which ultimately causes death of adult worms. Usually, the contraindications of benzimidazole anthelmintic drugs are rare; however, few symptoms may appear such as sporadic pain in the abdomen, diarrhoea, nausea, vertigo and headache. The drugs are not advised for infants and pregnant women as they cause embryotoxicity and teratogenicity (Montresor et al. 2002a, b, 2003). Barring some minor impacts, the drugs have considerable advantages as repeated chemotherapy (deworming) in high-risk populations at regular intervals has generally been successful. The impacts have been more obvious and remarkable in children with the heaviest infections than in adults. Significant improvement has been observed in motor and language development as well as nutritional status in treated preschool children (Stoltzfus et al. 2004). Such children also showed an increase in height compared to nontreated infected children. Treatment of school-age children had led to better appetite and nutrient uptake (Stoltzfus et al. 2004), improved physical growth (Stephenson et al. 2000) and intellectual abilities (Stephenson et al. 1989; Awasthi et al. 2000; Drake et al. 2000). In Indonesia, the children treated for *A. lumbricoides* had reportedly performed better in cognitive tests due to improvement in learning and eye-hand coordination (Hadidjaja et al. 1998). However, contrary to that observation, a school-based deworming trial in Sri Lanka could not yield significant improvement in the cognitive test scores of treated children (Ebenezer et al. 2013). The effects on growth in light infections may not be marked in the children with poor nutritional status.

8.20 Root Cause of Problem Globally

Of the various routes of transmission, the five Fs, faeces, fields, fingers, flies and food, are considered to be the most important. *Ascaris* and *Trichuris* have been largely reported from urban environments (Phiri et al. 2000) especially urban slum; however, hookworms are more prevalent in poor rural areas (Albonico et al. 1997).

These parasites usually withstand extremes of environmental conditions, and their eggs are adhesive to adhere to a variety of objects (Raisanen 1985). A global review of the available literature between 1980 and 2001 of cities of Africa, Asia and the Americas as done by Sobsey (2002) revealed an increase in morbidity in households due to water-related diseases. The reason of the problem is the deterioration of water quality that largely is due to inadequately stored water.

In the light of available information, there are several reasons for contamination of water, *viz.*:

(a) **Contamination of source water**

- Location of latrines close to or uphill of the water source, faecal contamination due to open defecation in the adjoining areas, through nearby septic tanks and garbage pits or due to intensive grazing in the area next to the source
- Water accumulated at or near water source due to inadequate drainage or lack of fencing of the water source providing access to animals, etc.
- Improper maintenance of dug wells; buckets used in wells or reservoirs with bases soiled
- Infrequent cleaning and treatment of source water either due to nonoperational plant or inadequate maintenance and supervision largely because of ill-trained operators, unavailability of treatment chemicals and inadequate record keeping
- Rainwater harvesting tanks without covers/inadequately covered, contaminated with faeces of birds and small animals, water tanks cracked or vents improperly sealed, allowing the entry of insects and small animals, with inadequate or poorly maintained filters
- Groundwater source inadequately protected from contamination
- Cracked pipes showing leaking due to erosion or poor construction
- Dirty water accumulated around tap points or open defecation near tap stands

(b) **Contamination of water during the transport and storage**

- Careless handling of commonly used transport and storage containers, their infrequent cleaning or washing with contaminated hands or cloths
- Improper household latrines without handwashing facilities and poor public awareness result in more faeces in the vicinity
- Wide-mouth storage containers or uncovered containers may allow contamination through hands and cups/ladles
- Storage containers which often are easily accessible to children and animals

- The above problems occur due to poor awareness about hygiene and because of not setting the minimum safe distance (MSD) between sources of contamination and main water source. Also excessive use of wastewater and excreta in both agriculture and aquaculture is another major reason of contamination of usable water. In rural and poor urban areas of many developing countries, good water quality records are not available because either data are not recorded or recorded sporadically or recorded without specific parameter. Hence a routine monitoring is largely lacking.

- Other factors that may facilitate the dispersion of the parasites or may offer suitable conditions for its sustenance are stated hereunder.

8.21 Climatic Conditions

STHs require adequate warmth and moisture. Thus, rainfall patterns and climatic conditions usually determine the patterns of prevalence of STH. As *Ascaris* and *Trichuris* eggs tolerate drier conditions better than hookworm infective larvae, the latter's transmission rates increase during the rainy season (Mark 1975; Udonsi et al. 1980). However, *A. lumbricoides* prevalence increases by 10% with a minimum of 1400 mm annual rainfall (Prost 1987; Brooker and Michael 2000). *Ascaris* eggs require a minimum of 80% relative humidity to embryonate (Brooker and Michael 2000), whereas the larvae of *S. stercoralis* show nictation for their dispersal even if the relative humidity decreases. Moisture also plays a crucial role in vertical migration of hookworm larvae to a height of 30–40 cm (Komiya and Yasuraoka 1966). The prevalence declined with increasing altitude (Appleton and Gouws 1996; Jemaneh 1998; Flores et al. 2001). *A. lumbricoides* and *T. trichiura* are not found in areas with land surface temperature (LST) above 37 °C (Brooker et al. 2002a, b).

8.22 Global Target

Though not considered fatal, STH infections kill 135,000 people a year. The global targets have been set to eliminate STH infections in children by 2020 by regularly treating at least 75% of the children in endemic areas of the world. Other tasks include reducing transmission and controlling the neglected tropical diseases by improving basic sanitation, providing safe drinking water and promoting health and education to control vector population. Thrust is, therefore, given on WASH (water, sanitation and hygiene), a strategy to counter soil-transmitted helminths.

8.22.1 Anthelmintic Therapy

The first important aspect is the diagnosis of infection in the community. Due to the insidious nature, a faecal examination is advised on the basis of the local epidemiology or country of origin or persistent eosinophilia. Common tests, viz. Kato-Katz faecal-thick smear and the McMaster methods (Dunn and Keymer 1986; Santos et al. 2005) to measure the intensity of infection, are based on the count of helminth eggs per gram stool sample of infected individual (Booth et al. 2003; Knopp et al. 2008). To increase the reliability, greater number of slides must be prepared from a single sample, or multiple stool samples should be examined (Knopp et al. 2008; Steinmann et al. 2008).

Chemotherapy is the prime strategy to control STH infections. The WHO opted for preventive chemotherapy with periodic doses of the highest-risk populations (WHO 2006). The strategy can be helpful in reducing the intensity of infections besides eliminating associated morbidities (Gabrielli et al. 2011). Despite certain limitations the chemotherapy has even been advised during pregnancy after the first trimester (Larocque and Gyorkos 2006). The drugs – albendazole (400 mg) and mebendazole (500 mg) – are donated to national ministries of health through WHO for the treatment of all school-going children. Two other drugs, viz. levamisole and pyrantel pamoate, have also been recommended by WHO against soil-transmitted helminths. Continued use of a drug may lead to development of resistance by the parasite; hence drug combinations are considered a better strategy that could delay the emergence of drug resistance.

8.22.2 Monitoring and Surveillance Systems

It involves the regular inspection and examination of water supply sources. Awareness is created among people about water quality and about the consequences of drinking unsafe water. It involves development and management of community treatment systems and maintenance of household treatment systems. UNICEF programmes, therefore, not only direct resources such as surveillance agencies to ensure the effective transmission of water quality messages to consumers but also involve local government bodies and NGOs so that such programmes should not become only futile exercises of data collection.

STH infections can be prevented by reducing contamination of soil and water on a wide scale; however, the execution of such programmes is not easy in poor and deprived countries because of high costs involved (Asaolu and Ofoezie 2003). The adequate disposal of human excreta is necessary in this regard. Improved sanitation can keep the helminth infections away and hence prevent learning impairment. Non-chemotherapeutic measures also help preventing the infections. The concept of sanitation, the right education and motivation can work towards prevention of the parasitic worms (Brown et al. 2013). The positive impact has been reported in a review article where improved sanitation reduced the chances of infection by *T. trichiura*, hookworm and *A. lumbricoides* by 42%, 40% and 46%, respectively (Ziegelbauer et al. 2012). In Brazil, the households with better sanitation showed 40% decrease in the prevalence of *A. lumbricoides* compared to the households lacking drains and sewage system (Moraes et al. 2004).

8.23 Community Capacity Building

Programmes of support in the area of sanitation and hygiene and water quality must be designed for capacity building. Health education camps create awareness among the masses about the disease caused by STHs and the measures to reduce its transmission and reinfection. A change in perception and attitude of the community can be brought by such campaigns to make people more enthusiastic towards deworming.

Programmes for prompting people for latrine construction and health awareness decrease the incidence of helminth infections, although much time is required for sanitation to be effective (Brooker et al. 2004). Such educational campaign launched in Northern Bangladesh resulted in a 64% drop of prevalence of disease in children (Northrop-Clewes et al. 2001). People have started realizing the importance of health practices and norms and have generally adopted them except few ethnic minority groups with staunch cultural beliefs (McMullin et al. 2005; Bóia et al. 2006; Ribera et al. 2009; Vandemark et al. 2010). A recent study from Hunan Province showed considerable reductions in *A. lumbricoides* transmission in the people availing educational health (Bieri et al. 2013).

8.24 Vaccine Development

Vaccination is considered the most inexpensive and competent procedure for controlling diseases and hence has also been tried to counter STH infections (Hotez et al. 2003, 2010). Helminths are found to be evading and modifying the mammalian immune response which may be exploited for therapeutic purposes (Zaccone et al. 2006). Helminth infections have been found to be inversely linked to immune functions, and a reduction in helminth infections may result in rising rates of autoimmunity and atopy (Cooper et al. 2003; Cooper 2006). The individuals infected with STHs were found to be less susceptible to allergic disorders (Feary et al. 2010; Kim and Drake-Lee 2003).

Vaccination seems to be still far from reach because of lack of correct data on disease burden and degree of resistance to anthelmintic drug. Some other limitations are lack of good experimental model and a poor understanding of the suppression of host immune response by the parasite for its long-term survival. Although Na-ASP-2 hookworm vaccine consisting of the recombinant larval antigen ASP2 has been a success on animal model (Bethony et al. 2005; Goud et al. 2004, 2005; Mendez et al. 2005a, b), however, it requires proper verification with human clinical trials as hookworm antigens show amino acid sequences partially homologous with mammalian proteins. For the development and manufacture of vaccines, the infrastructure and the high levels of technological ambience require enormous funding (Hotez 2001; Broder et al. 2002) which may not give good returns due to their large application in poor countries.

Acknowledgement The author is grateful to the Department of Science and Technology, New Delhi, for financial support. Also, the help by those who have contributed in any way in preparation of this article is gratefully acknowledged.

References

ACC/SCN (2000) Fourth report on the world nutrition situation: nutrition throughout the lifecycle. United Nations Administrative Committee on Coordination, Sub-Committee on Nutrition, Geneva

Albonico M, Chwaya HM, Montresor A, Stolfzfus RJ, Tielsch JM, Alawi KS, Savioli L (1997) Parasitic infections in Pemba Island school children. East Afr Med J 74:29–298

Albonico M, Allen H, Chitsulo L, Engels D, Gabrielli AF (2008) Controlling soil transmitted helminthiasis in pre-school age children through preventive chemotherapy. PLoS Negl Trop Dis 2:126

Anderson RM, Gordon DM (1982) Processes influencing the distribution of parasite numbers within host populations with special emphasis on parasite-induced host mortalities. Parasitology 85:373–398

Anderson RM, May RM (1985) Vaccination and herd immunity to infectious diseases. Nature 318:323–329

Anderson TJ, Romero-Abal ME, Jaenike J (1993) Genetic structure and epidemiology of *Ascaris* populations: patterns of host affiliation in Guatemala. Parasitology 107:319–334

Appleton CC, Gouws E (1996) The distribution of common intestinal nematodes along an altitudinal transect in Kwa-Zulu Natal, South Africa. Ann Trop Med Parasitol 90:181–188

Asano T, Burton FL, Leverenz H, Tsuchihashi R, Tchobanoglous G (2007) Water reuse: issues, technologies and applications. McGraw-Hill, New York

Asaolu SO, Ofoezie IE (2003) The role of health education and sanitation in the control of helminth infections. Acta Trop 86:283–294

Ash LR, Schacher JF (1971) Early life cycle and larval morphogenesis of *Wuchereria bancrofti* in the jird, *Meriones unguiculatus*. J Parasitol 57:1043–1051

Ashton F, Schad G, Li J (1999) Chemo- and thermosensory neurons: structure and function in animal parasitic nematodes. Vet Parasitol 84:297–316

Awasthi S, Pande VK, Fletcher RH (2000) Effectiveness and cost effectiveness of albendazole in improving nutritional status of preschool children in urban slums. Indian J Pediatr 37:19–29

Ayoya MA, Spiekermann-Brouwer GM, Traore AK, Stoltzfus RJ, Garza C (2006) Determinants of anemia among pregnant women in Mali. Food Nutr Bull 27:3–11

Baron S, Cross J (1996) Medical microbiology: parasitology. The University of Texas Medical Branch of Galveston, Galveston

Becker SL, Sieto B, Silué KD, Adjossan L, Koné S (2011) Diagnosis, clinical features, and self-reported morbidity of *Strongyloides stercoralis* and hookworm infection in a co-endemic setting. PLoS Negl Trop Dis 5:e1292

Behnke JM, De Clercq D, Sacko M, Gilbert FS, Ouattara DB, Vercruysse J (2000) The epidemiology of human hookworm infections in the southern region of Mali. Trop Med Int Health 5:343–354

Beigal Y, Greenburg Z, Ostfeld I (2000) Letting the patient off the hook. N Engl J Med 342:1658–1661

Bethony J, Loukas A, Smout M, Brooker S, Mendez S, Olieskatt J, Goud G, Bottazzi ME, Zhan B, Wang Y, Williamson A, Lustigman S, Correa-Oliveira R, Xiao S, Hotez PJ (2005) Antibodies against a secreted protein from hookworm larvae reduce intensity of hookworm infection in humans and vaccinated laboratory animals. FASEB J 19:1743–1745

Bethony J, Brooker S, Albonico M, Geiger SM, Loukas A (2006) Soil-transmitted helminth infections: ascariasis, trichuriasis and hookworm. Lancet 367:1521–1532

Bieri FA, Gray DJ, Williams GM, Raso G, Li Y-S, Yuan L, He Y, Li RS, Guo F-Y, Li S-M, McManus DP (2013) Health-education package to prevent worm infections in Chinese school children. N Engl J Med 368:1603–1612

Bina JC, Tavares-Neto J, Prata A, Azevedo ES (1978) Greater resistance to development of severe schistosomiasis in Brazilian Negroes. Hum Biol 50:41–49

Blumenthal DS, Schultz MG (1975) Incidence of intestinal obstruction in children infected with *Ascaris lumbricoides*. Am J Trop Med Hyg 24:801–805

Bogitsh BJ, Cheng TC (1998) Human parasitology. Academic, New York

Bóia MN, Carvalho-Costa FA, Sodré FC, Eyer-Silva WA, Lamas CC, Lyra MR, Pinto-Junior VL, Cantalice Filho JP, Oliveira AL, Carvalho LM, Gross JB, Sousa AL, Moraes TI, Bermudez-Aza EH, Martins EB, Coura JR (2006) Mass treatment for intestinal helminthiasis control in an Amazonian endemic area in Brazil. Rev Inst Med Trop Sao Paulo 48:189–195

Booth M, Vounatsou P, N'Goran EK, Tanner M, Utzinger J (2003) The influence of sampling effort and the performance of the Kato-Katz technique in diagnosing *Schistosoma mansoni* and hookworm co-infections in rural Côte d'Ivoire. Parasitology 127:525–531

Bradley JE, Jackson JA (2004) Immunity, immunoregulation and the ecology of trichuriasis and ascariasis. Parasite Immunol 26:429–441

Broder S, Hoffman SL, Hotez PJ (2002) Cures for the third world's problems: the application of genomics to the diseases plaguing the developing world may have huge medical and economic benefits for those countries and might even prevent armed conflict. EMBO Rep 3:806–812

Brooker S (2010) Estimating the global distribution and disease burden of intestinal nematode infections: adding up the numbers- a review. Int J Parasitol 40:1137–1144

Brooker S, Michael E (2000) The potential of geographical information systems and remote sensing in the epidemiology and control of human helminth infections. Adv Parasitol 47:245–288

Brooker S, Beasley NMR, Ndinaromtan M, Madjiouroum EM, Baboguel M, Djenguinabe E, Hay SI, Bundy DAP (2002a) Use of remote sensing and a geographical information system in a national helminth control programme in Chad. Bull World Health Organ 80:783–789

Brooker S, Hay SI, Bundy DAP (2002b) Tools from ecology: useful for evaluating infection risk models? Trends Parasitol 18:70–74

Brooker S, Bethony J, Hotez PJ (2004) Human hookworm infection in the 21st century. Adv Parasitol 58:197–288

Brooker S, Clements AC, Hotez PJ (2006) The co-distribution of *Plasmodium falciparum* and hookworm among African school children. Malar J 5:99. https://doi.org/10.1186/1475-2875-5-99

Brown KA, Khanafer N, Daneman N, Fisman DN (2013) Meta-analysis of antibiotics and the risk of community-associated *Clostridium difficile* infection. Antimicrob Agents Chemother 57:2326–2332

Brusca RC, Brusca GJ (2003) Invertebrates. 1990, 2nd edn. Sinauer Associates, New York

Bundy DAP (1986) Epidemiological aspects of *Trichuris* and trichuriasis in Caribbean communities. Trans R Soc Trop Med Hyg 80:706–718

Bundy DAP (1988) Population ecology of intestinal helminth infections in human communities. Philos Trans R Soc Ser B 321:405–420

Bundy DAP (1994) The global burden of disease due to intestinal nematode disease. Am J Trop Med Hyg 88:259–261. https://doi.org/10.1016/0035-9203(94)90069-8

Bundy DAP (1995) Epidemiology and transmission of intestinal helminths. In: Farthing MJG, Keusch GT, Wakelin D (eds) Enteric infection 2, intestinal helminths. Chapman & Hall Medical, London, pp 5–24

Burkhart CN, Burkhart CG (2005) Assessment of frequency, transmission and genitourinary complications of enterobiasis (pinworms). Int J Dermatol 44:837–840

Cairncross S, Muller R, Zagaria N (2002) Dracunculiasis (Guinea worm disease) and the eradication initiative. Clin Microbiol Rev 15:223–246

Carr A, Frei B (1999) Does vitamin C act as a pro-oxidant under physiological conditions? FASEB J 13:1007–1024
Carson-DeWitt R (1999) Gale encyclopedia of medicine, 1st edn. Gale Research, Detroit, p 1475
CDC (2009) Parasites and health. (On-line). Filariasis
Chan MS (1997) The global burden of intestinal nematode infections-fifty years on. Parasitol Today 13:438–443
Chan MS, Medley GF, Jamison D, Bundy DAP (1994) The evaluation of potential global morbidity attributable to intestinal nematode infections. Parasitology 109:373–387
Changhua L, Xiaorong Z, Dongchuan Q, Shuhua X, Hotez P (1999) Epidemiology of human hookworm infections among adult villagers in Hejiang and Santai Counties, Sichuan Province. China Acta Trop 73:243–249
Cheng TC (1986) General parasitology. Academic Press College Division, Orlando
Chitwood BG, Chitwood MB (1950) Introduction to nematology. University Park Press, Baltimore
Chong Y (2003) Ascaris lumbricoides. (On-line). Web Atlas of Medical Parasitology
Coles GC (1985) Allergy and immunopathology of ascariasis. In: Crompton DWT, Nesheim MC, Pawlowski ZS (eds) Ascariasis and its public health significance. Taylor & Francis, London, pp 167–184
Conway DJ, Hall A, Anwar KS, Rahman MI, Bundy DA (1995) Household aggregation of *Strongyloides stercoralis* infection in Bangladesh. Trans R Soc Trop Med Hyg 89:258–261
Cook GC, Zumla-Alimuddin I (2009) Manson's tropical diseases. In: Cook GC, Manson SP, Zumla-Alimuddin I (eds) Soil-transmitted helminths (geohelminths), 22nd edn. Elsevier Health Sciences, Germany, pp 1515–1519
Cooper ES (1995) Epidemiology and transmission of intestinal helminths. In: Farthing MJG, Keusch GT, Wakelin D (eds) Enteric infection 2. Intestinal helminths. Chapman & Hall Medical, London, pp 201–212
Cooper PJ (2006) The potential impact of early exposures to geohelminth infections on the development of atopy. Clin Rev Allergy Immunol 26:5–14
Cooper PJ, Chico ME, Rodrigues LC, Ordonez M, Strachan D, Griffin GE (2003) Reduced risk of atopy among school-age children infected with geohelminth parasites in a rural area of the tropics. J Allergy Clin Immunol 111:995–1000
Cox F, Chappell L (1993) Modern parasitology: a textbook of parasitology. Blackwell Science Ltd, Cambridge, MA
Crofton HD (1971a) A quantitative approach to parasitism. Parasitology 62:179–194
Crofton HD (1971b) A model of host-parasite relationships. Parasitology 63:343–364
Crompton DWT (1989) Prevalence of ascariasis. In: Crompton DWT, Nesheim MC, Pawlowski ZS (eds) Ascariasis and its prevention and control. Taylor & Francis Ltd, London, pp 45–69
Crompton DWT, Nesheim MC (1982) Nutritional science and parasitology: a case for collaboration. Bioscience 32:677–680
Crompton DWT, Nesheim MC (2002) Nutritional impact of intestinal helminthiasis during the human life cycle. Annu Rev Nutr 22:35–59
de Almeida A, Freedman D (1999) Epidemiology and immunopathology of bancroftian filariasis. Microbes Infect 1:1015–1022
de Silva NR, Brooker S, Hotez PJ, Montresor A, Engels D (2003) Soil-transmitted helminth infections: updating the global picture. Trends Parasitol 19:547–551
Despommier D, Gwadz R, Hotez P (1994) Parasitic diseases. Springer, New York
Dessein AJ, Hillaire D, Elwali NEMA, Marquet S, Mohamed-Ali Q, Mirghani A, Henri S, Abdelhameed AA, Saeed OK, Magzoub MMA, Abel L (1999) Severe hepatic fibrosis in *Schistosoma mansoni* infection is controlled by a major locus that is closely linked to the interferon-gamma receptor gene. Am J Hum Genet 65:709–721
Donowitz L (1999) Infection control in the child care center and preschool. Lippincott, Williams and Wilkins, Philadelphia
Dora-Laskey A, Ezenkwele U, Weiss E (2009) Ascaris lumbricoides. (On-line). eMedicine- Medscape

Drake L, Korchev Y, Bashford L (1994) The major secreted product of the whipworm, *Trichuris*, is a pore forming protein. Proc R Soc Lond B 257:255–261

Drake LJ, Jukes MCH, Sternberg RJ, Bundy DAP (2000) Geohelminth infections (ascariasis, trichuriasis and hookworm): cognitive and developmental impacts. Sem Pediatr Infect Dis 11:245–251

Dundas KC, Calder AA, Alyusuf R (1999) *Enterobius vermicularis* threadworm infestation of paraovarian tissue in a woman who has had a hysterectomy. Br J Obstet Gynaecol 106:605–607

Dunn A, Keymer A (1986) Factors affecting the reliability of the McMaster technique. J Helminthol 60:260–262

Durán-Álvarez JC, Jiménez-Cisneros B (2014) Beneficial and negative impacts on soil by the reuse of treated/untreated municipal wastewater for agricultural irrigation – a review of the current knowledge and future perspectives. In: Hernandez-Soriano MC (ed) Environmental risk assessment of soil contamination. https://doi.org/10.5772/57226

Ebenezer R, Gunawardena K, Kumarendran B, Pathmeswaran A, Jukes MCH, Drake LJ, de Silva N (2013) Cluster-randomised trial of the impact of school-based deworming and iron supplementation on the cognitive abilities of schoolchildren in Sri Lanka's plantation sector. Trop Med Int Health 18:942–951

Ellis MK, Raso G, Li Y-S, Rong Z, Chen H-G, McManus DP (2007a) Familial aggregation of human susceptibility to co- and multiple helminth infections in a population from the Poyang Lake region. China Int J Parasitol 37:1153–1161

Ellis JS, Smith JD, Braham S, Lock M, Barlow K, Zambon MC (2007b) Design and validation of an H5 TaqMan real-time one-step reverse transcription-PCR and confirmatory assays for detection and verification of influenza A virus H5 infections in humans. J Clin Microbiol 45:1535–1543

Eppig C, Fincher CL, Thornhill R (2010) Parasite prevalence and the worldwide distribution of cognitive ability. Proc R Soc B 277:3801–3808

Faulkner H, Turner J, Kamgno J, Pion SD, Boussinesq M, Bradley JE (2002) Age- and infection intensity-dependent cytokine and antibody production in human trichuriasis: the importance of IgE. J Infect Dis 185:665–672

Feary J, Britton J, Leonardi-Bee J (2010) Atopy and current intestinal parasite infection: a systematic review and meta-analysis. Allergy 66(4):569–578

Fenwick A, Figenschou BH (1972) The effect of *Schistosoma mansoni* infection of the productivity of cane cutters on a sugar estate in Tanzania. Bull World Health Organ 47:567–572

Finkelman FD, Shea-Donohue T, Goldhill J (1997) Cytokine regulation of host defense against parasitic gastrointestinal nematodes: lessons from studies with rodent models. Ann Rev Immunol 15:505–533

Flores A, Esteban JG, Angles R, Mas-Coma S (2001) Soil-transmitted helminth infections at very high altitude in Bolivia. Trans R Soc Trop Med Hyg 95:272–277

Forrester JE, Scott ME, Bundy DAP, Golden MHN (1988) Clustering of *Ascaris lumbricoides* and *Trichuris trichiura* infections within households. Trans R Soc Trop Med Hyg 82:282–288

Gabrielli M, Bonazzi P, Scarpellini E, Bendia E, Lauritano EC, Fasano A, Ceravolo MG, Capecci M, Bentivoglio AR, Provinciali L, Tonali PA, Gasbarrini A (2011) Prevalence of small intestinal bacterial overgrowth in Parkinson's disease. Mov Disord 26:889–892

Garcia LS, Bruckner DA (1997) Diagnostic medical parasitology, 3rd edn. ASM Press, Washington, DC, p 937

Gargano R, Di Legami R, Maresi E, Restivo S (2003) Chronic sialoadenitis caused by *E. vermicularis*: case report. Acta Otorhinolaryngol Ital 23:319–321

Gershwin LJ, Gunther RA, Anderson ML, Woolums AR, McArthur-Vaughan K, Randel KE, Boyle GA, Friebertshauser KE, McInturff PS (2000) Bovine respiratory syncytial virus specific IgE is associated with interleukin-2 and -4, and interferon-gamma expression in pulmonary lymph of experimentally infected calves. Am J Vet Res 61:291–298

Ghosh K, Ghosh K (2007) *Strongyloides stercoralis* septicaemia following steroid therapy for eosinophilia: report of three cases. Trans R Soc Trop Med Hyg 101:1163–1165

Gilgen D, Mascie-Taylor CGN, Rosetta L (2001) Intestinal helminth infections, anaemia and labour productivity of female tea pluckers in Bangladesh. Trop Med Int Health 6:449–457

Gill GV, Welch E, Bailey JW, Bell DR, Beeching NJ (2004) Chronic *Strongyloides stercoralis* infection in former British far east prisoners of war. Q J Med 97:789–795

Goud GN, Zhan B, Ghosh K, Loukas K, Haedon J, Dobardzic A, Deumic V, Liu S, Dobardzic R, Zook RC, Qun J, Liu YY, Hoffman L, Chung-Debose D, Patel R, Mendez S, Hotez PJ (2004) Cloning, yeast expression, isolation and vaccine testing of recombinant Ancylostoma secreted protein 1 (ASP-1) and ASP-2 from *Ancylostoma ceylanicum*. J Infect Dis 189:919–929

Goud GN, Bottazzi ME, Zhan B, Mendez S, Deumic V, Pleiskatt J, Lin S, Wang Y, Bueno L, Fujiwara R, Samuel A, Ahn SY, Solanki M, Asoja O, Wen J, Saul A, Bethony JM, Loukas A, Roy M, Hotez PJ (2005) Expression of the *Necator americanus* hookworm larval antigen Na-ASP-2 in *Pichia pastoris* and purification of the recombinant protein for use in human clinical trials. Vaccine 23:4754–4764

Greenaway C (2004) Dracunculiasis (guinea worm disease). Can Med Assoc J 170:495–500

Grove DI (1989) Clinical manifestations. In: Grove DI (ed) Strongyloidiasis a major roundworm infection of man. Taylor and Francis, London, pp 155–174

Hadidjaja P, Bonang E, Suyardi MA, Abidin SA, Ismin IS, Margano SS (1998) The effect of intervention on nutritional status and cognitive function of primary school children infected with *Ascaris lumbricoides*. Am J Trop Med Hyg 59:791–795

Hadju V, Stephenson LS, Abadi K, Mohammed HO, Bowman DD, Parker RS (1996) Improvement in appetite and growth in helminth-infected schoolboys three and seven weeks after a single dose of pyrantel pamoate. Parasitology 113:497–504

Haswell-Elkins MR, Elkins DB, Anderson RM (1987) Evidence for predisposition in humans to infection with *Ascaris*, hookworm, *Enterobius* and *Trichuris* in a South Indian fishing community. Parasitology 95:323–337

Haswell-Elkins MR, Kennedy MW, Maizels RM (1989) The antibody recognition profiles of humans naturally infected with *Ascaris lumbricoides*. Parasite Immunol 11:615–627

Helmby H (2009) Helminths and our immune system: friend and foe? Parasitol Int 58:121–127

Hlaing T, Saw T, Lwin M (1987) Reinfection of people with *Ascaris lumbricoides* following single 6-month and 12-month interval mass chemotherapy in Okpo village, rural Burma. Trans R Soc Trop Med Hyg 81:140–146

Hong ST, Choi MH, Chai JY, Kim YT, Kim MK, Kim KR (2002) A case of ovarian enterobiasis. Korean J Parasitol 40:149–151

Hopkins DR, Richards FO, Ruiz-Tiben E Jr, Emerson P, Withers PC Jr (2008) Dracunculiasis, onchocerciasis, schistosomiasis and trachoma. Ann N Y Acad Sci 1136:45–52

Hotez PJ (1989) Hookworm disease in children. Pediatr Infect Dis J 8:516–520

Hotez PJ (2000) Pediatric geohelminth infections: trichuriasis, ascariasis and hookworm infections. Semin Pediatr Infect Dis 11:236–244

Hotez PJ (2001) Vaccines as instruments of foreign policy. EMBO Rep 2:862–868

Hotez PJ, Zhan B, Bethony JM (2003) Progress in the development of a recombinant vaccine for human hookworm disease: the human hookworm vaccine initiative. Int J Parasitol 33:1245–1258

Hotez PJ, Molyneux DH, Fenwick A, Ottesen E, Ehrlich Sachs S (2006) Incorporating a rapid-impact package for neglected tropical diseases with programs for HIV/AIDS, tuberculosis and malaria. PLoS Med 3:e102

Hotez PJ, Molyneux DH, Fenwick A, Kumaresan J, Ehrlich Sachs S (2007) Control of neglected tropical diseases. N Engl J Med 357:1018–1027

Hotez PJ, Engels D, Fenwick A, Savioli L (2010) Africa is desperate for praziquantel. Lancet 376:496–498

Hotez PJ, Savioli L, Fenwick A (2012) Neglected tropical diseases of the Middle East and North Africa: review of their prevalence, distribution and opportunities for control. PLoS Negl Trop Dis 6:e1475

Howard SC, Donnelly CA, Kabatereine NB, Ratard RC, Brooker S (2002) Spatial and intensity dependent variations in associations between multiple species helminth infections. Acta Trop 83:141–149

Igra-Siegman Y, Kapila R, Sen P, Kaminski ZC, Louria DB (1981) Syndrome of hyperinfection with *Strongyloides stercoralis*. Rev Infect Dis 3:397–407

Jain SK, Dwivedi A, Shrivastava A, Vijayananth P, Vidyavardhini R, Venkatesh S (2016) Prevalence of soil-transmitted helminthic infection in India in current scenario: a systematic review. J Commun Dis 48:24–35

Jardim-Botelho A, Raff S, Rodrigues Rde A, Hoffman HJ, Diemert DJ, Corrêa-Oliveira R, Bethony JM, Gazzinelli MF (2008) Hookwork, Ascaris lumbrocoides infection and polyparasitism associated with poor cognitive performance in Brazilian school children. Trop Med Int Health 13:994–1004

Jemaneh L (1998) Comparative prevalences of some common intestinal helminth infections in different altitudinal regions in Ethiopia. Ethiop Med J 36:1–8

Jiménez B, Asano T (2008) Water reclamation and reuse around the world. In: Jiménez B, Asano T (eds) Water reuse: an international survey of current practice, issues and needs. IWA Publishing, London, pp 3–26

Jones JE (1983) Parasites in Kentucky: the past seven decades. J Ky Med Assoc 81:621–626

Kappus KD Jr, Lundgren RG, Juranek DD, Roberts JM, Spencer HC (1994) Intestinal parasitism in the United States: update on a continuing problem. Am J Trop Med Hyg 50:705–713

Katona P, Katona-Apte J (2008) The interaction between nutrition and infection. Clin Infect Dis 46:1582–1588

Keddy PA (2000) Wetland ecology: principles and conservation. Cambridge University Press, Cambridge, p 614

Keller AE, Leathers WS, Knox JC (1937) The present status of hookworm infestation in North Carolina. Am J Epidemiol 26:437–454

Kennedy MW (1992) Genetic control of the antibody response to parasite allergens. In: Moqbel R (ed) Allergy and immunity to helminths. Taylor & Francis, London, pp 63–80

Keraita BN, Jiménez B, Drechsel P (2008) Extent and implications of agricultural reuse of untreated, partly treated and diluted wastewater in developing countries. CAB Rev Perspect Agric Vet Sci Nutr Nat Resour 3:1–15

Kettler HE, Modi R (2001) Building local research and development capacity for the prevention and cure of neglected diseases: the case of India. Bull World Health Organ 79:742–747

Khuroo MS (1996) Ascariasis. Gastroenterol Clin N Am J 25:553–577

Kim AB, Drake-Lee AB (2003) Infection, allergy and the hygiene hypothesis: historical perspective. J Laryngol Otol 117:946–950

King CL, Connelly M, Alpers MP, Bockarie M, Kazura JW (2001) Transmission intensity determines lymphocyte responsiveness and cytokine bias in human lymphatic filariasis. J Immunol 166:7427–7436

Knopp S, Mgeni AF, Khamis IS, Steinmann P, Stothard JR (2008) Diagnosis of soil-transmitted helminths in the era of preventive chemotherapy: effect of multiple stool sampling and use of different diagnostic techniques. PLoS Negl Trop Dis 2:e331

Komiya Y, Yasuraoka Y (1966) The biology of hookworms. Prog Med Parasitol Jpn 3:4–114

Kramer JM, French RP, Park EC, Johnson JJ (1990) The *Caenorhabditis elegans* rol-6 gene, which interacts with the sqt-1 collage gene to determine organismal morphology, encodes a collagen. Mol Cell Biol 10:2081–2089

Lammie PJ, Hitch WL, Allen EMW, Hightower W, Eberhard ML (1991) Maternal filarial infection as risk factor for infection in children. Lancet 337:1005–1006

Larocque R, Gyorkos TW (2006) Should deworming be included in antenatal packages in hookworm-endemic areas of developing countries? Can J Public Health 97:222–224

Lewthwaite P, Gill GV, Hart CA, Beeching NJ (2005) Gastrointestinal parasites in the immunocompromised. Curr Opin Infect Dis 18:427–435

Litt E, Baker MC, Molyneux D (2012) Neglected tropical diseases and mental health: a perspective on comorbidity. Trends Parasitol 28:195–201

Lobo DA, Velayudhan R, Chatterjee P, Kohli H, Hotez PJ (2011) The neglected tropical diseases of India and South Asia: review of their prevalence, distribution and control or elimination. PLoS Negl Trop Dis 5:e1222

Loeffler W (1956) Transient lung infiltrations with blood eosinophilia. Int Arch Allergy Appl Immunol 8:54–59

Lozano R et al (2012) Global and regional mortality from 235 causes of death for 20 age groups in 1990 and 2010: a systematic analysis for the Global Burden of Disease Study 2010. ScienceDirect 380:2095–2128

Lustigman S, Prichard RK, Gazzinelli A, Grant WN, Boatin BA (2012) A research agenda for helminth diseases of humans: the problem of helminthiases. PLoS Negl Trop Dis 6:e1582

Maggini J, Mirkin G, Bognanni I, Holmberg J, Piazzón IM, Nepomnaschy I (2010) Mouse bone marrow-derived mesenchymal stromal cells turn activated macrophages into a regulatory-like profile. PLoS One 5:e9252

Maizels RM, Holland MJ, Falcone FH, Zang XX, Yazdanbakhsh M (1999) Vaccination against helminth parasites: the ultimate challenge for vaccinologists? Immunol Rev 171:125–147

Malafaia G, Serafim TD, Silva ME, Pedrosa ML, Rezende SA (2009) Protein-energy malnutrition decreases immune response to *Leishmania chagasi* vaccine in BALB/c mice. Parasite Immunol 31:41–49

Marcos LA, Terashima A, Dupont HL, Gotuzzo E (2008) *Strongyloides* hyperinfection syndrome: an emerging global infectious disease. Trans R Soc Trop Med Hyg 102:314–318. https://doi.org/10.1016/j.trstmh.2008.01.020

Mark DL (1975) Survival of *Ancylostoma caninum* on bluegrass pasture. J Parasitol 61:484–488

Martorell R, Scrimshaw NS (1995) The effects of improved nutrition in early childhood. The Institute of Nutrition of Central America and Panama (INCAP). Follow-up study. J Nutr 125:4

May RM, Anderson RM (1979) Population biology of infectious diseases. Part II Nat Lond 280:455–461

McDonald GSA, Hourihane DOB (1972) Ectopic *Enterobius vermicularis*. Gut 13:621–626

McMullin MF, Bareford D, Campbell P, Green AR, Harrison C, Hunt B, Oscier D, Polkey MI, Reilly JT, Rosenthal E, Ryan K, Pearson TC, Wilkins B (2005) Guidelines for the diagnosis, investigation and management of polycythaemia/erythrocytosis. Br J Haematol 130:174–195

Mendez S, Valenzuela J, Wu WH, Hotez P (2005a) Host cytokine production, lymphoproliferation and antibody responses during the coarse of *Ancylostoma ceylanicum* infection in the golden Syrian hamster. Infect Immun 73:3402–3407

Mendez S, Zhan B, Goud G, Ghosh K, Dobardzic A, Wu WH, Liu S, Deumic V, Dobardzic R, Liu YY, Bethony J, Hotez PJ (2005b) Effect of combining the larval antigens *Ancylostoma* secreted protein 2 (ASP-2) and metalloprotease 1 (MTP-1) in protecting hamsters against hookworm infection and disease caused by *Ancylostoma ceylanicum*. Vaccine 23:3123–3130

Mohammed-Ali Q, Nasr-Eldin MAE, Abdelhameed AA, Mergani A, Rahoud S, Elagib KE, Saeed OK, Abel L, Magzoub MMA, Dessein A (1999) Susceptibility to periportal (Symmers) fibrosis in human *Schistosoma mansoni* infections: evidence that intensity and duration of infection, gender and inherited factors are critical in disease progression. J Infect Dis 180:1298–1306

Molyneux DH, Hotez PJ, Fenwick A (2005) Rapid-impact interventions: how a policy of integrated control for Africa's neglected tropical diseases could benefit the poor. PLoS Med 2:e336

Montresor A, Crompton D, Gyorkos T, Savioli L (2002a) Helminth control in school-age children: a guide for managers of control programmes. World Health Organization, Geneva

Montresor A, Stoltzfus RJ, Albonico M, Tielsch JM, Rice AL, Chwaya HM, Savioli (2002b) Is the exclusion of children under 24 months from anthelmintic treatment justifiable? Trans R Soc Trop Med Hyg 96:197–199

Montresor A, Awasthi S, Crompton DW (2003) Use of benzimidazoles in children younger than 24 months for the treatment of soil transmitted helminthiasis. Acta Trop 86:223–232

Moraes LRS, Azevedo Cancio J, Cairncross S (2004) Impact of drainage and sewerage on intestinal nematode infections in poor urban areas in Salvador, Brazil. Trans R Soc Trop Med Hyg 98:197–204

Morgan DO, Jarnagin K, Roth RA (1986) Purification and characterization of the receptor for insulin-like growth factor I. Biochemistry 25:5560–5564

Mupfasoni D, Karibushi B, Koukounari A, Ruberanziza E, Kaberuka T, Kramer MH (2009) Polyparasite helminth infections and their association to anaemia and undernutrition in Northern Rwanda. PLoS Negl Trop Dis 3:e517

Murray CJ, Lopez AD (1994) Quantifying disability: data, methods and results. Bull World Health Organ 72:481–494

Murray CJ, Lopez AD, Jamison DT (1994) The global burden of disease in 1990: summary results, sensitivity analysis and future directions. Bull World Health Organ 72:495–509

Mwangi TW, Bethony J, Brooker S (2006) Malaria and helminth interactions in humans: an epidemiological viewpoint. Ann Med Trop Parasitol 100:551–570

Nagpal K, Abboudi M, Manchanda C, Vats A, Sevdalis N, Bicknell C, Vincent C, Moorthy K (2013) Improving postoperative handover: a prospective operational study. Am J Surg 206:494–501

Napier L (1994) Filariasis due to *Wuchereria bancrofti*. Medicine 23:149–180

Nematode and Neglected Genomics (2008) *Wuchereria bancrofti*. Accessed 22 Feb 2010 at http://www.nematodes.org/fgn/pnb/wuchban.html

Northrop-Clewes CA, Rousham EK, Mascie-Tayor CGN, Lunn PG (2001) Anthelmintic treatment of rural Bangladeshi children: effect on host physiology, growth and biochemical status. Am J Clin Nutr 73:53–60

Nunesmaia HG, Azevedo ES, Arandas EA, Widmer CG (1975) Racial composition and ahaptoglobinemia in carries of the hepatosplenic form of *Schistosomiasis mansoni*. Revesta do Instuto de Medicina Tropical de Sao Paulo 17:160–163

Ogunrinade A, Boakye D, Merriweather A, Unnasch TR (1999) Distribution of the blinding and nonblinding strains of *Onchocerca volvulus* in Nigeria. J Infect Dis 179:1577–1579

Ohta N, Waikagul J (2007) Disease burden and epidemiology of soil-transmitted helminthiases and schistosomiasis in Asia: the Japanese perspective. Trends Parasitol 23:30–35

Ottesen EA (1992) Infection and disease in lymphatic filariasis- an immunological perspective. Parasitology 104:71–79

Palmer DR, Bradley M, Bundy DA (1996) IgG4 responses to antigens of adult *Necator americanus*: potential for use in large-scale epidemiological studies. Bull World Health Organ 74:381–386

Pasricha SR, Caruana SR, Phuc TQ, Cassey GJ, Jolley D, Kingsland S (2008) Anemia, iron deficiency, meat consumption and hookworm infection in women of reproductive age in northwest Vietnam. Am J Trop Med Hyg 78:375–381

PATH (2016) Soil-transmitted helminths: the business case for new diagnostics. PATH, Seattle

Petro M, Iavu K, Minocha A (2005) Unusual endoscopic and microscopic view of *Enterobius vermicularis*: a case report with a review of the literature. South Med J 98:927–929

Phiri K, Whitty CJ, Graham SM, Ssembatya-Lule G (2000) Urban/rural differences in prevalence and risk factors for intestinal helminth infection in southern Malawi. Ann Trop Med Parasitol 94:381–387

Porter RS, Kaplan JL (1999) The Merck manual of diagnosis and therapy, 17th edn. White house station, Kenilworth

Prata A (1992) Influence of the host related factors in the development of the hepatosplenic form of schistosomiasis mansoni. Mem Inst Oswaldo Cruz 87:39–44

Pritchard DI, Quinnell RJ, Walsh EA (1995) Immunity in humans to *Necator americanus*: IgE, parasite weight and fecundity. Parasite Immunol 17:71–75

Prost A (1987) L'ascaridiose en Afrique de l'ouest. Revue épidémiologique. Annals de Parasitologie Humaine et Comparée 62:434–455

Pullan R, Brooker S (2008) The health impact of polyparasitism in humans: are we underestimating the burden of parasitic diseases? Parasitology 135:783–794

Quinnell RJ, Woolhouse ME, Walsh EA, Pritchard DI (1995) Immunoepidemiology of human necatoriasis: correlations between antibody responses and parasite burdens. Parasite Immunol 17:313–318

Raisanen M (1985) Studies on the synthesis and excretion of bufotenine and N, N-dimethyltryptamine in man. Academic dissertation, University of Helsinki, Helsinki

Raschid-Sally L, Jayakody P (2008) Drivers and characteristics of wastewater agriculture in developing countries: results from a global assessment, Colombo, Sri Lanka', IWMI Research Report 127. International Water Management Institute, Colombo

Raso G, Luginbühl A, Adjoua CA, Tian-Bi NT, Silue KD, Matthys B, Vounatsou P, Wang Y, Dumas M-E, Holmes E, Singer BH, Tanner M, N'Goran EK, Utzinger J (2004) Multiple parasite infections and their relationship to self-reported morbidity in a community of rural Côte d'Ivoire. Int J Epidemiol 33:1092–1102

Ribera JM, Grietens KP, Toomer E, Hausmann-Muela S (2009) A word of caution against the stigma trend in neglected tropical disease research and control. PLoS Negl Trop Dis 3:e445

Roberts L, Janovy J (2000) Foundations for parasitology. McGraw Hill Companies, Boston

Roche M, Layrisse M (1966) The nature and causes of "hookworm anemia". Am J Trop Med Hyg 15:1030–1100

Romstad A, Gasser RB, Nansen P, Polderman AM, Chilton NB (1998) *Necator americanus* (Nematoda: Ancylostomatidae) from Africa and Malaysia have different ITS-2 rDNA sequences. Int J Parasitol 28:611–615

Salomon JA, Wang J, Freeman MK, Vos T, Flaxman AD, Lopez AD, Murray CJL (2012) Healthy life expectancy for 187 countries, 1990–2010: a systematic analysis for the global burden disease study 2010. Lancet 380:2144–2162. https://doi.org/10.1016/S0140-6736(12)61690-0

Santos FL, Cerqueira EJ, Soares NM (2005) Comparison of the thick smear and Kato-Katz techniques for diagnosis of intestinal helminth infections. Laboratório de Parasitologia Clínica, Faculdade de Farmácia, Universidade Federal da Bahia, Salvador, BA. Revista da Sociedade Brasileira de Medicina Tropical 38:196–198

Schierenberg E (1997) Nematodes: the roundworms. In: Gilbert SF, Raunio AM (eds) Embryology: constructing the organism. Sinauer Associates, Sunderland, pp 131–147

Scrimshaw N, SanGiovanni JP (1997) Synergism of nutrition, infection and immunity: an overview. Am J Clin Nutr 66:464–477

Secor WE (2012) The effects of schistosomiasis on HIV/AIDS infection, progression and transmission. Curr Opin HIV AIDS 7:254–259

Siddiqui AA, Berk SL (2001) Diagnosis of *Strongyloides stercoralis* infection. Clin Infect Dis 33:1040–1047

Smillie WG, Augustine DL (1925) Intensity of hookworm infestation in Alabama. Its relationship to residence, occupation, age, sex and race. J Am Med Assoc 85:1958–1963

Smolyakov R, Talalay B, Yanai-Inbar I, Pak I, Alkan M (2003) *Enterobius vermicularis* infection of female genital tract: a report of three cases and review of the literature. Eur J Obstet Gynecol Reprod Biol 107:220–222

Smyth JD (1994) Introduction to animal parasitology, 3rd edn. Cambridge University Press, Cambridge/New York, pp 50–95

Sobsey MD (2002) Managing water in the home: accelerated health gains from improved water supply. The World Health Organisation (WHO/SDE/WSH/02.07), Geneva

Steinmann P, Du ZW, Wang LB, Wang XZ, Jiang JY, Li LH, Marti H, Zhou XN, Utzinger J (2008) Extensive multiparasitism in a village of Yunnan Province, People's Republic of China, revealed by a suite of diagnostic methods. Am J Trop Med Hyg 78:760–769

Steinmann P, Usubalieva J, Imanalieva C, Minbaeva G, Stefiuk K (2010) Rapid appraisal of human intestinal helminth infections among schoolchildren in Osh oblast. Kyrgyzstan Acta Trop 116:178–184

Stephenson LS, Latham MC, Kurz KM, Kinoti SM, Brigham H (1989) Treatment with a single dose of albendazole improves growth of Kenyan school children with hookworm, *Trichuris trichiuria* and *Ascaris lumbricoides* infections. Am J Trop Med Hyg 41:78–87

Stephenson LS, Latham MC, Ottesen EA (2000) Malnutrition and parasitic helminth infections. Parasitology 121:23–38
Stoltzfus RJ, Dreyfuss ML, Chwaya HM, Albonico M (1997) Hookworm control as a strategy to prevent iron deficiency. Nutr Rev 55:223–232
Stoltzfus RJ, Chwaya HM, Montresor A, Tielsch JM, Jape JK, Albonico M, Savioli L (2004) Low dose daily iron supplementation improves iron status and appetite but not anemia whereas quarterly anthelminthic treatment improves growth, appetite and anemia in Zanzibari preschool children. J Nutr 134:2
Supali T, Verweij JJ, Wiria AE, Djuardi Y, Hamid F, Kaisar MM, Wammes LJ, van Lieshout L, Luty AJ, Sartono E, Yazdanbakhsh M (2010) Polyparasitism and its impact on the immune system. Int J Parasitol 40:1171–1176
Tavares-Neto J, Prata A (1989) Family occurrence of schistosomal hepatosplenomegaly and maternal effect. Rev Soc Bras Med Trop 22:13–18
Trouiller P, Olliaro P, Torreele E, Orbinski J, Laing R, Ford N (2002) Drug development for neglected diseases: a deficient market and a public health policy failure. Lancet 359:2188–2194
Udonsi JK, Nwosu ABC, Anya AO (1980) *Necator americanus*: population structure, distribution and fluctuations in population densities of infective larvae in contaminated farmlands. Z Parasitenkd 63:251–259
Uga S, Matsuo J, Kimura D, Rai SK, Koshino Y, Igarashi K (2000) Differentiation of *Toxocara canis* and *T. cati* eggs by light and scanning electron microscopy. Vet Parasitol 92:287–294
Utzinger J, de Savigny D (2006) Control of neglected tropical diseases: integrated chemotherapy and beyond. PLoS Med 3:e112
Utzinger J, Keiser J (2004) Schistosomiasis and soil-transmitted helminthiasis: common drugs for treatment and control. Expert Opin Pharmacother 5:263–285
Utzinger J, Becker SL, Knopp S, Blum J, Neumayr AL (2012) Neglected tropical diseases: diagnosis, clinical management, treatment and control. Swiss Med Wkly 142:w13727
van Eijk AM, Lindblade KA, Odhiambo F, Peterson E, Rosen DH, Karanja D (2009) Geohelminth infections among pregnant women in rural western Kenya; a cross-sectional study. PLoS Negl Trop Dis 3:e370
Vandemark LM, Jia T-W, Zhou X-N (2010) Social science implications for control of helminth infections in Southeast Asia. Adv Parasitol 73:137–170
Villamizar E, Mendez M, Bonilla E, Varon H, de Onatra S (1996) *Ascaris lumbricoides* infestation as a cause of intestinal obstruction in children: experience with 87 cases. J Paediatr Surg 31:201–204
Weiss M (2008) Stigma and the social burden of neglected tropical diseases. PLoS Negl Trop Dis 2:e237
Williams D, Burke G, Hendley JO (1974) Ascariasis: a family disease. J Pediatr 84:853–854
World Health Organization (2006) Schistosomiasis and soil-transmitted helminth infections-preliminary estimates of the number of children treated with albendazole or mebendazole. World Health Organization, Geneva, pp 145–164
World Health Organization (2012) Wkly Epidemiol Rec 87:17–28
World Health Organization (2015) Investing to overcome the global impact of neglected tropical diseases. Third WHO report on neglected tropical diseases. WHO Press, Geneva, p 190
World Health Organization (2017) Wkly Epidemiol Rec 92:749–760.
Xue HC, Wang Y, Xiao SH, Liu S, Wang Y, Shen GJ, Wu WT, Zhan B, Drake L, Feng Z, Hotez PJ (2000) Epidemiology of human ancylostomiasis among rural villagers in Nanlin County (Zhongzhou Village), Anhui Province, China: II. Seroepidemiological studies of the age relationships of serum antibody levels and infection status. Southeast Asian J Trop Med Public Health 31:736–741
Zaccone P, Fehervari Z, Phillips JM, Dunne DW, Cooke A (2006) Parasitic worms and inflammatory diseases. Parasite Immunol 28:515–523
Ziegelbauer K, Speich B, Mäusezahl M, Bos R, Keiser J, Utzinger J (2012) Effect of sanitation on soiltransmitted helminth infection: systematic review and meta-analysis. PLoS Med 9:el001162
Zygmunt DJ (1990) Strongyloides stercoralis. Infect Control Hosp Epidemiol 11:495–497.

Chapter 9
Neglected Tropical Diseases: A Biosocial Perspective

Abhiruchi Galhotra and Abhisek Mishra

Abstract Neglected tropical diseases (NTDs) are a medically diverse group of communicable diseases that prevail in tropical and subtropical climates, globally. The World Health Organization has identified 18 diseases as NTDs. These diseases flourish under conditions of poverty and filth, where housing is substandard, drinking water is unsafe, sanitation is poor, insect vectors are rampant, and there is restricted or nearly nonexistent access to health care. NTDs are disablers rather than killers and are a proxy for poverty and disadvantage. The biological and medical diversity of NTDs signifies the fact that their control or elimination strategies also need to be very diverse. There is a necessity of multiple approaches and techniques for control and elimination, which include specialized drugs, preventive chemotherapy, mass drug administration, and vector control (which limit or eradicate insects, e.g., flies and bugs, which transmit the infectious pathogens). Despite an encouraging progress to tackle the menace of NTDs, a large number of people still need high-quality free treatments, care, and much more. There is a need to build capacity of health-care providers in order to sustain implementation of efficient control programs.

Keywords Neglected tropical diseases (NTDs) · Poverty · Social stigma · Sustainable Development Goals (SDGs)

9.1 What Are NTDs?

Neglected tropical diseases (NTDs) are a medically diverse group of communicable diseases that prevail in tropical and subtropical climates. Globally, these diseases affect more than one billion people and cost developing economies billions of dollars every year (Molyneux 2010; Hotez et al. 2009). The various international

A. Galhotra (✉)
Department of Community and Family Medicine, All India Institute of Medical Sciences, Raipur, India

A. Mishra
Department of Community and Family Medicine, All India Institute of Medical Sciences, Patna, India

P. P. Singh (ed.), *Infectious Diseases and Your Health*,
https://doi.org/10.1007/978-981-13-1577-0_9

agencies have defined the NTDs differently. The World Health Organization (WHO) defines NTDs as "Chronically endemic and epidemic-prone tropical diseases," which have an exceptionally large negative impact on the lives of poor populations and remain critically neglected in the global public health agenda (WHO 2007). According to the Global Network for Neglected Tropical Diseases (2017), NTDs are defined as follows: "The neglected tropical diseases are a group of 13 parasitic and bacterial infections that affect over 1.4 billion individuals, majority of whom live on less than US $1.25 per day." The Public Library of Science Neglected Tropical Diseases defines NTDs as "a group of poverty-promoting chronic infectious diseases, which primarily occur in rural areas and poor urban areas of low-income and middle-income countries" (Madon et al. 2014). They are poverty-promoting because of their impact on child health and development, pregnancy, and worker productivity as well as their stigmatizing features. According to the United States Agency for International Development, "These diseases disproportionately impact the poor and rural populations, who lack access to safe water, sanitation, and essential medicines" (Armah et al. 2015). They (NTDs) cause disorder and disability, compromise children's mental and physical development, and result in blindness and severe disfigurement.

The World Health Organization (WHO) has identified 18 diseases as NTDs, viz., Buruli ulcer, Chagas disease, dengue and chikungunya, dracunculiasis (Guinea worm disease), echinococcosis, foodborne trematodiases, human African trypanosomiasis (sleeping sickness), leishmaniasis, leprosy, lymphatic filariasis, onchocerciasis (river blindness), rabies, schistosomiasis, soil-transmitted helminthes, taeniasis/cysticercosis, trachoma, yaws (endemic treponematoses), and mycetoma. Geographical overlap patterns between the highest-prevalence NTDs (soil-transmitted helminths, schistosomiasis, onchocerciasis and lymphatic filariasis, and trachoma), and malaria, HIV are seen, exhibiting a high degree of coinfection. New evidence highlights that NTDs can affect the disease progression of HIV and AIDS, tuberculosis (TB), and malaria. A complex combination of epidemiological, immunological, and clinical factors sums up to these interactions and add to a worsening prognosis for people affected by HIV/AIDS, TB, and malaria. Soil-transmitted helminthic infections have had a long hidden contributing effect on the AIDS epidemic (Simon 2016).

Dengue: A viral infection where mosquito acts as a vector causing flu-like illness that may develop into severe dengue and have deadly consequences.

Rabies: A preventable viral illness transmitted to humans through the bites of infected canines, which is perpetually deadly once symptoms develop.

Trachoma: An infection caused by *Chlamydia*, transmitted through direct contact with infectious eye or nasal discharge or through indirect contact with unsafe living conditions and hygiene practices, which, if left untreated, causes irreversible corneal opacities and blindness.

Buruli ulcer: Mycobacterial skin infection often debilitating causing severe destruction of the skin, bone, and soft tissue.

Yaws: A chronic bacterial infection which affects mainly the skin and bone, caused by a bacteria.

Leprosy: A complex disease caused by bacterial infection primarily affecting the skin, peripheral nerves, mucosa of the upper respiratory tract, and eyes.
Chagas disease: A life-threatening disease transmitted to humans through contact with vector insects (triatomine bugs), ingestion of contaminated food, infected blood transfusions, congenital transmission, organ transplantation, and laboratory accidents.
Human African trypanosomiasis (sleeping sickness): A parasitic illness spread by the bites of tsetse flies which is nearly 100% fatal without prompt diagnosis and treatment to prevent the parasitic invasion into the central nervous system.
Leishmaniasis: A disease transmitted through the bites of infected female sandflies, which in its most severe (visceral) form invades the internal body parts and in its most common (cutaneous) form causes ulcers on face, disfiguring scars, and disability.
Taeniasis and neurocysticercosis: This disease is caused by adult tapeworms in human intestines; cysticercosis results when humans ingest tapeworm eggs which hatch as larvae in host tissues.
Dracunculiasis (Guinea worm disease): A nematode infection transmitted solely by drinking water contaminated with parasite-infected water.
Echinococcosis: This infection is due to the larval stages of tapeworms which later form pathogenic cysts in humans and transmitted following ingestion of eggs predominately shed in the feces of dogs and wild animals.
Foodborne trematodiases: Consumption of fish, vegetables, and crustaceans contaminated with larval parasites causes it.
Lymphatic filariasis: Infection transmitted by mosquitoes which causes abnormal enlargement of appendages and genitals due to blockade of lymphatic flow. The adult worms inhabit and reproduce in the lymphatic system.
Onchocerciasis (river blindness): Infected blackflies transmit this infection. Some of the symptoms are severe itching and eye lesions, as the adult worm produces larvae leading to visual impairment and permanent blindness.
Schistosomiasis: This trematodal disease is transmitted to human when larval forms are released by freshwater snails which penetrate human skin during contact with infested water.
Soil-transmitted helminthiases: These nematode infections are transmitted through soil contaminated by human feces. They cause anemia, vitamin A deficiency, stunted growth, malnutrition, intestinal obstruction, and impaired development.
Mycetoma: A chronic, progressively destructive inflammatory skin ailment which commonly affects the lower limbs. Infection is thought to be caused by the inoculation, through a thorn prick or skin damage, of fungi or bacteria into the subcutaneous tissue.

The 10th meeting of the Strategic and Technical Advisory Group (in 2017) for Neglected Tropical Diseases received proposals for the addition of diseases, and, pursuant to the required procedures, chromoblastomycosis, other deep mycoses, scabies and other ectoparasites, and snakebite envenoming have been added to the NTD portfolio.

The transmission patterns of NTDs are also diverse and can take place via:

- Flies, fomites (e.g., skin cells, hair, clothing, or bedding), and fingers (trachoma)
- Mosquitoes (dengue fever and filariasis)
- Tsetse flies (sleeping sickness)
- Sandflies (leishmaniasis)
- Blackflies (onchocerciasis)
- Snails, which release infective larvae into water, which, in turn, penetrate human skin (e.g., schistosomiasis)
- Feco-oral route (e.g., soil-transmitted helminths) or via food products

9.2 Who Is Affected by NTDs?

Individuals living in poverty, without adequate sanitation, and in proximity with infectious vectors, domestic animals, and livestock are chiefly affected. These diseases grow vigorously under conditions of poverty and filth, where housing is substandard, drinking water is unsafe, sanitation is poor, insect vectors are rampant, and access to health care is restricted or nonexistent (Feasey and Wansbrough-Jones 2010). Neglected tropical diseases cause immense human suffering and death. NTDs are a serious deterrent to poverty reduction and socioeconomic development (Fig. 9.1).

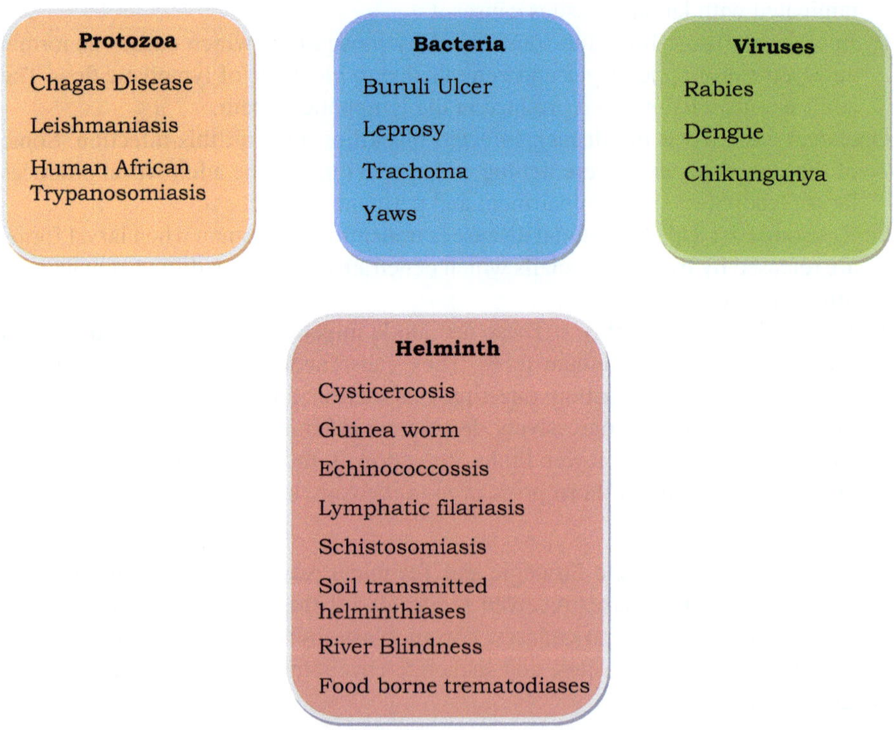

Fig. 9.1 Infectious agents responsible for NTDs

Globally, 149 nations and territories are affected by at least one NTD. There is a high potential for NTD risk globally, as over 40% of the human population (about three billion people) currently live in the tropics. There is additionally a high potential for an increase in NTD prevalence, particularly, in the absence of enhanced public health efforts (Gubler 2002; Dujardin et al. 2008). These diseases kill an estimated 534,000 individuals worldwide every year (Neglected Tropical Disease WHO 2009). NTDs are a noteworthy reason of disease burden, resulting in approximately 57 million years of life lost due to premature disability and death. Individuals are often afflicted with more than one parasite or infection. Treatment cost for most NTD mass drug administration programs is estimated at less than US 50 cents per person per year (Neglected Tropical Diseases-Fast facts CDC 2017. para 1).

9.3 Why the Term Neglected for These Diseases?

NTDs are a proxy for poverty and disadvantage (World Bank Report 2012). The neglected tropical diseases have their hotbeds in the places left farthest behind by financial advance, where substandard housing, lack of access to safe and clean drinking water and sanitation, chronic hunger, filthy environments, and abundant insects and other vectors contribute to their efficient transmission (Hotez et al. 2009).

The NTDS have a strong presence in rural, remote, vulnerable, and marginalized populations. They are found most in populations who have limited access to sustainable, adequate, and affordable water supply and sanitation facilities. The poor territories, therefore, endure the maximum burden of multiple NTDs at any given time (WHO 2015). HIV, malaria, and tuberculosis have been in focus because of their higher mortality. There has been an underestimation of the contribution of NTDs to mortality due to the asymptomatic and long incubation period that is characteristic of many of these diseases and a lack of interest in developing (nonprofitable) treatments by pharmaceutical companies. Therefore, the term neglected is used for these diseases. Historically, the NTDs have been connected with witchcraft and other beliefs such as bad behavior, and these diseases are seen as divine punishment. The NTDs usually result in stigma, silence, and hiding, with people usually resorting to traditional healers and other forms of informal treatments and self-treatment.

9.4 What Differentiates NTDs from Non-neglected Diseases?

It is a fact that NTDs are disablers instead of killers. NTD infections are co-endemic; an individual might be infected with more than one NTD in addition to other diseases such as HIV, tuberculosis, and malaria. For example, the parasitic infection with schistosomiasis increases susceptibility in females to HIV infection. This parasite also saps micronutrients and iron from developing children to stunt their growth, thereby contributing to increased absenteeism in schools. Lymphatic filariasis (a chronic helminth parasitic infection) causes severe swelling (lymphedema) in populations affected, rendering them socially stigmatized and unfit to work. Various other NTDs can be portrayed by chronic disabilities, increased susceptibility to infectious and noninfectious diseases, social stigma, and an economic burden on the individual, the family, and the country (Cotton 2014; Perera et al. 2007; Weiss 2008).

This biological and medical diversity of NTDs signifies the fact that the control or elimination strategies also need to be very diverse. They require different approaches and strategies for control/elimination, which include specialized medicines, preventive chemotherapy, mass drug administration, and vector control (which limit or eradicate insects, e.g., flies and bugs, that transmit the infectious pathogens).

9.5 Why Are NTDs Receiving Increased International Recognition?

As the NTDs have wide range of medical signs and symptoms and transmission patterns, it is challenging to concentrate the world's attention on these very diverse diseases which required an equally diverse range of different interventions. Though medically diverse, NTDs have certain characteristics in common, and some of them are given below:

- They affect the poorest of poor – those with limited access to safe and clean drinking water, proper sanitation, and even basic health services.
- High-income groups are rarely affected.
- Most of NTDs are chronic, gradually progressive conditions which become worse if undetected and untreated and cause an irreversible damage.
- They cause severe pain, agony and lifelong disabilities, with long-term consequences for the person and their caregivers.
- People with NTDs are often stigmatized and excluded from society, which might affect their psychological well-being (Molyneux 2013).

Therefore, by recognizing what NTDs have in common, and grouping them together under the NTD "brand," it became conceivable to convince the concerned parties for an international action. This was supported by good evidence: that addressing NTDs is cost-effective in terms of economic rates of returns on investment of health dollars, leading to "More health, for more people, for fewer dollars." Further, the NTDs have a social and economic impact, which, in turn, has significant bearing on equity, equality, and development issues. These issues fall within the mandate of development organizations accordingly justifying both technical and financial support (Molyneux 2004).

9.6 The Social and Economic Impact of NTDs

It is now well accepted fact that prevalence of NTDs is a strong and effective indicator of poverty. These diseases are an after effect of destitution, and they thus contribute to further poverty in affected people (Durrheim et al. 2004; WHO 2012). The social and economic impacts of NTDs are:

- Loss of ability to undertake traditional farming practices, which is the main source of livelihood in rural areas, especially those thriving on agrarian economy.
- Loss of capacity to play an economic and social role(s) within the family and community.
- Due to lack of awareness and inaccessibility of proper treatment for NTDs, people usually go in for inappropriate treatment (e.g., traditional healers), which augments the cycle of poor health and poverty.
- Losses of educational opportunities, as children are compelled to act as caregivers for their parents, which creates a generation of people with little or no education.
- Poor mental health of the patient and the caregiver, particularly chronic depression.
- Women typically tend to have poorer access to health care than men (Courtright and Lewallen 2009). When women become ill, they are less able to do work such as growing vegetables, fetching water and fuel, providing care for older people and children, and ensuring that family members wash their hands or wear shoes which reduce the transmission of NTDs. Hence, the impact of NTDs on the unpaid work provided by women in the community is difficult to quantify/measure.

9.7 How to Reduce the Burden of NTDs?

WHO recommends an integrated approach to overcoming the global impact of NTDs through following five interventions:

- Innovative and intensified disease management
- Preventive chemotherapy
- Vector ecology and management
- Veterinary public health services
- Provision of safe water, sanitation, and hygiene

There is a need to coordinate and bolster approaches and strategies to enhance access to interventions for the prevention, diagnosis, treatment, care, control, elimination, and eradication of NTDs, including zoonotic diseases to all those in need. Further, there is a need to strengthen and maintain national control programs with clearly defined responsibilities in order to coordinate essential functions such as situation analysis, strategic planning, budgeting, prevention, diagnosis, treatment,

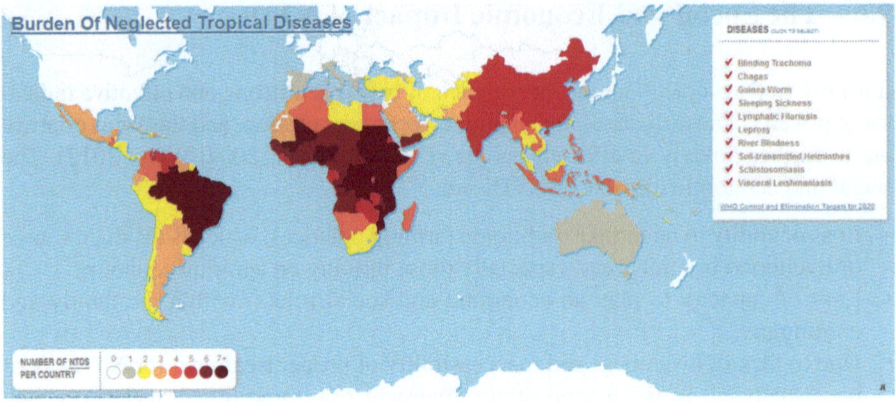

Fig. 9.2 Burden of Neglected Tropical Diseases globally. (Source -https://neglectedtropicaldiseases.wordpress.com/category/uncategorized/page/2/)

surveillance, and capacity building, timely distribution of medicines, and supervision of tasks at all levels of the national health systems. The results are more effective when these approaches are combined and delivered locally (Neglected Tropical Diseases 2017) (Fig. 9.2).

9.8 What Is Being Done to Control NTDs?

The first NTD global partner meeting was convened by WHO in 2007, with commitments from member states to donate large quantities of high-quality medicines to suppress common tropical parasitic and bacterial infections. Since 2007, various local and international partners have worked alongside ministries of health in endemic countries to deliver quality-assured medicines and provide individuals with care and long-term management. In 2012, partners embraced a WHO NTD roadmap, which committed additional support and resources to eliminate ten of the most common NTDs (WHO 2012).

The first WHO report on NTDs (2010) set the scene by exhibiting proof for how these interventions had produced results. The second report (2013) assessed the progress made in deploying them and detailed the obstacles to their implementation (WHO 2013). The third report (Investing to overcome the global impact of neglected tropical diseases) analyzed for the first time the investments needed (finances needed) to accomplish the targets of the targets of the WHO Roadmap on NTDs and universal coverage against NTDs. WHO's fourth global report on neglected tropical diseases – Integrating neglected tropical diseases into global health and

Table 9.1 Major milestones regarding steps taken by WHO for the elimination of NTDs

2017 – WHO hosts 2nd Global Partners Meeting amid unprecedented progress. Pledges of more than US$ 800 million dollars are made over the next 5–7 years to accelerate the elimination and eradication of NTDs
2015 – NTD interventions reach almost a billion of the poorest people
2013 – WHO's decision-making body – the World Health Assembly – passes Resolution 66.12 calling on countries to accelerate interventions to eliminate NTDs
2012 – WHO publishes the NTD Roadmap with 2015 and 2020 targets. Inspired by the Roadmap, partners endorse the London Declaration which committed renewed support to eliminating 10 NTDs
2010 – WHO publishes its first global NTD report, triggering renewed and additional medicine and in-kind donations
2008 – WHO publishes its Global Plan to Combat Neglected Tropical Diseases 2008–2015
2007 – WHO hosts the first Global Partners' Meeting, heralding a close multi-stakeholder collaboration

development – was launched to celebrate "Collaborate. Accelerate. Eliminate" in April 2017 ("Uniting to Combat" 2016). Due to limited financial resources, ineffective surveillance, disruptive conflicts, and other barriers to access the needed health services, the NTD programs continue to struggle. The need of the hour is integrated approaches (coordinated methodologies) for simplification, cost-effectiveness, and streamlined efficiency ("World Health assembly adopts resolution on NTDs" 2013) (Table 9.1).

9.9 NTDs and Sustainable Development Goals (SDGs)

NTDs have wide cross-cutting and cross-sectoral linkages and effects. An effort to mitigate the impact of NTDs provides a chance to mitigate poverty and to have a direct bearing on the achievements of SDGs. The NTDs have the greatest relevance for SDG 3 (the Health Goal). NTDs influence and are influenced by many of other developmental areas covered in the agenda 2030.

The SDG 1 targets the ending of poverty in all its forms everywhere; NTD programs can go a long way toward achieving that target as NTDs are considered as proxy indicators of poverty and disadvantage. The SDG 2 (Zero Hunger), 4 (Quality Education), 6 (Clean Water and Sanitation), 11 (Sustainable Cities and Communities), and 17 (Partnerships for the Goals) are in consonance with the NTD programs. Therefore it is important to integrate the NTD control programs into the broader health systems, based on principles of Universal Health Coverage which is the focus of SDG health agenda (Fig. 9.3).

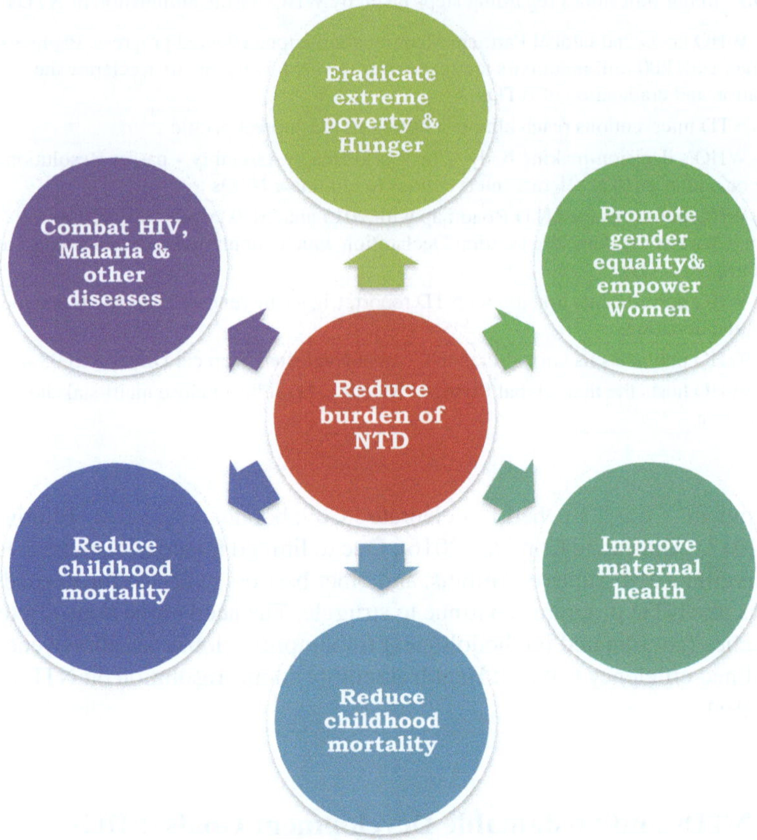

Fig. 9.3 Integrated approach to overcome the impact of NTDs through five interventions as per WHO

9.10 Conclusions

NTDs handicap, incapacitate, and propagate cycles of poverty, school dropouts, parents out of work, and dampening hope of any chance of an economic future. Medical interventions alone cannot solve the problem of NTDs. A broader attack on the social, environmental, and economic determinants of health is needed. Ignorance is the primary fight that must be fought in this war. Community engagement has a noteworthy potential to create grassroots demand for treatment and reduce stigma because these diseases are so deeply feared by affected populations. Despite an encouraging progress to tackle the menace of NTDs, millions of people still need free high-quality treatments, and millions more still need care and treatment. There is a need to build capacity of health-care providers in order to sustain implementation of efficient control programs. Strengthened epidemiological surveillance systems should be in place for effective monitoring and evaluation of the programs. The health systems need to develop beyond the health center for the successful integration of NTDs into the mainstream health care. The financial resources should be coordinated through effective plans and budgets at national and district levels. To succeed in the elimination of NTDs, we need to look beyond mass drug administration to the removal of the primary risk factors for NTDs (i.e., poverty and exposure) by ensuring access to safe and clean drinking water, basic sanitation, improvement in vector control, integration of the NTDs into poverty reduction schemes and vice versa, and building stronger and equitable health systems in endemic areas.

References

Armah FA, Quansah R, Luginaah I, Chuenpagdee R, Hambati H, Campbell G (2015) Historical perspective and risk of multiple neglected tropical diseases in coastal Tanzania: compositional and contextual determinants of disease risk. PLoS Negl Trop Dis 9:e0003939

Cotton R (2014) Neglected tropical diseases: progress and challenges for the post 2015 development era: special edition on infectious disease. Retrieved from sitn.hms.harvard.edu/flash/special-edition-on-infectiousdisease/2014/neglected-tropical-diseases. Progress and challenges

Courtright P, Lewallen S (2009) Why are we addressing gender issues in vision loss? Community Eye Health J 22:17–19

Dujardin J-C, Campino L, Cañavate C, Dedet J-P, Gradoni L, Soteriadou K et al (2008) Spread of vectorborne diseases and neglect of Leishmaniasis. Eur J Emerg Infect Dis 14:1013–1018

Durrheim DN, Wynd S, Liese B, Gyapong JO (2004) Lymphatic filariasis endemicity: an indicator of poverty? Trop Med Int Health 8:43–45

Feasey N, Wansbrough-Jones M, Mabey DCW, Solomon AW (2010) Neglected tropical diseases. Br Med Bull 93:179–200. Retrieving from. https://doi.org/10.1093/bmb/ldp046

Global Network for Neglected Tropical Diseases (2017) About neglected tropical diseases. Retrieved from http://gnntdc.sabin.org/about-ntds

Gubler DJ (2002) Epidemic dengue/dengue hemorrhagic fever as a public health, social and economic problem in the 21st century. Trends Microbiol 10:100–103

Hotez PJ, Fenwick A, Savioli L, Molyneux DH (2009) Rescuing the bottom billion through control of neglected tropical diseases. Lancet 373:1570–1575. Retrieved from. https://doi.org/10.1016/S0140-6736(09)60233-6

Madon S, Amaguru JO, Malecela MN, Michael E (2014) Can mobile phones help control neglected tropical diseases? Experiences from Tanzania. Soc Sci Med 102:103–110 pmid:24565147

Molyneux DH (2004) "Neglected" diseases but unrecognized successes-challenges and opportunities for infectious disease control. Lancet 364:380–383

Molyneux DH (2010) Neglected tropical diseases-beyond the tipping point? Lancet 375:3–4

Molyneux D (2013) Neglected tropical diseases. Community Eye Health J 26:21–24

Neglected tropical Diseases Definition in Public Library of Sciences (PLOS) Neglected Tropical Diseases (2017) Retrieved from http://www.plosntds.org/static/scope.action

Neglected tropical Diseases Fast Facts-Centers for Disease Control and Prevention (2017) Retrieved from https://www.cdc.gov/globalhealth/ntd/fastfacts.html

Neglected Tropical Diseases. World Health Organization (2017) Neglected tropical diseases. (n.d.) in neglected tropical diseases WHO report on neglected tropical diseases. Retrieved from http://www.who.int/neglected_diseases/en/

Perera M, Whitehead M, Molyneux D, Weerasooriya M, Gunatilleke G (2007) Neglected patients with a neglected disease? A qualitative study of lymphatic filariasis. PLoS Negl Trop Dis 1:e128

Simon GG (2016) Impacts of neglected tropical disease on incidence and progression of HIV/AIDS, tuberculosis, and malaria: scientific links. Int J Infect Dis 42:54–57

Uniting to combat neglected tropical disease (2016) Reaching the unreached. Fourth progress report of the London declaration. Retrieving from http://unitingtocombatntds.org/report/fourth-report-reaching-unreached

Uniting to Combat Neglected Tropical Diseases: Translating the London Declaration into Action (2012) Meeting report. The World Bank, Washington, DC Retrieved from www.unitingtocombatntds.org

Weiss MG (2008) Stigma and the social burden of neglected tropical diseases. PLoS Negl Trop Dis 2:e237

World Health Organization (2012) Accelerating work to overcome the global impact of neglected tropical diseases: a roadmap for implementation. Geneva. Retrieved from www.who.int/neglected_diseases/NTD_RoadMap_2012_Fullversion.pdf

World Health Organization (2013) Sustaining the drive to overcome the Global impact of Neglected tropical diseases, Second WHO Report on NTDs, Geneva. Retrieved from www.who.int/iris/bitstream/10665/77950/1/9789241564540_eng.pdf.

World Health Organization (2015) Water, Sanitation and Hygiene for accelerating and sustaining progress on Neglected Tropical Diseases-A global strategy, 2015–2020, Geneva, Switzerland. Retrieved from http://www.who.int/water_sanitation_health/publications/wash-and-ntd-strategy/en/

World Health Organization (2017a) Neglected tropical diseases. Summary. Retrieved from http://www.who.int/neglected_diseases/diseases/summary/en/

World Health Organization (2017b) Neglected tropical diseases – gravis learning agenda drives change for rabies. Retrieved from http://www.who.int/neglected_diseases/zoonoses/en/

World Health Organization (2017c) Integrating neglected tropical diseases into global health and development: fourth WHO report on neglected tropical diseases. Executive summary. Geneva. Retrieved from http://www.who.int/neglected_diseases/resources/9789241565448/en/

World Health Organization (2017d) Neglected tropical diseases – world health assembly approves comprehensive global approach against vector borne diseases. Retrieved from http://www.who.int/neglected_diseases/vector_ecology/en/

World Health Organization Neglected Tropical Diseases-Neglected tropical diseases Hidden success, Emerging opportunities (2009) World Health Organization, Geneva. Available at www.who.int/neglected_diseases/diseases/NTD_Report_APPMG.pdf

World Health Organization Neglected Tropical Diseases-world health assembly adopts resolution on Neglected Tropical Diseases (2013) Retrieved from www.who.int/neglected_diseases/WHA_66_seventh_day_resolution_adopted/en/

World Health Organization WHO-Global plan to combat neglected tropical diseases 2008–2015 (2007) World Health Organization, Geneva. Available at whqlibdoc.who.int/hq/2007/WHO_CDS_NTD_2007.3_eng.pdf

Part II
Bacterial Infections

Chapter 10
Autophagy: A Potential Antibacterial Therapeutic Target

Madhu Puri, Trinad Chakraborty, and Helena Pillich

Abstract Bacterial infections caused by pathogenic bacteria, like tuberculosis by *Mycobacterium tuberculosis*, listeriosis by *Listeria monocytogenes*, and gastroenteritis by *Salmonella typhimurium*, are on the rise. With the increase in pathogen resistance to antibiotics, novel approaches are required for therapeutic interventions to treat bacterial infections. Autophagy is an essential host defense mechanism against infections and, in recent times, has shown promising potential as a therapeutic target in this regard. This article reviews the role of autophagy during infection with pathogenic bacteria and recent studies which highlight the importance of autophagy as a prospective therapeutic target.

Keywords Autophagy · Bacterial infections · Therapeutic target

10.1 Introduction

Autophagy is an evolutionarily conserved cellular defense mechanism which involves the cloistering of cargo molecules (viz., damaged cellular organelles, protein aggregates, or pathogens) in a double-membrane vacuole, known as an autophagosome, which are ultimately degraded by lysosomal hydrolases. Autophagy can be triggered by a variety of factors, such as damaged cellular organelles, withdrawal of growth factors, amino acid deprivation, oxidative stress, hypoxia, endoplasmic reticulum stress, low cellular energy levels, and infection (Lin and Baehrecke 2015). The autophagy of cellular organelles and protein aggregates is an essential part of the maintenance of cellular homeostasis, whereas that of pathogens acts as a defense mechanism against infections (termed xenophagy) (Samson 1981).

M. Puri (✉)
Institute of Medical Microbiology, Justus Liebig University, Giessen, Germany

School of Life Sciences, Jawaharlal Nehru University, New Delhi, India

T. Chakraborty · H. Pillich
Institute of Medical Microbiology, Justus Liebig University, Giessen, Germany

P. P. Singh (ed.), *Infectious Diseases and Your Health*,
https://doi.org/10.1007/978-981-13-1577-0_10

Autophagy can be classified into different types. In selective autophagy, molecules called autophagy adaptors or cargo receptors recognize and bind to a specific target cargo molecule and subsequently lead to its degradation by autophagy. Nonselective autophagy involves the indiscriminate entrapment of cargo into developing autophagosomes (Moy and Cherry 2013). Autophagy adaptors like optineurin (OPTN), sequestosome 1 (SQSTM1), neighbor of BRCA1 gene 1 (NBR1), nuclear dot protein 52 (NDP52), Toll-interacting protein (Tollip), TAX1-binding protein 1 (TAX1BP1), and nuclear receptor subfamily 1, group D, member 1 (NR1D1) have been identified, and most of them have been shown to be involved in xenophagy (Bjørkøy et al. 2005; Kirkin et al. 2009; Thurston et al. 2009; Zheng et al. 2009; Dupont et al. 2009; Ogawa et al. 2011; Osawa et al. 2011; Newman et al. 2012; Khweek et al. 2013; Lu et al. 2014; Chandra et al. 2015). Autophagy can also be further divided into three categories: macroautophagy, microautophagy, and chaperone-mediated autophagy. Macroautophagy is often referred to as conventional autophagy, wherein cytoplasmic cargo is cloistered into autophagosomes, which is followed by fusion of lysosomes with autophagosomes to form autolysosomes and subsequent degradation of cargo by lysosomal hydrolases (Fig. 10.1). Microautophagy involves the uptake of cytosomal components directly

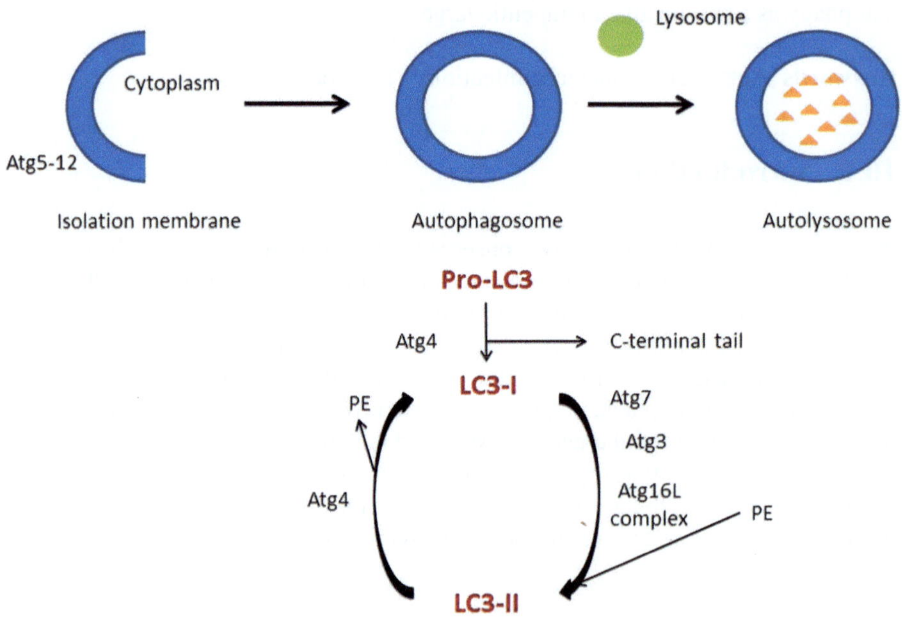

Fig. 10.1 Stages in autophagy. The process of autophagosomal membrane formation starts when the autophagy-related genes (Atg) Atg5 and Atg12 conjugate to form an isolation membrane. The membrane then envelops cargo molecules and closes to form an autophagosome. Lysosomes fuse with autophagosomes to form autolysosomes, wherein the autophagic cargo is degraded via lysosomal hydrolases. In the LC3 pathway, Atg4 cleaves the C-terminal tail of sequence of the pro-LC3 molecule to generate LC3-I (cytosolic). Atg7 activates LC3-I and is conjugated to Atg3. This Atg3-LC3-I conjugate binds to the Atg16L complex, and LC3-I binds to phosphatidylethanolamine (PE), thereby generating LC3-II (membrane-bound). The PE of LC3-II is subsequently cleaved by Atg4 to produce LC3-I. (Modified from Noda et al. 2009)

by the lysosomes by invagination, without forming intermediate autophagosomal structures. Chaperone-mediated autophagy is mediated by chaperone proteins recognized by the lysosomal membrane receptor lysosome-associated membrane protein 2A, which form a complex with cargo and are translocated across the lysosomal membrane (Glick et al. 2010).

At the molecular level, autophagy is mediated by autophagy-related genes (Atg), which are present both in yeast and mammals (Mizushima et al. 2011). The induction of autophagy is responsible for the activation of the Unc-51-like autophagy activating kinase 1 (ULK1), which, in turn, activates Beclin-1/Atg6 (Russell et al. 2013). The class III phosphatidylinositol 3-phosphate kinase Vps34 phosphorylates phosphatidylinositol to produce phosphatidylinositol 3-phosphate (PtdIns(3)P) which provides a docking site for WD-repeat protein which then interacts with phosphoinositides (WIPI) protein family (Proikas-Cezanne et al. 2015). Atg12 binds to Atg5 and then to Atg16L1 to form a complex which binds and activates Atg3 (Hanada et al. 2007). Atg3 attaches Atg8 (microtubule-associated protein 1 light chain 3 [LC3]), which is first processed by Atg4, to phosphatidylethanolamine (PE) on the surface of autophagosomes leading to the closure of autophagosomes (Fujita et al. 2008; Kirisako et al. 2000). The fusion of lysosomes with closed autophagosomes results in cargo degradation.

Autophagy is both pro- and antibacterial during infections. This review discusses the role of autophagy during bacterial infections and also if autophagy can act as a target for therapeutic interventions during bacterial infections.

10.2 Role of Autophagy in Bacterial Pathogenesis

Several pathogenic bacteria are known to induce autophagy during infection, and some have also devised various strategies to evade autophagic recognition (Table 10.1). The following sections discuss these aspects in greater detail.

Table 10.1 Bacteria and bacterial factors that are involved in the induction or evasion of autophagy

Bacterium	Bacterial factor(s)	Autophagy induction	Autophagy evasion
Listeria monocytogenes	LLO, ActA, InlK	Yes	Yes
Salmonella typhimurium	SipB, TTSS, SseL	Yes	Yes
Mycobacterium tuberculosis	EspB, EIS	Yes	Yes
Shigella flexneri	IcsB, IcsA	Yes	Yes
Legionella pneumophila	RavZ, LpSpl	Yes	Yes
Streptococcus pyogenes	SLO, NADase, SpeB	Yes	Yes
Streptococcus pneumoniae	PLY	Yes	n.d.
Pseudomonas aeruginosa	Phycocyanin, TpIE	Yes	n.d.
Francisella tularensis	dipA, O-antigen	Yes	Yes

10.2.1 Mechanisms of Autophagy Induction by Bacteria During Infection

10.2.1.1 Virulence Factors of Bacteria

Bacterial virulence factors play an essential role in mediating the recognition of pathogenic bacteria by the host autophagy machinery. The type III secretion system (TTSS) of *Salmonella typhimurium* ruptures the cytosolic compartments in which intracellular *S. typhimurium* are contained (termed as *Salmonella*-containing vacuoles [SCV]), and the entrapped bacteria are ubiquitinated and subsequently targeted by autophagy (Birmingham et al. 2006). The key virulence factor of the Gram-positive bacterium *Listeria monocytogenes* reported to be involved in autophagy induction is the pore-forming toxin listeriolysin O (LLO) (Py et al. 2007). Amino acid starvation can also be triggered by LLO-dependent phagosomal lysis during *L. monocytogenes* infection, which can result in induction of autophagy (Tattoli et al. 2013). LC3-associated phagocytosis (LAP) is also induced by LLO and facilitates the formation of spacious *Listeria*-containing phagosomes (SLAPs: LC3-positive *L. monocytogenes*-containing phagosomes). These LC3-positive single-membrane compartments allow listerial survival and their slow replication (Lam et al. 2013; Birmingham et al. 2008).

10.2.1.2 Regulation of Host Autophagy Signaling

Autophagy can also be induced when bacteria regulate host signaling pathways during infection. Autophagy is activated when amino acid starvation is triggered by the infection of epithelial cells with *S. typhimurium* (Tattoli et al. 2012). It is already established that macrophage scavenger protein apoptosis inhibitor of macrophages (AIM) enhances the mycobactericidal activity of macrophages by increasing the levels of processed LC3 form and Beclin 1 (Sanjurjo et al. 2013). It has also been reported that during infection of macrophages with *Mycobacterium tuberculosis*, the cytosolic DNA sensor cyclic GMP-AMP synthase triggers STING/TBK1/IRF3-dependent interferon production (Watson et al. 2015). Autophagy is regulated by eukaryotic microRNAs including miR-155 in macrophages. Thus, during infection with *M. tuberculosis*, miR-155 enhances bacterial elimination and, via binding to the Ras homologue enriched in brain (Rheb), a negative regulator of autophagy, accelerates autophagy (Wang et al. 2013). It is well established that infection with *L. monocytogenes* induces autophagy in host cells (Rich et al. 2003). Toll-like receptor 2 (TLR2) and Nod-like receptors 1 and 2, acting via the downstream extracellular signal-regulated kinases, have been shown to play a crucial role for autophagy in *Listeria*-infected cells (Anand et al. 2011). The role of histone deacetylase 6

(HDAC6) during *L. monocytogenes* infection has been studied recently, and HDAC6 has been reported to control innate immune and autophagy responses to TLR-mediated signaling during infection with *L. monocytogenes* (Morenzo-Gonzalo et al. 2017). Additionally, Gluschko et al. (2018) have very recently reported that the in vivo infection of macrophages by *L. monocytogenes* leads to their interaction with the β-2 integrin macrophage-1 antigen (Mac-1), which activates Nox2 and induces the production of reactive oxidation species that subsequently leads to the recruitment of LC3 to *L. monocytogenes*-containing phagosomes.

10.2.1.3 Recruitment of Autophagy Receptors

Numerous studies have reported on the recruitment of autophagy receptors to intracellular bacteria in order to mediate their recognition by the host autophagy machinery. Intracellular *L. monocytogenes* is ubiquitinated and detected by the autophagy receptors SQSTM1 and NDP52 (Yoshikawa et al. 2009; Mostowy et al. 2011). In response to *M. tuberculosis* infection, SQSTM1 is phosphorylated by TBK1 which also coordinates the assembly and function of the autophagic machinery. The transmembrane protein STING recognizes *M. tuberculosis* extracellular DNA which is ubiquitinated, and the autophagy receptors SQSTM1 and NDP52 are recruited to it (Watson et al. 2012). The autophagy receptors SQSTM1 and NDP52 have been shown to be recruited to intracellular *S. typhimurium* independently of each other and with similar kinetics (Zheng et al. 2009; Thurston et al. 2009). The depletion of either of the receptors hampers autophagy. It has also been reported that SQSTM1 and NDP52 have convergent roles in mediating antibacterial autophagy (Cemma et al. 2011). Moreover, NDP52 has been reported to target bacteria to autophagosomes and thereby promote the maturation of *Salmonella*-containing autophagosomes by binding to LC3A, LC3B, GABARAPL2, and myosin VI (Verlhac et al. 2015). Thurston et al. (2012) have reported that galectin-8 (a danger receptor) recruits NDP52 to damaged SCVs and restricts the growth of *S. typhimurium* by autophagy. We have recently reported the involvement of another autophagy receptor, OPTN, in the growth inhibition of *L. monocytogenes* and that OPTN phosphorylation by TBK1 enhances the growth restriction of intracellular *L. monocytogenes* in an LLO-dependent manner (Puri et al. 2017). Moreover, OPTN and TAX1BP1 restrict the growth of *S. typhimurium* (Wild et al. 2011; Tumbarello et al. 2015). It has also been shown that the expression of the autophagy receptor NR1D1 increases the number of acidic vacuoles and the levels of processed LC3 and also modulates lysosome biogenesis during *M. tuberculosis* infection (Chandra et al. 2015).

10.2.2 Mechanisms of Autophagy Evasion by Bacteria During Infection

Several infection-causing bacteria have also devised strategies to evade autophagy during infection. *S. typhimurium* produces the virulence protein SseL which deubiquitinates *S. typhimurium*-induced aggregates which accumulate at SCV (Thomas et al. 2012). Another mechanism is the suppression of the overall autophagy by acting on the Akt-mTOR signaling pathway (Owen et al. 2014). *L. monocytogenes* expresses two phospholipases C, PlcA and PlcB, which allow escape from autophagosomes (Birmingham et al. 2007; Py et al. 2007). Additionally, it produces the surface-located protein actin assembly-inducing protein (ActA) which binds to host cell actin machinery. This, on the one hand, allows bacterial intracellular movement and, on the other hand, disguises the pathogen as a host cell organelle and thereby allows autophagosomal evasion (Yoshikawa et al. 2009). In the absence of ActA, *L. monocytogenes* harbors another protein, internalin K (InlK), which camouflages the pathogen from autophagic recognition as it interacts with the major vault protein (MVP) (Dortet et al. 2011). An interesting study by Mitchell et al. (2018) has reported that upon *L. monocytogenes* infection, noncanonical autophagy is activated, whereas growth-restricting xenophagy is inhibited in a FIP200- and TBK1-dependent manner. *M. tuberculosis* is capable of evading autophagy by various mechanisms including the expression of the early secretory antigenic target 6 (ESAT-6) system 1 (ESX-1) secretion-associated protein B (EspB) of *M. tuberculosis* which suppresses LC3B expression and autophagosome formation (Huang and Bao 2016). *M. tuberculosis* blocks also phagosomal maturation (via IL-27 induction) and can promote the intracellular growth of *M. tuberculosis* by the inhibition of IFN-γ- and starvation-induced autophagy (Sharma et al. 2014). *M. tuberculosis* growth is facilitated by the inhibition of autophagy by the overexpression of miR-30A (Chen et al. 2015). Another mechanism includes the enhanced intracellular survival (EIS) gene-dependent upregulation of IL-10 which acts, via acetylation of histone H3, on mTOR pathway and thereby suppresses autophagy (Duan et al. 2016). The phospholipase A2-dependent phagosome escape by some strains of *M. tuberculosis* is crucial because of their reduced capacity to tolerate phagosomal stresses, and it serves as a "virulence-rescue" mechanism which favors suppression of autophagy in macrophages (Jamwal et al. 2016).

10.2.3 Role of Autophagy in Crohn's Disease

Genome-wide association studies have implicated autophagy as an essential part in the pathogenesis of Crohn's disease (Hampe et al. 2007; Barrett et al. 2008). In particular, Rioux et al. 2007 have reported that the autophagy gene ATG16L1 is expressed in intestinal epithelial cells and its knockdown revokes the autophagy of *S. typhimurium*. Moreover, mice deficient in Nod2 have decreased expression of α-defensins associated with Paneth cells and a severe defect in handling orally administered *L. monocytogenes* (Kobayashi et al. 2005). However, Atg16l1 hypomorphic mice are not deficient in handling *L. monocytogenes* despite differences in Paneth cell granule structure and composition (Cadwell et al. 2008). ATG16L1 T300A variant-transfected epithelial cells show impaired capture of internalized *Salmonella* within autophagosomes (Kuballa et al. 2008).

10.3 Autophagy as a Potential Therapeutic Target

With a plethora of studies on bacterial infections and autophagy, the current research should focus on the potential of autophagy as a therapeutic target for bacterial infections. A promising strategy in this direction could be to target bacterial factors that antagonize the functions of autophagy or autophagy factors. The inhibition of bacterial virulence factors which enable intracellular bacteria to escape autophagic recognition – such as ActA of *L. monocytogenes* or EspB of *M. tuberculosis* – could possibly enhance the xenophagic degradation of these bacteria and thereby provide an adjuvant therapy against bacterial infection. This approach may prove to be more specific and effective in the treatment of infections as it avoids the potential drawbacks associated with the manipulation of autophagy itself. Another strategy that can be employed to control bacterial infections is to exploit the autophagy receptor-bacteria interaction. Several autophagy receptors are known to bind to ubiquitinated bacteria and deliver them to autophagosomes like SQSTM1, NDP52, NBR1, optineurin, and TAX1BP1. Therefore, approaches that increase the interaction of autophagy receptors and bacteria, and also which augment certain modifications of autophagy receptors, like the phosphorylation of OPTN by TBK1 increases its LC3-binding affinity, may prove to be effective in enhancing the autophagic degradation of intracellular pathogens. Deciphering the molecular mechanisms of how autophagy receptors function could provide new avenues for the development of compounds that selectively enhance microbial autophagy. However, the downside of this strategy is that most autophagy receptors also mediate the selective autophagy of damaged cellular organelles and have other autophagy-independent functions; therefore, such manipulating these receptors may have undesired repercussions for the host.

Another plausible approach to target autophagy for antibacterial therapy involves the identification of novel autophagy-inducing compounds. Toward this end, chemical compounds can be screened on the basis of measurements of autophagosomal fluorescence (green fluorescent protein-LC3-positive puncta) by live-cell imaging methods, and the total LC3 levels can be determined by FACS analysis. Proteomic mapping methods like spatially restricted enzymatic tagging in living cells can be employed for the identification of autophagy-specific regulatory steps (Rhee et al. 2013). A caveat for compounds which modulate autophagy is that they usually induce other effects which may be unrelated to autophagy, thereby making it difficult to determine the contribution of the autophagy to their therapeutic effects. It is known that some autophagy-inducing agents fail to induce their beneficial effects in host organisms lacking autophagy genes (Levine et al. 2015). The upregulation of autophagy has been shown to have promising effects in preclinical models of diseases, viz., trifluoperazine in *Salmonella* infection and statins in *M. tuberculosis* infection (Conway et al. 2013; Parihar et al. 2014). It is unknown whether the therapeutic effects shown by the clinically recommended concentrations of these agents correspond to considerable increase in autophagy induction.

Several compounds have been shown to modulate the autophagy of pathogenic bacteria. The treatment with isoniazid has been shown to activate autophagy and decrease the pro-inflammatory responses induced by *M. tuberculosis* in macrophages (Kim et al. 2012). The intracellular growth of *M. tuberculosis* has also been shown to be inhibited by autophagy induction upon treatment with the anti-protozoan drug nitazoxanide and its active metabolite tizoxanide (Lam et al. 2012). Lieberman and Higgins (2009) have shown that a small molecule called pimozide, which promotes autophagy (Zhang et al. 2007) and is used as an antipsychotic drug, inhibits *L. monocytogenes* infection. They have also reported that the antipsychotic drug thioridazine, also known to induce autophagy (Chen et al. 2015), inhibits vacuolar escape and the intracellular growth of *L. monocytogenes* in murine macrophages (Lieberman and Higgins 2010). It is, therefore, imperative to further examine the connection between antipsychotic drugs and their antibacterial and pro-autophagy effects. Simvastatin, a drug known to modulate cholesterol turnover and to enhance autophagy, also prevents the phagosomal escape of *L. monocytogenes* and thereby decreases infection in mice (Parihar et al. 2013).

10.4 Conclusions

Autophagy is an integral part of the host defense mechanism against infections. With current antibiotics being prone to drug resistance, alternate strategies should be adopted for the treatment of bacterial infections. Targeting autophagy as an additional novel therapeutic target apart from conventional antibacterial therapy has a promising potential, and that should be the focus of upcoming research in the field of pathogenic bacterial infections.

Acknowledgments M.P. is thankful to T.C. for guidance, laboratory facilitites and financial assistance. This work was supported by the LISTRESS and PROANTILIS networks of the EU ERANet pathogenomics program to T.C.

References

Anand PK, Tait SW, Lamkanfi M, Amer AO, Nunez G, Pagès G, Pouysségur J et al (2011) TLR2 and RIP2 pathways mediate autophagy of *Listeria monocytogenes* via extracellular signal-regulated kinase (ERK) activation. J Biol Chem 286:42981–42991

Barrett JC, Hansoul S, Nicolae DL, Cho JH, Duerr RH, Rioux JD et al (2008) Genome-wide association defines more than 30 distinct susceptibility loci for Crohn's disease. Nat Genet 40:955–962

Birmingham CL, Smith AC, Bakowski MA, Yoshimori T, Brumell JH (2006) Autophagy controls *Salmonella* infection in response to damage to the *Salmonella*-containing vacuole. J Biol Chem 281:11374–11383

Birmingham CL, Canadien V, Gouin E, Troy EB, Yoshimori T, Cossart P, Higgins DE, Brumell JH (2007) *Listeria monocytogenes* evades killing by autophagy during colonization of host cells. Autophagy 3:442–451

Birmingham CL, Canadien V, Kaniuk NA, Steinberg BE, Higgins DE, Brumell JH (2008) Listeriolysin O allows *Listeria monocytogenes* replication in macrophage vacuoles. Nature 451:350–354

Bjørkøy G, Lamark T, Brech A, Outzen H, Perander M, Overvatn A, Stenmark H, Johansen T (2005) p62/SQSTM1 forms protein aggregates degraded by autophagy and has a protective effect on huntingtin-induced cell death. J Cell Biol 171:603–614

Cadwell K, Liu JY, Brown SL, Miyoshi H, Loh J, Lennerz JK et al (2008) A key role for autophagy and the autophagy gene Atg16l1 in mouse and human intestinal Paneth cells. Nature 456:259–263

Cemma M, Kim PK, Brumell JH (2011) The ubiquitin-binding adaptor proteins p62/SQSTM1 and NDP52 are recruited independently to bacteria-associated microdomains to target *Salmonella* to the autophagy pathway. Autophagy 7:341–345

Chandra V, Bhagyaraj E, Nanduri R, Ahuja N, Gupta P (2015) NR1D1 ameliorates *Mycobacterium tuberculosis* clearance through regulation of autophagy. Autophagy 11:1987–1997

Chen Z, Wang T, Liu Z, Zhang G, Wang J, Feng S et al (2015) Inhibition of autophagy by miR-30A induced by *Mycobacteria tuberculosis* as a possible mechanism of immune escape in human macrophages. Jpn J Infect Dis 68:420–424

Conway KL, Kuballa P, Songh JH, Patel KK, Castoreno AB, Yilmaz OH et al (2013) Atg16l1 is required for autophagy in intestinal epithelial cells and protection of mice from *Salmonella* infection. Gastroenterology 145:1347–1357

Dortet L, Mostowy S, Samba-Louaka A, Gouin E, Nahori MA, Wiemer EA, Dussurget O, Cossart P (2011) Recruitment of the major vault protein by InlK: a *Listeria monocytogenes* strategy to avoid autophagy. PLoS Pathog 7:e1002168

Duan L, Yi M, Chen J, Li S, Chen W (2016) *Mycobacterium tuberculosis* EIS gene inhibits macrophage autophagy through up-regulation of IL-10 by increasing the acetylation of histone H3. Biochem Biophys Res Commun 473:1229–1234

Dupont N, Lacas-Gervais S, Bertout J, Paz I, Freche B, Van Nhieu GT, van der Goot FG, Sansonetti PJ, Lafont F (2009) *Shigella* phagocytic vacuolar membrane remnants participate in the cellular response to pathogen invasion and are regulated by autophagy. Cell Host Microbe 6:137–149

Fujita N, Itoh T, Omori H, Fukuda M, Noda T, Yoshimori T (2008) The Atg16L complex specifies the site of LC3 lipidation for membrane biogenesis in autophagy. Mol Biol Cell 19:2092–2100

Glick D, Barth S, Macleod KF (2010) Autophagy: cellular and molecular mechanisms. J Pathol 221:3–12

Gluschko A, Herb M, Wiegmann K, Krut O, Neiss WF, Utermohlen O et al (2018) The β2 integrin Mac-1 induces protective LC3-associated phagocytosis of *Listeria monocytogenes*. Cell Host Microbe 23:324–337

Hampe J, Franke A, Rosenstiel P, Till A, Teuber M, Huse K et al (2007) A genome-wide association scan of nonsynonymous SNPs identifies a susceptibility variant for Crohn disease in ATG16L1. Nat Genet 39:207–211

Hanada T, Noda NN, Satomi Y et al (2007) The Atg12-Atg5 conjugate has a novel E3-like activity for protein lipidation in autophagy. J Biol Chem 282:37298–37302

Huang D, Bao L (2016) *Mycobacterium tuberculosis* EspB protein suppresses interferon-γ-induced autophagy in murine macrophages. J Microbiol Immunol Infect 49:859–865

Jamwal SV, Mehrotra P, Singh A, Siddiqui Z, Basu A, Raoy KV (2016) Mycobacterial escape from macrophage phagosomes to the cytoplasm represents an alternate adaptation mechanism. Sci Rep 6:23089. https://doi.org/10.1038/srep23089

Khweek AA, Caution K, Akhter A, Abdulrahman BA, Tazi M, Hassan H, Majumdar N et al (2013) A bacterial protein promotes the recognition of the *Legionella pneumophila* vacuole by autophagy. Eur J Immunol 43:1333–1344

Kim JJ, Lee HM, Shin DM, Kim W, Yuk JM, Jin HS et al (2012) Host cell autophagy activated by antibiotics is required for their effective antimycobacterial drug action. Cell Host Microbe 11:457–468

Kirisako T, Ichimura Y, Okada H, Kabeya Y, Mizushima N, Yoshimori T, Ohsumi M, Takao T, Noda T, Ohsumi Y (2000) The reversible modification regulates the membrane-binding state of Apg8/Aut7 essential for autophagy and the cytoplasm to vacuole targeting pathway. J Cell Biol 151:263–276

Kirkin V, Lamark T, Sou YS, Bjørkøy G, Nunn JL, Bruun JA, Shvets E et al (2009) A role for NBR1 in autophagosomal degradation of ubiquitinated substrates. Mol Cell 33:505–516

Kobayashi KS, Chamaillard M, Ogura Y, Henegariu O, Inohara N, Nunez G et al (2005) Nod2-dependent regulation of innate and adaptive immunity in the intestinal tract. Science 307:731–734

Kuballa P, Huett A, Rioux JD, Daly MJ, Xavier RJ (2008) Impaired autophagy of an intracellular pathogen induced by a Crohn's disease associated ATG16L1 variant. PLoS One 3(10):e3391. https://doi.org/10.1371/journal.pone.0003391 Epub 2008 Oct 13

Lam KK, Zheng X, Forestieri R, Balgi AD, Nodwell M, Vollett S et al (2012) Nitazoxanide stimulates autophagy and inhibits mTORC1 signaling and intracellular proliferation of *Mycobacterium tuberculosis*. PLoS Pathog 8:e1002691. https://doi.org/10.1371/journal.ppat.1002691

Lam GY, Cemma M, Muise AM, Higgins DE, Brumell JH (2013) Host and bacterial factors that regulate LC3 recruitment to *Listeria monocytogenes* during the early stages of macrophage infection. Autophagy 9:985–995

Levine B, Packer M, Codogno P (2015) Development of autophagy inducers in clinical medicine. J Clin Invest 125:14–24

Lieberman LA, Higgins DE (2009) A small-molecule screen identifies the antipsychotic drug pimozide as an inhibitor of *Listeria monocytogenes* infection. Antimicrob Agents Chemother 53:756–764

Lieberman LA, Higgins DE (2010) Inhibition of *Listeria monocytogenes* infection by neurological drugs. Int J Antimicrob Agents 35:292–296

Lin L, Baehrecke EH (2015) Autophagy, cell death, and cancer. Mol Cell Oncol 2:e985913. https://doi.org/10.4161/23723556.2014.985913

Lu K, Psakhye I, Jentsch S (2014) A new class of ubiquitin-Atg8 receptors involved in selective autophagy and polyQ protein clearance. Autophagy 10:2381–2382

Mitchell G, Cheng MI, Chen C, Nguyen BN, Whiteley AT, Kianian S, Cox JS et al (2018) *Listeria monocytogenes* triggers noncanonical autophagy upon phagocytosis, but avoids subsequent growth-restricting xenophagy. Proc Natl Acad Sci USA 115:E210–E217

Mizushima N, Yoshimori T, Ohsumi Y (2011) The role of Atg proteins in autophagosome formation. Annu Rev Cell Dev Biol 27:107–132

Morenzo-Gonzalo O, Ramirez-Huresca M, Blas-Rus N, Cibrian D, Saiz ML, Jorge I, Camafeita E et al (2017) HDAC6 controls innate immune and autophagy responses to TLR-mediated signalling by the intracellular bacteria *Listeria monocytogenes*. PLoS Pathog 13:e1006799. https://doi.org/10.1371/journal.ppat.1006799 eCollection 2017 Dec

Mostowy S, Sancho-Shimizu V, Hamon MA, Simeone R, Brosch R, Johansen T, Cossart P (2011) p62 and NDP52 proteins target intracytosolic *Shigella* and *Listeria* to different autophagy pathways. J Biol Chem 286:26987–26995

Moy RH, Cherry S (2013) Antimicrobial autophagy: a conserved innate immune response in *Drosophila*. J Innate Immun 5:444–455

Newman AC, Scholefield CL, Kemp AJ, Newman M, McIver EG, Kamal A, Wilkinson S (2012) TBK1 kinase addiction in lung cancer cells is mediated via autophagy of Tax1bp1/Ndp52 and non-canonical NF-κB signaling. PLoS One 7:e50672

Noda T, Fujita N, Yoshimori T (2009) The late stages of autophagy: how does the end begin? Cell Death Differ 16:984–990

Ogawa M, Yoshikawa Y, Mimuro H, Hain T, Chakraborty T, Sasakawa C (2011) Autophagy targeting of *Listeria monocytogenes* and the bacterial countermeasure. Autophagy 7:310–314

Osawa T, Mizuno Y, Fujita Y, Takatama M, Nakazato Y, Okamoto K (2011) Optineurin in neurodegenerative diseases. Neuropathol Off J Japan Soc Neuropathol 31:569–574

Owen KA, Meyer CB, Bouton AH, Casanova JE (2014) Activation of focal adhesion kinase by *Salmonella* suppresses autophagy via an Akt/mTOR signaling pathway and promotes bacterial survival in macrophages. PLoS Pathog 10:e1004159. https://doi.org/10.1371/journal.ppat.1004159 eCollection 2014

Parihar SP, Guler R, Lang DM, Harukazu S, David Marais A, Brombacher F, Ratner AJ (2013) Simvastatin enhances protection against Listeria monocytogenes infection in mice by counteracting listeria-induced Phagosomal escape. PLoS One 8(9):e75490

Parihar SP, Guler R, Khutlang R, Lang DM, Hurdayal R, Mhlanga MM et al (2014) Statin therapy reduces the *Mycobacterium tuberculosis* burden in human macrophages and in mice by enhancing autophagy and phagosome maturation. J Infect Dis 209:754–763

Proikas-Cezanne T, Takacs Z, Dönnes P, Kohlbacher O (2015) WIPI proteins: essential PtdIns3P effectors at the nascent autophagosome. J Cell Sci 128:207–217

Puri M, La Pietra L, Mraheil MA, Lucas R, Chakraborty T and Pillich H. 2017. Listeriolysin O regulates the expression of optineurin, an autophagy adaptor that inhibits the growth of *Listeria monocytogenes*. Toxins (Basel) 9:273. pii: E273. https://doi.org/10.3390/toxins9090273

Py BF, Lipinski MM, Yuan J (2007) Autophagy limits *Listeria monocytogenes* intracellular growth in the early phase of primary infection. Autophagy 3:117–125

Rhee HW, Zou P, Udeshi ND, Martell JD, Mootha VK, Carr SA et al (2013) Proteomic mapping of mitochondria in living cells via spatially restricted enzymatic tagging. Science 339:1328–1331

Rich KA, Burkett C, Webster P (2003) Cytoplasmic bacteria can be targets for autophagy. Cell Microbiol 5:455–468

Rioux JD, Xavier RJ, Taylor KD, Silverberg MS, Goyette P, Huett A et al (2007) Genome-wide association study identifies new susceptibility loci for Crohn disease and implicates autophagy in disease pathogenesis. Nat Genet 39:596–604

Russell RC, Tian Y, Yuan H et al (2013) ULK1 induces autophagy by phosphorylating Beclin-1 and activating VPS34 lipid kinase. Nat Cell Biol 15:741–750

Samson E (1981) Xenophagy. Br Dent J 150:136

Sanjurjo L, Amézaga N, Vilaplana C, Cáceres N, Marzo E, Valeri M et al (2013) The scavenger protein apoptosis inhibitor of macrophages (AIM) potentiates the antimicrobial response against *Mycobacterium tuberculosis* by enhancing autophagy. PLoS One 8:e79670. https://doi.org/10.1371/journal.pone.0079670 eCollection 2013

Sharma G, Dutta RK, Khan MA, Ishaq M, Sharma K, Malhotra H (2014) IL-27 inhibits IFN-γ induced autophagy by concomitant induction of JAK/PI3 K/Akt/mTOR cascade and up-regulation of Mcl-1 in *Mycobacterium tuberculosis* H37Rv infected macrophages. Int J Biochem Cell Biol 55:335–347. https://doi.org/10.1016/j.biocel.2014.08.022 Epub 2014 Sep 4

Tattoli I, Sorbara MT, Vuckovic D, Ling A, Soares F, Carneiro LA, Yang C, Emili A, Philpott DJ, Girardin SE (2012) Amino acid starvation induced by invasive bacterial pathogens triggers an innate host defense program. Cell Host Microbe 11:563–575

Tattoli I, Sorbara MT, Yang C, Tooze SA, Philpott DJ, Girardin SE (2013) *Listeria* phospholipases subvert host autophagic defenses by stalling pre-autophagosomal structures. EMBO J 32:3066–3078

Thomas M, Mesquita FS, Holden DW (2012) The DUB-ious lack of ALIS in *Salmonella* infection: a *Salmonella* deubiquitinase regulates the autophagy of protein aggregates. Autophagy 8:1824–1826

Thurston TL, Ryzhakov G, Bloor S, von Muhlinen N, Randow F (2009) The TBK1 adaptor and autophagy receptor NDP52 restricts the proliferation of ubiquitin-coated bacteria. Nat Immunol 10:1215–1221

Thurston TL, Wandel MP, von Muhlinen N, Foeglein A, Randow F (2012) Galectin 8 targets damaged vesicles for autophagy to defend cells against bacterial invasion. Nature 482:414–418

Tumbarello DA, Manna PT, Allen M, Bycroft M, Arden SD, Kendrick-Jones J, Buss F (2015) The autophagy receptor TAX1BP1 and the molecular motor myosin VI are required for clearance of *Salmonella typhimurium* by autophagy. PLoS Pathog 11:e1005174. https://doi.org/10.1371/journal.ppat.1005174 eCollection 2015

Verlhac P, Viret C, Faure M (2015) Dual function of CALCOCO2/NDP52 during xenophagy. Autophagy 11:965–966. https://doi.org/10.1080/15548627.2015.1046672

Wang J, Yang K, Zhou L, Minhaowu WY, Zhu M et al (2013) MicroRNA-155 promotes autophagy to eliminate intracellular mycobacteria by targeting Rheb. PLoS Pathog 9:e1003697. https://doi.org/10.1371/journal.ppat.1003697 Epub 2013 Oct 10

Watson RO, Manzanillo PS, Cox JS (2012) Extracellular *M. tuberculosis* DNA targets bacteria for autophagy by activating the host DNA-sensing pathway. Cell 150:803–815. https://doi.org/10.1016/j.cell.2012.06.040

Watson RO, Bell SL, MacDuff DA, Kimmey JM, Diner EJ, Olivas J (2015) The cytosolic sensor cGAS detects *Mycobacterium tuberculosis* DNA to induce type I interferons and activate autophagy. Cell Host Microbe 17:811–819. https://doi.org/10.1016/j.chom.2015.05.004 Epub 2015 Jun 2

Wild P, Farhan H, McEwan DG, Wagner S, Rogov VV, Brady NR, Richter B et al (2011) Phosphorylation of the autophagy receptor optineurin restricts *Salmonella* growth. Science 333:228–233

Yoshikawa Y, Ogawa M, Hain T, Yoshida M, Fukumatsu M, Kim M, Mimuro H et al (2009) *Listeria monocytogenes* ActA-mediated escape from autophagic recognition. Nat Cell Biol 11:1233–1240

Zhang L, Yu J, Pan H, Hu P, Hao Y, Cai W, Zhu H, Yu AD, Xie X, Ma D, Yuan J (2007) Small molecule regulators of autophagy identified by an image-based high-throughput screen. Proc Natl Acad Sci USA 104:19023–19028

Zheng YT, Shahnazari S, Brech A, Lamark T, Johansen T, Brumell JH (2009) The adaptor protein p62/SQSTM1 targets invading bacteria to the autophagy pathway. J Immunol 183:5909–5916

Chapter 11
Antibiotics and Antimicrobial Resistance

Varsha Gupta and Lipika Singhal

Abstract Antibiotics fundamentally revolutionized the practice of medicine and patient care, shifting the approach from diagnosis without intervening to a treatment-focused approach. However, the irrational use of antibiotic over the last seven decades has led to alarming clinical, environmental, and economic consequences. The judicious use of antibiotics limits the spread of antibiotic resistance as this strategy minimizes any unnecessary, inappropriate, or irrational use of antimicrobials. Most of the treating doctors hold good intentions of providing the best possible care to patients, and prescription of antibiotics is often considered a routine activity for achieving this by both patients and treating doctors. Researchers have found out that physician with high-volume practices and those who are in practice for longer are more likely to prescribe antibiotics inappropriately. On the other hand, patients have habit of repeating the prescription on their own, purchasing antibiotics without any prescription, or purchasing fewer units, all of which contribute to emergence of resistance. It would be rewarding to invest in seeking proper medical advice and further microbiological investigation in a patient who is suspected to have an infective etiology (bacterial) before instituting antibiotics. We have outlined the importance of rational use of antibiotics and good practices on part of both the doctor and the patients for effective containment of antimicrobial resistance.

Keywords Antibiotics · Antimicrobial resistance

11.1 Antibiotics

The accidental discovery of penicillin "the magic bullet" by Fleming in the 1930s paved way for the era of antibiotics. Antibiotics fundamentally revolutionized the practice of medicine and patient care, shifting the approach from diagnosis without means to interfere to a treatment-focused approach. The most fundamental impact of the introduction of antibiotics was a dramatic decline in death from bacterial

V. Gupta (✉) · L. Singhal
Department of Microbiology, Government Medical College Hospital, Chandigarh, India

P. P. Singh (ed.), *Infectious Diseases and Your Health*,
https://doi.org/10.1007/978-981-13-1577-0_11

infections of all types. Many communicable diseases have been contained since then. However, the potential for misuse of antibiotics was recognized shortly after introduction of antibiotics in 1940. As early as 1945, in an interview with the *New York Times*, Fleming called for stopping the overuse of penicillin in order to slow the development of resistance (Spellberg et al. 2013). The irrational use of antibiotic over the last seven decades has led to alarming clinical, environmental, and economic consequences.

11.2 What Is "Rational" Use of Medicines?

The World Health Organization (WHO) defines rational use of medicines as "Patients receive medications appropriate to their clinical needs, in doses that meet their own individual requirements, for an adequate period of time, and at the lowest cost to them and their community" (WHO 1985).

11.3 Common Types of Irrational Antibiotic Use Are (WHO 2002; Pathak et al. 2012)

(a) *Incorrect choice of antibiotics*:

- Use of too many medicines per patient (polypharmacy) or use of broad-spectrum antibiotics to cover unusual/many organisms when a narrow-spectrum drug would be sufficient. This approach is often chosen as a substitute for appropriate diagnostic evaluation which may be either due to inadequate knowledge of treating physician about the utility and availability of the diagnostic tests or due to high cost and unaffordability of diagnostic tests in resourcc-limited country like ours.
- Tendency to use newly introduced and expensive antibiotics, when there is no evidence supporting better drug susceptibility of the newer drug over an older one.

(b) *Inappropriate use of antibiotics*:

- Prescribing antibiotics to any patients with fever, taking it as a sign of bacterial infection.
- Use of antibiotics for self-limiting bacterial diseases that do not benefit from use of antimicrobials.
- Prescribing antibiotics for non-bacterial infections as viral upper respiratory tract infections (URTI) and diarrhea.
- Use of antibiotics for minor superficial skin infections which may be more suitably treated with local antiseptics.
- Overuse of injections when oral formulations would be more appropriate.

- Use of antibiotics for abscesses which primarily need surgical drainage, and antibiotics are often not required if adequate drainage has been done.

(c) *Inappropriate dosing and duration*:

- Antibiotics are often inappropriately given/taken in higher doses for longer duration or in insufficient doses or duration. Often drugs of low potency and effectiveness are used because of poor manufacture (Taylor et al. 1995). This may be due to the insufficient knowledge or financial constraints.
- Antibiotics used for prophylaxis are often inappropriately chosen and administered. Thirty percent of antibiotics used in hospital are for prophylaxis, and more than 80% are given inappropriately for >48 h duration.

11.4 Philosophy of Antimicrobial Use

Most of the treating doctors hold good intentions of providing the best possible care to patients, and prescription of antibiotics is often considered a routine activity for achieving this by both patients and treating doctors. It is more of a psychological and philosophical problem than a scientific exercise. Moreover, many a times doctors face pressure from patients for prescribing antibiotics. Doctors feel the need to give into this pressure due to time constraints (avoidance of time-consuming patient education) or to avoid losing the patient to another practitioner or at times to prevent claims of negligence. Researchers have found out that physician with high-volume practices and those who are in practice for longer are more likely to prescribe antibiotics inappropriately (Cadieux et al. 2007; Butler et al. 1998). As a leading role in containment of resistance is played by the prescribers, it should be understood that even if patients do expect a prescription for trivial conditions as cold, they may not necessarily be expecting a prescription of antibiotics. Physicians should carefully explore if a perceived wish for antibiotics exists or whether the consulting patient seeks symptomatic relief or just reassurance. Sometimes, malpractice, easy solutions, and incentives provided by pharmaceutical campaigns also contribute to rampant misuse of antibiotics.

Findings from various studies (Sivagnanam et al. 2004; Kotwani et al. 2010; Ghafur et al. 2013) have revealed that clinicians prescribe antibiotics mainly in the following contexts: (1) diagnostic uncertainty due to the inadequacy of the microbiology facility, (2) perceived demand or expectations from patient to get "capsules," (3) less time to interact with patients, (4) under the influence of medical representatives, and (5) oversupplied and near-expiry antibiotics. Doctors have also pointed out that once a prescription is written for a symptom, the same prescription is used for self-medication or given to near and dear ones with same symptoms to save money and time required for repeated consultations.

This behavior of repeating the prescription, acquiring antibiotics without a prescription, and purchasing fewer units contributes to emergence of resistance. It would be rewarding to invest in seeking proper medical advice and further microbiological investigation in a patient who is suspected to have an infective etiology (bacterial) before instituting antibiotics.

11.5 Why Do We Need to Rationally Use Available Antibiotics?

- At present, antibiotics are one of the most commonly prescribed drugs being responsible for 30–50% of hospital's total drug budget.
- Bacteria may not be the sole cause for all fevers which are due to infections, and infections can also be caused by microorganisms apart from bacteria. In a study conducted to determine the epidemiology of fever in the intensive care unit (ICU), it was observed that less than 50% febrile episodes were attributed to infection (Circiumaru et al. 1999).
- According to WHO Policy Perspective on Medicines (2002),"Worldwide more than 50 per cent of all medicines are prescribed, dispensed, or sold inappropriately, while 50 per cent of the patients fail to take them correctly" (Dellit et al. 2007).
- Approximately one third (~30%) of all antibiotic prescribing is for prophylaxis, mainly in surgical services. More than 80% of patients received antibiotics for longer than 48 h, which is a cause of concern that should be considered for possible intervention (Pathak et al. 2012).
- In India, there is unrestricted dispensing of "over-the-counter" (OTC) antibiotics for consumption without professional control. The antibiotics are available in retail without prescriptions though they belong to "prescription-only" medicines and/or are prescribed by practitioners from alternative medical branches and healers (Ghafur et al. 2013; Smith et al. 1996).
- Emergence of antibiotic resistance (AMR): Antibiotic use both appropriate and inappropriate drives resistance. Antibiotic usage can lead to the development and spread of antibiotic-resistant bacteria by at least two mechanisms: (Spellberg et al. 2013) by applying selective pressure, selects out strains of antibiotic-resistant bacteria, and (World Health Organization 1985) eliminates normal bacterial flora in human hosts, which promotes colonization and spread of existing antibiotic-resistant strains. Regardless whether antibiotic treatment is appropriate, the bacteria inhabiting the gut and other parts of the body are affected by antibiotics and will evolve genetic material coding for resistance that they pass on to other – even unrelated – bacteria.
- Bad Bugs, No Drugs: An "Impending Disaster": Antibiotic resistance is depleting the number of effective antibiotic agents. Research and development of any antibiotic is a colossal investment for the big pharmaceutical industry (Boucher et al. 2009). Antibiotics have a poor return on investment because they are taken for a short period and cure their target disease. In contrast, drugs that treat chronic illness, such as high blood pressure, are taken daily for decades and often for the

rest of a patient's life. Lack of profitability has led to stagnation in the field of anti-infective research (Shlaes and Moellering Jr 2002; Boucher et al. 2009; Braine 2011).

- The increasing number of multidrug-resistant infections and the diminishing number of new antibiotics in development with the potential to treat these infections represent one of the world's greatest health threats. The WHO has supported this premise, identifying antimicrobial resistance as one of the three greatest threats to human health. The Infectious Diseases Society of America (IDSA) launched the "10 X '20 initiative" which calls for the development of ten novel, safe and effective, systemic antibiotics by 2020 (ISDA 2010).
- The high prevalence of extended spectrum β-lactamase (ESBL)- and carbapenemase-producing microorganisms makes majority of antibiotics ineffective. Increasing carbapenem resistance forces increased usage of colistin, currently the last line of defense, with a potential for colistin-resistant and pan-drug-resistant bacterial infections and a world without antibiotics (Marchaim et al. 2011; Datta et al. 2012).
- Strict supervision on the usage of higher-end antibiotics such as colistin, tigecycline, vancomycin, linezolid, etc. which are currently the most precious antibiotics in an era of pan-resistance is mandatory.
- The burden of infectious disease in India is among the highest in the world. Overcrowding, poor sanitation, and a warm and humid environment result in hurriedly spreading the deadly "superbugs". These superbugs are putting us at increased risk of not only healthcare-associated and community-acquired infections but also impose devastating threats (bioterrorism, pandemics) that can affect our nation's security.

  (a) Healthcare-associated infections (HCAI): These result in prolonged hospital stays, long-term disability, increased resistance of microorganisms to antibiotics, and massive additional costs for patients, their families, and health systems and cause preventable deaths.
  (b) Community-acquired infections: Once confined to the inpatient setting, resistant bacteria like MRSA, VISA, ESBL, and MDR *Clostridium difficile* are now common in community-acquired infections as well (Datta et al. 2012).

- The inappropriate use of antibiotics is not just restricted to the hospital setting or community but to other fields like veterinary and agriculture. Antibiotics are widely used in the food animals as growth promoters and to prevent and treat any kind of infections in them. There are no regulatory provisions regarding the use of antibiotics in livestock.
- Currently, there are no accepted definite guidelines for appropriate use of antimicrobials for diseases of public health importance as enteric fever, diarrheal disease, and respiratory infections except for those where there is a specific national health program, e.g., RNTCP, National AIDS Control Program, etc.
- Lack of enforcement of legislation at all levels of the drug delivery and disposal system to improve antibiotic use and prevent pharmaceutical contamination of the environment.

Containment of antimicrobial resistance, although not easy, is not impossible. The significance of the judicious use of antibiotics in restraining the spread of antibiotic resistance cannot be overstressed as this strategy minimizes any unnecessary, inappropriate, or irrational use of antimicrobials. Policies and regulations that encourage more appropriate and rational use of antimicrobials are key long-term interventions for preventing or slowing the progression of resistance. There is at present no functioning national antibiotic policy or a national policy to contain antimicrobial resistance in India. The National Centre for Disease Control, under the Directorate General of Health Services, Ministry of Health and Family Welfare, Government of India, published "The National Policy for Containment of Antimicrobial Resistance" in 2011 (The National Policy for Containment of Antimicrobial Resistance, India 2011), which has been put on hold due to nonavailability of major recommendations (Ghafur et al. 2013; Ganguly et al. 2011). Being one of the largest countries with more than 20,000 hospitals, more than a billion population, wide cultural diversity, socioeconomic disparity, and a large medical community of more than three fourths of a million doctors, India finds translating recommendations into national action plans very difficult task to implement.

This requires a stepwise implementation of strategies at various levels. The first step at hospital level is to establish an Infection Control Team (ICT) and to initiate antimicrobial stewardship (Voss and Ghafur 2013).

Antimicrobial stewardship (i.e., coordinated interventions designed to improve appropriate use of antimicrobial drugs, including preventing inappropriate antimicrobial use and limiting antimicrobial exposure) is a perilous tool to protect antibiotics from being misused or overused (Boucher et al. 2009). The medical director and/or hospital manager should ensure that the prevention and control of HCAI which includes a mandatory official hospital body should be involved in the formation of an antibiotic policy.

11.6 The Antibiotic Management Team

It is a multidisciplinary team with experts from all major departments: infectious diseases, internal medicine, intensive care, surgery, pediatrics, clinical microbiology, pharmacology, and hospital pharmacy. A multidisciplinary approach offers the best potential for sustained improvement in both clinical and economic outcomes and increases the general acceptability. However, membership of an antibiotic committee may diverge according to local conditions and needs.

11.7 Main Tasks of Antibiotic Management Team

The main tasks of an antibiotic committee as also recommended by the IDSA in 1987 are the following:

- To establish guidelines for antibiotic use which lead to manufacturing of an antibiotic formulary restricted to the minimal number of agents needed for most effective therapy
- To formulate guidelines for prophylaxis and therapy of infection (which is the responsibility of the clinicians)
- To conduct periodic reviews on antibiotic use
- To eliminate duplicate agents within the same class
- To restrict certain agents for special indications, toxicity, and costs
- To consider local prevalence and susceptibility data and to provide feedback to the clinicians (on appropriate usage)
- To be responsible for education and dissemination of information
- To work closely with the ICT in the hospital

11.7.1 Antibiotic Formulary Must Be Restricted to the Minimal Number of Agents Needed for Most Effective Therapy

An essential medicine list should be developed.

(i) First choice antibiotics: Can be prescribed by all doctors
(ii) Restricted list of antibiotics: Only after permission from HOD or AMT representative
(iii) Reserve antibiotics: Only after AMT member's permission

The formulary should include side effects, contraindications, dose information, and references for the antibiotics.

- *De-escalation policy*: Presumptive therapy should only be applicable for a maximum of 48 h after which the therapy needs to be de-escalated based on clinical or microbiological evidence. AMT should interact with the unit that prescribes more than two antibiotics and satisfy them to the correctness of the regimen.
- *Prophylactic antibiotic policy*: Prophylactic antibiotics given for a short duration should be free of side effects and relatively inexpensive and should not be used as a routine therapy.
- The antibiotic committee will make rational choices among "equivalent drugs" and classes of drugs to select the least expensive, most effective agents. Cost should determine the selection, when microbiological, pharmacological, and other relevant properties are similar.
- *Therapeutic substitution and streamlining process*
  - Involves substitution of a member of a class of drugs with a less expensive therapeutic equivalent and change to a different class to switch over from parenteral to oral therapy.
  - Limiting availability of agents: It is the most direct and simple method to stop use of newer, more expensive antibiotics in favor of older, cheaper, and equally effective ones.

11.7.2 Indian Government's Vision

The Chennai Declaration "A roadmap to tackle the challenge of antimicrobial resistance" was the first ever joint meeting of medical societies in India addressing antibiotic resistance, held in Chennai in August 2012 (Ghafur et al. 2013). Several strategies for regulating antibiotic prescribing practices have been recommended, few of which are:

- An Infection Control Team (ICT) should be made mandatory in all hospitals.
- Regulatory authorities and accreditation agencies such as the National Accreditation Board for Hospitals and Healthcare Providers (NABH) should insist on strict implementation of hospital antibiotic and infection control policy, during hospital licensing and accreditation process.
- Hospitals without compliance with the policy should not be given accreditation.
- The Drugs Controller General of India will need to formulate and implement a policy on rationalizing antibiotic usage in the country, both in hospitals and over the counters.
- Those hospitals with an existing ICT and an antibiotic policy should expand efforts to increase compliance to the policy.

Good Practices for Patients and Treating Physicians to Follow

- Bacteria may not be the sole cause for all fevers which are due to infections, and infections can also be caused by microorganisms apart from bacteria.
- Most infections seen in general practice are of viral origin, and antibiotics can neither treat viral infections nor prevent secondary bacterial infections in these patients.
- Consider whether the patient requires an antibiotic.
- Even where a bacterial etiology is established, an antibiotic may not always be necessary:
  - Many bacterial infections resolve spontaneously.
  - Minor superficial skin infections may be more suitably treated with local antiseptics.
  - Collections of pus should be drained surgically, and if drainage is adequate, antibiotics are often not required.
- The outcome of successful therapy depends very much on the choice of the antibacterial agent.
- In the process of selecting an antibiotic, the etiological agent, the patient, and the antibiotic are the three main factors which need to be considered.
- Bacterial isolates from the specimens to be cultured may represent normal flora, colonizers, or contaminants rather than true pathogens.
- Microbiology laboratory reports should always be correlated with the clinical findings. Every attempt should be made to avoid unnecessary treatment of isolates that represent *colonization* rather than infection.

- It is necessary to interpret the bacteriology report as the importance of etiological agent depends on correlation clinical acumen and the laboratory support.
- In many instances, empirical therapy has to be given. Local epidemiologic susceptibility patterns (analyzed by a microbiology laboratory) need to be considered in selecting an appropriate antimicrobial agent.
- Once microbiological report is available, there should be immediate shift to susceptible first-line drug.
- The patient's compliance to antimicrobials is a crucial factor for consideration in the choice of antibiotics.
- While selecting a particular antibiotic, many patient factors have to be considered

  (a) Age

    (i) The patients in extremes of ages are found to be more prone to the adverse effects of the antibiotics.
    (ii) Neonates due to immature liver and renal functions have altered ability to metabolize or excrete antibiotics.
    (iii) Antibiotics and their metabolites can affect the growing tissues and organs in children adversely.
    (iv) Elderly patients suffer from nephrotoxicity and allergic reactions more often.

- Patients with hepatic or renal impairment need dosage modifications.
- The doctor should be aware of the drug-drug interactions since many antibiotics can interact with other non-antibiotic drugs.
- Beta-lactam antibiotics and erythromycin are probably the safest antibiotics if needed during pregnancy.
- A history of allergy to antibiotics should always be informed to the doctor.
- Monotherapy is always preferred to polypharmacy, which increases antibiotic costs, potential for antibiotic side effects, and potential for drug-drug interactions.
- If the clinical condition indicates improvement, antibiotic therapy need not be changed.
- If no clinical response is seen within 72 h of clinical diagnosis, the antibiotic needs to be changed, and a secondary infection should be reconsidered.
- Duration of antibiotic therapy has to be reviewed periodically, and usage of such antibiotics should be stopped after 5 days.

References

Boucher HW, Talbot GH, Bradley JS, Edwards JE, Gilbert D, Rice LB et al (2009) Bad bugs, no drugs: no ESKAPE! An update from the Infectious Diseases Society of America. Clin Infect Dis 48:1–12

Braine T (2011) Race against time to develop new antibiotics. Bull World Health Organ 89:88–89

Butler CC, Rollnick S, Pill R, Maggs-Rapport F, Stott N (1998) Understanding the culture of prescribing: qualitative study of general practitioners' and patients' perceptions of antibiotics for sore throats. Br Med J 317:637–642

Cadieux G, Tamblyn R, Dauphinee D, Libman M (2007) Predictors of inappropriate antibiotic prescribing among primary care physicians. Can Med Assoc J 177:877–883

Circiumaru B, Baldock G, Cohen J (1999) A prospective study of fever in the intensive care unit. Intensive Care Med 25:668–673

Datta S, Wattal C, Goel N, Oberoi JK, Raveendran R, Prasad KJ (2012) A ten year analysis of multi-drug resistant blood stream infections caused by *Escherichia coli* & *Klebsiella pneumoniae* in a tertiary care hospital. Indian J Med Res 135:907–912

Dellit TH, Owens RC, McGowan JE Jr, Gerding DN, Weinstein RA, Burke JP et al (2007) Infectious Diseases Society of America and the Society for Healthcare Epidemiology of America guidelines for developing an institutional program to enhance antimicrobial stewardship. Clin Infect Dis 44:159–177

Ganguly NK, Arora NK, Chandy SJ, Fairoze MN, Gill JP, Gupta U et al (2011) Rationalizing antibiotic use to limit antibiotic resistance in India. Indian J Med Res 134:281–294

Ghafur A, Mathai D, Muruganathan A, Jayalal JA, Kant R, Chaudhary D et al (2013) The Chennai declaration: a roadmap to tackle the challenge of antimicrobial resistance. Indian J Cancer 50:71–73

Infectious Diseases Society of America (2010) The 10 x '20 initiative: pursuing a global commitment to develop 10 new antibacterial drugs by 2020. Clin Infect Dis 50:1081–1083

Kotwani A, Wattal C, Katewa S, Joshi PC, Holloway K (2010) Factors influencing primary care physicians to prescribe antibiotics in Delhi India. Fam Pract 27:684–690

Marchaim D, Chopra T, Pogue JM, Perez F, Hujer AM, Rudin S et al (2011) Outbreak of colistin-resistant, carbapenem-resistant *Klebsiella pneumoniae* in metropolitan Detroit. Mich Antimicrob Agents Chemother 55:593–599

Pathak A, Mahadik K, Dhaneria SP, Sharma A, Eriksson B, Lundborg CS (2012) Surveillance of antibiotic consumption using the "focus of infection" approach in 2 hospitals in Ujjain, India. PLoS One 7:e38641 Epub 2012/06/21

Shlaes DM, Moellering RC Jr (2002) The United States Food and Drug Administration and the end of antibiotics. Clin Infect Dis 34:420–422

Sivagnanam G, Thirumalaikolundusubramanian P, Mohanasundaram J, Raaj AA, Namasivayam K, Rajaram S (2004) A survey on current attitude of practicing physicians upon usage of antimicrobial agents in southern part of India. Medscape Gen Med 6:1

Smith RD, Coast J, Millar MR (1996) Over-the-counter antimicrobials: the hidden costs of resistance. J Antimicrob Chemother 37:1031–1032

Spellberg B, Bartlett JG, Gilbert DN (2013) The future of antibiotics and resistance. N Engl J Med 368:299–302

Taylor RB, Shakoor O, Behrens RH (1995) Drug quality, a contributor to drug resistance? Lancet 346:122

The National Policy for Containment of Antimicrobial Resistance, India (2011) The National Centre for Disease Control, Director General of Health Services, Ministry of Health and Family Welfare, Government of India. Available from: nicd.nic.in/ab_policy.pdf

Voss A, Ghafur A (2013) "The Chennai declaration" – Indian doctors' fight against antimicrobial resistance. Antimicrob Resist Infect Control 2:7

World Health Organization (1985) The rational use of drugs report of the conference of experts [Internet]

World Health Organization (2002) Perspectives on medicines. Promoting rational use of medicines: core components [Internet]. Available from: http://www.who.int/medicines/

Chapter 12
Nontuberculous Mycobacteria: An Update on Infections Caused, Laboratory Identification and their Treatment

Swetarka Das, Tanu Garg, Sidharth Chopra, and Arunava Dasgupta

Abstract Nontuberculous mycobacteria (NTM) or mycobacteria other than tuberculosis (MOTT) are opportunistic environmental pathogens capable of causing different kinds of infections to humans starting from hospital acquired to pulmonary and soft tissue infections. They share the common genera *Mycobacterium tuberculosis* (Mtb) and *Mycobacterium leprae*, the causative agents of two major diseases, tuberculosis (TB) and leprosy, respectively. Although NTMs or MOTT are similar to Mtb and *M. leprosy* in terms of mycolic acid-containing cell wall, acid fastness and their capability of causing pulmonary and extrapulmonary diseases, they are highly dissimilar in terms of growth rate and their antibiotic resistance profile. Several clinically relevant NTM strains are inherently resistant to different classes of antibiotics including the first line of drugs (e.g. isoniazid, rifampicin and pyrazinamide) used for treatment of TB. Currently, NTM infections have become a major concern to mankind in terms of mortality and morbidity in immunocompromised individuals. In addition, lack of antibacterial molecules, specific diagnostic tools and their intrinsic resistance to common antibiotics have turned these historically neglected pathogens to serious threats. In this chapter, we have discussed about infections caused by different NTMs, their laboratory identification and the antibiotics available to treat these infections.

Keywords Drug resistance · *Mycobacterium* · Mycobacteria other than tuberculosis (MOTT) · Nontuberculous mycobacteria

---

Authors Swetarka Das and Tanu Garg have equally contributed to this chapter.

S. Das · T. Garg · S. Chopra · A. Dasgupta (✉)
Division of Microbiology, CSIR-Central Drug Research Institute, Lucknow, India
e-mail: a.dasgupta@cdri.res.in

P. P. Singh (ed.), *Infectious Diseases and Your Health*,
https://doi.org/10.1007/978-981-13-1577-0_12

12.1 Introduction

The genus *Mycobacterium* is composed of more than 100 species characterized by complex lipid-rich cell walls, which confer the acid-fast staining amongst other properties including resistance to clinically utilized drugs as well as to desiccation. This genus is broadly divided into two groups based on the growth rate: slow and fast growers. The slow growers exhibit an incubation period generally greater than 7 days and include human and animal pathogens such as *M. tuberculosis* (Mtb), *M. leprae*, *M. bovis*, *M. avium* complex (MAC) and *M. ulcerans*, while the fast growers exhibit an incubation period generally less than 7 days and include *M. kansasii*, *M. abscessus* and *M. fortuitum*.

The nontuberculous mycobacteria (NTM) comprises of frequently opportunistic pathogens that are mostly isolated from environment including water and soil and comprise species of mycobacterium other than Mtb, *M. leprae* and *M. bovis* (Brown-Elliott and Wallace 2002; Falkinham 2003). These bacteria share many common properties, such as acid fastness and the ability to cause pulmonary and extrapulmonary granulomatous disorders but also exhibit great dissimilarities at molecular and biochemical level when compared to slow-growing mycobacteria (Soni et al. 2016).

Owing to the fact that, historically, a lot of attention of healthcare professionals has been to study and contain the infection caused due to slow-growing mycobacteria, there has been a significant neglect of NTMs which is reflected in a serious lack of potent active antimicrobials against as well as lack of sensitive and specific diagnostic tools available to contain the spread of these infectious agents. This is alarming since there is a significant upward trend in infections caused worldwide due to NTMs with >80% of clinical isolates being either *M. fortuitum*, *M. chelonae* or *M. abscessus* (Brown-Elliott and Wallace 2002; Johnson and Odell 2014). The lack of proper diagnostic tools and bactericidal antimicrobials is one of the major reasons for delays in diagnosis and the long treatment regimens for most NTM infections with sub- to moderate efficacy leading to high mortality (Brown-Elliott et al. 2012). As can be seen in Fig. 12.1, a picture of a patient suffering from NTM infection is depicted after CO_2 surfacing.

The depth of knowledge regarding the infection caused due to Mtb owes a lot to the identification of its genome sequence (Cole et al. 1998). Similar efforts for most of the NTMs are lacking although there is some progress (Bryant et al. 2013; Tettelin et al. 2014; Shallom et al. 2013; Davidson et al. 2014); however accompanying in-depth biological validation is currently lacking. NTMs share limited physiologic characteristics with slow-growing mycobacteria with significant differences in generation time and metabolic capabilities, which is reflected in their inherent drug-resistant status as well as possessing myriad antimicrobial resistance mechanisms (Griffith et al. 2007). Additionally, there is a serious lack of animal models mimicking human pathology, which is a very significant prerequisite for drug discovery; this situation is slowly improving with a number of new animal models being developed mimicking human pathology (Obregón-Henao 2015).

The current treatment of NTMs has been standardized to a great extent based upon the recommendations set forth in the guidelines jointly issued by the American

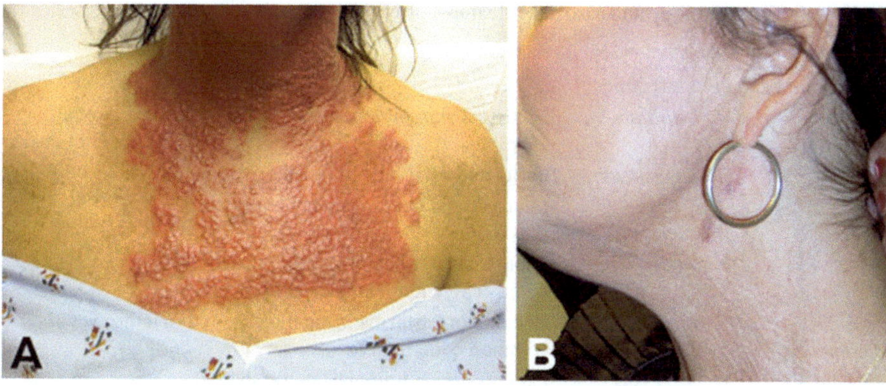

Fig. 12.1 (**a**) The neck and chest of a 53-year-old woman 14 days after fractionated CO_2 laser resurfacing, showing nontuberculous mycobacterial infection. (**b**) The neck of the patient after 5 months of multidrug therapy and pulsed dye laser (Culton et al. 2013)

Thoracic Society (ATS) and Infectious Disease Society of America (IDSA) (Griffith et al. 2007). The antimicrobial therapy against rapidly growing NTM primarily consists of macrolides, aminoglycosides, fluoroquinolones, oxazolidinones, tigecycline, carbapenems and cephalosporins, while for the slow-growing NTM species, rifampicin, macrolides, ethambutol and amikacin are the front-line agents.

Diagnosis of NTM infections remains challenging for clinicians owing mostly to a lack of clear guidelines, and depending on the extent of disease and species involved, a cure may be difficult to achieve as all these therapies have varying efficacies in vivo (Maurer et al. 2014). Thus, identification of NTM is of significant clinical relevance.

In this chapter, we provide an overview of the infections caused by NTM along with the currently utilized tools for their identification and the clinically utilized treatment options.

12.2 Infections Caused due to NTM

The infections caused due to NTM have been seen in most industrialized countries; incidence rates vary from 1.0 to 1.8 cases per 100,000 persons (REF). Typically, MAC is the most common NTM species followed by *M. fortuitum*, *M. kansasii*, *M. chelonae* or *M. abscessus*. The most common clinical manifestation of NTM disease is pulmonary disease although lymphatic, localized skin and soft tissue and disseminated disease are increasingly reported. In majority of the cases, the reservoir of the infection is the environment although reservoir for pulmonary disease is unknown (Griffith et al. 2007).

The infections along with their epidemiology and associated features are tabulated in Table 12.1.

Table 12.1 Epidemiology and associated features of NTM infections

Infection	Causative NTM	Associated features	Refs.
Nonhospital acquired infections			
Footbath-associated folliculitis	*M. fortuitum*	Persistent skin infections below the knee	Cooksey et al. (2004), Gira et al. (2004), Winthrop et al. (2004), (2009), Viana-Niero et al. (2008) and Vugia et al. (2005)
	M. mageritense, *M. cosmeticum*	Furunculosis of hair follicles	
	M. abscessus	Microtrauma due to shaving legs prior to pedicures	
		Footbath water heavily contaminated with NTM; filters not cleaned properly	
Anti-TNF therapy-associated infections	*M. avium* complex	Rheumatoid arthritis is the most common complaint	Winthrop et al. (2009)
	M. abscessus	Patients are on immunosuppressive therapies	
	M. chelonae		
	M. fortuitum		
INF-gamma-/interleukin-12-associated infections	*M. avium* complex	Multifocal infections affecting the lymph nodes, osteoarticular tissue, lungs, skin and/or soft tissues	Valour et al. (2016), Czaja et al. (2014) and Havlir et al. (1996)
	M. abscessus,		
	M. fortuitum	>90% cases in Asian born population	
Hospital-acquired infections			
Surgical infections	*M. fortuitum*	Infections related to catheters, sternal wounds following cardiac bypass surgery, infected augmentation mammaplasty sites, prosthetic devices, lens implants, artificial knees and hips and metal rods inserted to stabilize bones following fractures	Brown-Elliott and Wallace (2002), Gubler et al. (1992) and Tiwari et al. (2003)
	M. chelonae		
	M. abscessus		
	M. immunogenum		
		Infections associated with various water sources	
Postinjection abscesses	*M. abscessus*	Large multi-year outbreaks	Tiwari et al. (2003), Villanueva et al. (1997) and Galil et al. (1999)
Catheter-associated bacteraemia	*M. phocaicum*	Large outbreak	Cooksey et al. (2008)
Post-video laparoscopy infections	*M. abscessus* subsp. *massiliense*	Isolates tolerant to 2% glutaraldehyde disinfectants	Matsumoto et al. (2012)
		Strain isolated from sputum, BAL fluid and urine from patients not related to outbreak	
		Geographically and temporally widespread	
Cardiothoracic surgery	*M. wolinskyi*	Most likely infection sources were medical equipment utilized during surgery	Nagpal et al. (2014) and Dupont et al. (2016)
Lung transplant infections	*M. abscessus*	Infections associated with various water sources	Baker et al. (2015)

12.3 Laboratory Identification of NTM

Laboratory identification of the causative NTM is extremely important owing to inherent drug resistance as well as to tailor the antimicrobial chemotherapy accordingly. The current identification falls into either phenotypic methods or, more recently, molecular methods. It should be noted that recent identification of a large number of new pathogenic species has exposed the limitation of phenotypic methods.

12.4 Phenotypic Testing

The typical testing of NTM is primarily based on growth rate, gram staining, colony morphology, acid fastness, the presence or absence of pigmentation and arylsulfatase test (Adékambi and Drancourt 2004; Brown-Elliott and Wallace 2002). The major limitation with these tests is their inability to identify NTM at the species level, which is absolutely required for guiding the antimicrobial chemotherapy, for appropriate diagnosis as well as for containment and management of disease outbreaks. Thus, these should be correlated with identification at the molecular level.

12.5 High-Performance Liquid Chromatography (HPLC)

HPLC is a practical, rapid, sophisticated and reliable method for identifying many species of NTM, especially the slowly growers. Primary cultures of mycobacteria that are grown in BACTEC 7H12B medium (Becton Dickinson) can be directly analysed by HPLC. HPLC can also be used in the identification of MAC directly from samples with AFB smear-positive results. Although HPLC may be suitable for separating organisms into complexes or groups, it lacks the specificity needed for full species-level identification (Brown-Elliott and Wallace 2002; Tortoli 2014). For example, recognition of some newer species within the *M. simiae* complex has been reported to be challenging along with not being able to separate some species within *M. abscessus* and *M. chelonae*. Additionally, it is typically utilized in large reference labs due to its high cost as well as requirement for experienced personnel for operation. Because of these limitations, HPLC analysis has been overtaken by molecular methods for identification of NTMs.

12.5.1 Molecular Testing

PCR restriction fragment analysis

Attributes	Comments
Viable cells required	No
Technology	Restriction fragment length polymorphism analysis of 441 bp portion (Telenti fragment) of hsp65 sequence is used to identify NTM
Utilized in	Reference labs only but adaptable to clinical labs with extensive in-house validation
Caveats	Not evaluated for identification of newly described species
	Some taxa may require additional endonucleases for species identification
Availability of commercial systems	No
Replaced by	Gene sequencing and other molecular methods
References	Zelazny et al. (2009), Tortoli (2003), (2006), Steingrube et al. (1995) and Telenti et al. (1993)

Matrix-assisted laser deionization-time of flight mass spectrometry (MALDI-TOF MS)

Attributes	Comments
Viable cells required	No
Technology	One of the newest methods being used to identify NTM
	Identifies based on unique spectral fingerprints produced by extracted ribosomal proteins by MALDI-TOF MS
	Relatively uncomplicated, accurate, rapid and less expensive per test
Utilized in	Typically reference labs
Caveats	Initial expense for the MS instrument (~$200,000)
	Species extraction and inactivation methods are required
	Limited MS commercial databases; need to enrich in-house database for better species identification
	Difficulty in differentiating closely related species. For example, clinically significant species and subspecies of *M. abscessus* subsp. *abscessus*, *M. abscessus* subsp. *massiliense* and *M. abscessus* subsp. *bolletii* have been difficult to differentiate by MALDI-TOF
Availability of commercial systems	Yes
Replaced by	Gene sequencing and other molecular methods
References	Tortoli (2014), Buckwalter et al. (2016), Saleeb et al. (2011), Rodríguez-Sánchez et al. (2016) and Wilen et al. (2015)

Genotypic methods for identification of NTM-16S rRNA gene sequencing

Attributes	Comments
Viable cells required	No
Technology	One of the most reliable methods being used to identify NTM
	Identifies based on 16S rRNA gene signature sequence which is species specific
	Typically, region A is sequenced, while region B could be sequenced for additional confirmation
	16s rRNA genes sequences vary by 95.7–99.7%
Utilized in	Typically reference labs but can be adapted to clinical labs
Caveats	Species extraction and inactivation methods are required
	Limited MS commercial databases; need to enrich in-house database for better species identification
	Difficulty in differentiating closely related species especially *M. chelonae* and *M. abscessus*
	Complex and cost-prohibitive method
Availability of commercial systems	Yes
References	Turenne et al. (2001), Wayne (2011) and Forbes et al. (2008)

Hsp65 gene sequencing

Attributes	Comments
Viable cells required	No
Technology	Identifies based on hsp65 hypervariable regions which are species specific
	Like PRA, 441 bp region of the hsp65 gene is analysed
	Advantageous over 16s rRNA gene sequence for differentiating closely related species
Utilized in	Typically reference labs but can be adapted to clinical labs
Caveats	Species extraction and inactivation methods are required
	Limited MS commercial databases; need to enrich in-house database for better species identification
	Complex and cost-prohibitive method
	Hsp65 is less conserved than 16S rRNA
Availability of commercial systems	Yes
References	Telenti et al. (1993) and McNabb et al. (2004)

rpoB gene sequencing

Attributes	Comments
Viable cells required	No
Technology	Identifies based on rpoB gene which are species specific
	723 bp fragment of region V of the rpoB gene is analysed
	Advantageous over 16s rRNA gene sequence for differentiating closely related species due to its small size and discriminating nature
	rpoB genes sequences vary by 84.3–96.6%
Utilized in	Typically reference labs but can be adapted to clinical labs
Caveats	Species extraction and inactivation methods are required
	Limited MS commercial databases; need to enrich in-house database for better species identification
	Complex and cost-prohibitive method
Availability of commercial systems	Yes
References	Adékambi et al. (2003), (2004), (2006), Adékambi and Drancourt (2004) and Kim et al. (2008)

In addition to the genes listed above, *dnaJ* gene, 32 kDa protein gene, superoxide dismutase (*sod*) gene, 16S-23S rRNA internal transcribed spacer, *secA1* gene, *recA* gene and erythromycin ribosomal resistance methylase (erm) genes have been proposed for identification of NTM but typically suffer from lack of enriched databases (Adékambi and Drancourt 2004).

Nucleic acid probes (Inno-LiPA multiplex probe assay)

Attributes	Comments
Viable cells required	No
Technology	Biotinylated PCR product corresponding to 16S-23S internal transcribed spacer region is hybridized with specific oligonucleotide probes immobilized as parallel lines on a membrane strip
	Major advantage is that a large variety of species can be identified by a single probe without necessitating the selection of a specific probe for each species
	Only approved for use in Europe
Utilized in	Typically reference labs but can be adapted to clinical labs
Caveats	Not currently approved by FDA
	Cross-reactivity amongst closely related species (most notably the *M. fortuitum* group)
	Inability to differentiate isolates of *M. chelonae* from *M. abscessus*
Availability of commercial systems	Yes
References	Tortoli et al. (2001), (2010) and Richter et al. (2006)

Whole-genome sequencing

Attributes	Comments
Viable cells required	No
Technology	Whole-genome sequence allows the study of multiple genes at once
	Allows linkage of genes to various organismal attributes like virulence, pathogenicity, etc.
Utilized in	Not currently widely used in reference and clinical labs but has potential to be employed in both
Caveats	Technology is in its infancy
	Complex and cost-prohibitive method
Availability of commercial systems	Yes
References	Tettelin et al. (2014), Bryant et al. (2013), Davidson et al. (2014), Ngeow et al. (2012a, b, c) and Chan et al. (2012)

12.6 Antimicrobial Susceptibility and the Treatment of NTM Infections

Unlike most other infections, the current antimicrobial treatment regimens for NTM infections are based upon their unique in vitro susceptibility patterns and have not been established by clinical trials with the exception of clarithromycin for *M. chelonae* (Wallace et al. 2002, 2004, 2005; Brown-Elliott and Wallace 2002; Griffith et al. 2007). Typically, NTM is not susceptible to the front-line anti-Mtb drugs and requires specialized susceptibility testing according to CLSI guidelines (Wayne 2011, Clinical and Laboratory Standards Institute). Due to extensive inherent drug resistance exhibited by most NTM, drug combination therapy is typically recommended for most of the NTM infections.

The currently utilized antimicrobials against NTM are tabulated below.

NTM	Antibiotics typically utilized	Comments
M. fortuitum	Amikacin, cefoxitin, imipenem, tigecycline, clarithromycin, ciprofloxacin, levofloxacin, trimethoprim-sulfamethoxazole, moxifloxacin, doxycycline, linezolid	Typically most susceptible NTM
		Express inducible erythromycin gene (*erm*)
M. abscessus	Amikacin, cefoxitin, imipenem, tigecycline, clarithromycin, azithromycin, doxycycline, ciprofloxacin, moxifloxacin, linezolid	Most drug-resistant NTM
		Express inducible erythromycin gene (*erm*)
		Lung disease due to *erm1*-positive isolates is difficult to eradicate
M. chelonae	Tobramycin, amikacin, imipenem, tigecycline, doxycycline, ciprofloxacin, linezolid, clarithromycin, azithromycin	Most drug-resistant NTM
		Typically highly resistant to cephalosporins
		No functional *erm* gene

Due to increasing drug resistance and the fact that there is a distressing lack of bactericidal antibiotics, there has been a spurt in the identification of novel antimicrobials acting against NTM, typically repurposed FDA-approved drugs. These are listed below with their antimicrobial spectrums and caveats, if any.

Drug	Attributes	Comments
Linezolid	Belongs to oxazolidinone class	MIC 4–8 μg/ml against *M. fortuitum* and *M. chelonae* (Wallace et al. 2001)
	Active against *M. fortuitum* and *M. chelonae*	
	M. abscessus expresses variable susceptibility	
Tigecycline	Belongs to glycylcycline class	MIC ≤1 μg/ml against all species
	Active against *M. fortuitum, M. chelonae* and *M. abscessus* including tetracycline-resistant *M. fortuitum*	CLSI breakpoint not available
Clofazimine	Belongs to riminophenazine class	Synergizes with amikacin against *M. fortuitum*
	Active against *M. chelonae* and *M. abscessus*	CLSI breakpoint not available
Bedaquiline	Belongs to diarylquinolone class	CLSI breakpoint not available
	Proposed to be active against all NTM	Clinical studies are lacking

As can be seen, there is a significant need for novel antibiotics or therapeutic options to improve the outcome of patients infected with NTMs, especially with *M. abscessus* (Maurer et al. 2014).

12.7 Antimicrobial Resistance and NTMs

As has been shown for multiple species, most of the NTMs are inherently resistant to a number of antimicrobial drugs including front-line anti-TB drugs. This has been extensively documented by multiple investigators (Brown-Elliott et al. 2002, 2012; Soni et al. 2016) and again highlights the lack of active antimicrobials available against NTMs.

12.8 Conclusion

Taken together, the data presented above delineates the troika of problems surrounding NTM: because the infections due to NTM are not reportable to public health authorities, the infections are typically underreported and there is lack of proper diagnostic tools and lack of bactericidal antibiotics, which significantly impacts the therapeutic outcome of the disease. Coupled with the fact that there is a serious lack of animal models mimicking human pathology, lack of utilizable vaccine and concomitantly increasing drug resistance, it presents the perfect vortex of human misery. Keeping these factors in mind, there is an urgent need to deeply and seriously study the host-pathogen interaction with NTM, develop better diagnostic tools and discover novel drugs which exhibit potent bactericidal activity to positively impact the therapeutic outcome of NTM infections.

Acknowledgements This manuscript bears CSIR-CDRI communication number 104/2018/AD.

References

Adékambi T, Drancourt M (2004) Dissection of phylogenetic relationships among 19 rapidly growing Mycobacterium species by 16S rRNA, hsp65, sodA, recA and rpoB gene sequencing. Int J Syst Evol Microbiol 54:2095–2105

Adékambi T, Colson P, Drancourt M (2003) rpoB-based identification of nonpigmented and late-pigmenting rapidly growing mycobacteria. J Clin Microbiol 41:5699–5708

Adékambi T, Reynaud-Gaubert M, Greub G, Gevaudan MJ, La Scola B et al (2004) Amoebal coculture of "*Mycobacterium massiliense*" sp. nov. from the sputum of a patient with hemoptoic pneumonia. J Clin Microbiol 42:5493–5501

Adékambi T, Berger P, Raoult D, Drancourt M (2006) rpoB gene sequence-based characterization of emerging non-tuberculous mycobacteria with descriptions of Mycobacterium bolletii sp. nov., Mycobacterium phocaicum sp. nov. and Mycobacterium aubagnense sp. Nov. Int J Syst Evol Microbiol 56:133–143

Baker AW, Lewis SS, Alexander BD, Isaacs PJ, Pickett LC, Wallace RJ Jr et al (2015) A cluster of Mycobacterium abscessus among lung transplant patients: investigation and mitigation. Open Forum Infectious Disease 2:627

Brown-Elliott BA, Wallace RJ Jr (2002) Clinical and taxonomic status of pathogenic nonpigmented or late-pigmenting rapidly growing mycobacteria. Clin Microbiol Rev 15:716–746

Brown-Elliott BA, Nash KA, Wallace RJ Jr (2012) Antimicrobial susceptibility testing, drug resistance mechanisms, and therapy of infections with nontuberculous mycobacteria. Clin Microbiol Rev 25:545–582

Bryant JM, Grogono DM, Greaves D, Foweraker J, Roddick I, Inns T (2013) Whole-genome sequencing to identify transmission of *Mycobacterium abscessus* between patients with cystic fibrosis: a retrospective cohort study. Lancet 381:1551–1560

Buckwalter SP, Olson SL, Connelly BJ, Lucas BC, Rodning AA, Walchak RC et al (2016) Evaluation of matrix-assisted laser desorption ionization-time of flight mass spectrometry for identification of Mycobacterium species, Nocardia species, and other aerobic actinomycetes. J Clin Microbiol 54:376–384

Chan J, Halachev M, Yates E, Smith G, Pallen M (2012) Whole genome sequence of the emerging pathogen *Mycobacterium abscessus* strain 47J26. J Bacteriol 194:549

Cole ST, Brosch R, Parkhill J, Garnier T, Churcher C, Harris D et al (1998) Deciphering the biology of *Mycobacterium tuberculosis* from the complete genome sequence. Nature 11;393(6685):537–544

Cooksey RC, de Waard JH, Yakrus MA, Rivera I, Chopite M, Toney SR, Morlock GP, Butler WR (2004) Mycobacterium cosmeticum sp. nov., a novel rapidly growing species isolated from a cosmetic infection and from a nail salon. Int J Syst Evol Microbiol 54:2385–2391

Cooksey RC, Jhung MA, Yakrus MA, Butler WR, Adékambi T, Morlock GP et al (2008) Multiphasic approach reveals genetic diversity of environmental and patient isolates of Mycobacterium mucogenicum and Mycobacterium phocaicum associated with an outbreak of bacteremias at a Texas hospital. Appl Environ Microbiol 74:2480–2487

Culton DA, Lachiewicz AM, Miller BA, Miller MB, Mackuen C, Groben P, White B, Cox GM, Stout JE (2013) Nontuberculous mycobacterial infection after fractionated CO(2) laser resurfacing. Emerg Infect Dis 19(3):365–370

Czaja CA, Merkel PA, Chan ED, Lenz LL, Wolf ML, Alam R et al (2014) Rituximab as successful adjunct treatment in a patient with disseminated nontuberculous mycobacterial infection due to acquired anti-interferon-γ autoantibody. Clin Infect Dis 58:e115–e118

Davidson RM, Hasan NA, Reynolds PR, Totten S, Garcia B, Levin A et al (2014) Genome sequencing of Mycobacterium abscessus isolates from patients in the United States and comparisons to globally diverse clinical strains. J Clin Microbiol 52:3573–3582

Dupont C, Terru D, Aguilhon S, Frapier J-M, Paquis M-P, Morquin D et al (2016) Source-case investigation of *Mycobacterium wolinskyi* cardiac surgical site infection. J Hosp Infect 93:235–239

Falkinham JO (2003) The changing pattern of nontuberculous mycobacterial disease. Can J Infect Dis 14:281–286

Forbes BA, Banaiee N, Beavis KG, Brown-Elliott BA, Della Latta P, Elliott LB et al (2008) Laboratory detection and identification of mycobacteria. Approved guideline, CLSI document M48-A. CSLI, Wayne

Galil K, Miller LA, Yakrus MA, Wallace RJ Jr, Mosley DG, England B et al (1999) Abscesses due to *Mycobacterium abscessus* linked to injection of unapproved alternative medication. Emerg Infect Dis 5:681–687

Gira AK, Reisenauer AH, Hammock L, Nadiminti U, Macy JT, Reeves A, Burnett C, Yakrus MA, Toney S, Jensen BJ, Blumberg HM, Caughman SW, Nolte FS (2004) Furunculosis due to *Mycobacterium mageritense* associated with footbaths at a nail salon. J Clin Microbiol 42:1813–1817

Griffith DE, Aksamit T, Brown-Elliott BA, Catanzaro A, Daley C, Gordin F et al (2007) An official ATS/IDSA statement: diagnosis, treatment, and prevention of nontuberculous mycobacterial diseases. Am J Respir Crit Care Med 175:367–416

Gubler JGH, Salfinger M, von Graevenitz A (1992) Pseudoepidemic of nontuberculous mycobacteria due to a contaminated bronchoscope cleaning machine. Report of an outbreak and review of the literature. Chest 101:1245–1249

Havlir DV, Dubé MP, Sattler FR, Forthal DN, Kemper CA, Dunne MW, California Collaborative Treatment Group et al (1996) Prophylaxis against disseminated *Mycobacterium avium* complex with weekly azithromycin, daily rifabutin, or both. N Engl J Med 335:392–398

Johnson MM, Odell JA (2014) Nontuberculous mycobacterial pulmonary infections. J Thorac Dis 6:210–220

Kim H-Y, Kook Y, Yun Y-J, Park CG, Lee NY, Shim TS et al (2008) Proportions of *Mycobacterium massiliense* and *Mycobacterium bolletii* strains among Korean *Mycobacterium chelonae-Mycobacterium abscessus* group isolates. J Clin Microbiol 46:3384–3390

Matsumoto CK, Chimara E, Ramos JP, Campos CED, De Suza Caldas PC, Lima KVB et al (2012) Rapid tests for the detection of the *Mycobacterium abscessus* subsp. *bolletii* strain responsible for an epidemic of surgical-site infections in Brazil. Mem Inst Oswaldo Cruz 107:969–970

Maurer FP, Bruderer VL, Ritter C, Castelberg C, Bloemberg GV, Böttger EC (2014) Lack of antimicrobial bactericidal activity in *Mycobacterium abscessus*. Antimicrob Agents Chemother 58:3828–3836

McNabb A, Eisler D, Adie K, Amos M, Rodrigues M, Stephens G et al (2004) Assessment of partial sequencing of the 65-kilodalton heat shock protein gene (hsp65) for routine identification of *Mycobacterium* species isolated from clinical sources. J Clin Microbiol 42:3000–3011

Nagpal A, Wentink JE, Berbari EF, Aronhalt KC, Wright AJ, Krageschmidt DA et al (2014) A cluster of *Mycobacterium wolinskyi* surgical site infections at an academic medical center. Infect Control Hosp Epidemiol 35:1169–1175

Ngeow YF, Wee WY, Wong YL, Tan JL, Ongi CS, Ng KP et al (2012a) Genomic analysis of *Mycobacterium abscessus* strain M139, which has an ambiguous subspecies taxonomic position. J Bacteriol 194:6002–6003

Ngeow YF, Wong YL, Lokanathan N, Wong GJ, Ong CS, Ng KP et al (2012b) Genomic analysis of *Mycobacterium massiliense* strain M115, an isolate from human sputum. J Bacteriol 194:4786

Ngeow YF, Wong YL, Tan JL, Arumugam R, Wong GJ, Ong CS et al (2012c) Genome sequence of *Mycobacterium massiliense* M18, isolated from a lymph node biopsy specimen. J Bacteriol 194:4125

Obregón-Henao A, Arnett KA, Henao-Tamayo M, Massoudi L, Creissen E et al (2015) Susceptibility of *Mycobacterium abscessus* to antimycobacterial drugs in preclinical models. Antimicrob Agents Chemother 59:6904–6912

Richter E, Rüsch-Gerdes S, Hillemann D (2006) Evaluation of the GenoType Mycobacterium assay for identification of mycobacterial species from cultures. J Clin Microbiol 44:1769–1775

Rodríguez-Sánchez B, Ruiz-Serrano MJ, Ruiz A, Timke M, Kostrzewa M, Bouza E (2016) Evaluation of MALDI biotyper mycobacterial library v3.0 for identification of nontuberculous mycobacteria. J Clin Microbiol 54:1144–1147

Saleeb PG, Drake SK, Murray PR, Zelazny AM (2011) Identification of mycobacteria in solid-culture media by matrix-assisted laser desorption ionization-time of flight mass spectrometry. J Clin Microbiol 49:1790–1794

Shallom SJ, Gardina PJ, Myers TG, Sebastian Y, Conville P, Calhoun LB et al (2013) New rapid scheme for distinguishing the subspecies of the *Mycobacterium abscessus* group and identifying *Mycobacterium massiliense* isolates with inducible clarithromycin resistance. J Clin Microbiol 51:2943–2949

Soni I, De Groote MA, Dasgupta A, Chopra S (2016) Challenges facing the drug discovery pipeline for non-tuberculous mycobacteria. J Med Microbiol 65:1–8

Steingrube VA, Gibson JL, Brown BA, Zhang Y, Wilson RW, Rajagopalan M et al (1995) PCR amplification and restriction endonuclease analysis of a 65-kilodalton heat shock protein gene sequence for taxonomic separation of rapidly growing mycobacteria. J Clin Microbiol 33:149–153. (Erratum, 33:1686.)

Telenti A, Marchesi F, Balz M, Bally F, Böttger EC, Bodmer T (1993) Rapid identification of mycobacteria to the species level by polymerase chain reaction and restriction enzyme analysis. J Clin Microbiol 31:175–178

Tettelin H, Davidson RM, Agrawal S, Aitken ML, Shallom S, Hasan NA et al (2014) High-level relatedness among *Mycobacterium abscessus subsp. massiliense* strains from widely separated outbreaks. Emerg Infect Dis 20:364–371

Tiwari TSP, Ray B, Jost KC Jr, Rathod MK, Zhang Y, Brown-Elliott BA et al (2003) Forty years of disinfectant failure: outbreak of post-injection *Mycobacterium abscessus* infection caused by contamination of benzalkonium chloride. Clin Infect Dis 36:954–962

Tortoli E (2003) Impact of genotypic studies on mycobacterial taxonomy: the new mycobacteria of the 1990s. Clin Microbiol Rev 16:319–354

Tortoli E (2006) The new mycobacteria: an update. FEMS Immunol Med Microbiol 48:159–178

Tortoli E (2014) Microbiological features and clinical relevance of new species of the genus Mycobacterium. Clin Microbiol Rev 27:727–752

Tortoli E, Nanetti A, Piersimoni C, Cichero P, Farina C, Mucignat G et al (2001) Performance assessment of new multiplex probe assay for identification of mycobacteria. J Clin Microbiol 39:1079–1084

Tortoli E, Pecorari M, Fabio G, Messinò M, Fabio A (2010) Commercial DNA probes for mycobacteria incorrectly identify a number of less frequently encountered species. J Clin Microbiol 48:307–310

Turenne CY, Tschetter L, Wolfe J, Kabani A (2001) Necessity of quality-controlled 16S rRNA gene sequence databases: identifying nontuberculous *Mycobacterium* species. J Clin Microbiol 39:3637–3648

Valour F, Perpoint T, Sénéchal A, Kong X-F, Bustamante J, Ferry T, Lyon TB Study Group et al (2016) Interferon-γ autoantibodies as predisposing factor for nontuberculous mycobacterial infection. Emerg Infect Dis 22:1124–1126

Viana-Niero C, Lima KVB, Lopes ML, Rabello MC, Marsola LR, Brilhante VCR et al (2008) Molecular characterization of *Mycobacterium massiliense* and *Mycobacterium bolletii* in isolates collected from outbreaks of infections after laparoscopic surgeries and cosmetic procedures. J Clin Microbiol 46:850–855

Villanueva A, Calderon RV, Vargas BA, Ruiz F, Aguero S, Zhang Y et al (1997) Report on an outbreak of post-injection abscesses due to *Mycobacterium abscessus*, including management with surgery and clarithromycin therapy and comparison of strains by random amplified polymorphic DNA polymerase chain reaction. Clin Infect Dis 24:1147–1153

Vugia DJ, Jang Y, Zizek C, Ely J, Winthrop KL, Desmond E (2005) Mycobacteria in nail salon whirlpool footbaths, California. Emerg Infect Dis 11:616–618

Wallace RJ Jr, Brown-Elliott BA, Ward SC, Crist CJ, Mann LB, Wilson RW (2001) Activities of linezolid against rapidly growing mycobacteria. Antimicrob Agents Chemother 45:764–767

Wallace RJ Jr, Brown-Elliott BA, Hall L, Roberts G, Wilson RW, Mann LB et al (2002) Clinical and laboratory features of *Mycobacterium mageritense*. J Clin Microbiol 40:2930–2935

Wallace RJ Jr, Brown-Elliott BA, Wilson RW, Mann L, Hall L, Zhang Y et al (2004) Clinical and laboratory features of *Mycobacterium porcinum*. J Clin Microbiol 42:5689–5697

Wallace RJ Jr, Brown-Elliott BA, Brown J, Steigerwalt AG, Hall L, Woods G et al (2005) Polyphasic characterization reveals that the human pathogen *Mycobacterium peregrinum* type II belongs to the bovine pathogen species *Mycobacterium senegalense*. J Clin Microbiol 43:5925–5935

Wayne PA (2011) Clinical and laboratory standards institute. Susceptibility testing of mycobacteria, nocardiae, and other aerobic actinomycetes. Approved standard, 2nd ed. CLSI document M24-A2.

Wilen CB, McMullen AR, C-AD B (2015) Comparison of sample preparation methods, instrumentation platforms, and contemporary commercial databases for identification of clinically relevant mycobacteria by matrix-assisted laser desorption ionization-time of flight mass spectrometry. J Clin Microbiol 53:2308–2315

Winthrop KL, Albridge K, South D, Albrecht P, Abrams M, Samuel MC, Leonard W, Wagner J, Vugia DJ (2004) The clinical management and outcome of nail salon-acquired Mycobacterium fortuitum skin infection. Clin Infect Dis 38:38–44

Winthrop KL, Chang E, Yamashita S, Iademarco MF, LoBue PA (2009) Nontuberculous mycobacteria infections and anti-tumor necrosis factor-alpha therapy. Emerg Infect Dis 15:1556–1561

Zelazny AM, Root JM, Shea YR, Colombo RE, Shamputa IC et al (2009) Cohort study of molecular identification and typing of *Mycobacterium abscessus, Mycobacterium massiliense*, and *Mycobacterium bolletii*. J Clin Microbiol 47:1985–1995

Chapter 13
Drugs Under Preclinical and Clinical Testing for the Treatment of Infections Caused due to *Staphylococcus aureus*: An Update

Manjulika Shukla, Isha Soni, Arunava Dasgupta, and Sidharth Chopra

Abstract *Staphylococcus aureus* is a significant pathogen of interest worldwide owing to its increasing drug resistance and dwindling antimicrobial armamentarium available against it. Although it typically causes uncomplicated infections, recently increasingly it has been linked to serious community and nosocomial infections ranging from boils to bloodstream infections to endocarditis. Even though there are a number of antibiotics available for the treatment of uncomplicated *S. aureus* infections, the advent of drug resistance has complicated the picture due to decreasing options available for the treatment coupled with increased transmission of drug-resistant strains in the community. Even though the drug discovery pipeline has recently been augmented with the discovery of some new molecules active against *S. aureus*, the true status is anemic owing to lack of molecules acting via new mechanism of action. This chapter describes the various molecules which are in preclinical and clinical development against *S. aureus* and depicts the various challenges as well as lacunae in them.

Keywords Staphylococcus · Preclinical · Drugs · MRSA

13.1 Introduction

Staphylococcus aureus, a typically commensal microflora, is also one of the most notorious pathogens among the so-called "ESKAPE" pathogens (*Enterobacter* sp., *S. aureus*, *Klebsiella pneumoniae*, *Acinetobacter baumannii*, *Pseudomonas aeruginosa*, and *Enterococcus* sp.) as defined by IDSA and other organizations responsible for causing a variety of serious acute and chronic community as well as

Authors Manjulika Shukla and Isha Soni have equally contributed to this chapter.

M. Shukla · I. Soni · A. Dasgupta · S. Chopra (✉)
Division of Microbiology, CSIR-Central Drug Research Institute, Lucknow, India
e-mail: a.dasgupta@cdri.res.in

P. P. Singh (ed.), *Infectious Diseases and Your Health*,
https://doi.org/10.1007/978-981-13-1577-0_13

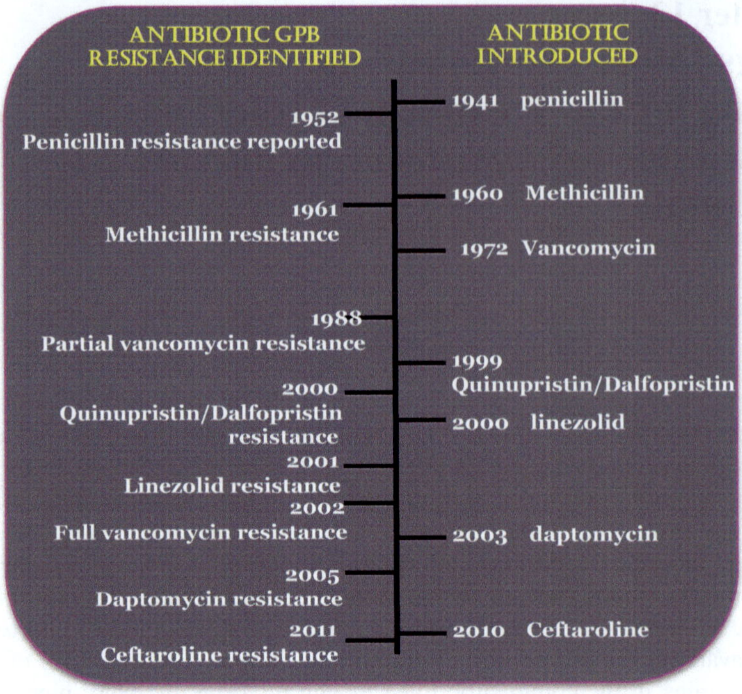

Fig. 13.1 Diagrammatic representation of the introduction of the antibiotic into the clinic and subsequent identification of antibiotic resistance

hospital-acquired infections (HAI) (Boucher et al. 2009). These infections range from skin and skin structure infections such as boils and abscesses to bacteremia, pneumonia, osteomyelitis, endocarditis, meningitis, and toxic shock syndrome. Figure 13.1 (A) depicts a scanning electron micrograph of MRSA (A), as well as cutaneous infection caused due to MRSA in (B).

The treatment of infections caused due to *S. aureus* is typically uncomplicated, but due to the increasing advent of antimicrobial resistance exhibited by *S. aureus* clinical isolates, the discovery and development of new molecules active against *S. aureus* has become a worldwide scientific and medical high priority. This is reflected in the increasingly populated drug discovery pipeline targeting *S. aureus*, although majority of the molecules are from already known classes of drugs (Kumar and Chopra 2013). This is disturbing as increasingly drug-resistant strains of *S. aureus* are being isolated worldwide, thus decreasing the therapeutic options which are available to the attending infectious disease physicians and are directly responsible for increasing morbidity and mortality associated with staphylococcal infections.

Drug resistance in *S. aureus* is not a new feature. As seen in Fig. 13.2, the identification of resistance usually happens within a couple of years of the commercial release of the drug. This trend started with the introduction of penicillin for the treatment of *S. aureus* infections in 1941, and its resistance was identified in 1952;

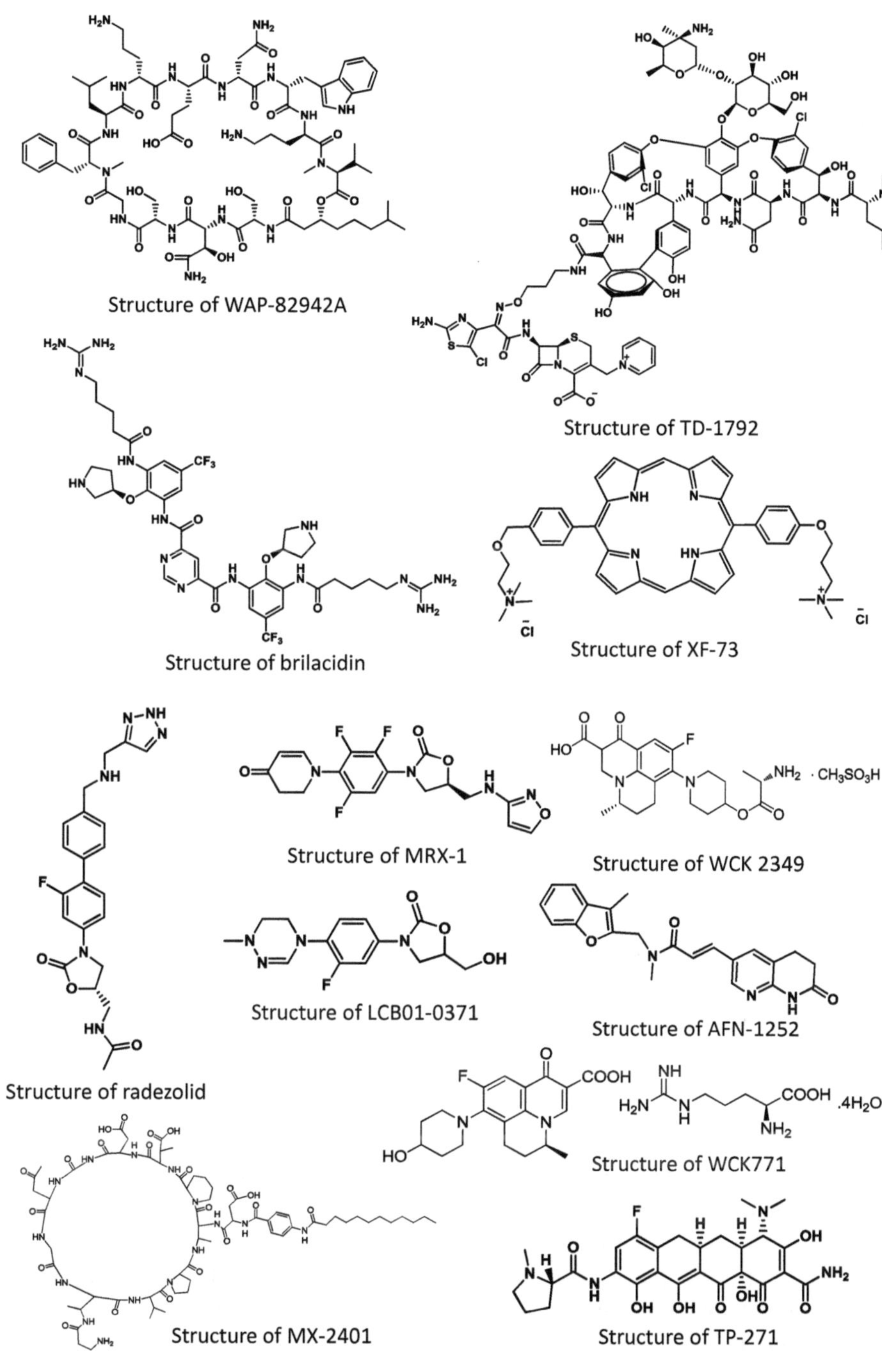

Fig. 13.2 Structure of drugs under preclinical and clinical assessment against *S. aureus*

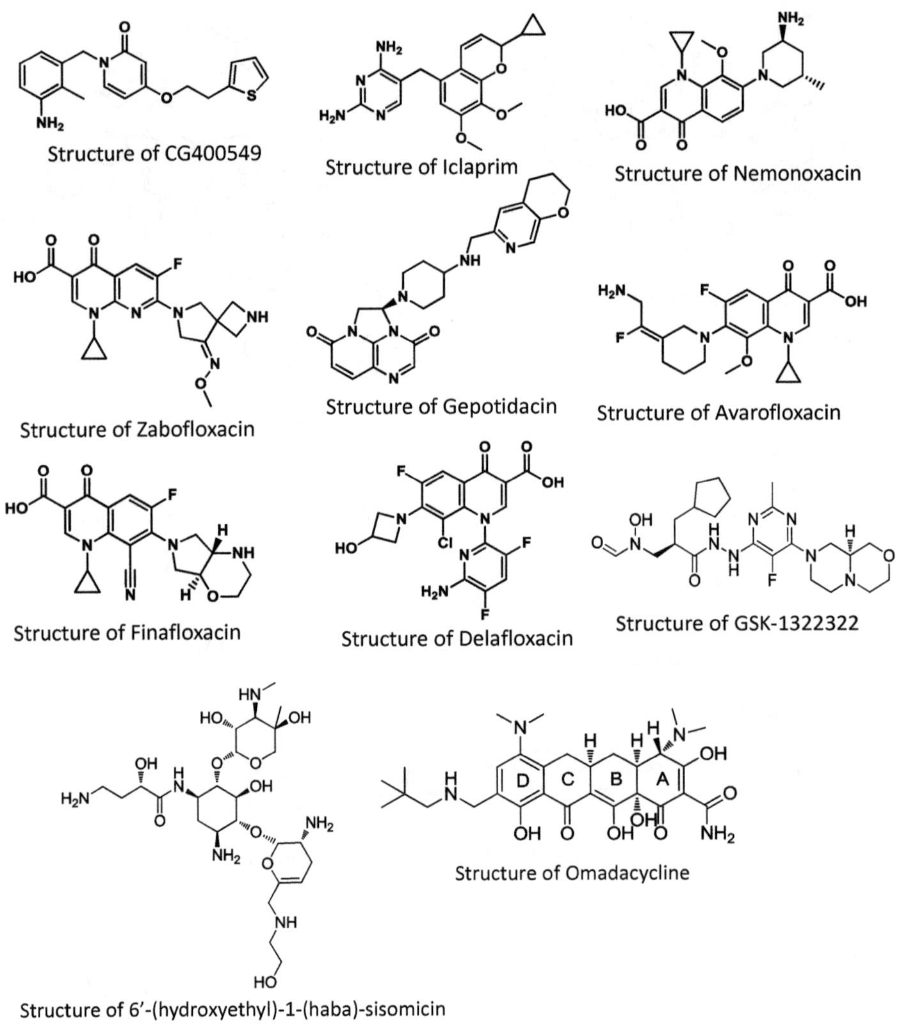

Fig. 13.2 (continued)

methicillin was introduced in 1960, and resistance to methicillin was identified in 1961, the so-called MRSA, followed by identification of vancomycin-resistant *S. aureus* (VRSA) in 1988. The latest casualty of this trend was ceftaroline with its introduction in 2010 and the first reported resistance in 2011. In addition, MRSA is listed a high-priority pathogen for which novel drugs are urgently required by WHO in 2017. Taken together, it makes a powerful case for a dedicated drug discovery and development pipeline to continuously address this issue.

The aim of this article is to give an update of drugs in the preclinical and clinical pipeline active against *S. aureus*. In this report, agents are mainly categorized and subcategorized on the basis of their mode of action and chemical class, respectively. Their structures are depicted in Fig. 13.3.

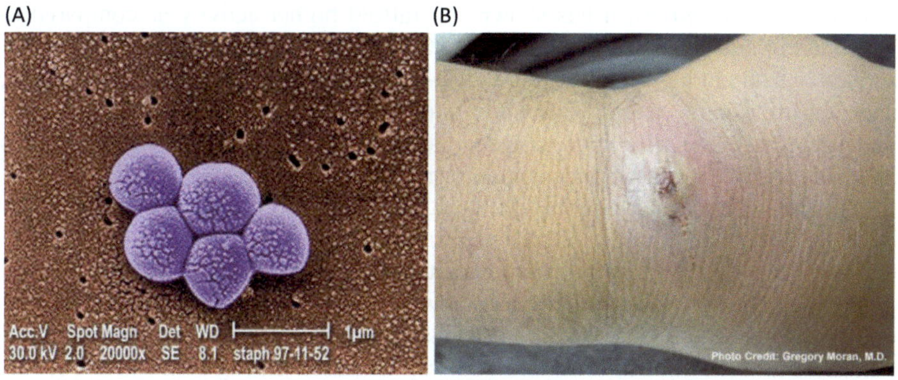

Fig. 13.3 EM of *S. aureus* and cutaneous abscess caused due to methicillin-resistant *S. aureus* (**a**) Colorized SEM depicts MRSA (Magnified 20,000X). (Courtesy: CDC, USA) (**b**) Cutaneous abscess on the foot post packing (front view), caused due to MRSA. (Courtesy: CDC, USA)

13.2 Molecules Acting on Peptidoglycan Biosynthesis and Cell Membrane

Lipopeptide MX-2401 (Migenix Incorporated), a novel semisynthetic analog of amphomycin, is an expanded-spectrum calcium-dependent cyclic lipopeptide possessing bactericidal activity against several clinically relevant Gram-positive bacteria including *Enterococcus*, *S. pneumoniae*, and MRSA (Craig et al. 2010). Despite being structurally similar to daptomycin and possessing a similar mechanism of action by binding to undecaprenyl phosphate (C_{55}-P) and inhibiting cell wall biosynthesis, MX-2401 is not inhibited by lung surfactant. This improves its efficacy for the treatment of pulmonary infections (Rubinchik et al. 2011). Although still in preclinical development phase, MX-2401 seems to be a promising candidate for further evaluation for the treatment of Gram-positive infections, including hospital-acquired pneumonia.

Lotilibcin (WAP-8294A_2; aRigen Pharmaceuticals) is a naturally occurring cyclic lipodesipeptide isolated from Gram-negative bacteria *Lysobacter staphylocidin* that exhibits potent activity against MRSA through selective interaction with bacterial membrane phospholipids (Chen et al. 2015). Lotilibcin has rapidly bactericidal nature and is specific to MRSA, thus expressing a narrow spectrum of action. It has shown 14-fold higher antibacterial activity as compared to vancomycin and has shown potent activity in animal infection models of MRSA. It is being developed for the treatment of ABSSI and is currently under Phase I clinical trials.

Cefilavancin (TD-1792; Theravance Biopharma) is a bactericidal multivalent glycopeptide-cephalosporin heterodimer antibiotic that possesses properties of both a glycopeptide and a beta-lactam and a synergized mode of action (Leuthner et al. 2010). In neutropenic murine thigh infection model and subcutaneous thigh infection model of *S. aureus*, it has shown severalfold higher activity as compared to vancomycin and linezolid (Hedge et al. 2007). In a randomized, double-blinded, active-controlled Phase II clinical trial, TD-1792 was shown to be safe, efficacious, and non-inferior to vancomycin when injected IV (Stryjewski et al. 2012). It is undergoing a Phase III clinical trial for the treatment of Gram-positive infections and SSI (http://www.theravance.com/bacterial).

TD-1607 (Theravance Biopharma) is a glycopeptide-cephalosporin heterodimer developed for the treatment of serious Gram-positive infections. It is structurally distinct from cefilavancin but demonstrates a similar mode of action, in vitro potency, and bactericidal profile. It has shown potent in vitro activity against numerous contemporary clinical MRSA isolates and has completed Phase I clinical trial where it has been shown to be safe and well tolerated among healthy individuals (https://www.jmilabs.com/data/posters/ICAAC2014/F-970.PDF).

13.3 Cell Membrane Inhibitors

Brilacidin (PMX-30063; Cellceutix) is an arylamide foldamer designed to mimic the amphiphilic properties of antimicrobial peptides and is being developed for the treatment of ABSSSI as IV. In Phase II clinical trials, it has demonstrated clinical response rates comparable to those of daptomycin and was shown to be safe and well tolerated (NCT01211470). Brilacidin causes a dose-dependent depolarization of the *Staphylococcus aureus* membrane comparable to that of daptomycin, thus representing a new class of antibiotics (Mensa et al. 2014).

XF-73 (Destiny Pharma) is a novel dicationic porphyrin that rapidly kills a wide range of Gram-positive bacteria including MRSA by interfering cell membrane, resulting in the loss of vital components from the cell, without bursting the cells themselves. Due to its targeting the bacterial membrane, XF-73 kills bacteria in all growth phases, including nongrowing cultures and bacteria within biofilms (Farrell et al. 2011a, b). Owing to its target, there is a less likely possibility of generating XF-73-resistant mutants. It is being currently investigated for topical use in the treatment of MRSA infections and decolonization and has successfully completed Phase I trial of a nasal formulation for safety and local tolerability for decolonization of *S. aureus.*

13.4 Protein Synthesis Inhibitors

Fusidic acid (CEM-102; Cempra Pharmaceuticals), approved as a topical antibacterial, is being developed as an oral formulation for use in the USA for the treatment of ABSSSI and joint infections. It inhibits protein synthesis by preventing the release of elongation factor-G (EF-G) from the ribosome. In Phase II clinical trial, fusidic acid was shown to be non-inferior to linezolid. In a double-blinded Phase III trial, the oral formulation has shown non-inferiority to linezolid for the treatment of MRSA infection (https://www.investor.cempra.com/releasedetail.cfm?ReleaseID=1014395). Although approved since the 1960s, there is a significant lack of resistance to fusidic acid among *S. aureus* clinical isolates (McLaws et al. 2011).

13.5 Macrolide

Solithromycin (CEM-101; Cempra Pharmaceuticals), a next-generation semisynthetic ketolide, has been developed for the treatment of CABP and differs from telithromycin in having fluorine substitution at C-2 and an aminophenyl moiety instead of the pyridine moiety, which ameliorates the nicotinic acetylcholine receptor blockage. Solithromycin inhibits protein synthesis by binding to the domain II and domain V of 23S rRNA. Additionally, because of its substitutions, it interacts at a third site on the ribosome and thus is especially active against macrolide-resistant strains (Fernandes et al. 2016). It has successfully completed Phase III trials with both oral and IV formulations, and a NDA has been accepted by the FDA.

13.6 Oxazolidinone

The oxazolidinones, a totally synthetic class of novel antibiotics, have strong activity against nearly all Gram-positive organisms, including those resistant to other drugs. They inhibit protein synthesis by binding to domain V of the 23S rRNA, thereby blocking formation of the initiation complex.

Radezolid (RX-1741; Melinta Therapeutics) is a novel biaryloxazolidinone having improved activity, even against linezolid-resistant strains, and binds to 23S rRNA in the peptidyl transferase center, thus inhibiting protein synthesis by obstructing binding of incoming aminoacyl-tRNA to the A site (Lawrence et al. 2008; Leach et al. 2007; Colca et al. 2003; Ippolito et al. 2008). It is being developed as an oral and topical formulation for the treatment of CAP and uSSSI. It has shown good cure rates and was well tolerated in two completed Phase II trials against uSSSI and CAP. The most commonly reported adverse effects in clinical trial were gastrointestinal symptoms.

MRX-1 (MRX-I; MicuRx Pharmaceuticals), an analog of linezolid, is being developed as an oral formulation to specifically reduce myelotoxicity and monoamine oxidase inhibition, the signature toxicities associated with linezolid (Gordeev and Yuan 2014). MRX-1 has completed Phase II clinical trials with excellent antibacterial activities against MRSA, outpacing linezolid, and demonstrated no significantly adverse side effects. Phase III trials have begun enrolling patients with ABSSSI to compare MRX-1 with linezolid.

LCB01-0371 (LegoChem Biosciences), a novel oxazolidinone with cyclic amidrazone, has been developed for the treatment of infections caused by MRSA, VRE, and MDR-TB. LCB01-0371 demonstrates potency comparable to torezolid in mouse systemic infection models. It has completed Phase I clinical trial with potency better than linezolid and improved safety in terms of lower myelosuppression and monoamine oxidase inhibition. LCB01-0371 can be administered both parenterally and IV. Combined with its excellent spectrum of activity and multiple routes of administration makes it a good candidate for long-term therapy (Jeong et al. 2010, http://adisinsight.springer.com/drugs/800035856). A Phase II study to explore the EBA, safety, and pharmacokinetics of orally administered LCB01-0371 is about to be initiated.

LCB01-0699 (LegoChem Biosciences) is a new oxazolidinone under preclinical studies for Gram-positive organisms including MRSA, VRE, MDR-TB, and linezolid-resistant strains. It is under preclinical studies and shares most of its properties with LCB01-0371 (http://adisinsight.springer.com/drugs/800027159).

13.7 Aminomethylcyclines

Omadacycline (PTK-0796, Paratek Pharmaceuticals) is a novel, broad-spectrum aminomethyl tetracycline developed for the treatment of ABSSSI, CABP, and cUTI caused due to Gram-positive bacteria. It was designed to circumvent two key mechanisms of tetracycline resistance: ribosome protection and tetracycline efflux, which is reflected in its potent antimicrobial activity against a wide range of drug-resistant clinical isolates. Omadacycline was shown to be efficacious, well tolerated, and non-inferior to linezolid for the treatment of ABSSSI in a Phase II clinical trial (Noel et al. 2012). In a Phase III trial, omadacycline was reported to be safe, efficacious, and non-inferior to linezolid for ABSSSI. Two more Phase III trials are under way, one to compare once-daily oral dose of omadacycline to twice-daily oral dose of moxifloxacin for the treatment of ABSSSI and second to compare safety and efficacy of omadacycline with linezolid for CABP patients.

Eravacycline (TP-434, Tetraphase Pharma) is a broad-spectrum synthetic fluorocycline which retains its activity in the presence of tetracycline efflux pumps and ribosome protection proteins. In vitro studies have shown potent activity against numerous clinical isolates and MDR bacteria such as MRSA (Xiao et al. 2012; Chopra and Dasgupta 2014). It has recently completed a Phase II clinical trial for its use in cIAI when compared with ertapenem where it demonstrated equal efficacy at comparatively lesser dosage. Eravacycline is currently being examined in a multicenter Phase III study to assess the efficacy, safety, and pharmacokinetics of eravacycline compared with meropenem in the treatment of cIAIs (NCT02784704). It is also being examined for its efficacy, safety, and pharmacokinetics as compared to ertapenem and levofloxacin in treating cUTI patients (NCT03032510 and NCT01978938).

TP-271 (Tetraphase Pharma) is a novel broad-spectrum synthetic fluorocycline developed for the treatment of respiratory tract infections caused by susceptible and antibiotic-resistant public health pathogens, including *Francisella tularensis*, *Yersinia pestis*, and *Bacillus anthracis* (Liu and Myers 2016 (http://adisinsight.springer.com/drugs/800044518)). It has received the QIDP designation from FDA, and Phase I trial to assess the safety, tolerability, and pharmacokinetics of TP-271 has been initiated.

13.8 Aminoglycosides

Plazomicin (ACHN-490; Achaogen Incorporated), a neoglycoside derived from sisomicin, has been developed to overcome the effects of aminoglycoside-modifying enzymes (AMEs), which are the primary mode of resistance to aminoglycosides (Tenover et al. 2011). It has shown potent efficacy against MDR Gram-negative bacteria as well as MRSA. In a Phase II clinical trial for the treatment of cUTI, plazomicin was shown to be safe and non-inferior to levofloxacin. Similarly, in Phase III clinical trial termed EPIC (Evaluating Plazomicin in cUTI), it has met the primary goal of non-inferiority to meropenem.

13.9 Peptide Deformylase Inhibitors

Lanopepden (GSK 1322322; GlaxoSmithKline) is a selective peptide deformylase inhibitor with potent activity against MRSA (Ross et al. 2011). In a randomized, double-blinded, placebo-controlled Phase I clinical trial, it was shown to be safe and well tolerated among healthy individuals. In another Phase II clinical trial, it was shown to be safe, well tolerated, and efficacious for the treatment of confirmed cases of ABSSSI. However, GlaxoSmithKline has terminated the lanopepden program (http://adisinsight.springer.com/drugs/800030124).

13.10 DNA Synthesis Inhibitors

13.10.1 Fluoroquinolones

Delafloxacin (Baxdela, Melinta Therapeutics) is an anionic fluoroquinolone that acts on both DNA topoisomerase II and topoisomerase IV and thus is designed to exhibit potency against MDR pathogens (Remy et al. 2012). It is being developed for the treatment of ABSSSI and is available as both oral and IV formulations. In a

Phase II clinical trial, where it was compared to tigecycline, delafloxacin met its primary and secondary efficacy end points and is currently being evaluated in a Phase III clinical trial versus vancomycin and aztreonam for the treatment of ABSSSI; no results are posted yet (NCT01984684). It is also undergoing a Phase III trial to evaluate the safety and efficacy of delafloxacin compared to moxifloxacin in the treatment of adult patients with CAP (NCT02679573).

Finafloxacin (Xtoro, MerLion Pharmaceuticals) is a novel fluoroquinolone designed to be activated under acidic conditions, thus eliminating the reservoirs of intracellular infection (Stubbings et al. 2011). Under acidic conditions, finafloxacin demonstrates superior activity when compared to moxifloxacin. A finafloxacin suspension was recently approved by FDA for the treatment of acute *otitis externa* or swimmer's ear caused due to *P. aeruginosa* and *S. aureus*. Various formulations are in Phase I and II evaluation for the treatment of uUTI, cUTI, and pyelonephritis.

Avarofloxacin (JNJ-Q2, Johnson and Johnson) is an aminoethylidenylpiperidine fluoroquinolone with a zwitterion structure that demonstrates antibacterial effect against numerous MDR Gram-negative and Gram-positive bacteria including MRSA and is being developed as both oral and IV formulations (Farrell et al., 2011a, b). During one Phase II trial conducted to study its safety and efficacy for treating ABSSSI caused by MRSA, JNJ-Q2 was found to be highly active, well tolerated, and non-inferior to linezolid.

Gepotidacin (GSK 2140944; GlaxoSmithKline) is a novel triazaacenaphthylene DNA topoisomerase II inhibitor (Bouchillon et al. 2013). Its mechanism is different from fluoroquinolone binding to DNA gyrase; thus it demonstrates efficacy against fluoroquinolone-resistant isolates. In addition, it is especially active against various MDR isolates of *N. gonorrhoeae* (Farrell et al. 2017). It is currently being evaluated for bioavailability, food effect, and pharmacokinetics of gepotidacin in a Phase I trial in healthy subjects (NCT02853435).

Nemonoxacin (TG-873870) is a C-8-methoxy nonfluorinated broad-spectrum quinolone. The C-8-methoxy substituent on the quinolone ring increases antibacterial effectiveness against Gram-positives and reduces selection of resistant mutants (Guo et al. 2012). When tested in vitro against clinical fluoroquinolone-resistant isolates of Gram-negative and Gram-positive bacteria including staphylococci, streptococci, enterococci, *N. gonorrhoeae*, and *H. influenzae*, nemonoxacin demonstrates better activity than ciprofloxacin, levofloxacin, and moxifloxacin (http://adisinsight.springer.com/drugs/800022726). It has been tested in a Phase II to determine the safety and efficacy of nemonoxacin in diabetic foot infections (NCT00685698). It has also been evaluated in Phase II trial for the treatment of CAP in comparison to levofloxacin (NCT00434291). It has been tested against levofloxacin to determine its efficacy and safety in CAP patients (NCT01529476).

Zabofloxacin (DW-224a, Dong Wha) is a broad-spectrum fluoroquinolone active against fluoroquinolone-resistant bacteria including MRSA and *Neisseria gonorrhoeae* (Kim et al. 2004). In a double-blinded, double-dummied, randomized, parallel-grouped Phase III clinical trial, zabofloxacin was shown to be non-inferior to moxifloxacin (DW224a-II-1; KCT0001343). It has also been evaluated for the treatment of CAP in a Phase II trial, but the studies have been terminated due to financial considerations (NCT01081964). It also demonstrates excellent activity against intracellular pathogens such as *Legionella pneumophila* (http://adisinsight.springer.com/drugs/800019661).

WCK 771/WCK 2349 (Wockhardt) is a broad-spectrum bactericidal fluoroquinolone drug derived from benzoquinolizine levonadifloxacin. It targets bacterial DNA gyrase and DNA topoisomerase IV and inhibits *NorA* and thus exhibits potent activity against levofloxacin- and moxifloxacin-resistant MRSA. In various animal infection models, it has shown superior activity as compared to other quinolones (Patel et al. 2004). In Phase I clinical trial, it was shown to be safe and well tolerated among healthy individuals (http://adisinsight.springer.com/drugs/800016734).

13.11 Dihydrofolate Reductase (DHFR) Inhibitors

Iclaprim (AR-100 and RO-48-2622, Arpida Limited) is a novel diaminopyrimidine and has been shown to possess strong antagonistic activity against the bacterial DHFR enzyme. It is also active against trimethoprim-resistant MRSA (Schneider et al. 2003). In Phase II and Phase III trial studies, iclaprim was shown to be equally safe and non-inferior to vancomycin and linezolid for treating ABSSSI and cSSSI when compared to linezolid (NCT00299520) caused by MRSA, respectively.

13.12 Fatty Acid Synthesis Inhibitors

CG400549 (Crystal Genomics) belongs to a novel structural scaffold that targets the fatty acid biosynthesis enzyme called FabI within the FASII pathway, a critical enzyme in generating bacterial cell membrane. It has displayed superior in vitro efficacy and four- to eightfold higher potency when compared with linezolid, daptomycin, and vancomycin in treating MRSA, VRSA, and VISA (Park et al. 2007a, b). It has successfully completed a Phase II trial for the treatment of cABSSSI caused due to MRSA and has been demonstrated to be more potent in comparison to linezolid and vancomycin (Kim and Sohn 2011; Yum et al. 2007; Park et al. 2007a, b) (NCT01593761), but there is no further information available publically as to its future.

Debio1450 (AFN-1720; Debiopharm), the prodrug of Debio1452 (AFN-1252), is a highly potent and selective agent against several staphylococcal species and MDR strains such as MRSA and VISA. Both chemical entities inhibit the enzyme FabI enoyl reductase, the acyl carrier protein, resulting in a bacteriostatic mode of action (Kaplan et al. 2012). A preclinical study of Debio1452 in several murine models of infection has validated its effectiveness. It has completed a Phase II clinical trial evaluating the efficacy of both intravenous (IV) and oral formulations in comparison with IV vancomycin switched to oral linezolid for the treatment of ABSSSI (NCT02426918). It is predicted that this entity will preserve human microbiota and reduce adverse effects associated with antibiotics such as *Clostridium difficile* overgrowth because of its specificity for staphylococcal species.

13.13 Summary

This chapter provides a detailed overview of current anti-MRSA drug pipeline and summarizes the excitements and loopholes in the current experimental therapies, which are tabulated in Table 13.1. It is important to keep in mind that the drug pipeline for most of the infectious disease pathogens including *S. aureus* is facing a severe deficit of putative drug candidates. In the past decades, the overly misuse of antibiotics has led to severe healthcare crises as the cases of MRSA are growing in both developed and developing countries with very little options available for their treatment. Recent Phase III trials for investigative drugs including MRX-1, omadacycline, plazomicin, zabofloxacin, and iclaprim have raised the hope for their promising effects against a range of the clinical isolates of MRSA. There have been some bright spots such as the discovery and development of TD-1792 and TD1607 and other novel modes of action antibiotics; however much still needs to be done to stem this burgeoning crisis.

Apart from these although the drug pipeline against MRSA has multiple potent candidates, there is a massive void which still exists in the discovery of new antibacterial chemical classes. Post-marketing surveillance and judicious use of new antimicrobial agents are necessary to ensure the effectiveness and longevity of the currently approved drugs. Presently, the drug pipeline is occupied with the focus on the single target. In the future, focusing on multiple targets and other avenues including combination therapy and identification of newer targets is the need of time for the generation of effective antimicrobials to combat resistance.

The discovery and development of molecules exhibiting anti-MRSA activity is a highly active field with a number of agents under preclinical and clinical development. Despite its appearance, the drug discovery pipeline is extremely anemic and requires a continuous resupply effort to be able to ameliorate the antimicrobial resistance crises.

Table 13.1 The various compounds active against *S. aureus* are listed along with their status above

Drug name	Company	Chemical class	Indication	Development phase	Mechanism of action (known or novel)
Cell wall and cell membrane inhibitors					
MX-2401	Migenix Incorporated	Lipopeptide	Gram-positive infections	Preclinical	Known
WAP-8294A2 (lotilibcin)	aRigen Pharmaceutical	Lipopeptide	ABSSI	Phase I	Known
TD-1792	Theravance	Cephalosporin/vancomycin heterodimer	ABSSI	Phase III	Known
TD-1607	Theravance	Cephalosporin/glycopeptide heterodimer	Gram-positive infections	Phase I	Known
Brilacidin	Cellceutix Corporation	Defensin mimetic	ABSSI, mucositis, stomatitis	Phase II	Known
XF-73	Destiny Pharmaceutical	Porphyrin	Gram-positive infections	Phase I	Novel
Protein synthesis inhibitors					
Fusidic acid	Cempra Pharmaceutical	Cholestadienes	ABSSI and joint infections	Phase III	Known
Solithromycin	Cempra Pharmaceutical	Macrolide	CABP	Phase III	Known
Radezolid	Melinta Therapeutics	Oxazolidinone	uSSSCI; CABP	Phase II	Known
MRX-I	MicuRx Pharmaceutical	Oxazolidinone	ABSSI	Phase III	Known
LCB01-037	LegoChem Biosciences	Oxazolidinone	Gram-positive infections; TB	Phase II	Known
Omadacycline	Paratek	Aminomethylcycline	ABSSI	Phase III	Known

(continued)

Table 13.1 (continued)

Drug name	Company	Chemical class	Indication	Development phase	Mechanism of action (known or novel)
Eravacycline	Tetraphase Pharmaceutical	Aminomethylcycline	cIAI	Phase III	Known
TP-271	Tetraphase Pharmaceutical	Aminomethylcycline	Respiratory infections	Phase I	Known
Plazomicin	Achaogen	Aminoglycosides	cUTI and pyelonephritis	Phase III	Known
Lanopepden	GSK	Cyclopentanes	cSSSI	Phase II	Known
Nucleic acid inhibitors					
Delafloxacin	Melinta Therapeutics	Fluoroquinolone	ABSSSI (Melinta)	Phase III	Known
Finafloxacin	MerLion Pharmaceutical	Fluoroquinolone	cUTI and pyelonephritis	Phase II	Known
Avarofloxacin	Furiex	Fluoroquinolone	ABSSSI	Phase II	Known
Gepotidacin	GSK	Fluoroquinolone	Gram-positive infections and *N. gonorrhoeae*	Phase I	Known
Nemonoxacin	TaiGen Biotech	Fluoroquinolone	ABSSSI and CABP	Phase III	Known
Zabofloxacin	Dong Wha Pharma	Fluoroquinolone	CABP	Phase III	Known
Iclaprim	Arpida Ltd.	Dihydrofolate reductase inhibitors	ABSSSI	Phase III	Known
Fatty acid inhibitors					
CG400549	Crystal genomics	Fatty acid synthesis inhibitor	ABSSSI	Phase II	Known
Debio1450	Debio/Nobelex	Benzodiazepine	ABSSSI	Phase II	Known

Acknowledgments This manuscript CSIR-CDRI communication number 9662.

References

Boucher HW, Talbot GH, Bradley JS, Edwards GD, Rice LB et al (2009) Bad bugs, no drugs: no ESKAPE! An update from the Infectious Diseases Society of America. Clin Infect Dis 48:1–12

Bouchillon SK, Hackel M, Miller LA, Scangarella-Oman NE (2013) Abstract F-1216. 53rd annual interscience conference on antimicrobial agents and chemotherapy; September, Denver, CO

Chen H, Olson AS, Dussault AH, Du L (2015) Fatty acyl incorporation in the biosynthesis of WAP-8294A, a group of potent anti-MRSA cyclic lipodepsipeptides. RSC Adv 5:105753–105759

Chopra S, Dasgupta A (2014) Eravacycline: a monograph. Drugs Future 39:247

Colca JR, McDonald WG, Waldon DJ et al (2003) Cross-linking in the living cell locates the site of action of oxazolidinone antibiotics. J Biol Chem 278:21972–21979

Craig WA, Andes DR, Stamstad T (2010) In vivo pharmacodynamics of new lipopeptide MX-2401. Antimicrob Agents Chemother 54:5092–5098

Farrell DJ, Liverman LC, Biedenbach DJ et al (2011a) JNJ-Q2, a new fluoroquinolone with potent in vitro activity against *Staphylococcus aureus*, including methicillin- and fluoroquinolone-resistant strains. Antimicrob Agents Chemother 55:3631–3634

Farrell DJ, Robbins M, Rhys-Williams W et al (2011b) Investigation of the potential for mutational resistance to XF-73, retapamulin, mupirocin, fusidic acid, daptomycin, and vancomycin in methicillin-resistant *Staphylococcus aureus* isolates during a 55-passage study. Antimicrob Agents Chemother 55:1177–1181

Farrell DJ, Sader HS, Rhomberg PR, Scangarella-Oman NE, Flamm RK (2017) In vitro activity of Gepotidacin (GSK2140944) against *Neisseria gonorrhoeae*. Antimicrob Agents Chemother 61:e02047–e02016

Fernandes P, Martens E, Bertrand D, Pereira D (2016) The solithromycin journey-it is all in the chemistry. Bioorg Med Chem 24:6420–6428

Gordeev MF, Yuan ZY (2014) New potent antibacterial oxazolidinone (MRX-I) with an improved class safety profile. J Med Chem 57:4487–4497

Guo B, Wu X, Zhang Y, Shi Y, Yu J, Cao G, Zhang J (2012) Safety and clinical pharmacokinetics of nemonoxacin, a novel non-fluorinated quinolone, in healthy Chinese volunteers following single and multiple oral doses. Clin Drug Investig 32(7):475–486

Hedge SS et al (2007) Efficacy of Td-1792, a novel multivalent glycopeptide, against Gram-positive pathogens in the neutropenic thigh model, abstr. A-42. Abstr. 47th intersci. conf. antimicrobial agents and chemotherapy. Chicago, IL

http://adisinsight.springer.com/drugs/800016734. Last accessed 18 June 2017

http://adisinsight.springer.com/drugs/800019661. Last accessed 18 June 2017

http://adisinsight.springer.com/drugs/800022726. Last accessed 18 June 2017

http://adisinsight.springer.com/drugs/800027159. Last accessed 18 June 2017

http://adisinsight.springer.com/drugs/800030124. Last accessed 18 June 2017

http://adisinsight.springer.com/drugs/800035856. Last accessed 18 June 2017

http://adisinsight.springer.com/drugs/800044518. Last accessed 18 June 2017

http://www.theravance.com/bacterial. Last accessed 18 June 2017

http://www.who.int/mediacentre/news/releases/2017/bacteria-antibiotics-needed/en/. Last accessed on 18 June 2017

http://www.who.int/mediacentre/news/releases/2017/bacteria-antibiotics-needed/en/

https://www.investor.cempra.com/releasedetail.cfm?ReleaseID=1014395. Last accessed 18 June 2017

https://www.jmilabs.com/data/posters/ICAAC2014/F-970.PDF. Last accessed 18 June 2017

Ippolito JA, Kanyo ZF, Wang D et al (2008) Crystal structure of the oxazolidinone antibiotic linezolid bound to the 50S ribosomal subunit. J Med Chem 51:3353–3356

Jeong J-W, Jung S-J, Lee H-H, Kim Y-Z, Park T-K, Cho Y-L et al (2010) In vitro and in vivo activities of LCB01-0371, a new oxazolidinone. Antimicrob Agents Chemother 12:5359–5362

Kaplan N, Albert M, Awrey D et al (2012) Mode of action, in vitro activity, and in vivo efficacy of AFN-1252, a selective antistaphylococcal FabI inhibitor. Antimicrob Agents Chemother 56:5865–5874

Kim BY, Sohn YT (2011) Solid state of CG-400549, a novel FabI inhibitor: characterization, dissolution, transformation. Arch Pharm Res 34:775–779

Kim EJ, Shin WH, Kim KS, Han SS (2004) Safety pharmacology of DW-224a, a novel fluoroquinolone antibiotic agent. Drug Chem Toxicol 27:295–307

Kumar K, Chopra S (2013) New drugs for methicillin-resistant *Staphylococcus aureus*: an update. J Antimicrob Chemother 68:1465–1470

Lawrence L, Danese P, DeVito J et al (2008) In vitro activities of the Rx-01 oxazolidinones against hospital and community pathogens. Antimicrob Agents Chemother 52:1653–1662

Leach KL, Swaney SM, Colca JR et al (2007) The site of action of oxazolidinone antibiotics in living bacteria and in human mitochondria. Mol Cell 26:393–392

Leuthner KD, Vidaillac C, Cheung CM et al (2010) In vitro activity of the new multivalent glycopeptide-cephalosporin antibiotic TD-1792 against vancomycin-nonsusceptible *Staphylococcus* isolates. Antimicrob Agents Chemother 54:3799–3803

Liu F, Myers AG (2016) Development of a platform for the discovery and practical synthesis of new tetracycline antibiotics. Curr Opin Chem Biol 32:48–57

McLaws FB, Larsen AR, Skov RL et al (2011) Distribution of fusidic acid resistance determinants in methicillin-resistant *Staphylococcus aureus*. Antimicrob Agents Chemother 55:1173–1176

Mensa B, Howell G, Scott R, DeGrad WF (2014) Comparative mechanistic studies of brilacidin, daptomycin, and the antimicrobial peptide LL16. Antimicrob Agents Chemother 58:5136–5145

Noel GJ, Draper MP, Hait H et al (2012) A randomized, evaluator-blind, phase 2 study comparing the safety and efficacy of omadacycline to those of linezolid for treatment of complicated skin and skin structure infections. Antimicrob Agents Chemother 56:5650–5654

Park HS, Yoon YM, Jung SJ, Kim CM, Kim JM, Kwak J-H (2007a) Antistaphylococcal activities of CG400549, a new bacterial enoyl-acyl carrier protein reductase (FabI) inhibitor. J Antimicrob Chemother 60:568–564

Park HS, Yoon YM, Jung SJ et al (2007b) Antistaphylococcal activities of CG400549, a new bacterial enoyl-acyl carrier protein reductase (FabI) inhibitor. J Antimicrob Chemother 60:568–564

Patel MV, De Souza NJ, Gupte SV, Jafri MA, Bhagwat SS, Chugh Y et al (2004) Antistaphylococcal activity of WCK 771, a tricyclic fluoroquinolone, in animal infection models. Antimicrob Agents Chemother 48:4754–4761

Remy JM, Tow-Keogh CA, McConnell TS et al (2012) Activity of delafloxacin against methicillin-resistant *Staphylococcus aureus*: resistance selection and characterization. J Antimicrob Chemother 67:2814–2820

Ross JE, Scangarella-Oman NE, Miller LA et al (2011) Determination of disk diffusion and MIC quality control ranges for GSK1322322, a novel peptide deformylase inhibitor. J Clin Microbiol 49:3928–3930

Rubinchik E, Schneider T, Elliott M et al (2011) Mechanism of action and limited cross-resistance of new lipopeptide MX-2401. Antimicrob Agents Chemother 55:2743–2754

Schneider P, Hawser S, Islam K (2003) Iclaprim, a novel diaminopyrimidine with potent activity on trimethoprim sensitive and resistant bacteria. Bioorg Med Chem Lett 13:4217–4221

Stryjewski ME, Potgieter PD, Li Y-P et al (2012) TD-1792 versus vancomycin for treatment of complicated skin and skin structure infections. Antimicrob Agents Chemother 56:5476–5483

Stubbings W, Leow P, Yong GC et al (2011) In vitro spectrum of activity of finafloxacin, a novel, pH-activated fluoroquinolone, under standard and acidic conditions. Antimicrob Agents Chemother 55:4394–4397

Tenover FC, Tickler I, Armstrong ES et al (2011) Activity of ACHN-490 against meticillin-resistant *Staphylococcus aureus* (MRSA) isolates from patients in US hospitals. Int J Antimicrob Agents 38:352–354

Xiao XY, Hunt DK, Zhou J et al (2012) Fluorocyclines. 1. 7-Fluoro-9-pyrrolidinoacetamido-6-demethyl-6-deoxytetracycline: a potent, broad spectrum antibacterial agent. J Med Chem 55:597–595

Yum JH, Kim CK, Yong D et al (2007) In vitro activities of CG400549, a novel FabI inhibitor, against recently isolated clinical staphylococcal strains in Korea. Antimicrob Agents Chemother 51:2591–2593

Chapter 14
Recurrent Vulvovaginal Infections: Etiology, Diagnosis, Treatment and Management

Jatinder Singh, Namarta Kalia, and Manpreet Kaur

Abstract Abnormal vaginal discharge is a representative attribute of vulvovaginal infections (VVI). Frequent episodes of these infections are referred to as recurrent VVI (RVVI). The most common types of RVVI are vulvovaginal candidiasis (VVC) and bacterial vaginosis (BV). These infections are a significant cause of HIV acquisition and reproductive morbidity and hence have become a leading public health concern. The disturbances in vaginal ecosystem are the key events in the development of these disease conditions. The development of RVVI has been shown to be influenced by various host-related and behavioural risk factors. Traditional techniques used for detection of RVVI pathogens are based on the physiological and morphological characteristics. However, precise recognition of different clinical isolates is often found lengthy and complex. Therefore, many automatic as well as manual systems have been developed for the rapid identification of these opportunistic pathogens, though some of these methods were found to have considerable sensitivity issues. These limitations have been addressed by the development of various molecular typing techniques, which permits the early detections of these pathogens. However, incomplete understanding of pathophysiology of RVVI is a significant hurdle in the development of optimal treatment and prevention approaches. This chapter provides a comprehensive compilation of the present status of knowledge regarding etiology, diagnosis, treatment and management of RVVI. Systematic investigations in this particular area can provide better understanding of RVVI, further contributing to novel target recognition for more proficient therapeutic advances against these clinically relevant infections.

Keywords Antimycotics · Anti-RVVI vaccines · Mixed infections · Molecular approaches · Probiotics · Vaginal microbiota

---

J. Singh · N. Kalia
Department of Molecular Biology and Biochemistry, Guru Nanak Dev University, Amritsar, India

M. Kaur (✉)
Department of Human Genetics, Guru Nanak Dev University, Amritsar, India

P. P. Singh (ed.), *Infectious Diseases and Your Health*,
https://doi.org/10.1007/978-981-13-1577-0_14

14.1 Introduction

Abnormal vaginal discharge and itching are the frequent complaints of reproductive age women seeking gynaecological care. Vulvovaginal infections (VVI) are the most commonly documented cause of these complaints. Vulvovaginal candidiasis (VVC), bacterial vaginosis (BV) and trichomoniasis/trichomonas vaginitis (TV) are the most prevalent types of VVI. Out of these, BV and VVC result from the disturbance in vaginal microflora, while TV is a sexually transmitted disease (STD). BV is a poly-bacterial syndrome, which involves the replacement of the healthy Gram-positive *Lactobacilli* microbiota by diverse range of anaerobic bacteria. VVC is characterised by the overgrowth of *Candida* species, accompanied by yeast to hyphal form transition. In addition to vaginal microflora disturbances, several high-risk sexual behaviours, such as having multiple partners, are also associated with the development of these disease conditions. However, the status of these diseases as STDs is still ambiguous (Morris et al. 2001; CDC 2011). Moreover, some women experience repeated events of vaginal infections, clinically designated as recurrent VVI (RVVI) (Powell and Nyirjesy 2014). These include recurrent vulvovaginal candidiasis (RVVC) and recurrent bacterial vaginosis (RBV) and are the focus of this chapter. Different criteria have been described to define these recurrent infections. Occurrence of symptomatic acute VVC for ≥ four times in 1 year is denoted as RVVC (Sobel 2006), whereas repeated events of BV within 3 months, with recurrence rate 30–50%, are referred to as RBV (Verstraelen and Verhelst 2009). RVVI may affect 15–39% of women of all ages, leading to considerable healthcare costs. Untreated common infections give rise to increased possibility of preterm birth, infertility, pelvic inflammatory disease, vulvovaginal inflammation, risk of spreading invasively and acquiring other infectious diseases (Ralph et al. 1999; Atashili et al. 2008; Toth et al. 2015). Hence, prevention, precise diagnosis and timely management/treatment, particularly amongst risk groups, are necessary to avert these complications (Nwadioha et al. 2010). In spite of theranostic advancements, the fundamental mechanisms and causative factors of RVVI remain poorly understood.

Though VVI has been thought as uncomplicated and simple to treat by many practitioners, still cases of recurrent infections and treatment failures reported commonly. Same ineffective regimens are frequently being offered to patients with such infections. However, various advancements have been made in assessment and management of RVVI women that can contribute to better managements of these disease conditions. Ironically, the clinical manifestations caused by different pathogenic species are indistinguishable from each other. For instance, symptoms of BV overlap with TV and symptoms of VVC caused by non-*albicans Candida* species have common characteristics as caused by *C. albicans*. However, susceptibility of these pathogenic species varies to different antifungal agents and further shows differential sensitivity to frequently used drugs (Johnson et al. 1995). Hence, it is extremely important that the causative agent should be recognised at its species level for better clinical diagnosis, prophecy of possible drug susceptibility and to direct management. Many automatic as well as manual systems have been developed to identify these organisms, some of these methods were found to have considerable sensitivity issues. To surmount these limitations, advanced techniques based on molecular typing are coming up that permit specific and early diagnosis of these pathogens.

To understand the treatment failures in RVVI, consideration of ecosystem of the vagina in health as well as in disease is important. Parallel re-establishment of both vaginal epithelium and protective healthy microflora is much needed to shun recurrent infections. There is a need to develop prophylactic and treatment strategies that can prevent or specifically treat this enigmatic disorder. Thus, an effort is made in this chapter to provide comprehensive and current status of knowledge about the etiology, risk factors, diagnosis, treatment and management of RVVI.

14.2 Microbiome of Healthy Vagina

The human vaginal microbiome is a complex environment containing abundance of *Lactobacilli* including *L. iners*, *L. gasseri*, *L. jensenii* and *L. crispatus*, along with low number of dimorphic *Candida* species as well as obligate and facultative anaerobic bacteria (Fig. 14.1). The acidic pH (3.5–4.5) of the vagina is maintained by

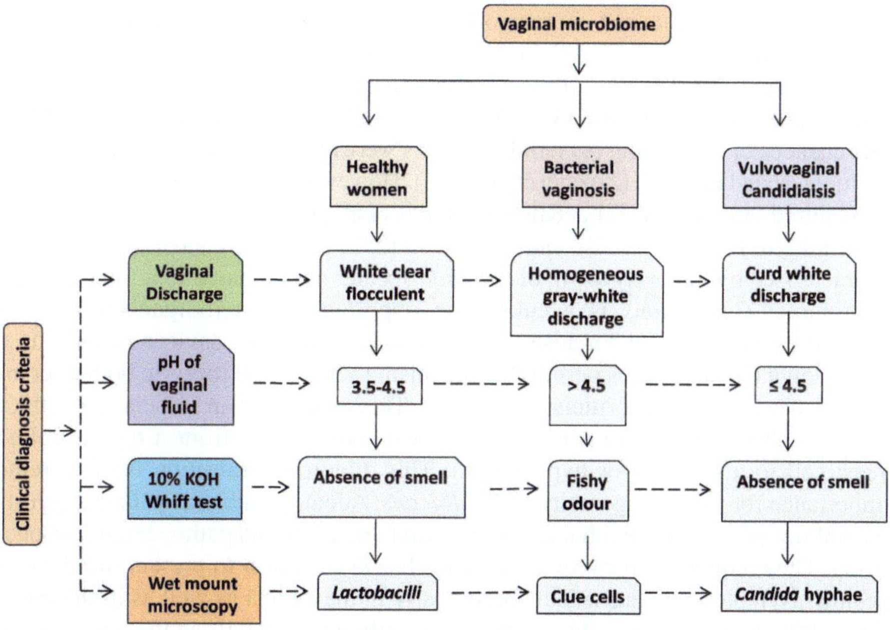

Fig. 14.1 Differential clinical diagnosis criteria based on vaginal microbiome in health and disease

lactic acid, produced by *Lactobacilli* as a result of fermentation of carbohydrates accumulated in the vaginal epithelium. In addition to this, *Lactobacilli* also produce a bactericidal compound that plays an essential role in the host immunity. This host defence, i.e. low pH and bactericidal compound, tends to suppress the growth of many obligate and facultative anaerobic pathogens. In addition, it also inhibits the transition of *Candida* from avirulent yeast form to virulent hyphal form (Mayer et al. 2013). In addition, *Lactobacilli* release fatty acids and hydrogen peroxide that hampers overgrowth of *Candida* and its transition to hyphal form (Boris et al. 1998). The constancy of vaginal ecosystem in many women is significantly affected by menstrual cycle due to hormonal changes, which further varies considerably after menopause (Santoro and Randolph 2011). The pH along with the number of anaerobic species, which are normally present in vaginal microbiome, increases with concomitant decreases in *Lactobacilli* during menstrual flow. While, pH along with the number of anaerobic species decreases with concomitant increase in *Lactobacilli* on termination of menstrual flow (Onderdonk et al. 2016).

14.3 Etiology of RVVI

In spite of these vaginal defences against microbial pathogens, microbiota disruption by physiological variations of the menstrual cycle, along with other risk factors (discussed later), leads to the development of these recurrent infections. BV is characterised by predominance of anaerobic organisms and low number of hydrogen peroxide producing *Lactobacilli* in the vagina. Obligately anaerobic species of genera *Mobiluncus*, *Prevotella*, *Atopobium* and *Sneathia* are more frequently found in symptomic women with BV (Van de wijgert et al. 2014; Onderdonk et al. 2016). The chances of developing bacterial vaginosis were also seemed to increase with the multiple sexual partners (Zozaya et al. 2016). However, the status of BV as sexually transmitted infection is still debatable (Morris et al. 2001).

Similar to BV, VVC is also caused by disturbance in the vaginal microbiota. It is characterised by the overgrowth of *Candida* species as compared to normal vaginal microbiota. *C. albicans* is a eukaryotic opportunistic microorganism with an unusual capability to acclimatise to diverse host niches and environments. These exceptional characteristics permit this organism to live a dual lifestyle both as commensal as well as an opportunistic pathogen. This duality has an association with *C. albicans* ability to endure a morphological transformation from a typical round yeast cell to a mycelial or hyphal form. This dimorphic transition is of extreme importance for the pathogenicity of *C. albicans*. Adequate substantial data suggests the linking of yeast and hyphal form with commensalism and pathogenicity, respectively. The commensal form of *C. albicans* has been shown to present in >50% of healthy asymptomatic individuals particularly in the vagina as well as in the intestine, while the hyphal form has been consistently detected only in the pathological conditions including VVC or RVVC. Host cells endure yeast cells and employ diverse mechanisms to maintain their low number on the vaginal epithelial surface and inhibit its conversion to the hyphal form. How yeast form of *Candida* provides benefit to the host, either by safeguarding of local immune homeostasis or by balancing microbiota composition, is still debatable (Wang 2009). However, *Candida* has a capacity to liberate ammonia through amino acid metabolism that neutralises excess acid and thus opposes extreme pH variations in vaginal environment. Amusingly, the same refined mechanism autoinduces hyphal growth and expansion (Vylkova et al. 2011). The hyphae further invade the outermost layer of the vaginal epithelium subsequent to adhering and forming a strong biofilm layer (Wang 2009; Harriott et al. 2010). Finally, a curdy white vaginal discharge is formed as a result of vaginal cell wreckage, separation and loosening of hyphae from vaginal epithelium, conscripted inflammatory cells and vaginal fluid. Non-pathogenic *Candida* strains were found to be incapable of producing biofilm and undergoing dimorphic transition (Lo et al. 1997; Peters et al. 2014). Moreover, allergic responses against components of *C. albicans* in the yeast form were also shown to cause VVC or RVVC (Neves et al. 2005). However, some studies have suggested no reliable vaginal microbiome distinctions amongst VVC patients and healthy women (Vylkova et al. 2011; Liu et al. 2013).

14.4 Prevalence

BV is the most frequent cause of abnormal vaginal discharge in reproductive age women and infrequently found in the menopausal women and children (Oliveira et al. 2007; Akinbiyi et al. 2008; Kalia et al. 2015). Worldwide prevalence of BV varies from 5% to 56%. African and American blacks are reported to have higher prevalence (45–55%) as compared to Caucasians (5–15%). The prevalence of BV was 29% in the USA (3.1 times greater prevalence in African-Americans relative to whites), 44% in high-risk HIV women and 56% amongst injection drug users (Warren et al. 2001; Plitt et al. 2005). The prevalence of BV is not studied systematically in Asian women, but generally it ranged from 20% to 30% (Sherrard et al. 2011). In India, BV prevalence varies from 32% to 51% (Bhalla et al. 2007; Baruahet al. 2014; Kalia et al. 2015).

After BV, the second main prevalent cause of RVVI is VVC (Sobel 1988; Holland et al. 2003). About 75% of the reproductive age women experience a minimum of one episode of VVC in their lifetime. Although VVC affects women globally, data regarding its prevalence in ethnically diverse regions is not available. The prevalence of VVC in India varies from 17.7% to 31% (Vijaya et al. 2014; Kalia et al. 2015). *C. albicans* has been recognised as the foremost causative agent of VVC, but the prevalence of non-*albicans Candida* (NAC) species has increased in the last few decades (Guzel et al. 2011; Hamad et al. 2014; Hedayati et al. 2015; Kalia et al. 2015). This can be explained by the decreased susceptibility of NAC species to the commonly used antifungal drugs. A prevalence of 10–30% of NAC species in VVC patients has been suggested in last decade by many studies (Vermitsky et al. 2008; Weissenbacher et al. 2009; Zeng et al. 2011). A recent study has indicated slightly higher prevalence of overall NAC species (53%) relative to *C. albicans* (47%) in the Indian population (Kalia et al. 2015). Amongst NAC species, *C. tropicalis* was found to be the largely prevailing species after *C. glabrata*, *C. krusei* and *C. dubliniensis* (Sharma and Solanki 2014; Kalia et al. 2015). However, another study showed *C. glabrata* to be the most prevalent NAC species (Holland et al. 2003).

14.5 Risk Factors

Although VVI develops in healthy women intermittently, this infection is often accredited to the existence of risk factors that disturb the vaginal environment, promoting RVVI (Tables 14.1 and 14.2). As explained above, vaginal flora is vastly dynamic with a confined microbial system including abundant *Lactobacilli* and small numbers of *Candida* species. The composition of this microbiota can be troubled by many physiological or nonphysiological factors. These disturbances include various lifestyle factors, psychological factors and genetic predispositions. This may create favourable conditions for the development of numerous facultative and obligate anaerobic microbes with pathogenic prospective and for the transition of *Candida* yeast to hyphal form.

14.5.1 Racial Disparities and Genetic Predispositions

One of the important risk factors responsible for susceptibility to BV and VVC is racial disparity. Black women were found to have lesser number of *lactobacilli*, higher vaginal pH and more diverse population of BV-associated bacteria including *Atopobium*, *Prevotella*, *Gardnerella*, *Megasphera* and *Mobiluncus* relative to white women (Ravel et al. 2011; Fettweis et al. 2014). Also, normal microbiota consisting of *Lactobacillus* species has been observed more commonly in White women than African-American women (Fettweis et al. 2014). Black race has been reported as an absolute analyst of BV after controlling for lifestyle and demographic factors (Ness et al. 2003). Studies have found higher prevalence of VVC in African-Americans with higher possibility of *Candida* colonisation than white and Hispanic women (Geiger and Foxman 1996; Cotch et al. 1998). Additionally, *Lactobacilli* that form protective barrier against *Candida* were found to have lower prevalence in black women (Antonio et al. 1999). Biologic, behavioural and genetic determinants along with diversity in socioeconomic status were suggested as potential factors for these disparities. Development of RVVI in many women lacking any recognised disposing factor suggests the involvement of genetic predispositions for these disease conditions (Bradford et al. 2013). Genetic polymorphisms in various immune-modulatory molecules including toll-like receptors, cytokines, etc. may describe some of the racial inconsistencies in the vaginal microbiota and possibility of developing RVVI (Kalia et al. 2017, 2018).

Table 14.1 Risk factors associated with bacterial vaginosis

Risk factor	Effect	References
Smoking	Smoking may suppress the immune system facilitating infection	Jonsson et al. (1997)
Racial origin	Black ethnic groups have the highest prevalence of bacterial vaginosis. Vaginal douching is more commonly practised by African-American women, which has been independently associated with acquiring bacterial vaginosis	Goldenberg et al. (1996) and Hawes et al. (1996)
Chronic stress	Chronic stress has been linked to alterations in the systemic and the vaginal immune response	Culhane et al. (2001), Ashcraft and Bonneau (2008), and Dhabhar (2014)
Pregnancy	Sex hormones were proposed to inhibit aspects of both innate and adaptive immunity at systemic or local level	Nelson and Macones (2002)
Viral co-infections/ immunosuppression	*Herpes simplex* virus (HSV) and HIV effect innate and adaptive immunity	Lieberman et al. (2008) and Esber et al. (2015)
Genetic predispositions	Polymorphisms in mannose-binding lectin (MBL), IL-1β, corticotrophin-releasing hormone, corticotrophin-releasing hormone receptor-2, toll-like receptor 2, MTHFR, protein kinase C, FMS-like tyrosine kinase 1and IL-6 genes	Goepfert et al. (2005), Cauci et al. (2007), Giraldo et al. (2007), Ryckman et al. (2009), Fang et al. (2010), Gomez et al. (2010), and Taylor et al. (2014)
Oral contraceptive pills/ intrauterine device	These pills/devices increase the risk of infections	Avonts et al. (1990) and Shoubnikova et al. (1997)
Poor personal hygiene	Inappropriate personal hygiene can change the normal balance of vaginal microflora possibly contributing to development of BV	Chiaffarino et al. (2004) and Zozaya et al. (2016)
Sexual activity (higher numbers of sexual partners, lower age of first intercourse)	Changes in the vaginal environment induced by sexual intercourse with a new partner may increase susceptibility to abnormal colonisation in certain women. Women aged under 25 report higher numbers of sexual partners and higher rates of partner change	Nillson et al. (1997), Ralph et al. (1999), and Wilson et al. (2000)
Tight or poorly ventilated clothing	Poorly ventilated clothing and/or synthetic underwear has been attributed to increased perineal moisture levels and temperature which may contribute towards BV	Chiaffarino et al. (2004)

Table 14.2 Risk factors for vulvovaginal candidiasis

Risk factors	Effects	References
Pregnancy	Excessive proliferation of *Candida* species, inhibit aspects of both innate and adaptive immunity at systemic or local level	Kalo-Klein and Witkin (1991), Zhang et al. (2000), Spacek et al. (2007), Kamath et al. (2013), and Vijaya et al. (2014)
Antibiotics	Excess antibiotic use leads to the depletion of the vaginal bacterial microflora, which represents the dominant vaginal defence mechanism against *Candida*	Gibbs (1987), Geiger and Foxman (1996), Ahmad and Khan (2009), and Guzel et al. (2011);
Immunosuppression	RVVC may have a dysfunction in the normal protective immune response, acquired from early exposure to *Candida* organisms	Nwadioha et al. (2010)
Uncontrolled diabetes	Impairing basic mechanisms of host defence and by promoting *Candida* adhesion to vaginal cells	Hostetter et al. (1990) and Grigoriou et al. (2006)
Genetic predispositions	Genetic polymorphisms in mannose-binding lectin (MBL), dectin-1 stop-codon, interleukin-4 (IL-4) and NLRP3 genes	Rosentul et al. (2009) and Bradford et al. (2013)
Oral contraceptive pills	Oral contraceptive pills add constant levels of hormones that act in the host and *Candida* microorganisms, possibly contributing to VVC	Egbe et al. (2011)
Intrauterine device	Due to the *Candida* adhesion and biofilm formation on the surface of the device	Auler et al. (2010) and Guzel et al. (2011)
Condoms	Spermicidal compound, nonoxynol-9 (N9), has bactericidal activity resulting in depletion of *Lactobacillus* species. It is metabolised by *Candida* resulting in altered surface characteristics, increasing *Candida* adhesion to epithelial tissues	McGroarty et al. (1990), Watts et al. (1999), and Cetin et al. (2007)
Poor personal hygiene and sexual habit	Inappropriate personal hygiene lead to increase in the number of *Candida* spores in the vaginal environment triggering epithelial invasion possibly contributing to VVC	Ahmad and Khan (2009)
Tight or poorly ventilated clothing	Poorly ventilated clothing and/or synthetic underwear lead to increased perineal moisture levels and temperature which may contribute to *Candida* proliferation	Sobel (2014)

14.5.2 Chronic Stress

Studies have shown effect of chronic stress on vaginal and systemic immune responses (Ashcraft and Bonneau 2008; Dhabhar 2014). Psychological stress has been linked with disparities in mucosal immunity, signifying an association with VVC and BV (Wouters et al. 2016).

14.5.3 Smoking

Cigarette smoking has been suggested to affect cervicovaginal immunity. Elevated levels of IL-10 have been found in cervicovaginal secretions of smokers compared to non-smokers (Gravitt et al. 2003; Simhan et al. 2005). However, one study showed significantly lower IL-10 levels, while another study showed no significant difference (Scott et al. 2006; Lieberman et al. 2008). Also a study has shown a lower prevalence of *Lactobacilli* in smokers relative to non-smokers (Brotman et al. 2014).

14.5.4 Pregnancy and Uncontrolled Diabetes

Pregnancy has been suggested as a significant influencing risk factor of VVC due to its high prevalence observed in pregnant than non-pregnant women. From the past last years, many studies of VVC carried out in India were based on pregnant women only (Neerja et al. 2006; Ahmad and Khan 2009; Vijaya et al. 2014). This can be attributed to high birth rate in India, making pregnancy-related diseases as a major focus of the epidemiological studies. The elevated secretion of sex hormones in pregnancy has been suggested as the main reason for high prevalence of VVC in pregnancy. Even if symptomatic VVC episodes are common all throughout the pregnancy, its prevalence was found to be higher in the last trimester, when the hormone levels are prominent (Nelson et al. 2013; Aslam et al. 2017). One proposed mechanism of high incidence of VVC in pregnancy is concomitant rise in glycogen levels in vaginal epithelium, along with load of oestrogen and progesterone. Storage of glycogen in the epithelial cells, lining the vaginal wall, is influenced by the hormonal levels. Thus the status of the vaginal mucosa emulates the situation of female hormones in different stages of life span (Dennerstein and Ellis 2001). High glycogen content provides a dietary source of carbon for growth of *Candida* and hence probably add to the abundance of *Candida* species (Dennerstein and Ellis 2001; Spacek et al. 2007). In addition, sex hormones are also proposed to inhibit both adaptive and innate immunity at local as well as systemic level. As for instance, progesterone has shown to suppress mice neutrophil anti-*Candida* activity (Nohmi et al. 1995). In studies involving vaginal epithelial cells, progesterone and oestrogen have been shown to inhibit lymphocyte responses against *Candida*, while estradiol significantly reduced production of antimicrobial molecules (Kalo-Klein and Witkin 1991; Wira and Fahey 2008). In addition, estradiol leads to increased growth of *C. albicans* and directly stimulates its dimorphic changeover from yeast to hyphal form, along with increase in hyphae length (Cheng et al. 2006).

Moreover, diabetics were found to have advanced rate of RVVC development and *Candida* colonization than nondiabetics (Bohannon 1998; Gunther et al. 2014). This can be due to more availability of carbon, which is the nutritive substrate of *Candida*, promoting its adhesion to vaginal epithelial wall. A high percentage of VVC in diabetic women is attributed to NAC group, particularly *C. glabrata*, which has been detected in 50–61% of VVC patients with diabetes, while *C. albicans* was isolated in 29–36% of these cases (Peer et al. 1993; Goswami et al. 2006; Ray et al. 2007). Adding up, most type 2 diabetic participants have been found to have vaginal colonization with *C. glabrata* (54%), whereas *C. albicans* (56%) were most frequently found amongst type 1 diabetic patients (De Leon et al. 2002). As NAC species are innately less vulnerable to the conventional antifungal treatment, its high prevalence in the women with diabetes may have significant inferences for VVC therapy.

14.5.5 Antibiotics

The unchecked use of antibiotics, both systemic as well as local, has also been suggested as a risk factor for VVC development (Geiger and Foxman 1996). Excessive antibiotic use can lead to reduction in vaginal *Lactobacilli* that corresponds to the central vaginal defence system against *Candida* (Gibbs 1987). Additionally, *Lactobacilli* have higher affinity for epithelial receptors and thus remain in competition with *Candida* for adhesion sites (Boris et al. 1998). Reduced adherence of *C. albicans* to vaginal epithelial cells was observed in the presence of *Lactobacilli*, in contrast to the absence of *Lactobacilli* (Boris et al. 1998). *Lactobacilli* are also shown to secrete biosurfactants that physically decrease *Candida* binding. Also, *C. albicans* adherence to silicone rubber roofed with surlactin (biosurfactant) was shown to decrease by 50% relative to adherence observed on surlactin free silicone rubber (Velraeds et al. 1998). Additionally, pentocin TV35b, a peptide isolated from *L. pentosus*, has been shown to hamper growth of *C. albicans* similar to other bacteriocins (Okkers et al. 1999).

14.5.6 Condoms

Where condoms are suggested as a preventive measure against BV and TV, the same has been suggested as possible risk factor of VVC (Cetin et al. 2007; Amouri et al. 2011). This relationship amongst spermicides/condoms and VVC development was linked with a component known as nonoxynol-9 (N9), which is a spermicidal compound found in many spermicidal measures including condoms and spermicide foams/creams (McGroarty et al. 1990). This non-ionic compound leads to immobilization of many viruses and bacteria by breaking their cell membrane and operates likewise on sperms (North 1988). However, the same compound disturbs the ecological stability of vaginal microbiota by inhibiting vaginal commensal organisms including *Lactobacillus* species. In addition, this compound can be metabolised by *Candida* species, resulting in distorted vaginal epithelial cells surface, increasing *Candida* adhesion to epithelial tissues (McGroarty et al. 1990). In addition to N-9, several other factors have been postulated to explain the link between condoms/spermicide use and increased risk of VVC. These include vaginal epithelial cell sensitivity to vaginal microtraumas and condoms and allergic responses to latex, with subsequent disordering of the vaginal ecosystem (Schreiber et al. 2006). In addition to these, other associated risk factors of RVVI include use of oral contraceptives, intrauterine devices, immunosuppression, sexual activity and hygiene and clothing habits.

14.6 Diagnosis

As traditional techniques used for RVVI diagnosis are lengthy and less responsive, hence, molecular and immunodiagnostic methods are suggested for specific and early diagnosis. These different methods are discussed in detail below.

14.6.1 Bacterial Vaginosis

14.6.1.1 Laboratory-Based Diagnosis

Diverse clinical criteria have been documented for the diagnosis of BV (Amsel et al. 1983; Spiegel et al. 1983; Srinivasan and Fredricks 2009). But the most widely used criterion for BV diagnosis was given by Amsel et al. (1983). According to this criteria, clinical diagnosis of BV is made by the presence of at least three of the following four features, i.e. homogeneous discharge, vaginal pH > 4.5 and presence of fishy amine odour on potassium hydroxide treatment and clue cells (bacteria laden squamous epithelial cells) to clinically diagnose BV (Fig. 14.1). As the microbiological disturbance is thought to be the main reason for BV development, gram staining (Spiegel et al. 1983) has been extensively used as a technique to evaluate vaginal health. However, results of gram staining varied widely, founded on the expertise of the individual examining the smear. Hence a standardised evaluation method (Nugent score) was reported based on 0–10 scoring system (Nugent et al. 1991). This has been used as a gold standard for BV diagnosis and is based on the estimation of comparative proportion of different gram stained bacteria in vaginal smear to give a rank between 0 and 10 (Fig. 14.2). However, except bacterial vaginosis, it does not take other anomalous flora categories into consideration. Therefore, a simpler version called Hay-Ison criteria was described, in which vaginal flora was divided into three different categories, i.e. normal (grade I), intermediate (grade II) and BV (grade III), depending on the relative amount of *Lactobacillus* morphotypes as compared to the *Gardnerella* morphotypes (Ison and Hay 2002). Two new grades were also added, i.e. grade 0, with epithelial cells only and grade IV with Gram-positive cocci only. This criterion reflects difference in flora better than Nugent score. Grade IV given by Ison and Hay represents aerobic vaginitis, i.e. abnormal flora other than BV, which gives evidence that the two conditions are different and

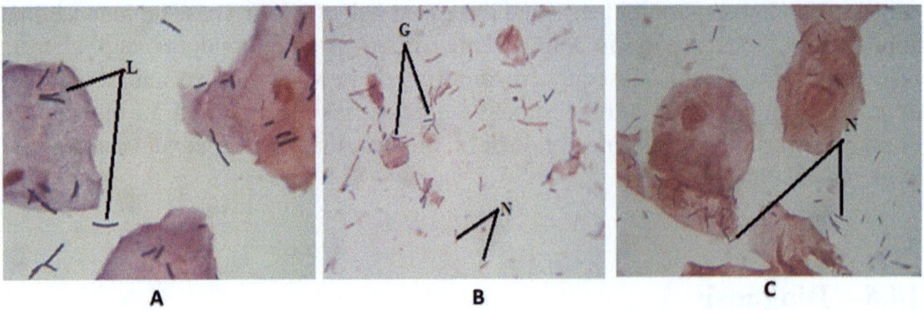

Fig. 14.2 Gram staining of vaginal fluid smears: L – *Lactobacillus* morphotype; G – *Gardnerella* morphotype; N-small Gram-negative rods; (**a**) vaginal squamous epithelial cells and only 3+ large Gram-positive rods (*Lactobacillus* morphotype), clinical examination was normal. (**b**) Mixed vaginal flora including 2+ small Gram-positive rods (*Gardnerella* morphotype) and 4+ large Gram-negative rods. No *Lactobacillus* morphotype was present; clinical diagnosis was BV. (**c**) Vaginal squamous epithelial cells with many 4+ large Gram-negative rods only, clinical diagnosis was BV. (Adapted from Kalia et al. 2015)

thus being used for the grading of aerobic vaginitis. This grade is also known as Donders' score (Donders et al. 2002). Though not each alleged case of BV was verified by criteria described by Amsel and Nugent, these criteria still holds a status for being the most popular approach for the clinical diagnosis of BV.

Most of the studies based on microbiology involves listing and comparative evaluation of the frequency of BV-associated specific species between women with BV and healthy women. Although percentage prevalence varies from study to study, all these studies showed overall high prevalence of BV-associated bacteria in symptomatic patients than healthy women or high prevalence of *Lactobacilli* in healthy women than BV symptomatic women.

14.6.2 Non-culture-Based Methods

Diagnosis of BV has not been limited only to the culture-based methods; instead molecular approaches are also being applied for the same. Quantitative polymerase chain reaction (qPCR) assays are being used for the diagnosis of specific organisms in the cervical vaginal lavage samples including *Lactobacillus* species, *Mycoplasma hominis* and *G. vaginalis*. However, for diagnosis of BV, qPCR for *M. hominis* and *G. vaginalis* was found to be better to *Lactobacillus* species (Beverly et al. 2005). Evaluation of 16S rRNA genes of eight different bacterial species has been carried out by qPCR assays. The analysis showed highest predictive values for *G. Vaginalis* and *Atopobium vaginae* with 99% specificity and 95% sensitivity for BV diagnosed with Nugent score (Menard et al. 2008). Alternatively, intravaginal microflora was also assessed using clone library analysis and 16S rRNA gene sequencing (Yoshimura et al. 2011). As organisms present in low numbers cannot be detected by 16S rRNA gene sequencing, a more sensitive method named as taxon-directed PCR assay has also been employed for recognition of BV-associated pathogens (Fredricks et al. 2007). Alternatively, another technique, i.e. DNA hybridization has been used to confirm findings of Nugent scores in women diagnosed with BV (Boggess et al. 2005). Microarray technology has offered another molecular tool for the diagnosis of BV (Dols et al. 2011). However, being expensive, the utility of all these methods for routine diagnostics has not been fully appreciated in the clinical practice. In addition, these molecular methods can only recognise the probable cause of vaginitis. Therefore, for accurate diagnosis of BV, clinical (Amsel's criteria) as well as laboratory diagnosis (Nugent criteria) followed by vigilant examination of pathogens at molecular level must be employed.

14.7 Vulvovaginal Candidiasis

14.7.1 Laboratory-Based Diagnosis

Clinical diagnosis of VVC includes normal vaginal pH, presence of a non-offensive curdy white discharge, an absence of odour on treatment of discharge with a KOH solution and presence of yeasts or pseudohypae, when observed microscopically using saline and 10% KOH (Fig. 14.1). Saline and KOH microscopy unfortunately have low sensitivity (40–70%) even in experienced hands. The fixed smear stained by Gram staining is still an extremely useful method that provides rapid confirmation of VVC. But these tests are now infrequently performed in an era where more expensive and complex tests, but not necessarily superior, have become diagnostic routine.

14.7.2 Culture-Based Methods

To remove doubt in the case of negative microscopy and normal vaginal pH, vaginal discharge culture should be obtained. The fungus-specific media commonly found in most laboratories can be used for growth of *Candida* species. The initial growth of *Candida* species is often carried out in Sabouraud dextrose agar (SDA) medium (Odds 1991). Due to its low pH, SDA medium does not allow the growth of many bacteria. *Candida* produces characteristic convex, pasty, smooth, creamy white colonies on SDA medium, which appear wrinkled on further incubation (Fig. 14.3a). The germ tube test is the primary test used for species identification in culture-based methods. This test is also known as Reynolds-Braude phenomenon after the name of researchers who discovered the formation of germ tube, a small cylindrical structure (Bradway and Levine 1993). This rapid test involves germ tube formation from *C. albicans* cells, when incubated in serum for 2 h at 37 °C, while other *Candida*

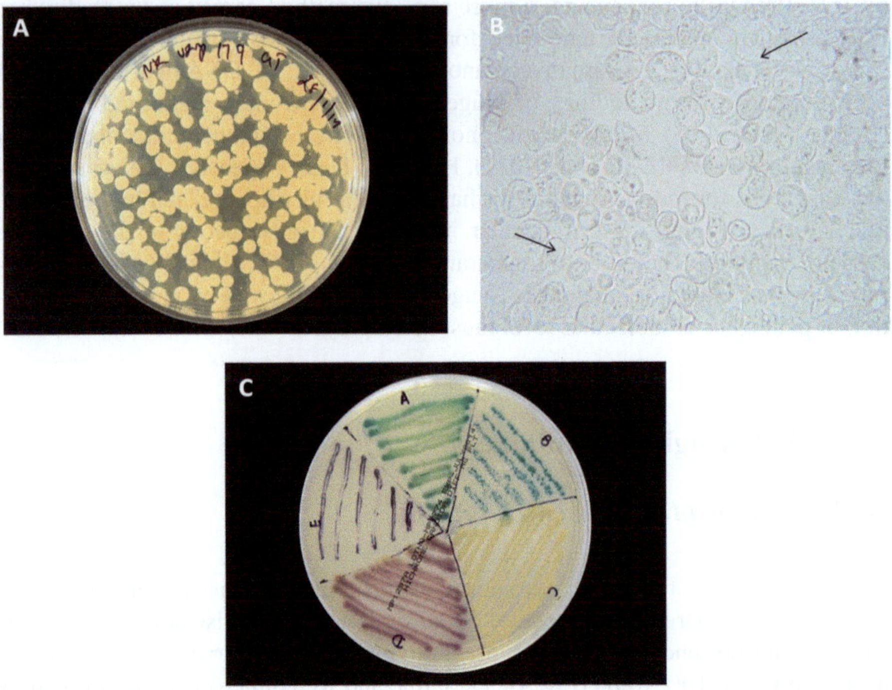

Fig. 14.3 (**a**) Sabouraud Dextrose Agar (SDA) showing creamy white smooth colonies of *Candida*. (**b**) Arrow showing germ tube, a characteristic feature of *C. albicans*. (**c**) Coloured streaks of *Candida* colonies on HiCrome Candida differential agar showing differential *Candida* species in sections *(A) C. albicans*, *(B) C. dubliniensis*, *(C) C. glabrata*, *(D) C. krusei*; and *(E) C. tropicalis*. (Adapted from Kalia et al. 2015)

species do not show the same structure formation (Fig. 14.3b). Other than this, a differential and selective medium known as chromogenic agar for *Candida* (manufactured by different companies such as Sigma-Aldrich, HiMedia, etc.) is used for identification and isolation of different *Candida* species including *C. albicans*, *C. dubliniensis*, *C. glabrata*, *C. tropicalis* and C. *krusei* (Horvath et al. 2003). Different *Candida* species produce enzymes that cleave chromogenic substrates present in medium resulting in contrasting colour colonies (Bauters and Nelis 2002) (Fig. 14.3c). This specific property of medium allows rapid detection of different *Candida* species in clinical samples (Powell et al. 1998). *C. krusei*, *C. tropicalis* and *C. albicans* grow on this medium with pink colour, dark blue-grey with a purple halo and leaf-green colonies, respectively. *C. glabrata* appear as cream to white colonies. However, slight colour variations have been observed in the colonies in chromogenic agar for *Candida* manufactured from different companies. Therefore, interpretation of different *Candida* colonies must be made as per manufacturer's specifications. Further, in contrast to the light green colonies of *C. albicans*, *C. dubliniensis* produce dark green colonies on chromogenic agar (Gutierrez et al. 2002; Campanha et al. 2005). However, a study has reported that repeated sub-culturing or

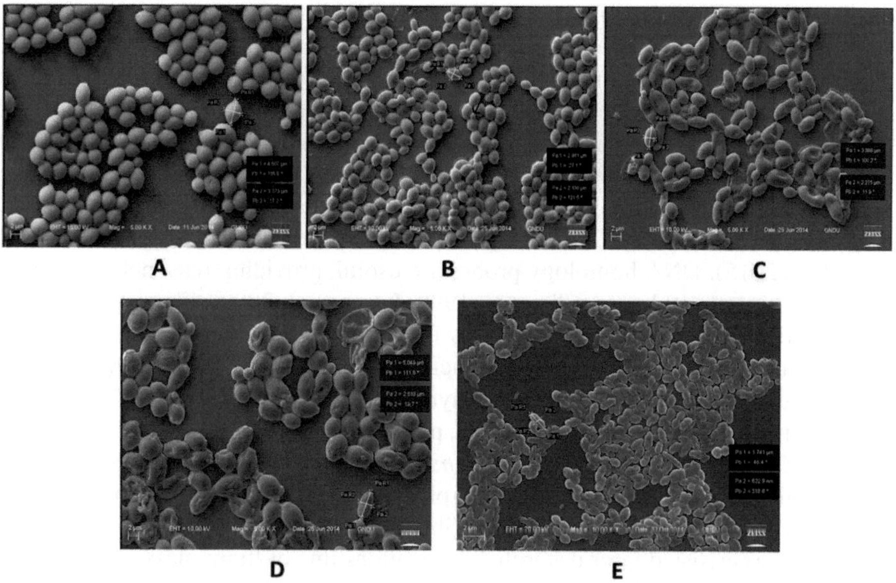

Fig. 14.4 Scanning electron micrographs showing various *Candida* species. (**a**) *C. albicans* colonies with smooth round spores (Length × Breath = 4.507 × 3.373 μm). A large arrow showing bud scar. (**b**) *C. glabrata* colonies having spherical shape with rough sides and pseudohypae, smaller than *C. albicans* (2.961 × 2.130 μm). (**c**) *C. krusei* colonies having oval to elongated spores with convoluted, rough, irregular and elevated surfaces (3.988 × 2.275 μm).(**d**) *C. dubliniensis* spores having extensively rough convoluted surface (5.049 × 2.613 μm). (**e**) *C. tropicalis* colonies showing oval spores with some pseudohypae. Smallest of all species studied (1.741 × 622.9 nm). Scale marker 2 μm is used to view each sample. (Adapted from Kalia et al. 2015)

storage at low temperature causes the fading of characteristic dark green colour of *C. dubliniensis* to light green colour of *C. albicans* colonies (Schoofs et al. 1997). This can be attributed to the property of *C. dubliniensis*, known as phenotypic switching, which allows it to exhibit characteristic attributes of *C. albicans* (Campanha et al. 2005). Alternatively, detection of *Candida species* based on their morphological features can also be carried out using scanning electron microscopy (SEM) (Fig. 14.4a–e). Although, distinguishing criteria between various morphotypes are still a matter of debate.

14.7.3 Non-culture-Based Methods

In addition to culture-based methods, molecular-based diagnostic methods are also available to diagnose *Candida* species (Cartwright et al. 2013; Sobel and Akins 2015). PCR was the basis and beginning of advanced molecular-based techniques for identification of different *Candida* species (Trtkova and Raclavsky 2006). In these PCR-based methods, DNA of pathogens is amplified and detected either directly from clinical samples or in the host DNA background (Kanbe et al. 2002; Trtkova and Raclavsky 2006; Schabereiter-Gurtner et al. 2007; Fricke et al. 2010). A variety of alterations in this technique allows its use in broad range of samples suggesting the flexibility of this technology. Amongst such alterations are the nested PCR (Bougnoux et al. 1999, Kanbe et al. 2002), real-time PCR (Souza et al. 2012) and multiplex PCR (Luo and Mitchell 2002; Lau et al. 2008). These methods have undoubtedly higher sensitivity than culture-based methods, but being expensive, they have not been shown to offer any advantage to the clinician in practice (Sobel and Akins 2015). DNA homology probes are useful, providing reasonably accurate results within several hours (Elie et al. 1998; Lowe et al. 2009). Fluorescent in situ hybridisation (FISH) is a convenient way that allows detection of yeasts even from impure culture with the help of oligonucleotide probes labelled with fluorescein. Peptide nucleic acid fluorescent in situ hybridization (PNA FISH assay) is another alternation of FISH with higher potential, particularly developed to distinguish *non-albicans Candida* species from *C. albicans* (Rigby et al. 2002). This technique uses high-affinity peptide nucleic acid (PNA) probes that target highly structured region of rRNA (Trtkova and Raclavsky 2006; Shepard et al. 2008). Alternatively, pyrolysis and matrix-assisted laser desorption ionization-time of flight mass spectrometry (MALDI-TOF MS) is slowly becoming an extensively accessible tool in the clinical microbiology laboratories and has emerged as an economical, dependable and quick alternative for *Candida* species identification (Dhiman et al. 2011; Spanu et al. 2012). Other methods based on β-d-(1, 3)-glucan and galactomannan detection of *Candida* in serum are also available (Ahmad and Khan 2012; Mikulska et al. 2012; Held et al. 2013).

In contrary to the above detection methods that target species-specific or universal sequences, strain typing can distinguish particular strains of a specified species amongst various clinical isolates. As, for instance, electrophoretic karyotyping employs pulse-field gel electrophoresis to assess chromosome-length polymorphisms. Due to chromosome breakage, uneven chromosomal reorganizations and healing, most of the fungal species exhibit chromosome-length polymorphisms. This technique can not only effectively distinguish different strains of *Candida* including *C. glabrata, C. tropicalis, C. parapsilosis, C. lusitaniae* and *C. albicans* but also differentiate phenotypically close species, i.e. C. *dubliniensis* from *C. albicans* (Merz et al. 1988; Sullivan and Coleman 1998; Espinel-Ingroff et al. 1999). In another technique called restriction analysis (REA), restriction endonuclease is used to cut genomic DNA that results in restriction fragment length polymorphism (RFLP) depending on the sequence. This resulted RFLP banding pattern is further visualised by commonly used agarose gel electrophoresis that separates the resulted restricted fragments based on size. The banding patterns are observed due to genomic variations between different strains, and these variations include insertion, deletions and single nucleotide polymorphisms in the sequence of restriction sites. Other than this, reports are available that have shown amplification of intergenic spacers, i.e. ITS1 and ITS2, followed by restriction analysis of these amplified products, for strain-specific detection and characterization of clinically relevant *Candida* species (Trost et al. 2004). Another typing technique that is most widely used in clinical mycology is random amplified polymorphic DNA (RAPD). This convenient tool has been constantly shown to be used for strain typing and species identification (Steffan et al. 1997; Bautista-Munoz et al. 2003). This PCR fingerprinting technique uses a randomly chosen short 10-mer primer sequence, rather than targeted specific complementary primer. An amplicon generates after successful PCR that further depends upon the chance event of binding of two same primer molecules in a proper orientation. The number of amplicons generated from the complex genomic DNA can be visualised on agarose gel. Another technique based on DNA fingerprinting by PCR is amplified fragment length polymorphism (AFLP) (Vos et al. 1995). This technique involves the use of two restriction enzymes, a tetra-cutter and a hexa-cutter, to cut genomic DNA. The ends of cleaved DNA were further ligated with double-stranded adapters. Further primers complementary to the sequences of restriction sites and adapters are used for the amplification of genomic DNA by PCR. At last, gel electrophoresis is used to separate and analyse amplified restriction fragments. Another variation of AFLP is fluorescent amplified fragment length polymorphism (FAFLP), in which computer-based automated sequence analyser allows the real-time analysis of gel electrophoretic patterns using fluorescent tag-labelled primers. Identification and typing of pathogenic *Candida* species have also been reported using this method (Borst et al. 2003; Ball et al. 2004). Multilocus sequence typing (MLST) is a potential technique that uses sequencing for typing purposes. Single nucleotide polymorphisms of numerous housekeeping genes are targeted in this method, and each permutation of single allele decides the strain type. This method was found to be stable and resolving for *C. albicans* in the clinical isolates (Bougnoux et al. 2002). Further, microarrays can be used to identify and measure the amount of sequence variations in internal transcribed spacers (ITSs), rRNA genes or other specific gene regions. Other than this, different sequencing projects on pathogenic yeast are in progress that will soon facilitate a simple and uncomplicated design of the whole-genome DNA microarrays (Bumgarner 2013).

14.8 Treatment

As over-the-counter drugs are easily available, self-diagnosis and self-treatment of VVI are usual practice. This lack of awareness about this chronic disease condition might consequently result in history of RVVI. It is important to confirm the diagnosis for guiding treatment and predicting the likely drug susceptibility. Available treatment strategies include oral or local use of antibiotics, azoles and steroids.

14.8.1 Treatment of BV

Clindamycin, tinidazole and metronidazole are the three antibiotics that are presently being used for the treatment of BV (Table 14.3). Metronidazole and tinidazole belong to the category of nitroimidazoles, having a 5-nitro group. The reduction of 5-nitro group results in formation of many products that is important for the activity of nitroimidazoles (Moreno and Docampo 1985). These reduction products include reduced reactive intermediates, i.e. hydroxylamine imidazoles or nitrosoimidazoles, nitroradical anions and reduction products produced as a result of single electron transfer. Out of tinidazole and metronidazole, tinidazole was found to have better tissue distribution, slower rate of elimination and longer half-life. However, for single-dose regimens, tinidazole is ten times more expensive than metronidazole, which considerably restricts its use. Both of these drugs were found to have similar efficacy. The combination of oral with local vaginal use was found to be providing better efficacy with 80–86% cure rates vs 75–86% when taken alone either orally or locally (Milani et al. 2003; Wang et al. 2008). The mechanism of clindamycin activity primarily involves its binding to 50s ribosomal subunit of bacteria. This molecule disrupts bacterial protein synthesis by interfering with the transpeptidation reaction, thereby inhibiting early chain elongation. Clindamycin is available as oral tablets, parenteral injection (intramuscular or intravenous) and topical and vaginal formulations. A culture is suggested for susceptibility testing, in cases of treatment failure.

14.8.2 Treatment of VVC

Once the diagnosis of RVVC is confirmed, accompanied by microbiological confirmation of an azole-sensitive strain of *Candida*, patients are recommended an induction regimen of fluconazole (Sobel et al. 2004; Seidman and Skokos 2005; Donders et al. 2008). A variety of formulations of numerous efficient topical azole drugs are available (Table 14.3). In general, the rate of cure of different topical azoles has been shown to range from 80% to 90%. However, no confirmation regarding the better efficacy of one azole over the other has been reported (Reef et al. 1995). Topical azoles were found to have marginally low or almost similar rate of cure than oral azoles. Similar efficacy of drugs was observed in a meta-analysis, when drugs were administered locally or through oral routes (Watson et al. 2002). However, use of oral azoles has been limited, due to its possible systemic side effects. *C. krusei*, being resistant to fluconazole, poses a therapeutic dilemma (Singh et al. 2002). Regardless of this, *C. krusei* has been observed to be susceptible to other azoles under in vitro conditions. Vaginal *C. krusei* strains were shown to be most susceptible to clotrimazole on in vitro susceptibility testing. Furthermore, azoles were documented to be inadequately efficient in VVI specifically caused by *C. glabrata* (Rodrigues et al. 2014). There are no published data on long-term maintenance

Table 14.3 Treatment for RVVI

Antibiotics	Route	Dose	References
Bacterial vaginosis			
Metronidazole	Per os	500 mg 2dd, 5 days	
	Per os	2 g stat, once	Mikamo et al. (1997)
	Vag dose	500 mg 1dd, 5 days	Brandt et al. (2008)
	Vag dose	2 g, once	Milani et al. (2003)
	Vag cream	0.75% 1dd, 5 days	Cunningham et al. (1994), Hanson et al. (2000)
Clindamycin	Per os	300 mg 2dd, 7 days	Greaves et al. (1988)
	Vag cream	2% once, 7 days	Paavonen et al. 2000
Tinidazole	Per os	1dd 1 g, 5 days	Livengood et al. (2007)
		2 g stat, once	Milani et al. (2003) and Livengood et al. (2007)
	Vag tablets	1dd 500 mg, 14 days	Heikkinen and Vuopala (1989)
Ornidazole	Vag tablets/sup	1dd 500 g, 7 days	Baloglu et al. (2003)
	Per os+ vag tablets	2dd	Wang et al. (2008)
		500 mg5 days +	
		1dd 500 mg, 5 days	
Secnidazole	Per os	1dd 1–2 g stat	Núñez and Gómez (2005)
	Per os+ vag tablets orni	2 g stat, once +	Wang et al. (2008)
		1dd 500 mg, 5 days	
Vulvovaginal candidiasis			
For *C. albicans*, *C. tropicalis*, *C. parapsilosis and C. krusei* (any of the below except fluconazole)			
Fluconazole	Per os	150 mg q 72 h for 3 doses	Sobel et al. (2004), Seidman and Skokos (2005), Donders et al. (2008)
		Maintenance regimen 150 mg q weekly for 6 months	
Itraconazole	Per os	200 mg twice daily, 3 days	Pitsouni et al. (2008)
		Maintenance regimen 100–200 mg/d for 6 months	Sobel (2016)
Clotrimazole	Vaginal cream	1% for 7 nights	Seidman and Skokos (2005)
		2% for 3 nights	
		10% once	
	Vaginal tablets	100 mg – one tablet per day for 7 days	Seidman and Skokos (2005)
		100 mg – two tablets per day for 3 days	
		500 mg – one tablet once	

(continued)

Table 14.3 (continued)

Antibiotics	Route	Dose	References
Miconazole	Vaginal cream	2% for 7 nights	Reef et al. (1995) and Sobel (2016)
		4% for 3 nights	
	Vaginal suppository	100 mg – one suppository per day for 7 days	Reef et al. (1995) and Sobel (2016)
		200 mg – one suppository per day for 3 days	
		1200 mg – one suppository once	
Tioconazole	Ointment	2% for 3 nights	Sobel (2016)
		6.5% for 1 night	
Terconazole	Vaginal cream	0.4% for 7 nights	Sobel (2016)
		0.8% for 3 nights	
	Vaginal suppository	80 mg for 3 nights	
Butoconazole	Vaginal cream	2% – once	Seidman and Skokos (2005)
For *C. glabrata* and for azole-resistant Candida species			
Boric acid	Vaginal suppository/ capsule	600 mg for 14 days	Guaschino et al. (2001)
Nystatin	Vaginal suppository	100,000 U per vagina for 14 days	Fan et al. (2015)

therapy for *C. glabrata*. The management of chronic and RVVC due to *C. glabrata* remains complex and frustrating for patients and practitioners alike. Azole agents both oral and topical frequently fail (Sobel 2007; Kennedy and Sobel 2010; Davies et al. 2013). Some success was achieved with daily dose of 0.6 g of vaginal boric acid for 1–2 weeks (60–70%). Additional successful efforts include intra-vaginal nystatin, flucytosine, amphotericin B and their combinations in order to eradicate *C. glabrata* in the symptomatic women (Fan et al. 2015). However, there are few studies on the efficacy, safety or even need for long-term maintenance regimen with boric acid or polyene agents (Guaschino et al. 2001; Sobel et al. 2003). Therefore, VVC involving *C. krusei* and *C. glabrata* should be monitored carefully to manage the chances of antifungal resistance.

14.9 Treatment of Poorly Defined Vaginal Infections

A variety of formulations are available for the management of poorly defined vaginal infections like aerobic vaginitis (AV), mixed vaginal infections and co-infections. In aerobic vaginitis, the vagina is colonised by commensal intestinal bacteria including *Escherichia coli*, *Staphylococcus aureus* and *Streptococcus agalactiae* (Donders et al. 2015). Its treatment is based on microscopic findings and mixed local vaginal treatment with oestrogens, steroids and antibiotics. If cultures or microscopy results show the presence of *Candida*, treatment with antifungals is recommended before other treatments. Though not long-standing, temporary and rapid relief of symptoms have been observed with vaginal rinsing with povidone iodine (Donders et al. 2015). Other than this, commonly used broad spectrum local antibiotics such as kanamycin are also recommended. Moreover, oral therapy with moxifloxacin or amoxyclav can be used, particularly in colpitis and deep dermal vulvitis with *Staphylococcus aureus or* group B *streptococci* to attain quick and instant relief from severe symptoms (Donders et al. 2015). A complete understanding of the definition of co-infections and mixed vaginal infections is necessary for the proper treatment of these less defined infections. Specific therapy is a mandate to suppress the parallel expression of all the pathogens in mixed vaginitis (Sobel et al. 2013). In co-infection, identification of likely pathogen that is responsible for the symptoms is necessary for the specific treatment (Sobel et al. 2013). These scenarios lead to increased demand of polytherapy containing many antimicrobials either pooled or in separate preparations. However, use of pooled antimicrobials for vaginitis has been prohibited in many countries because of their side effects, and it can only be indicated for confirmed mixed infections. Therefore, advancement in accessible investigative modalities is necessary that can make easy identification of mutual presence of multiple pathogens that can further be helpful in deciding the use of antimicrobial combination regimens.

14.10 Preventive/Prophylactic Measures

An additional interest is needed to reduce the load imposed by VVI in women with recurrent episodes. Care should be taken regarding nutrition, hygiene and clothing. In addition to this, some prophylactic measures including use of probiotics as well as vaccines are being evaluated to reduce the risk of RVVI.

14.10.1 Eating Habits and Medication

A balanced diet including non-fat dairy products, whole grains, vegetables and fruits must be taken. Eating foods containing *lactobacilli* can prevent RVVI. Proper management of blood sugar levels can decrease the RVVI risk. Unchecked and excessive use of antibiotics must be avoided, as normal vaginal microbiota can be altered, which further leads to uncontrolled colonisation of yeast.

14.10.2 Personal Hygiene and Clothing

The vaginal area must be kept clean by using mild, unscented soap and rinsed well with water. After using the toilet, anal area should be wiped to cut the dissemination of bacteria or yeast from anus to the urinary tract or vagina. While wearing underwear one good choice is underwear of cotton cloth that doesn't hold heat and dampness and helps keep genital area dry that could prevent infections. Clothing with tight fitting such as wet swimsuit, jeans and panty hose must be avoided. These may increase body heat and moisture in the genital area. Pads or tampons must be changed often. Douching or use of perfumes or powders, feminine sprays or deodorant tampons must be avoided as these items can change the normal vaginal microflora.

14.10.3 Probiotics and Vitamins

Treatment of BV, VVC or other types of VVI followed by application of probiotics in the vagina has efficiently been shown to prevent recurrences of infection over a 6-month period (Ozkinay et al. 2005; Falagas et al. 2006; Larsson et al. 2008). Vaginal application of vitamin C is proposed to restore the vaginal pH. Recently, a study that indicated the direct use of vitamin C in the vagina after menses for 6 days per month for 6 months was proficient to halve the risk of recurrent BV (Krasnopolsky et al. 2013).

14.10.4 Vaccination

Many research groups are concentrating on anti-RVVI vaccination, with major emphasis to RVVC (Cutler et al. 2007; Cassone 2013). Two such vaccines, i.e. Als3-alum and virosomal-Sap2, have been approved from phase 1 clinical trial for immunogenicity and safety; the former has also come into phase 2 clinical trial (Schmidt et al. 2012; Pietrella et al. 2013). Both vaccines though different in immunological mechanisms showed promising evidence of protection from *C. albicans* infection in in vivo models of vaginal infection, i.e. mouse and rat. The phase 1 clinical trial has shown the generation of vaginal antibodies after parenteral administration of both Als3 and Sap2 vaccines, signifying the protective effect of vaccination (Schmidt et al. 2012). Alternatively, many other vaccines against *Candida* have been developed, which are still pending for clinical trials, and have shown to be protective and immunogenic in in vivo investigational models. These vaccines involve glycoconjugate of polysaccharides of *Candida* cell wall and *C. albicans* attenuated strains (Saville et al. 2009; Cassone 2013). It is interesting to know that studies have suggested no promising role of antibodies in host immune defence against *Candida* that generate naturally, while specific neutralising antibodies generate against almost all of the anti-*Candida* vaccines (Cassone 2013). Prophylactic immunization with Gynevac, a Hungarian patented vaccine with five inactivated pathogenic strains of *Lactobacilli*, is suggested as a good alternative for prevention of bacterial vaginosis (BV) (Lazar and Varga 2011). Establishment of RVVI vaccine as part of routine vaccination can help to reduce the social, familial and community stigma associated with having or seeking treatment for a genital infection.

14.11 Conclusion

RVVI is an important public health concern. Vaginal environmental changes play a major role in transitions of opportunistic pathogen from commensal to pathogenic. Thus, investigations in this field will provide better understanding of RVVI that will further lead to the development of more proficient therapeutic advances by recognising new targets of this clinically important disease. Control and detection of risk factors that influence RVVI will provide better management of infection. However, the presence of risk factor does not necessarily direct towards infection, but the possibility of RVVI development in the future may increase. In the same way, if any risk factor is not present, it does not essentially keep away from the chance of getting RVVI. Therefore, diagnostic problems continue to prevail although the advances have been carried out by recognising the host genetic factors contributing to RVVI. However, present position of these molecular-based diagnoses is still needed to be authenticated though these techniques will surely have a role in the future. Since the last decade, no new formulations have been established. Although, both oral and topical regimens may provide effective therapy, the cure is often elusive. Uses of probiotics in the vagina, which can endure, adhere to epithelial cells and can further generate defence factors against pathogens, are the probable future goals of RVVI management. Anti-RVVI vaccines have been shown to be efficient in averting RVVI in animal models; however, studies regarding efficiency of these vaccines in humans are still in infancy. In spite of all these progressive investigations, numerous mechanisms in concern with RVVI pathogenesis are still awaited. Explication of the causal factors that enhance vulnerability to RVVI in women and use of this information to build effectual management and prevention approach will lead to the development of efficient diagnostics and prophylaxis of this inscrutable disorder.

Acknowledgements The authors express their sincere gratitude to the University Grants Commission, New Delhi, Govt. of India for providing Ph.D. fellowship to Namarta Kalia.

References

Ahmad A, Khan AU (2009) Prevalence of Candida species and potential risk factors for vulvovaginal candidiasis in Aligarh, India. Eur J Obstet Gynaecol Reprod Biol 144:68–71

Ahmad S, Khan Z (2012) Invasive candidiasis: a review of nonculture-based laboratory diagnostic methods. Indian J Med Microbiol 30:264

Akinbiyi AA, Watson R, Feyi-Waboso P (2008) Prevalence of Candida albicans and bacterial vaginosis in asymptomatic pregnant women in South Yorkshire, United Kingdom. Arch Gynecol Obstet 278:463–466

Amouri I, Sellami H, Borji N, Abbes S, Sellami A, Cheikhrouhou F et al (2011) Epidemiological survey of vulvovaginal candidosis in Sfax, Tunisia. Mycoses 54:e499–e505

Amsel R, Totten PA, Spiegel CA, Chen KC, Eschenbach D, Holmes KK (1983) Nonspecific vaginitis: diagnostic criteria and microbial and epidemiologic associations. Am J Med 74:14–22

Antonio MA, Hawes SE, Hillier SL (1999) The identification of vaginal *Lactobacillus* species and the demographic and microbiologic characteristics of women colonized by these species. J Infect Dis 180:1950–1956

Ashcraft KA, Bonneau RH (2008) Psychological stress exacerbates primary vaginal herpes simplex virus type 1 (HSV-1) infection by impairing both innate and adaptive immune responses. Brain Behav Immun 22:1231–1240

Aslam M, Hafeez R, Ijaz S (2017) Vulvovaginal candidiasis in pregnancy. Biomedica 24:54–56

Atashili J, Poole C, Ndumbe PM, Adimora AA, Smith JS (2008) Bacterial vaginosis and HIV acquisition: a meta-analysis of published studies. AIDS 22:1493–1501. https://doi.org/10.1097/QAD.0b013e3283021a37

Auler ME, Morreira D, Rodrigues FF, Abr Ão MS, Margarido PF, Matsumoto FE et al (2010) Biofilm formation on intrauterine devices in patients with recurrent vulvovaginal candidiasis. Med Mycol 48:211–216

Avonts D, Sercu M, Heyerick P, Vandermeeren I, Meheus A, Piot P (1990) Incidence of uncomplicated genital infections in women using oral contraception or an intrauterine device: a prospective study. Sex Transm Dis 17:23–29

Ball LM, Bes MA, Theelen B, Boekhout T, Egeler RM, Kuijper EJ (2004) Significance of amplified fragment length polymorphism in identification and epidemiological examination of Candida species colonization in children undergoing allogeneic stem cell transplantation. J Clin Microbiol 42:1673–1679

Baloğlu E, Özyazıcı M, Baloğlu A, Ova L (2003) A randomized controlled trial of a new ovule formulation of ornidazole for the treatment of bacterial vaginosis. J Clin Pharm Ther 28:131–136

Baruah FK, Sharma A, Das C, Hazarika NK, Agarwalla R (2014) Prevalence and epidemiological correlates of bacterial vaginosis among nonpregnant females at a tertiary care centre in Assam, India. Ann Trop Med Public Health 7:91–95

Bauters TG, Nelis HJ (2002) Comparison of chromogenic and fluorogenic membrane filtration methods for detection of four Candida species. J Clin Microbiol 40:1838–1839

Bautista-Munoz C, Boldo XM, Villa-Tanaca L, Hernández-Rodríguez C (2003) Identification of Candida spp. by randomly amplified polymorphic DNA analysis and differentiation between *Candida albicans* and *Candida dubliniensis* by direct PCR methods. J Clin Microbiol 41:414–420

Beverly ES, Chen HY, Wang QJ, Zariffard MR, Cohen MH, Spear GT (2005) Utility of Amsel criteria, Nugent score, and quantitative PCR for Gardnerella vaginalis, *Mycoplasma hominis*, and *Lactobacillus* spp. for diagnosis of bacterial vaginosis in human immunodeficiency virus-infected women. J Clin Microbiol 43:4607–4612

Bhalla P, Chawla R, Garg S, Singh MM, Raina U, Bhalla R et al (2007) Prevalence of bacterial vaginosis among women in Delhi, India. Indian J Med Res 125:167–172

Boggess KA, Trevett TN, Madianos PN, Rabe L, Hillier SL, Beck J et al (2005) Use of DNA hybridization to detect vaginal pathogens associated with bacterial vaginosis among asymptomatic pregnant women. Am J Obstet Gynecol 193:752–756. https://doi.org/10.1016/j.ajog.2005.01.068

Bohannon NJV (1998) Treatment of vulvovaginal candidiasis in patients with diabetes. Diabetes Care 21:451–456

Boris S, Suárez JE, Vázquez F, Barbés C (1998) Adherence of human vaginal lactobacilli to vaginal epithelial cells and interaction with uropathogens. Infect Immun 66:1985–1989

Borst A, Theelen B, Reinders E, Boekhout T, Fluit AC, Savelkoul PHM (2003) Use of amplified fragment length polymorphism analysis to identify medically important Candida spp., including *C. dubliniensis*. J Clin Microbiol 41:1357–1362

Bougnoux ME, Dupont C, Mateo J, Saulnier P, Faivre V, Payen D et al (1999) Serum is more suitable than whole blood for diagnosis of systemic candidiasis by nested PCR. J Clin Microbiol 37:925–930

Bougnoux ME, Morand S, d'Enfert C (2002) Usefulness of multilocus sequence typing for characterization of clinical isolates of *Candida albicans*. J Clin Microbiol 40:1290–1297

Bradford LL, Ravel J, Bruno V (2013) Understanding vulvovaginal candidiasis through a community genomics approach. Curr Fungal Infect Rep 7:126–131

Bradway SD, Levine MJ (1993) "Germ tube test" Reynolds-Braude phenomenon. Crit Rev Oral Biol Med 4:293–299

Brandt M, Abels C, May T, Lohmann K, Schmidts-Winkler I, Hoyme UB (2008) Intravaginally applied metronidazole is as effective as orally applied in the treatment of bacterial vaginosis, but exhibits significantly less side effects. Eur J Obstet Gynaecol Reprod Biol 141:158–162

Brotman RM, He X, Gajer P, Fadrosh D, Sharma E, Mongodin EF et al (2014) Association between cigarette smoking and the vaginal microbiota: a pilot study. BMC Infect Dis 14:471

Bumgarner R (2013) Overview of DNA microarrays: types, applications, and their future. Curr Potocols Mol Miol 22:1

Campanha NH, Neppelenbroek KH, Spolidório DMP, Spolidorio LC, Pavarina AC (2005) Phenotypic methods and commercial systems for the discrimination between C. albicans and *C. dubliniensis*. Oral Dis 11:392–398

Cartwright CP, Lembke BD, Ramachandran K, Body BA, Nye MB, Rivers CA, Schwebke JR (2013) Comparison of nucleic acid amplification assays with BD affirm VPIII for diagnosis of vaginitis in symptomatic women. J Clin Microbiol 51:3694–3699

Cassone A (2013) Development of vaccines for *Candida albicans*: fighting a skilled transformer. Nat Rev Microbiol 11:884–891

Cauci S, Di Santolo M, Casabellata G, Ryckman K, Williams SM, Guaschino S (2007) Association of interleukin-1b and interleukin-1 receptor antagonist polymorphisms with bacterial vaginosis in nonpregnant Italian women. MHR: Basic Sci Reprod Med 13:243–250

Centers for Disease Control and Prevention (2011) Sexually transmitted diseases treatment guidelines, 2010. Ann Emerg Med 58:67–68

Cetin M, Ocak S, Gungoren A, Ulvi Hakverdi A (2007) Distribution of Candida species in women with vulvovaginal symptoms and their association with different ages and contraceptive methods. Scand J Infect Dis 39:584–588

Cheng G, Yeater KM, Hoyer LL (2006) Cellular and molecular biology of *Candida albicans* estrogen response. Eukaryot Cell 5:180–191

Chiaffarino F, Parazzini F, De Besi P, Lavezzari M (2004) Risk factors for bacterial vaginosis. Eur J Obstet Gynecol Reprod Biol 117:222–226

Cotch MF, Hillier SL, Gibbs RS, Eschenbach DA, Prematurity Study Group (1998) Epidemiology and outcomes associated with moderate to heavy Candida colonization during pregnancy. Am J Obstet Gynaecol 178:374–380

Culhane JF, Rauh V, McCollum KF, Hogan VK, Agnew K, Wadhwa PD (2001) Maternal stress is associated with bacterial vaginosis in human pregnancy. Matern Child Health J 5:127–134

Cunningham FE, Kraus DM, Brubaker L, Fischer JH (1994) Pharmacokinetics of intravaginal metronidazole gel. J Clin Pharmacol 34:1060–1065

Cutler JE, Deepe GS Jr, Klein BS (2007) Advances in combating fungal diseases: vaccines on the threshold. Nat Rev Microbiol 5:13–28

Davies S, Johnson E, White D (2013) How to treat persistent vaginal yeast infection due to species other than *Candida albicans*. Sex Transm Infect 89:165–166

De Leon EM, Jacober SJ, Sobel JD, Foxman B (2002) Prevalence and risk factors for vaginal *Candida* colonization in women with type 1 and type 2 diabetes. BMC Infect Dis 2:1

Dennerstein GJ, Ellis DH (2001) Oestrogen, glycogen and vaginal candidiasis. Aust N Z J Obstet Gynaecol 41:326–328

Dhabhar FS (2014) Effects of stress on immune function: the good, the bad, and the beautiful. Immunol Res 58:193–210

Dhiman N, Hall L, Wohlfiel SL, Buckwalter SP, Wengenack NL (2011) Performance and cost analysis of matrix-assisted laser desorption ionization–time of flight mass spectrometry for routine identification of yeast. J Clin Microbiol 49:1614–1616

Dols JA, Smit PW, Kort R, Reid G, Schuren FH, Tempelman H et al (2011) Microarray-based identification of clinically relevant vaginal bacteria in relation to bacterial vaginosis. Am J Obstet Gynaecol 204:305–3e1

Donders GG, Vereecken A, Bosmans E, Dekeersmaecker A, Salembier G, Spitz B (2002) Definition of a type of abnormal vaginal flora that is distinct from bacterial vaginosis: aerobic vaginitis. BJOG Int J Obstet Gynaecol 109:34–43

Donders G, Bellen G, Byttebier G, Verguts L, Hinoul P, Walckiers R et al (2008) Individualized decreasing-dose maintenance fluconazole regimen for recurrent vulvovaginal candidiasis (ReCiDiF trial). Am J Obstet Gynaecol 199:613–619

Donders GG, Ruban K, Bellen G (2015) Selecting anti-microbial treatment of aerobic vaginitis. Curr Infect Dis Rep 17:24. https://doi.org/10.1007/s11908-015-0477-6

Egbe CA, Onwufor UC, Omoregie R, Enabulele OI (2011) Female reproductive tract infections among vaginal contraceptive users in Benin City, Nigeria. Genom Med Biomark Health Sci 3:49–52

Elie CM, Lott TJ, Reiss E, Morrison CJ (1998) Rapid identification of *Candida* species with species-specific DNA probes. J Clin Microbiol 36:3260–3265

Esber A, Vicetti Miguel RD, Cherpes TL, Klebanoff MA, Gallo MF, Turner AN (2015) Risk of bacterial vaginosis among women with herpes simplex virus type 2 infection: a systematic review and meta-analysis. J Infect Dis 212:8–17

Espinel-Ingroff A, Vazquez JA, Boikov D, Pfaller MA (1999) Evaluation of DNA-based typing procedures for strain categorization of Candida spp. Diagn Microbiol Infect Dis 33:231–239

Falagas ME, Betsi GI, Athanasiou S (2006) Probiotics for prevention of recurrent vulvovaginal candidiasis: a review. J Antimicrob Chemother 58:266–272

Fan S, Liu X, Wu C, Xu L, Li J (2015) Vaginal nystatin versus oral fluconazole for the treatment for recurrent vulvovaginal candidiasis. Mycopathologia 179:95–101

Fang X, Li H, Diao Y, Shan R, Dong J, Li H et al (2010) Polymorphisms in the MTHRF, VDR, MMP-9 and IL-β genes and the risk of premature rupture of membranes. Gynecol Obstet Investig 70:206–214

Fettweis JM, Serrano MG, Huang B, Brooks JP, Glascock AL, Sheth NU et al (2014) An emerging mycoplasma associated with trichomoniasis, vaginal infection and disease. PLoS One 9:e110943. https://doi.org/10.1371/journal.pone.0110943

Fredricks DN, Fiedler TL, Thomas KK, Oakley BB, Marrazzo JM (2007) Targeted PCR for detection of vaginal bacteria associated with bacterial vaginosis. J Clin Microbiol 45:3270–3276

Fricke S, Fricke C, Schimmelpfennig C, Oelkrug C, Schönfelder U, Blatz R et al (2010) A real time PCR assay for the differentiation of Candida species. J Appl Microbiol 109:1150–1158

Geiger AM, Foxman B (1996) Risk factors for vulvovaginal candidiasis: a case-control study among university students. Epidemiology 7:182–187

Gibbs RS (1987) Microbiology of the female genital tract. Am J Obstet Gynecol 156:491–495

Giraldo PC, Babula O, Gonçalves AKS, Linhares IM, Amaral RL, Ledger WJ et al (2007) Mannose-binding lectin gene polymorphism, vulvovaginal candidiasis, and bacterial vaginosis. Obstet Gynecol 109:1123–1128

Goepfert AR, Varner M, Ward K, Macpherson C, Klebanoff M, Goldenberg RL et al (2005) Differences in inflammatory cytokine and Toll-like receptor genes and bacterial vaginosis in pregnancy. Am J Obstet Gynaecol 193:1478–1485

Goldenberg RL, Klebanoff MA, Nugent R, Krohn MA, Hillier S, Andrews WW (1996) Bacterial colonization of the vagina during pregnancy in four ethnic groups. Am J Obstet Gynaecol 174:1618–1621

Gómez LM, Sammel MD, Appleby DH, Elovitz MA, Baldwin DA, Jeffcoat MK et al (2010) Evidence of a gene-environment interaction that predisposes to spontaneous preterm birth: a role for asymptomatic bacterial vaginosis and DNA variants in genes that control the inflammatory response. Am J Obstet Gynecol 202:386-e1. https://doi.org/10.1016/j.ajog.2010.01.042

Goswami D, Goswami R, Banerjee U, Dadhwal V, Miglani S, Lattif AA et al (2006) Pattern of *Candida* species isolated from patients with diabetes mellitus and vulvovaginal candidiasis and their response to single dose oral fluconazole therapy. J Infect 52:111–117

Gravitt PE, Hildesheim A, Herrero R, Schiffman M, Sherman ME, Bratti MC et al (2003) Correlates of IL-10 and IL-12 concentrations in cervical secretions. J Clin Immunol 23:175–183

Greaves WL, Chungafung J, Morris B, Haile A, Townsend JL (1988) Clindamycin versus metronidazole in the treatment of bacterial vaginosis. Obstet Gynecol 72:799–802

Grigoriou O, Baka S, Makrakis E, Hassiakos D, Kapparos G, Kouskouni E (2006) Prevalence of clinical vaginal candidiasis in a university hospital and possible risk factors. Eur J Obstet Gynaecol Reprod Biol 126:121–125

Guaschino S, De Seta F, Sartore A, Ricci G, De Santo D, Piccoli M, Alberico S (2001) Efficacy of maintenance therapy with topical boric acid in comparison with oral itraconazole in the treatment of recurrent vulvovaginal candidiasis. Am J Obstet Gynaecol 184:598–602

Gunther LSA, Martins HPR, Gimenes F, Abreu ALPD, Consolaro MEL, Svidzinski TIE (2014) Prevalence of *Candida albicans* and non-*albicans* isolates from vaginal secretions: comparative evaluation of colonization, vaginal candidiasis and recurrent vaginal candidiasis in diabetic and non-diabetic women. Sao Paulo Med J 132:116–120

Gutierrez J, Morales P, Gonzalez MA, Quindos G (2002) *Candida dubliniensis*, a new fungal pathogen. J Basic Microbiol 42:207–227

Guzel AB, Ilkit M, Akar T, Burgut R, Demir SC (2011) Evaluation of risk factors in patients with vulvovaginal candidiasis and the value of chromID Candida agar versus CHROMagar Candida for recovery and presumptive identification of vaginal yeast species. Med Mycol 49:16–25

Hamad M, Kazandji N, Awadallah S, Allam H (2014) Prevalence and epidemiological characteristics of vaginal candidiasis in the UAE. Mycoses 57:184–190

Hanson JM, McGregor JA, Hillier SL, Eschenbach DA, Kreutner AK, Galask RP et al (2000) Metronidazole for bacterial vaginosis. A comparison of vaginal gel vs. oral therapy. J Reprod Med 45:889–896

Harriott MM, Lilly EA, Rodriguez TE, Fidel PL Jr, Noverr MC (2010) *Candida albicans* forms biofilms on the vaginal mucosa. Microbiology 156:3635–3644

Hawes SE, Hillier SL, Benedetti J, Stevens CE, Koutsky LA, Wølner-Hanssen P et al (1996) Hydrogen peroxide – producing lactobacilli and acquisition of vaginal infections. J Infect Dis 174:1058–1063

Hedayati MT, Taheri Z, Galinimoghadam T, Aghili SR, Cherati JY, Mosayebi E (2015) Isolation of different species of Candida in patients with vulvovaginal candidiasis from Sari, Iran. Jundishapur J Microbiol 8:e15992. https://doi.org/10.5812/jjm.82015.15992

Heikkinen J, Vuopala S (1989) Anaerobic vaginosis: treatment with tinidazole vaginal tablets. Gynecol Obstet Investig 28:98–100

Held J, Kohlberger I, Rappold E, Grawitz AB, Häcker G (2013) Comparison of (1→ 3)-β-D-glucan, mannan/anti-mannan-antibodies and Cand-Tec Candida-antigen as serum biomarkers for Candidemia. J Clin Microbiol 51:1158–1164

Holland J, Young ML, Lee O, Chen SC (2003) Vulvovaginal carriage of yeasts other than *Candida albicans*. Sex Transm Infect 79:249–250
Horvath LL, Hospenthal DR, Murray CK, Dooley DP (2003) Direct isolation of *Candida* spp. from blood cultures on the chromogenic medium CHROMagar Candida. J Clin Microbiol 41:2629–2632
Hostetter MK, Lorenz JS, Preus L, Kendrick KE (1990) The iC3b receptor on *Candida albicans*: subcellular localization and modulation of receptor expression by glucose. J Infect Dis 161:761–768
Ison CA, Hay PE (2002) Validation of a simplified grading of gram stained vaginal smears for use in genitourinary medicine clinics. Sex Transm Infect 78:413–415
Johnson EM, Warnock DW, Luker J, Porter SR, Scully C (1995) Emergence of azole drug resistance in Candida species from HIV-infected patients receiving prolonged fluconazole therapy for oral candidosis. J Antimicrob Chemother 35:103–114
Jonsson M, Karlsson R, Rylander E, Gustavsson A, Wadell G (1997) The associations between risk behaviour and reported history of sexually transmitted diseases, among young women: a population-based study. Int J STD AIDS 8(8):501–505
Kalia N, Singh J, Sharma S, Kamboj SS, Arora H, Kaur M (2015) Prevalence of vulvovaginal infections and species specific distribution of vulvovaginal candidiasis in married women of North India. Int J Curr Microbiol App Sci 4:253–266
Kalia N, Singh J, Sharma S, Arora H, Kaur M (2017) Genetic and phenotypic screening of mannosebinding lectin in relation to risk of recurrent vulvovaginal infections in women of North India: a prospective cohort study. Front Microbiol 8:75
Kalia N, Kaur M, Sharma S, Singh J (2018) A comprehensive in Silico analysis of regulatory SNPs of human CLEC7A gene and its validation as genotypic and phenotypic disease marker in recurrent vulvovaginal infections. Front Cell Infect Microbiol 8:65
Kalo-Klein A, Witkin SS (1991) Regulation of the immune response to *Candida albicans* monocytes and progesterone. Am J Obstet Gynecol 164:1351–1354
Kamath P, Pais M, Nayak MG (2013) Risk of vaginal candidiasis among pregnant women. Int J Curr Microbiol App Sci 2:141–146
Kanbe T, Horii T, Arishima T, Ozeki M, Kikuchi A (2002) PCR based identification of pathogenic Candida species using primer mixes specific to Candida DNA topoisomerase II genes. Yeast 19:973–989
Kennedy MA, Sobel JD (2010) Vulvovaginal candidiasis caused by non-albicans Candida species: new insights. Curr Infect Dis Rep 12:465–470
Krasnopolsky VN, Prilepskaya VN, Polatti F, Zarochentseva NV, Bayramova GR, Caserini M et al (2013) Efficacy of vitamin C vaginal tablets as prophylaxis for recurrent bacterial vaginosis: a randomised, double-blind, placebo-controlled clinical trial. J Clin Med Res 5:309–315
Larsson PG, Stray-Pedersen B, Ryttig KR, Larsen S (2008) Human lactobacilli as supplementation of clindamycin to patients with bacterial vaginosis reduce the recurrence rate; a 6-month, double-blind, randomized, placebo-controlled study. BMC Womens Health 8:3. https://doi.org/10.1186/1472-6874-8-3
Lau A, Sorrell TC, Chen S, Stanley K, Iredell J, Halliday C (2008) Multiplex tandem PCR: a novel platform for rapid detection and identification of fungal pathogens from blood culture specimens. J Clin Microbiol 46:3021–3027
Lazar E, Varga R (2011) Gynevac-a vaccine, containing lactobacillus for therapy and prevention of bacterial vaginosis and related diseases. Akush Ginekol 50:36–42
Lieberman JA, Moscicki AB, Sumerel JL, Ma Y, Scott ME (2008) Determination of cytokine protein levels in cervical mucus samples from young women by a multiplex immunoassay method and assessment of correlates. Clin Vaccine Immunol 15:49–54
Liu MB, Xu SR, He Y, Deng GH, Sheng HF, Huang XM et al (2013) Diverse vaginal microbiomes in reproductive-age women with vulvovaginal candidiasis. PLoS One 8:e79812. https://doi.org/10.1371/journal.pone.0079812

Livengood IIICH, Ferris DG, Wiesenfeld HC, Hillier SL, Soper DE, Nyirjesy P et al (2007) Effectiveness of two tinidazole regimens in treatment of bacterial vaginosis: a randomized controlled trial. Obstet Gynecol 110:302–309

Lo HJ, Köhler JR, DiDomenico B, Loebenberg D, Cacciapuoti A, Fink GR (1997) Nonfilamentous *C. albicans* mutants are avirulent. Cell 90:939–949

Lowe NK, Neal JL, Ryan-Wenger NA (2009) Accuracy of the clinical diagnosis of vaginitis compared to a DNA probe laboratory standard. Obstet Gynaecol 113:89

Luo G, Mitchell TG (2002) Rapid identification of pathogenic fungi directly from cultures by using multiplex PCR. J Clin Microbiol 40:2860–2865

Mayer FL, Wilson D, Hube B (2013) *Candida albicans* pathogenicity mechanisms. Virulence 4:119–128

McGroarty JA, Soboh F, Bruce AW, Reid G (1990) The spermicidal compound nonoxynol-9 increases adhesion of *Candida* species to human epithelial cells in vitro. Infect Immun 58:2005–2007

Menard JP, Fenollar F, Henry M, Bretelle F, Raoult D (2008) Molecular quantification of Gardnerella vaginalis and Atopobium vaginae loads to predict bacterial vaginosis. Clin Infect Dis 47:33–43

Merz WG, Connelly C, Hieter P (1988) Variation of electrophoretic karyotypes among clinical isolates of *Candida albicans*. J Clin Microbiol 26:842–845

Mikamo H, Kawazoe K, Izumi K, Watanabe K, Ueno K, Tamaya T (1997) Comparative study on vaginal or oral treatment of bacterial vaginosis. Chemotherapy 43:60–68

Mikulska M, Furfaro E, Del Bono V, Raiola AM, Ratto S, Bacigalupo A et al (2012) Piperacillin/tazobactam (TazocinTM) seems to be no longer responsible for false-positive results of the galactomannan assay. J Antimicrob Chemother 67:1746–1748

Milani M, Barcellona E, Agnello A (2003) Efficacy of the combination of 2 g oral tinidazole and acidic buffering vaginal gel in comparison with vaginal clindamycin alone in bacterial vaginosis: a randomized, investigator-blinded, controlled trial. Eur J Obstet Gynaecol Reprod Biol 109:67–71

Moreno SN, Docampo R (1985) Mechanism of toxicity of nitro compounds used in the chemotherapy of trichomoniasis. Environ Health Perspect 64:199–208

Morris MC, Rogers PA, Kinghorn GR (2001) Is bacterial vaginosis a sexually transmitted infection? Sex Transm Infect 77:63–68

Neerja J, Aruna A, Paramjeet G (2006) Significance of *Candida* culture in women with vulvovaginal symptoms. J Obstet Gynecol India 56:139–141

Nelson DB, Macones G (2002) Bacterial vaginosis in pregnancy: current findings and future directions. Epidemiol Rev 24:102–108

Nelson M, Wanjiru W, Margaret MW (2013) Prevalence of vaginal candidiasis and determination of the occurrence of Candida species in pregnant women attending the antenatal clinic of Thika District Hospital, Kenya. Open J Med Microbiol 3:264–274

Ness RB, Hillier S, Richter HE, Soper DE, Stamm C, Bass DC et al (2003) Can known risk factors explain racial differences in the occurrence of bacterial vaginosis? J Natl Med Assoc 95:201–212

Neves NA, Carvalho LP, De Oliveira MAM, Giraldo PC, Bacellar O, Cruz AA et al (2005) Association between atopy and recurrent vaginal candidiasis. Clin Exp Immunol 142:167–171

Nilsson U, Hellberg D, Shoubnikova M, Nilsson S, Mardh PA (1997) Sexual behavior risk factors associated with bacterial vaginosis and *Chlamydia trachomatis* infection. Sex Transm Dis 24:241–246

Nohmi T, Abe S, Dobashi K, Tansho S, Yamaguchi H (1995) Suppression of anti-Candida activity of murine neutrophils by progesterone in vitro: a possible mechanism in pregnant women's vulnerability to vaginal candidiasis. Microbiol Immunol 39:405–409

North BB (1988) Vaginal contraceptives. Effective protection from sexually transmitted diseases for women? J Reprod Med 33:307–311

Nugent RP, Krohn MA, Hillier SL (1991) Reliability of diagnosing bacterial vaginosis is improved by a standardized method of gram stain interpretation. J Clin Microbiol 29:297–301
Núñez JT, Gómez G (2005) Low-dose secnidazole in the treatment of bacterial vaginosis. Int J Gynecol Obstet 88:281–285
Nwadioha SI, Egah DZ, Alao OO, Iheanacho E (2010) Risk factors for vaginal candidiasis among women attending primary health care centres in Jos, Nigeria. J Clin Med Res 2:111–113
Odds F (1991) Sabouraud(s) agar. J Med Vet Mycol 29:355–359
Okkers DJ, Dicks LMT, Silvester M, Joubert JJ, Odendaal HJ (1999) Characterization of pentocin TV35b, a bacteriocin-like peptide isolated from *Lactobacillus pentosus* with a fungistatic effect on *Candida albicans*. J Appl Microbiol 87:726–734
Oliveira FA, Pfleger V, Lang K, Heukelbach J, Miralles I, Fraga F et al (2007) Sexually transmitted infections, bacterial vaginosis, and candidiasis in women of reproductive age in rural Northeast Brazil: a population-based study. Mem Inst Oswaldo Cruz 102:751–756
Onderdonk AB, Delaney ML, Fichorova RN (2016) The human microbiome during bacterial vaginosis. Clin Microbiol Rev 29:223–238
Ozkinay E, Terek MC, Yayci M, Kaiser R, Grob P, Tuncay G (2005) The effectiveness of live lactobacilli in combination with low dose oestriol (Gynoflor) to restore the vaginal flora after treatment of vaginal infections. BJOG Int J Obstet Gynaecol 112:234–240
Paavonen J, Mangioni C, Martin MA, Wajszczuk CP (2000) Vaginal clindamycin and oral metronidazole for bacterial vaginosis: a randomized trial. Obstet Gynecol 96:256–260
Peer AK, Hoosen AA, Seedat MA, Van Den Ende J, Omar MA (1993) Vaginal yeast infections in diabetic women. S Afr Med J 83:727–729
Peters BM, Palmer GE, Nash AK, Lilly EA, Fidel PL, Noverr MC (2014) Fungal morphogenetic pathways are required for the hallmark inflammatory response during *Candida Albicans* vaginitis. Infect Immun 82:532–543
Pietrella D, Pandey N, Gabrielli E, Pericolini E, Perito S, Kasper L et al (2013) Secreted aspartic proteases of *Candida albicans* activate the NLRP3 inflammasome. Eur J Immunol 43:679–692
Pitsouni E, Iavazzo C, Falagas ME (2008) Itraconazole vs fluconazole for the treatment of uncomplicated acute vaginal and vulvovaginal candidiasis in nonpregnant women: a metaanalysis of randomized controlled trials. Am J Obstet Gynaecol 198:153–160
Plitt SS, Garfein RS, Gaydos CA, Strathdee SA, Sherman SG, Taha TE (2005) Prevalence and correlates of *Chlamydia trachomatis*, *Neisseria gonorrhoeae*, *Trichomonas vaginalis* infections, and bacterial vaginosis among a cohort of young injection drug users in Baltimore, Maryland. Sex Transm Dis 32:446–453
Powell AM, Nyirjesy P (2014) Recurrent vulvovaginitis. Best Pract Res Clin Obstet Gynaecol 28:967–976
Powell HL, Sand CA, Rennie RP (1998) Evaluation of CHROMagar Candida for presumptive identification of clinically important Candida species. Diagn Microbiol Infect Dis 32:201–204
Ralph SG, Rutherford AJ, Wilson JD (1999) Influence of bacterial vaginosis on conception and miscarriage in the first trimester: cohort study. Br Med J 319:220–223
Ravel J, Gajer P, Abdo Z, Schneider GM, Koenig SS, McCulle SL et al (2011) Vaginal microbiome of reproductive-age women. Proc Natl Acad Sci 108:4680–4687
Ray D, Goswami R, Banerjee U, Dadhwal V, Goswami D, Mandal P et al (2007) Prevalence of *Candida glabrata* and its response to boric acid vaginal suppositories in comparison with oral fluconazole in patients with diabetes and vulvovaginal candidiasis. Diabetes Care 30:312–317
Reef SE, Levine WC, McNeil MM, Fisher-Hoch S, Holmberg SD, Duerr A et al (1995) Treatment options for vulvovaginal candidiasis, 1993. Clin Infect Dis 20:S80–S90
Rigby S, Procop GW, Haase G, Wilson D, Hall G, Kurtzman C et al (2002) Fluorescence in situ hybridization with peptide nucleic acid probes for rapid identification of *Candida albicans* directly from blood culture bottles. J Clin Microbiol 40:2182–2186
Rodrigues CF, Silva S, Henriques M (2014) *Candida glabrata*: a review of its features and resistance. Eur J Clin Microbiol Infect Dis 33:673–688

Rosentul D, Delsing C, Joosten LA, van der Meer JW, Kullberg BJ, Netea MG (2009) Polymorphism in innate immunity genes and susceptibility to recurrent vulvovaginal candidiasis. J Mycol Méd/J Med Mycol 19:191–196

Ryckman KK, Simhan HN, Krohn MA, Williams SM (2009) Predicting risk of bacterial vaginosis: the role of race, smoking and corticotropin-releasing hormone-related genes. Mol Hum Reprod 15:131–137

Santoro N, Randolph JF (2011) Reproductive hormones and the menopause transition. Obstet Gynecol Clin 38:455–466

Saville SP, Lazzell AL, Chaturvedi AK, Monteagudo C, Lopez-Ribot JL (2009) Efficacy of a genetically engineered *Candida albicans* tet-NRG1 strain as an experimental live attenuated vaccine against hematogenously disseminated candidiasis. Clin Vaccine Immunol 16:430–432

Schabereiter-Gurtner C, Selitsch B, Rotter ML, Hirschl AM, Willinger B (2007) Development of novel real-time PCR assays for detection and differentiation of eleven medically important *Aspergillus* and *Candida* species in clinical specimens. J Clin Microbiol 45:906–914

Schmidt CS, White CJ, Ibrahim AS, Filler SG, Fu Y, Yeaman MR et al (2012) NDV-3, a recombinant alum-adjuvanted vaccine for *Candida* and *Staphylococcus aureus*, is safe and immunogenic in healthy adults. Vaccine 30:7594–7600

Schoofs A, Odds FC, Colebunders R, Ieven M, Goossens H (1997) Use of specialised isolation media for recognition and identification of *Candida dubliniensis* isolates from HIV-infected patients. Eur J Clin Microbiol Infect Dis 16:296–300

Schreiber CA, Meyn LA, Creinin MD, Barnhart KT, Hillier SL (2006) Effects of long-term use of nonoxynol-9 on vaginal flora. Obstet Gynaecol 107:136–143

Scott ME, Ma Y, Farhat S, Shiboski S, Moscicki AB (2006) Covariates of cervical cytokine mRNA expression by real-time PCR in adolescents and young women: effects of *Chlamydia trachomatis* infection, hormonal contraception, and smoking. J Clin Immunol 26:222–232

Seidman LS, Skokos CK (2005) An evaluation of butoconazole nitrate 2% site release® vaginal cream (Gynazole-1®) compared to fluconazole 150 mg tablets (Diflucan®) in the time to relief of symptoms in patients with vulvovaginal candidiasis. Infect Dis Obstet Gynaecol 13:197–206

Sharma M, Solanki A (2014) Prevalence of *Candida* infection in pregnant women with and without diabetes. Int J Curr Microbiol App Sci 3:605–610

Shepard JR, Addison RM, Alexander BD, Della-Latta P, Gherna M, Haase G et al (2008) Multicenter evaluation of the *Candida albicans/Candida glabrata* peptide nucleic acid fluorescent in situ hybridization method for simultaneous dual-color identification of *C. albicans* and *C. glabrata* directly from blood culture bottles. J Clin Microbiol 46:50–55

Sherrard J, Donders G, White D, Jensen JS (2011) European (IUSTI/WHO) guideline on the management of vaginal discharge, 2011. Int J STD AIDS 22:421–429

Shoubnikova M, Hellberg D, Nilsson S, Mårdh PA (1997) Contraceptive use in women with bacterial vaginosis. Contraception 55:355–358

Simhan HN, Caritis SN, Hillier SL, Krohn MA (2005) Cervical anti-inflammatory cytokine concentrations among first-trimester pregnant smokers. Am J Obstet Gynecol 193:1999–2003

Singh S, Sobel JD, Bhargava P, Boikov D, Vazquez JA (2002) Vaginitis due to *Candida krusei*: epidemiology, clinical aspects, and therapy. Clin Infect Dis 35:1066–1070

Sobel JD (1988) Pathogenesis and epidemiology of vulvovaginal candidiasis. Ann N Y Acad Sci 544:547–557

Sobel JD (2006) Management of recurrent vulvovaginal candidiasis: unresolved issues. Curr Infect Dis Rep 8:481–486

Sobel JD (2007) Vulvovaginal candidosis. Lancet 369:1961–1971

Sobel JD (2014) Genital candidiasis. Medicine 38:186–190

Sobel JD (2016) Recurrent vulvovaginal candidiasis. Am J Obstet Gynaecol 214:15–21

Sobel JD, Akins RA (2015) The role of PCR in the diagnosis of candida vulvovaginitis – a new gold standard? Curr Infect Dis Rep 17:488. https://doi.org/10.1007/s11908-015-0488-3

Sobel JD, Chaim W, Nagappan V, Leaman D (2003) Treatment of vaginitis caused by *Candida glabrata*: use of topical boric acid and flucytosine. Am J Obstet Gynaecol 189:1297–1300

Sobel JD, Wiesenfeld HC, Martens M, Danna P, Hooton TM, Rompalo A et al (2004) Maintenance fluconazole therapy for recurrent vulvovaginal candidiasis. N Engl J Med 351:876–883
Sobel JD, Subramanian C, Foxman B, Fairfax M, Gygax SE (2013) Mixed vaginitis – more than coinfection and with therapeutic implications. Curr Infect Dis Rep 15:104–108
Souza ACR, Ferreira RC, Gonçalves SS, Quindós G, Eraso E, Bizerra FC et al (2012) Accurate identification of *Candida parapsilosis* (sensu lato) by use of mitochondrial DNA and real-time PCR. J Clin Microbiol 50:2310–2314
Špaček J, Buchta V, Jílek P, Förstl M (2007) Clinical aspects and luteal phase assessment in patients with recurrent vulvovaginal candidiasis. Eur J Obstet Gynecol Reprod Biol 131:198–202
Spanu T, Posteraro B, Fiori B, D'Inzeo T, Campoli S, Ruggeri A et al (2012) Direct MALDI-TOF mass spectrometry assay of blood culture broths for rapid identification of Candida species causing bloodstream infections: an observational study in two large microbiology laboratories. J Clin Microbiol 50:176–179
Spiegel CA, Amsel R, Holmes KK (1983) Diagnosis of bacterial vaginosis by direct gram stain of vaginal fluid. J Clin Microbiol 18:170–177
Srinivasan S, Fredricks DN (2009) The human vaginal bacterial biota and bacterial vaginosis. Interdiscip Perspect Infect Dis 2008:750479. https://doi.org/10.1155/2008/750479
Steffan P, Vazquez JA, Boikov D, Xu C, Sobel JD, Akins RA (1997) Identification of Candida species by randomly amplified polymorphic DNA fingerprinting of colony lysates. J Clin Microbiol 35:2031–2039
Sullivan D, Coleman D (1998) Candida dubliniensis: characteristics and identification. J Clin Microbiol 36:329–334
Taylor BD, Darville T, Ferrell RE, Ness RB, Kelsey SF, Haggerty CL (2014) Cross-sectional analysis of Toll-like receptor variants and bacterial vaginosis in African–American women with pelvic inflammatory disease. Sex Transm Infect 90:563–566
Toth B, Würfel W, Bohlmann MK, Gillessen-Kaesbach G, Nawroth F, Rogenhofer N et al (2015) Recurrent miscarriage: diagnostic and therapeutic procedures. Guideline of the DGGG (S1-Level, AWMF Registry No. 015/050, December 2013). Geburtshilfe Frauenheilkd 75:1117–1129
Trost A, Graf B, Eucker J, Sezer O, Possinger K, Göbel UB et al (2004) Identification of clinically relevant yeasts by PCR/RFLP. J Microbiol Methods 56:201–211
Trtkova J, Raclavsky V (2006) Molecular-genetic approaches to identification and typing of pathogenic Candida yeasts. Biomed Pap-Palacky Univ Olomouc 150:51–61
van de Wijgert JH, Borgdorff H, Verhelst R, Crucitti T, Francis S, Verstraelen H et al (2014) The vaginal microbiota: what have we learned after a decade of molecular characterization? PloS One 9:e105998. https://doi.org/10.1371/journal.pone.0105998
Velraeds MM, Van de Belt-Gritter B, Van der Mei HC, Reid G, Busscher HJ (1998) Interference in initial adhesion of uropathogenic bacteria and yeasts to silicone rubber by a *Lactobacillus acidophilus* biosurfactant. J Med Microbiol 47:1081–1085
Vermitsky JP, Self MJ, Chadwick SG, Trama JP, Adelson ME, Mordechai E et al (2008) Survey of vaginal-flora Candida species isolates from women of different age groups by use of species-specific PCR detection. J Clin Microbiol 46:1501–1503
Verstraelen H, Verhelst R (2009) Bacterial vaginosis: an update on diagnosis and treatment. Expert Rev Anti-Infect Ther 7:1109–1124
Vijaya D, Dhanalakshmi TA, Kulkarni S (2014) Changing trends of vulvovaginal candidiasis. J Lab Phys 6:28–30
Vos P, Hogers R, Bleeker M, Reijans M, Lee TVD, Hornes M et al (1995) AFLP: a new technique for DNA fingerprinting. Nucleic Acids Res 23:4407–4414
Vylkova S, Carman AJ, Danhof HA, Collette JR, Zhou H, Lorenz MC (2011) The fungal pathogen *Candida albicans* autoinduces hyphal morphogenesis by raising extracellular pH. MBio 2:e00055–e00011. https://doi.org/10.1128/mBio.00055-11
Wang Y (2009) CDKs and the yeast-hyphal decision. Curr Opin Microbiol 12:644–649

Wang FM, Qian XD, Xu H, Yu HY (2008) Efficacy of 5-nitroimidazole derivatives in treatment of bacterial vaginosis. Zhonghua Yi Xue Za Zhi 88:2201–2203

Warren D, Klein RS, Sobel J, Kieke B, Brown W, Schuman P et al (2001) A multicenter study of bacterial vaginosis in women with or at risk for human immunodeficiency virus infection. Infect Dis Obstet Gynaecol 9:133–141

Watson MC, Grimshaw JM, Bond CM, Mollison J, Ludbrook A (2002) Oral versus intra-vaginal imidazole and triazole anti-fungal agents for the treatment of uncomplicated vulvovaginal candidiasis (thrush): a systematic review. BJOG Int J Obstet Gynaecol 109:85–95

Watts DH, Rabe L, Krohn MA, Aura J, Hillier SL (1999) The effects of three nonoxynol-9 preparations on vaginal flora and epithelium. J Infect Dis 180:426–437

Weissenbacher T, Witkin SS, Ledger WJ, Tolbert V, Gingelmaier A, Scholz C et al (2009) Relationship between clinical diagnosis of recurrent vulvovaginal candidiasis and detection of *Candida* species by culture and polymerase chain reaction. Arch Gynaecol Obstet 279:125–129

Wilson J, Ralph S, Rutherford A (2000) Rates of bacterial vaginosis in women with different types of infertility. In: STD conference, Baltimore, Maryland

Wira CR, Fahey JV (2008) A new strategy to understand how HIV infects women: identification of a window of vulnerability during the menstrual cycle. AIDS (London, England) 22:1909–1917

Wouters MM, Van Wanrooy S, Nguyen A, Dooley J, Aguilera-Lizarraga J, Van Brabant W et al (2016) Psychological comorbidity increases the risk for postinfectious IBS partly by enhanced susceptibility to develop infectious gastroenteritis. Gut 65:1279–1288

Yoshimura K, Morotomi N, Fukuda K, Nakano M, Kashimura M, Hachisuga T et al (2011) Intravaginal microbial flora by the 16S rRNA gene sequencing. Am J Obstet Gynaecol 205:235-e1–235.e9. https://doi.org/10.1016/j.ajog.2011.04.018

Zeng J, Zong LL, Mao T, Huang YX, Xu ZM (2011) Distribution of *Candida albican* genotype and Candida species is associated with the severity of vulvovagianl candidiasis. Nan Fang Yi Ke Da Xue Xue Bao 31:1649–1653

Zhang X, Essmann M, Burt ET, Larsen B (2000) Estrogen effects on *Candida albicans*: a potential virulence-regulating mechanism. J Infect Dis 181:1441–1446

Zozaya M, Ferris MJ, Siren JD, Lillis R, Myers L, Nsuami MJ et al (2016) Bacterial communities in penile skin, male urethra, and vaginas of heterosexual couples with and without bacterial vaginosis. Microbiome 4:16. https://doi.org/10.1186/s40168-016-0161-6

Chapter 15
Enterococci As Nosocomial Pathogen

Preeti Sharma, Sumanpreet Kaur, and Sukhraj Kaur

Abstract *Enterococcal* species are Gram-positive lactic acid bacteria that are ubiquitously present in environmental samples, plants, and gastrointestinal tracts of animals. As commensals they are known to benefit the host, but in recent years, they have earned dubious reputation as nosocomial pathogens. They are known to cause diseases like urinary tract infections, endocarditis, bacteremia, and intra-abdominal infections especially in immunocompromised patients that are subjected to prolonged antibiotic treatments. The various factors contributing to their status as nosocomial pathogens are their intrinsic and acquired resistance to various classes of antibiotics. They are known to persist on animate and inanimate surfaces for a long period of time which thus act as reservoirs for the spread of the infection in hospitals. Recent studies have shown that the nosocomial strains are genetically distinct from commensal *Enterococcal* strains. Thus, herein the various diseases caused by nosocomial *Enterococcus* spp., the problem of antibiotic resistance, and their treatment have been reviewed. Further, this chapter also discusses the various virulence factors contributing to its pathogenicity and highlights the genetic differences between pathogenic and commensal *Enterococcus* spp.

Keywords *Enterococcus faecium* · *E. faecalis* · Nosocomial · Pathogen · Virulence factors

15.1 Introduction

From the last three decades, *Enterococcus* spp. have emerged as one of the important etiological agents of hospital-associated infections in immunocompromised individuals and in patients with prolonged antibiotic treatment and hospital stay (Gilmore et al. 2013). The two species *Enterococcus faecalis* and *E. faecium* out of 50 known species of enterococci are particularly pathogenic to man; for example, *E. faecalis* accounts for 85–90%, and *E. faecium* accounts for 5–10% of all *Enterococcal*

P. Sharma · S. Kaur · S. Kaur (✉)
Department of Microbiology, Guru Nanak Dev University, Amritsar, India

P. P. Singh (ed.), *Infectious Diseases and Your Health*,
https://doi.org/10.1007/978-981-13-1577-0_15

diseases (Maki and Agger 1988). Other species that can also be pathogenic include *E. gallinarum*, *E. casseliflavus*, *E. durans*, and *E. mundtii*. In the USA and Europe, enterococci have been reported to be the second and third leading cause of nosocomial infections (Sievert et al. 2013; Zarb et al. 2012). In India, enterococci accounted for 2.3–9.7% of all nosocomial infections (Sreeja et al. 2012; Chakraborty et al. 2015; Karmarkar et al. 2004). Enterococci have immense genome plasticity, are prolific colonizers, and have the ability to persist in hospital environment for a long time owing to their sturdy nature. The various clinical manifestations of *Enterococcal* infection are urinary tract infections, bacteremia, endocarditis, and intra-abdominal infections (Murray 1990). Apart from *E. faecalis*, the incidence of nosocomial *E. faecium* infections is also on the rise, primarily due to antibiotic resistance in *E. faecium*. The ability to cause infection in *E. faecium* is mainly attributed to the increasing resistance to antibiotics, whereas *E. faecalis* has been reported to exhibit innate ability to cause infections irrespective of the antibiotic resistance (Mundy et al. 2000). To better understand the molecular mechanisms involved in the pathogenicity of *Enterococcus*, the whole genome sequences of various pathogenic (Qin et al. 2012) and commensal (Brede et al. 2011) *Enterococcus* spp. have been published and compared.

15.2 General Characterization of Enterococci

Enterococci are Gram-positive, facultative anaerobic bacteria, which are ovoid in shape and may occur in diploids, in chains, or as single cells (Fig. 15.1). They are catalase- and cytochrome c oxidase-negative. According to the classification by Sherman (1937), the bacteria belonging to *Enterococcus* genera possess

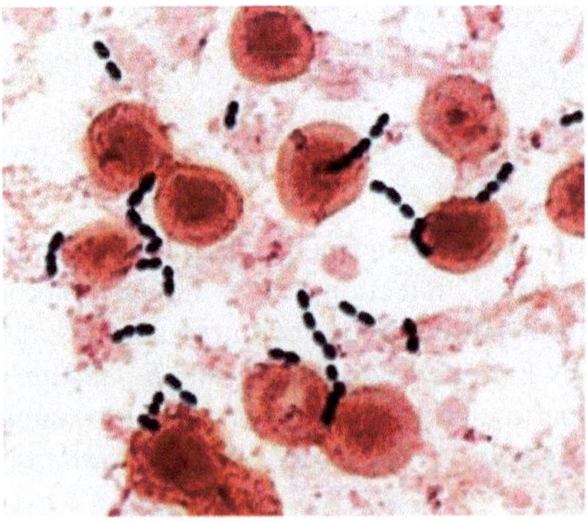

Scientific Classification

Domain: Bacteria

Division: Firmicutes

Class: Bacilli

Order: Lactobacillales

Family: Enterococcaceae

Genus: *Enterococcus*

A B

Fig. 15.1 (**a**) Photomicrograph showing Gram-positive coccus-shaped *Enterococcal* cells in the pulmonary tissue. (**b**) Scientific classification of *Enterococcus* genera

characteristic ability to grow at 10° and 45 °C, in media containing 6.5% NaCl, and do not produce gas from glucose (Schleifer and Kilpper-Bälz 1984). Hydrolysis of L-pyrrolidonyl-3-naphthylamide and group D antisera test is also one of the characteristic features of enterococci genera. The genera fall in low GC branch of bacteria, and G + C content ranges between 37 and 45 mol%. Species-level identification of *Enterococcus* genera is generally done by 16S rDNA sequencing or whole cell protein analysis. Till date approximately 50 species of *Enterococcus* have been reported. The prominent species present in the gut of animals and humans is *E. faecium* followed by *E. faecalis* (Silva et al. 2012).

15.3 Habitat

Enterococci are ubiquitous members of gastrointestinal tracts (GIT) of various organisms. In the 1960s and 1970s, enterococci were isolated from GIT of reptiles (85.7%), mammals (71.3%), insects (53%), and birds (31.8%) by Mundt et al. Enterococci have also been isolated from plants (Mundt 1963), soil (Mundt 1961), water, and fermented foods. It is postulated that enterococci may have been one of the earliest members of the GIT with their existence dating back from early Devonian period (Gilmore et al. 2013). Due to their role in highly evolved and extremely competitive environment of GIT, enterococci have well-adapted and reduced genomes ranging from 2.6 to 3.6 Mb. These bacteria are fastidious in nature and draw a number of vitamins and amino acids from their habitats (Niven and Sherman 1944). They have reduced their genome by eliminating genes necessary for the biosynthesis of amino acids and vitamins from simpler precursors.

15.4 Enterococcal Diseases

15.4.1 Endocarditis

Enterococcus spp. are known to cause endocarditis since 1899, when the first clinical description of the strain which was most certainly *E. faecalis* was published (MacCallum and Hastings 1899). Enterococci are the third main causative agents of infective endocarditis (IE) and prosthetic valve endocarditis worldwide after *Streptococcus* spp. and *Staphylococcus* spp. Worldwide approximately 10–15% IE cases are caused by enterococci (Murdoch et al. 2009). Frequency of *Enterococcal* IE is more in elderly debilitated patients with prior valvular damage, intracardiac devices, or a prosthetic valve. It occurs more in men than women (McDonald et al. 2005). Most common clinical findings of IE include subacute fever and the presence of a cardiac murmur. Almost half of the IE patients suffer from heart failure.

Mortality rate in *Enterococcal* IE is significant at 9–15%, but it is lower than IE caused by other pathogens such as *S. aureus* (McDonald et al. 2005; Rice et al. 1991). The first report of IE by vancomycin-resistant *E. faecium* (VRE) appeared in the late 1990s (Vijayvargiya and Veis 1996). The risk of VRE IE increases in the patients with the history of hemodialysis and organ transplantation and presence of a central venous catheter (Stevens and Edmond 2005). The study on clinical outcomes of IE by *E. faecalis* and *E. faecium* showed high mortality rates and longer persistence of bacteremia in *E. faecium* IE patients (Forrest et al. 2011).

15.4.2 Urinary Tract Infections (UTIs)

Enterococcus spp. are the third most common pathogen isolated from catheter-associated UTIs in hospitalized patients causing approximately 15% of the UTI cases in ICU patients. UTIs occur most frequently in elderly men. Vancomycin resistance was found to be 81% in *E. faecium* and 6% in *E. faecalis* urinary isolates in catheter-associated UTI (Hidron et al. 2008)

15.4.3 Bacteremia

At present, enterococci are the second leading causative agent of healthcare-related bacteremia (Hidron et al. 2008). The most common route of *Enterococcal* bloodstream infection (EBSI) is genitourinary tract and endovascular, intra-abdominal, or soft tissue infections. Other factors like old age, liver disease, male gender, renal impairment, hematologic transplant, diabetes, prior treatment with antibiotics, and malignancy have also been associated with EBSI (Noskin et al. 1995; Gray et al. 1994). In a study on patients diagnosed with acute leukemia, it was demonstrated that the risk of EBSI increases with increase in duration of hospital stay and with administrations of carbapenes and corticosteroids, diarrhea, and severe neutropenia (Ford et al. 2015). Twenty-five percent of EBSI cases have tendency to be polymicrobial in nature (Billington et al. 2014). Mortality rate in polymicrobial EBSI is approximately two times higher than monomicrobial EBSI (McKenzie 2006). Almost 9.5% and 82.6% *E. faecalis* and *E. faecium* isolated from blood stream infections are vancomycin resistant (Lautenbach et al. 1999). Unlike *S. aureus*-associated bacteremia, *Enterococcal* bacteremia rarely causes metastatic abscesses or seeds distant organs. Some studies have reported more mortality rate in *E. faecium* caused by bacteremia than that caused by *E. faecalis* (Noskin et al. 1995). Death rates can be up to 75% in patients with serious health conditions like diabetes mellitus, transplantation, heart diseases, or malignancy.

15.4.4 Intra-abdominal Infections

Intra-abdominal infections (IAIs) include various pathological conditions, such as fecal peritonitis, uncomplicated appendicitis, perforated viscus, and postoperative complications (Menichetti and Sganga 2009). In a 14-year-long study on recurrent and relapsing peritonitis in patients undergoing peritoneal dialysis, enterococci were found to be common causative agents of recurrent peritonitis (Szeto et al. 2009). The data collected at 68 medical centers demonstrated that enterococci were the most commonly isolated Gram-positive bacteria from IAI patients and constituted almost 12.9% of the isolates out of which 9.2% were *E. faecalis* and 3.7% were *E. faecium* (Sartelli et al. 2014). Further, it was reported that the increased intrinsic and acquired resistance in enterococci pose great challenges in the effective treatment of the diseases.

15.5 Nosocomial Transmission

Nosocomial *Enterococcal* infections are caused by increased colonization of GIT by resistant enterococci. Prolonged exposure of hospitalized patients to antibiotics alters gut microbiota which decreases the prevalence of Gram-negative microbes in the GIT, thus facilitating colonization by resistant Gram-positive bacteria such as enterococci. The primary mode of transmission of resistant enterococci in hospital setting is through the hands of healthcare workers (Hayden 2000). Studies have shown that enterococci have the ability to persist on the hands for approximately 1 h and for 4 months on inanimate surfaces, which thus act as reservoirs for transmission in the absence of regular sanitization (Kramer et al. 2006). Medical equipments like thermometers, stethoscopes, blood pressure cuffs, intravenous fluid pumps, bed rails, gowns, bedside tables, urinals, bedpans, and bed linens are readily contaminated with high densities of VRE (Bonilla et al. 1997; Hota 2004). Other factors posing VRE colonization risk include duration of patient's stay in hospital or intensive care unit and patient's proximity to other VRE-colonized patients (Fig. 15.2; Tornieporth et al. 1996). Strict surveillance of health workers is required to curb the spread of nosocomial *Enterococcal* infections. Further, various precautions, such as proper decontamination of inanimate surfaces, use of gloves, hand sanitization, etc., should be strictly followed.

15.6 Virulence Determinants of Enterococci

Virulence of enterococci is a multifactorial process, with the participation of several genes and their products. The virulence genes are mostly known to be present on the genome in special regions which are termed pathogenicity islands (PAI). However,

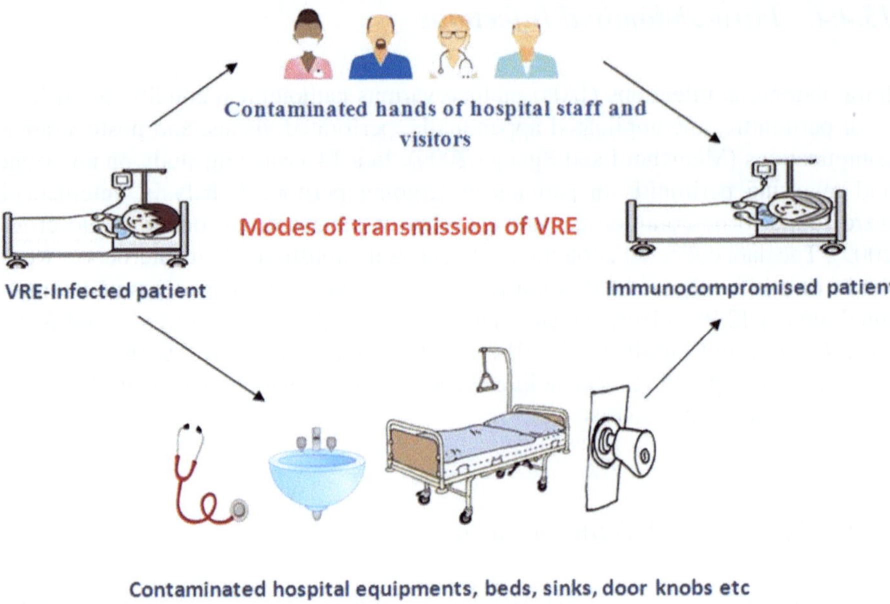

Fig. 15.2 Factors affecting the transmission of vancomycin-resistant enterococci (VRE) in nosocomial settings. The close proximity to VRE-infected patient passes on the contamination to hospital staff or hospital equipments and finally to the immunocompromised patient

some virulence factors may be present on plasmids also. *E. faecalis* are known to carry more numbers of virulence factors as compared to *E. faecium*.

15.7 Secreted Virulence Factors

15.7.1 Cytolysin

Cytolysin toxin belongs to the class of lantibiotic group of bacteriocins. Cytolysin production can be chromosomally encoded by PAI or encoded by a pheromone-responsive plasmid pAD1 (Ike et al. 1990). Cytolysin consists of two components CylLL and CylLS which are secreted as pro-peptides from the cell. These pro-peptides are proteolytically activated by CylM, CylB, and CylA.

Almost 30% of *E. faecalis* strains produce cytolysin, and its production has been associated with increased toxicity of *Enterococcal* infections. Cytolysin can lyse bacterial cells, mammalian erythrocytes, and other eukaryotic cells (Cox et al. 2005). In a rabbit endocarditis model, aggregation substance and cytolysin-positive strains caused 55% mortality as compared to 15% with only aggregation substance-positive strains (Chow et al. 1993).

15.7.2 Gelatinase

Gelatinase (Gel) is an extracellularly secreted Zn-metalloprotease which hydrolyzes casein, collagen, fibrinogen, complement components (C) 3 and C3a, and hemoglobin. Gel has been implicated as virulence factor by animal models and epidemiological studies. Gel expression is induced by *fsr* quorum-sensing system in response to high cell density (Qin et al. 2000). It mediates virulence by degrading host tissue and modulating host's immune response (Park et al. 2008). Gel can activate autolysin which is a peptidoglycan-degrading enzyme and leads to release of DNA and has a role in biofilm formation (Thomas et al. 2009).

15.8 Cell Surface-Associated Virulence Determinants

15.8.1 Aggregation Substance

Adherence to host tissue is important for pathogens to cause infection. In enterococci, aggregation substance (AS) and extracellular surface protein (esp) have been reported to play a role in adherence (Waters et al. 2004). In *E. faecalis*, AS is encoded by pheromone-responsive plasmid in response to pheromone (Clewell 1993). AS mediates aggregation of *Enterococcal* cells and facilitates transfer of plasmid during conjugation. It also has been reported to promote internalization of *Enterococcal* cells by intestinal cells, adhesion to renal tubular cells, and survival of *E. faecalis* cells inside polymorphonuclear neutrophils (Olmested et al. 1994; Kreft et al. 1992). In endocarditis model, destruction of pulmonary and myocardial tissue was found to be caused due to AS. Moreover, AS has been reported to mediate bacterial aggregation on cardiac valve resulting in increased pathogenesis in endocarditis (Schlievert et al. 1998).

15.8.2 Enterococcal *Surface Protein*

Esp is a large molecular weight protein expressed on the surface of *E. faecalis*. Esp is mainly associated with biofilm formation (Heikens et al. 2007) and plays a role in biofilm-associated infections such as UTI (Shankar et al. 2001), bacteremia, and endocarditis (Heikens et al. 2001) as demonstrated by studies with an *esp* deletion mutant. The *esp* gene located on PAI encodes a very large LPXTG-motif cell wall-anchored protein in both *E. faecalis* and *E. faecium*. It is widely present in *E. faecalis* strains, but *E. faecium* esp is predominantly found in hospital-associated isolates, suggesting its role in virulence.

15.8.3 Adhesion to Collagen Proteins

The ability of *E. faecalis* to adhere to extracellular matrix proteins like collagen, fibrinogen, laminin, lactoferrin, fibronectin, thrombospondin, and vitronectin has been reported in many studies (Rich et al. 1999; Nallapareddy and Murray 2008). The search for the gene encoding these adhesion molecules led to the discovery of *Ace*, an adhesin to collagen of *E. faecalis*. Most of these studies also found that only few isolates of *E. faecalis* exhibit adherence to extracellular matrix proteins under laboratory conditions (Woodford et al. 2001). Further experiments have found that most clinical isolates of *E. faecalis* do not express adherence phenotypes constitutively, but expression of these phenotypes is elicited under stress conditions or in the presence of host-derived factors (Nallapareddy and Murray 2008). The role of Ace was studied in experimental endocarditis rat model. Immunization with recombinant anti-Ace antibodies decreased endocarditis development (Singh et al. 2010).

Adherence studies showed that like *E. faecalis*, many isolates of *E. faecium* can bind to collagen following growth in brain-heart media, and the protein responsible is a cell wall-anchored Acm (adhesin to collagen of *E. faecium*; Nallapareddy et al. 2003). Collagen adherence of clinical isolates of *E. faecium* was found to be significantly more than the isolates from animal, community, or fecal origin, thus suggesting that it is an important factor responsible for infection-causing ability of *E. faecium*. Incorporation of *acm* gene in *acm*-negative mutants resulted in the expression of adherence phenotype. In *E. faecium* endocarditis mice model, *acm* mutant was less likely to develop endocarditis; thus it has been shown to be antigenic (Nallapareddy et al. 2008).

15.8.4 Endocarditis- and Biofilm-Associated Pili (ebp)

The ebp are surface-associated filamentous structures in *E. faecalis*. It plays a role in biofilm formation by binding to abiotic surfaces and adhering to platelets (Nallapareddy et al. 2011a). In animal models, it contributes to tissue colonization in IE and UTI (Sillanpaa et al. 2013). Deletion of the *ebp* locus resulted in a diminished capacity of *E. faecalis* OG1RF to colonize kidneys and bladders in a murine model of ascending UTI (Nallapareddy et al. 2011b) and in catheter-associated UTI (Nielsen et al. 2012).

15.9 Factors Contributing to *Enterococcus* Pathogenesis and Divergence from Commensals

Enterococci are part of normal GIT, and transition of commensals to pathogens is associated with change in ecology, acquisition of toxin genes, and mobile genetic elements carrying PAI. Comparative studies of the genomes of nosocomial and commensal strains show that nosocomial strains have acquired these virulent traits. Genome sequence analysis of V583, which is considered as representative strain of nosocomial lineage of *E. faecalis*, revealed that 26% of the genome comprised of mobile genetic elements, 38 insertion elements, remains of 3 integrative phages, 7 apparent phages, and 3 independently replicating plasmids (Paulsen 2003). Similar evolutionary sequences are found in case of nosocomial *E. faecium* strains which are found to possess large PAI carrying several IS and *esp* gene (Leavis et al. 2007). Further, these virulent genes synergize with each other to express the virulence phenotype; for example, in an endocarditis model, AS was found to synergize with cytolysin protein.

Further, the genome of V583 when compared with commensal strain OG1RF revealed the presence of CRISPR-*cas* in commensal strains which provides defense against plasmid acquisition and phage infection. The absence of CRISPR in V583 led to studies directed to find correlation between absence of CRISPR and accumulation of mobile genetic elements in nosocomial strains. Sixteen *E. faecalis* strains representing the deepest phylogenetic nodes were studied to identify polymorphism in location and content of CRISPR loci. Results suggested that CRISPR loci were highly conserved and influenced the movement of pheromone-responsive plasmids and phages. A complete absence of CRISPR-*cas* in MDR *Enterococcus* strains was also reported (Palmer and Gilmore 2010). Based on the whole genome data and the comparison of the 2113 core genes of the eight *E. faecium* strains, they were placed in two different clades. Clade A mainly comprised of nosocomial infection-causing strains, whereas clade B comprised of commensal strains. Clades A and B have similarity in the range 93.9–95.6% that shows the degree of divergence between the species. In case of *E. faecalis*, the commensal and nosocomial pathogens can also be similarly differentiated into different clonal complexes, but the division is overlapping in some cases (Gilmore et al. 2013).

15.10 Treatment

The treatment of infections caused by susceptible *Enterococcus* strains is mainly by using ß-lactam antibiotics. Monomicrobial *Enterococcal* infections that are susceptible are usually treated with ampicillin and penicillin alone. On the other hand, polymicrobial infections are treated with the combination of ampicillin and other antibiotics having broad spectrum of activity against aerobic, anaerobic, and Gram-negative bacteria. Combination drugs containing β-lactam along with β-lactamase inhibitor such as piperacillin-tazobactam or ampicillin-clavulanic acid can also be used. In case of penicillin allergies, a single agent like vancomycin or teicoplanin can be used to treat nonresistant *Enterococcal* infections. In case of penicillin-susceptible VRE infections like cystitis, oral application of nitrofurantoin, doxycycline, and fosfomycin can be considered (Heintz et al. 2010). Infections caused by VRE strains are mostly treated by antibiotic, linezolid, daptomycin, and quinupristin-dalfopristin (Q-D), but side effects are common in prolonged treatment.

For *E. faecium* Q-D is not used as it is resistant. UTI and skin and soft tissue infections caused by VRE are treated by fluoroquinolones or doxycycline depending upon the susceptibility patterns (Landman and Quale 1997). Adjunct measures including drainage of abscesses and removal of infected foci can also be used wherever possible.

Treatment of IE caused by *E. faecalis* mainly includes combination of ß-lactams, gentamicin, and aminoglycosides. In case of IE caused by *E. faecium*, treatment is challenging due to higher resistance to ß-lactams and aminoglycosides. Daptomycin and ceftaroline have also been found to be effective for the treatment (Pericás et al. 2015).

15.11 Antibiotic Resistance

Increasing antibiotic resistance in enterococci is a matter of concern because the increase in the cases of *Enterococcal* infections in humans is partially due to its ability to escape the action of commonly used antibiotics. Enterococci are known to be intrinsically resistant to many antibiotics due to which they can exist in environments enriched for antibiotic resistance (Mundy et al. 2000; Murray 1998). They are known to be intrinsically resistant to semisynthetic penicillins and cephalosporins due to the presence of low-affinity penicillin-binding proteins; for example, *E. faecalis* expresses PBP4, and *E. faecium* expresses PBP5. The minimum inhibitory concentrations (MICs) for penicillins are typically 2–8 μg/ml for *E. faecalis* and 8–16 μg/ml for *E. faecium* (Sifaoui et al. 2001). They also show decreased susceptibility to penicillin, ampicillin, aminoglycosides, and clindamycin.

Among aminoglycosides, the MIC of streptomycin and kanamycin in *E. faecalis* is 250 μg/ml, whereas that of tobramycin and gentamycin is 8–64 μg/ml (Chow 2000). Aminoglycoside resistance is due to the inability of the aminoglycoside to enter the thick cell wall (Aslangul et al. 2006). Therefore in the presence of cell wall-inhibiting antibiotics such as β-lactams, the enterococci become susceptible to aminoglycosides. The presence of aminoglycoside-converting enzymes such as acetyltransferases, phosphotransferases, and nucleotidyl transferases in *E. faecium* and *E. faecalis* also contributes to aminoglycoside resistance (Chow 2000; Miller et al. 2014). In *E. faecium* chromosomally encoded genes, rRNA methyltransferase (*efmM*) (Galimand et al. 2011) and a 6′-*N*-aminoglycoside acetyltransferase (*aac(6′)- Ii*) (Costa et al. 1993) are associated with intrinsic resistance to tobramycin and kanamycin. Resistance to clindamycin in *E. faecalis* is probably due to the presence of protein product of *lsa* (lincosamide and streptogramin A resistance) gene, i.e., ATP-binding cassette (ABC)-efflux pumps, i.e., ABC-23 (Singh et al. 2002). Similarly, putative ABC transporter was identified in all *E. faecium* isolates that result in resistance to erythromycin, and chloramphenicol resistance has been linked to efflux of the antibiotic out of the cell (Aakra et al. 2010).

Enterococcus spp. are known to acquire resistance to various antibiotics such as vancomycin, chloramphenicol, erythromycin, and fluoroquinolones and high-level resistance to aminoglycosides and penicillin. Acquired antibiotic resistance in a pathogen is due to either of the two mechanisms, i.e., either by mutation of the gene or by horizontal transfer of antibiotic resistance genes. However, in *Enterococcus* antibiotic resistance problem mainly stems from horizontal gene transfer. Enterococci exist in complex microbial environments where they come in contact with large diversity of genetic material, and in the presence of antibiotic pressure, they tend to acquire resistance to antibiotics. For example, there is a direct relationship between exposure to parenteral antibiotics, especially cephalosporins, antibiotics for anaerobes, and high-level gastrointestinal colonization by ampicillin-resistant *E. faecium* (Rice et al. 2004).

Enterococci were recognized as important agents of MDR nosocomial infections after vancomycin resistance was observed. In the USA, 87% *E. faecium* and 14% *E. faecalis* isolated from nosocomial infections are vancomycin resistant (Edelsberg et al. 2014). Vancomycin acts by binding to D-alanine-D-alanine moiety of peptide chain of peptidoglycans, thus preventing cross-linking of peptidoglycans. *Enterococcus* strains become resistant to vancomycin by altering peptidoglycan precursors from D-alanine-D-alanine to D-Ala-D-lactate or to D-Ala-D-serine. Nine gene clusters involved in vancomycin resistance are *vanA*, *vanB*, *vanD*, and *vanM* (Xu et al. 2010) leading to the formation of D-Ala-D-Lac and *vanC*, *vanE*, *vanG*, *vanL*, and *vanN* catalyzing D-Ala-D-Ser formation (Boyd et al. 2008; Lebreton et al. 2011). *VanA* and *vanB* are the most relevant vancomycin-resistant determinants and are located on transposons (Courvalin 2006). In the 1970s *Enterococcus* acquired high level of resistance against ampicillin due to specific mutations in *pbp* genes, which encode for penicillin-binding proteins (Galloway-Peña et al. 2011).

15.12 Conclusion

The emerging problem of hospital-associated *Enterococcal* infections appears to be man-made due to injudicious use of antibiotics especially for non-therapeutic purposes such as avoparcin which was employed as growth promoter in the animal feed for the first time in 1975. The overuse of avoparcin led to avoparcin resistance in the gut flora of animals. The avoparcin-resistant gut flora showed cross-resistance toward vancomycin (Mudd 2000). Thus, the first VRE was isolated from the gut flora of farm animals in the 1990s. Subsequently, the use of avoparcin was banned (Casewell et al. 2003) It is hypothesized that the VRE from farm animals were acquired by humans. This is a classic case that highlights the serious repercussions of overuse of antibiotics both for therapeutic and non-therapeutic purposes. To treat *Enterococcal* infections, antibiotics should be selected wisely based on various factors such as the age, immunocompromised status, type and location of infection, and antibiotic sensitivity spectrum of the *Enterococcal* pathogen. Further, general measures to prevent the spread of *Enterococcus* in the hospital settings should be strictly followed such as sanitization of hospital environment and equipments.

Acknowledgment This work was supported by research grant (grant number, 42-478/2013 SR) sponsored by the University Grants Commission (UGC), New Delhi, India. Ms. Preeti Sharma is thankful to the University of Potential for Excellence scheme of UGC for the fellowship. Ms. Sumanpreet Kaur is thankful to UGC for the fellowship.

References

Aakra A, Vebø H, Indahl U, Snipen L, Gjerstad O, Lunde M et al (2010) The response of *Enterococcus faecalis* V583 to chloramphenicol treatment. Int J Microbiol 2010:483048. https://doi.org/10.1155/2010/483048

Aslangul E, Massias L, Meulemans A, Chau F, Andremont A, Courvalin P et al (2006) Acquired gentamicin resistance by permeability impairment in *Enterococcus faecalis*. Antimicrob Agents Chemother 50:3615–3621

Billington EO, Phang SH, Gregson DB, Pitout JD, Ross T, Church DL et al (2014) Incidence, risk factors, and outcomes for *Enterococcus* spp. blood stream infections: a population-based study. Int J Infect Dis 26:76–82

Bonilla HF, Zervos MA, Lyons MJ, Bradley SF, Hedderwick SA, Ramsey MA et al (1997) Colonization with vancomycin-resistant *Enterococcus faecium*: comparison of a long-term-care unit with an acute-care hospital. Infect Control Hosp Epidemiol 18:333–339

Boyd DA, Willey BM, Fawcett D, Gillani N, Mulvey MR (2008) Molecular characterization of enterococcus faecalis N06-0364 with low-level vancomycin resistance harboring a novel D-Ala-D-Ser gene cluster, vanL. Antimicrob Agents Chemother 52:2667–2672

Brede DA, Snipen LG, Ussery DW, Nederbragt AJ, Nes IF (2011) Complete genome sequence of the commensal *Enterococcus faecalis* 62, isolated from a healthy Norwegian infant. J Bacteriol 193:2377–2378. https://doi.org/10.1128/JB.00183-11

Casewell M, Friis C, Marco E, McMullin P, Phillips I (2003) The European ban on growth-promoting antibiotics and emerging consequences for human and animal health. J Antimicrob Chemother 52:159–161

Chakraborty A, Pal NK, Sarkar S, Gupta MS (2015) Antibiotic resistance pattern of enterococci isolates from nosocomial infections in a tertiary care hospital in Eastern India. J Natl Sci Biol Med 6(2):394

Chow JW (2000) Aminoglycoside resistance in enterococci. Clin Infect Dis 31(2):586–589

Chow JW, Thal LA, Perri MB, Vazquez JA, Donabedian SM, Clewell DB et al (1993) Plasmid-associated hemolysin and aggregation substance production contribute to virulence in experimental *Enterococcal* endocarditis. Antimicrob Agents Chemother 37:2474–2477

Clewell DB (1993) Sex pheromones and the plasmid-encoded mating response in enterococcus faecalis. In: Clewell DB (ed) Bacterial conjugation. Springer, Boston

Costa Y, Galimand M, Leclercq R, Duval J, Courvalin P (1993) Characterization of the chromosomal aac (6′)-Ii gene specific for *Enterococcus faecium*. Antimicrob Agents Chemother 37:1896–1903

Courvalin P (2006) Vancomycin resistance in Gram-positive cocci. Clin Infect Dis 42(Supplement 1):S25–S34

Cox CR, Coburn PS, Gilmore MS (2005) *Enterococcal* cytolysin: a novel two component peptide system that serves as a bacterial defense against eukaryotic and prokaryotic cells. Curr Protein Pept Sci 6:77–84

Edelsberg J, Weycker D, Barron R, Li X, Wu H, Oster G et al (2014) Prevalence of antibiotic resistance in US hospitals. Diagn Microbiol Infect Dis 78:255–262

Ford CD, Lopansri BK, Haydoura S, Snow G, Dascomb KK, Asch J et al (2015) Frequency, risk factors, and outcomes of vancomycin-resistant *Enterococcus* colonization and infection in patients with newly diagnosed acute leukemia: different patterns in patients with acute myelogenous and acute lymphoblastic leukemia. Infect Control Hosp Epidemiol 36:47–53

Forrest GN, Arnold RS, Gammie JS, Gilliam BL (2011) Single center experience of a vancomycin resistant *Enterococcal* endocarditis cohort. J Infect 63:420–428

Galimand M, Schmitt E, Panvert M, Desmolaize B, Douthwaite S, Mechulam Y et al (2011) Intrinsic resistance to aminoglycosides in *Enterococcus faecium* is conferred by the 16S rRNA m5C1404-specific methyltransferase EfmM. RNA 17:251–262

Galloway-Pena JR, Rice LB, Murray BE (2011) Analysis of PBP5 of early US isolates of *Enterococcus faecium*: sequence variation alone does not explain increasing ampicillin resistance over time. Antimicrob Agents Chemother 55:3272–3277

Gilmore MS, Lebreton F, van Schaik W (2013) Genomic transition of enterococci from gut commensals to leading causes of multidrug-resistant hospital infection in the antibiotic era. Curr Opin Microbiol 16:10–16

Gray J, Marsh PJ, Stewart D, Pedler SJ (1994) *Enterococcal* bacteraemia: a prospective study of 125 episodes. J Hosp Infect 27:179–186

Hayden MK (2000) Insights into the epidemiology and control of infection with vancomycin-resistant enterococci. Clin Infect Dis 31:1058–1065

Heikens E, Singh KV, Jacques-Palaz KD, van Luit-Asbroek M, Oostdijk EA, Bonten MJ et al (2001) Contribution of the *Enterococcal* surface protein Esp to pathogenesis of *Enterococcus faecium* endocarditis. Microbes Infect 13:1185–1190

Heikens E, Bonten MJ, Willems RJ (2007) *Enterococcal* surface protein Esp is important for biofilm formation of *Enterococcus faecium* E1162. J Bacteriol 189:8233–8240

Heintz BH, Halilovic J, Christensen CL (2010) Vancomycin-resistant *Enterococcal* urinary tract infections. Pharmacother: J Hum Pharmacol Drug Ther 30:1136–1149

Hidron AI, Edwards JR, Patel J, Horan TC, Sievert DM, Pollock DA et al (2008) NHSN annual update: antimicrobial-resistant pathogens associated with healthcare-associated infections: annual summary of data reported to the National Healthcare Safety Network at the Centers for Disease Control and Prevention, 2006–2007. Infect Control Hosp Epidemiol 29:996–1011

Hota B (2004) Contamination, disinfection, and cross-colonization: are hospital surfaces reservoirs for nosocomial infection? Clin Infect Dis 39:1182–1189

Ike Y, Clewell DB, Segarra RA, Gilmore MS (1990) Genetic analysis of the pAD1 hemolysin/bacteriocin determinant in *Enterococcus faecalis*: Tn917 insertional mutagenesis and cloning. J Bacteriol 172:155–163

Karmarkar MG, Gershom ES, Mehta PR (2004) *Enterococcal* infections with special reference to phenotypic characterization & drug resistance. Indian J Med Res 119:22–25

Kramer A, Schwebke I, Kampf G (2006) How long do nosocomial pathogens persist on inanimate surfaces? A systematic review. BMC Infect Dis 6:130

Kreft B, Marre R, Schramm U, Wirth R (1992) Aggregation substance of *Enterococcus faecalis* mediates adhesion to cultured renal tubular cells. Infect Immun 60:25–30

Landman D, Quale JM (1997) Management of infections due to resistant enterococci: a review of therapeutic options. J Antimicrob Chemother 40:161–170

Lautenbach E, Bilker WB, Brennan PJ (1999) *Enterococcal* bacteremia: risk factors for vancomycin resistance and predictors of mortality. Infect Control Hosp Epidemiol 20:318–323

Leavis HL, Willems RJ, Van Wamel WJ, Schuren FH, Caspers MP, Bonten MJ (2007) Insertion sequence-driven diversification creates a globally dispersed emerging multiresistant subspecies of *E. faecium*. PLoS Pathog 26(3):e7

Lebreton F, Depardieu F, Bourdon N, Fines-Guyon M, Berger P, Camiade S et al (2011) D-Ala-D-Ser VanN-type transferable vancomycin resistance in *Enterococcus faecium*. Antimicrob Agents Chemother 55:4606–4612

MacCallum WG, Hastings TW (1899) A case of acute endocarditis caused by *Micrococcus zymogenes* (nov. spec.), with a description of the microorganism. J Exp Med 4:521–534

Maki DG, Agger WA (1988) *Enterococcal* bacteremia: clinical features, the risk of endocarditis, and management. Medicine 67:248

McDonald JR, Olaison L, Anderson DJ, Hoen B, Miro JM, Eykyn S et al (2005) *Enterococcal* endocarditis: 107 cases from the international collaboration on endocarditis merged database. Am J Med 118:759–766

McKenzie FE (2006) Case mortality in polymicrobial bloodstream infections. J Clin Epidemiol 59:760–761
Menichetti F, Sganga G (2009) Definition and classification of intra-abdominal infections. J Chemother 21:3–4
Miller WR, Munita JM, Arias CA (2014) Mechanisms of antibiotic resistance in enterococci. Expert Rev Anti-Infect Ther 12:1221–1236
Mudd AJ (2000) Vancomycin-resistant enterococci and use of avoparcin in animal feed: is there a link? Med J Aust 172:458–459
Mundt JO (1961) Occurrence of enterococci: bud, blossom, and soil studies. Appl Microbiol 9:541–544
Mundt JO (1963) Occurrence of enterococci on plants in a wild environment. Appl Microbiol 1:141–144
Mundy LM, Sahm DF, Gilmore MS (2000) Relationships between *Enterococcal* virulence and antimicrobial resistance. Clin Microbiol Rev 13:513–522
Murdoch DR, Corey GR, Hoen B, Miró JM, Fowler VG Jr, Bayer AS et al (2009) Clinical presentation, etiology, and outcome of infective endocarditis in the 21st century: the international collaboration on endocarditis-prospective cohort study. JAMA Intern Med 169:463–473
Murray BE (1990) The life and times of the *Enterococcus*. Clin Microbiol Rev 3:46–65
Murray BE (1998) Diversity among multidrug-resistant enterococci. Emerg Infect Dis 4:37
Nallapareddy SR, Murray BE (2008) Role played by serum, a biological cue, in the adherence of *Enterococcus faecalis* to extracellular matrix proteins, collagen, fibrinogen, and fibronectin. J Infect Dis 197:1728–1736
Nallapareddy SR, Weinstock GM, Murray BE (2003) Clinical isolates of *Enterococcus faecium* exhibit strain-specific collagen binding mediated by Acm, a new member of the MSCRAMM family. Mol Microbiol 47:1733–1747
Nallapareddy SR, Singh KV, Murray BE (2008) Contribution of the collagen adhesin Acm to pathogenesis of *Enterococcus faecium* in experimental endocarditis. Infect Immun 76:4120–4128
Nallapareddy SR, Sillanpää J, Mitchell J, Singh KV, Chowdhury SA, Weinstock GM et al (2011a) Conservation of Ebp-type pilus genes among enterococci and demonstration of their role in adherence of *Enterococcus faecalis* to human platelets. Infect Immun 79:2911–2920
Nallapareddy SR, Singh KV, Sillanpää J, Zhao M, Murray BE (2011b) Relative contributions of *Ebp* pili and the collagen adhesin *ace* to host extracellular matrix protein adherence and experimental urinary tract infection by *Enterococcus faecalis* OG1RF. Infect Immun 79:2901–2910
Nielsen HV, Guiton PS, Kline KA, Port GC, Pinkner JS, Neiers F et al (2012) The metal ion-dependent adhesion site motif of the *Enterococcus faecalis EbpA* pilin mediates pilus function in catheter-associated urinary tract infection. MBio 3:e00177-12. https://doi.org/10.1128/mBio.00177-12
Niven CF Jr, Sherman JM (1944) Nutrition of the enterococci. J Bacteriol 47:335–342
Noskin GA, Peterson LR, Warren JR (1995) *Enterococcus faecium* and *Enterococcus faecalis* bacteremia: acquisition and outcome. Clin Infect Dis 20:296–301
Olmested S, Dunny G, Erlandsen S, Wells C (1994) A plasmid encoded surface protein on *Enterococcus faecalis* augments its internalization by cultured intestinal epithelial cells. J Infect Dis 170:1549–1556
Palmer KL, Gilmore MS (2010) Multidrug-resistant enterococci lack CRISPR-cas. MBio 29(1):e00227–e00210
Park SY, Shin YP, Kim CH, Park HJ, Seong YS, Kim BS et al (2008) Immune evasion of *Enterococcus faecalis* by an extracellular gelatinase that cleaves C3 and iC3b. J Immunol 181:6328–6336
Paulsen IT (2003) Role of mobile DNA in the evolution of vancomycin-resistant enterococcus faecalis. Science 299(5615):2071–2074
Pericás J, Zboromyrska Y, Cervera C, Castañeda X, Almela M, Garcia-De-La-Maria C et al (2015) *Enterococcal* endocarditis revisited. Future Microbiol 10:1215–1240

Qin X, Singh KV, Weinstock GM, Murray BE (2000) Effects of *Enterococcus faecalis fsr* genes on production of gelatinase and a serine protease and virulence. Infect Immun 68:2579–2586
Qin X, Galloway-Peña JR, Sillanpaa J, Roh JH, Nallapareddy SR, Chowdhury S et al (2012) Complete genome sequence of *Enterococcus faecium* strain TX16 and comparative genomic analysis of *Enterococcus faecium* genomes. BMC Microbiol 12:135
Rice LB, Calderwood SB, Eliopoulos GM, Farber BF, Karchmer AW (1991) *Enterococal* endocarditis: a comparison of prosthetic and native valve disease. Rev Infect Dis 13:1–7
Rice LB, Lakticová V, Helfand MS, Hutton-Thomas R (2004) In vitro anti *Enterococcal* activity explains associations between exposures to antimicrobial agents and risk of colonization by multi resistant enterococci. J Infect Dis 190:2162–2166
Rich RL, Kreikemeyer B, Owens RT, LaBrenz S, Narayana SV, Weinstock GM et al (1999) Ace is a collagen-binding MSCRAMM from *Enterococcus faecalis*. J Biol Chem 274:26939–26945
Sartelli M, Catena F, Ansaloni L, Coccolini F, Corbella D, Moore EE et al (2014) Complicated intra-abdominal infections worldwide: the definitive data of the CIAOW study. World J Emerg Surg 9:37
Schleifer KH, Kilpper-Bälz R (1984) Transfer of *Streptococcus faecalis* and *Streptococcus faecium* to the genus *Enterococcus* nom. rev. as *Enterococcus faecalis* comb. nov. and *Enterococcus faecium* comb. nov. Int J Syst Evol Microbiol 34:31–34
Schlievert PM, Gahr PJ, Assimacopoulos AP, Dinges MM, Stoehr JA, Harmala JW et al (1998) Aggregation and binding substances enhance pathogenicity in rabbit models of *Enterococcus faecalis* endocarditis. Infect Immun 66:218–223
Shankar N, Lockatell CV, Baghdayan AS, Drachenberg C, Gilmore MS, Johnson DE (2001) Role of *Enterococcus faecalis* surface protein Esp in the pathogenesis of ascending urinary tract infection. Infect Immun 69:4366–4372
Sherman JM (1937) The streptococci. Bacteriol Rev 1:3–97
Sievert DM, Ricks P, Edwards JR, Schneider A, Patel J, Srinivasan A et al (2013) Antimicrobial-resistant pathogens associated with healthcare-associated infections: summary of data reported to the National Healthcare Safety Network at the Centers for Disease Control and Prevention, 2009–2010. Infect Control Hosp Epidemiol 34:1–14
Sifaoui F, Arthur M, Rice L, Gutmann L (2001) Role of penicillin-binding protein 5 in expression of ampicillin resistance and peptidoglycan structure in *Enterococcus faecium*. Antimicrob Agents Chemother 45:2594–2597
Sillanpää J, Chang C, Singh KV, Montealegre MC, Nallapareddy SR, Harvey BR et al (2013) Contribution of individual *Ebp* pilus subunits of *Enterococcus faecalis* OG1RF to pilus biogenesis, biofilm formation and urinary tract infection. PLoS One 8(7):e68813. https://doi.org/10.1371/journal.pone.0068813
Silva N, Igrejas G, Gonçalves A, Poeta P (2012) Commensal gut bacteria: distribution of *Enterococcus* species and prevalence of *Escherichia coli* phylogenetic groups in animals and humans in Portugal. Ann Microbiol 62:449–459
Singh KV, Weinstock GM, Murray BE (2002) An *Enterococcus faecalis* ABC homologue (Lsa) is required for the resistance of this species to clindamycin and quinupristin-dalfopristin. Antimicrob Agents Chemother 46:1845–1850
Singh KV, Nallapareddy SR, Sillanpää J, Murray BE (2010) Importance of the collagen adhesin ace in pathogenesis and protection against *Enterococcus faecalis* experimental endocarditis. PLoS Pathog 6(1):e1000716. https://doi.org/10.1371/journal.ppat.1000716
Sreeja S, Babu PRS, Prathab AG (2012) The prevalence and the characterization of the *Enterococcus* species from various clinical samples in a tertiary care hospital. J Clin Diagn Res 6:1486–1488
Stevens MP, Edmond MB (2005) Endocarditis due to vancomycin resistant enterococci: case report and review of the literature. Clin Infect Dis 41:1134–1142
Szeto CC, Kwan BC, Chow KM, Law MC, Pang WF, Chung KY et al (2009) Recurrent and relapsing peritonitis: causative organisms and response to treatment. Am J Kidney Dis 54:702–710
Thomas VC, Hiromasa Y, Harms N, Thurlow L, Tomich J, Hancock LE (2009) A fratricidal mechanism is responsible for eDNA release and contributes to biofilm development of *Enterococcus faecalis*. Mol Microbiol 72:1022–1036

Tornieporth NG, Roberts RB, John J, Hafner A, Riley LW (1996) Risk factors associated with vancomycin-resistant *Enterococcus faecium* infection or colonization in 145 matched case patients and control patients. Clin Infect Dis 23:767–772

Vijayvargiya R, Veis JH (1996) Antibiotic-resistant endocarditis in a hemodialysis patient. J Am Soc Nephrol 7:536–542

Waters CM, Hirt H, McCormick JK, Schlievert PM, Wells CL, Dunny GM (2004) An amino-terminal domain of *Enterococcus faecalis* aggregation substance is required for aggregation, bacterial internalization by epithelial cells and binding to lipoteichoic acid. Mol Microbiol 52:1159–1171

Woodford N, Soltani M, Hardy KJ (2001) Frequency of esp in *Enterococcus faecium* isolates. Lancet 358:584. https://doi.org/10.1016/S0140-6736(01)05726-9

Xu X, Lin D, Yan G, Ye X, Wu S, Guo Y et al (2010) vanM, a new glycopeptide resistance gene cluster found in *Enterococcus faecium*. Antimicrob Agents Chemother 54:4643–4647

Zarb P, Coignard B, Griskeviciene J, Muller A, Vankerckhoven V, Weist K et al (2012) The European Centre for Disease Prevention and Control (ECDC) pilot point prevalence survey of healthcare-associated infections and antimicrobial use. Euro Surveill 17:20316

Chapter 16
Dietary Antioxidants and Infectious Diseases

Jasleen Kaur, Rajvir Kaur, and Amarjeet Kaur

Abstract The role of antioxidants in helping to prevent diseases such as cancer, cardiovascular diseases and various neurodegenerative disorders arising due to oxidative stress has long been recognised. But recent researches have also highlighted their importance in progression and control of infectious diseases. Antioxidants can be endogenously produced by human body or externally supplied through foods and/or supplements. Dietary antioxidants have been related to modulate the host susceptibility or resistance to infectious diseases. This chapter focuses on the beneficiary effects of dietary antioxidants in limiting and preventing the progression of infectious diseases. Antioxidants can help in fighting infectious diseases in several ways. In addition to ameliorating the deteriorating and degenerative effects of reactive species, they also help in building or maintaining healthy immune cells to fight the pathogens. Many pathogens and their metabolites produce free radicals which help in establishing their virulence and pathogenicity. Antioxidants aid in limiting the progression by neutralising the free radicals. Some dietary antioxidants like phenolics and flavonoids also possess antimicrobial activities. Antimicrobial activities can be attributed to their ability to inhibit various enzymes and physiological processes in bacteria and viruses. Phenolics can also form complexes with proteins such as microbial adhesins, enzymes and cell envelope transport proteins and inactivate them. Alterations in membrane fluidity and quorum sensing inhibition by these molecules can also contribute to their antimicrobial activities.

Keywords Antimicrobial agents · Antioxidants · Free radicals · Infectious diseases

J. Kaur · R. Kaur · A. Kaur (✉)
Department of Microbiology, Guru Nanak Dev University, Amritsar, India

P. P. Singh (ed.), *Infectious Diseases and Your Health*,
https://doi.org/10.1007/978-981-13-1577-0_16

16.1 Introduction

Antioxidants are molecules that inhibit the oxidation of other molecules and prevent cell damage against free radicals. They are critical for maintaining optimum health in both humans and animals. The molecular species which are capable of independent existence and contain one unpaired electron in their outermost shell are regarded as free radicals. A variety of endogenous free radicals like superoxide (O^{2-}), hydroxyl (OH^*), nitric oxide (NO), peroxynitrite ($ONOO^*$), etc. are produced during the normal metabolic processes of the body or in pathophysiological conditions. During infection, immune cells of the body overproduce reactive oxygen species which are necessary for microbicidal activity, but at the same time, this overgeneration of free radicals damages the immune cells themselves as well as the host tissues. Free radicals are also produced by numerous pathogenic agents which help in establishing their virulence and pathogenicity by damaging the cells of the immune system (Knight 2000). Free radicals are highly reactive species, capable of damaging biologically relevant molecules such as DNA, proteins, carbohydrates, lipids, etc. and triggering a number of diseases like cancer, neurodegenerative disorders, etc. if produced in excess. The role of antioxidants in helping to prevent diseases such as cancer, cardiovascular diseases, Alzheimer's disease, etc. has long been recognised (Hajhashemi et al. 2010). But recent researches have also highlighted their importance in progression and control of infectious diseases (Elswaifi et al. 2009). The alarming increase in the emergence of antibiotic-resistant microorganisms has triggered a global concern for the need to search for alternative and effective strategies to fight this threat. Antioxidants can be used alone or in combination with antibiotics as a viable therapeutic strategy. Antioxidants can help in fighting infectious diseases in various ways. They help in building or maintaining healthy immune cells to fight the pathogens. The cell damages caused by the infections are also countered by antioxidants. And where the pathogenesis is aided by generation of free radicals, antioxidants aid in limiting the progression of disease by neutralising the free radicals. In addition to ameliorating the deteriorating and degenerative effects of reactive species, some antioxidants also have antimicrobial activities. Antioxidants can be endogenously produced by human body or externally supplied through foods and/or supplements. Dietary antioxidants have been recognised to modulate the immune system affecting the host susceptibility or resistance to infectious diseases. Besides, many dietary compounds like phenolics and flavonoids have been found to exhibit antibacterial and antiviral activities. This chapter focuses on the beneficiary effects of dietary antioxidants in limiting and preventing the progression of infectious diseases.

16.2 Classification of Antioxidants

Antioxidants can be categorised into exogenous and endogenous types. To maintain a balance between the production and inactivation of ROS, living organisms have evolved specific mechanisms for detoxification, comprising of enzymatic and

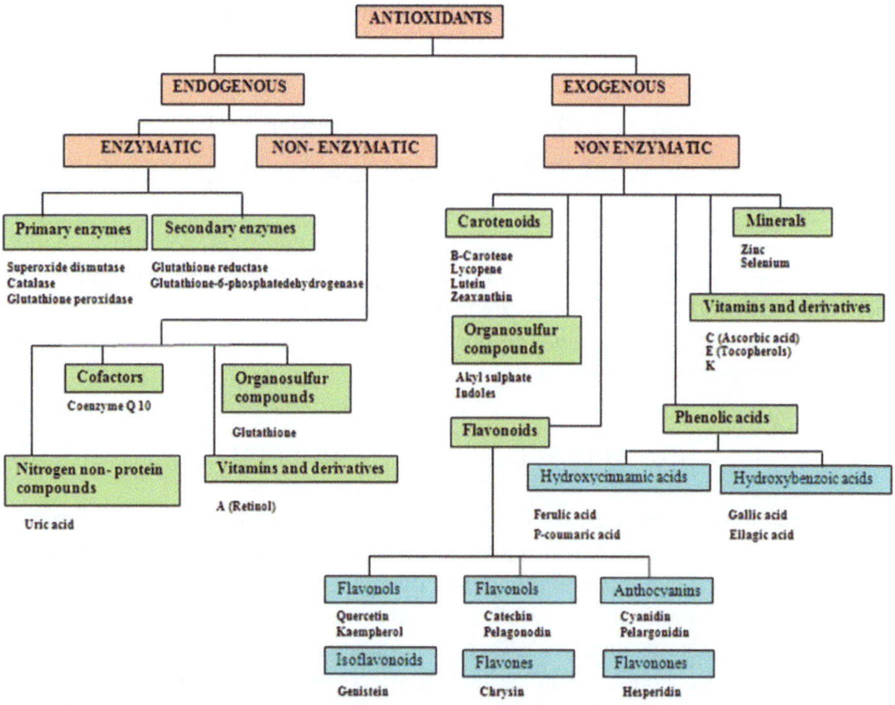

Fig. 16.1 Schematic representation of different types of antioxidants

non-enzymatic antioxidants. Endogenous antioxidants are produced by our body and include various enzymes like superoxide dismutase (SOD), glutathione, alpha lipoic acid (ALA), catalase and coenzyme Q10 (CoQ10). Exogenous antioxidants are obtained from our diet and supplements like vitamin C, vitamin E, flavonoids, etc. (Fig. 16.1). The role of these dietary antioxidants in infectious diseases is being discussed.

16.3 Role of Dietary Antioxidants in Infectious Diseases

16.3.1 Prevention Against Oxidative Stress

Oxidants are produced by various processes of immune system, e.g. neutrophil activation releases free radicals during respiratory burst for targeting foreign pathogens, but as it is unspecific, it also targets host tissue resulting in its damage. In phagocytosis**,** the cells phagocytose the bacteria and form a structure called phagosome. These phagosomes serve as centres for ROS production which is triggered by uptake of oxygen and activation of NADPH oxidase. This enzyme usually remains inactive in the cell. Activation of this enzyme oxidises NADPH to NADP+ and superoxide which destroys the pathogens. Superoxide dismutases convert superoxide radical into a nonradical form, H_2O_2 (Young and Woodside 2001).

$$2O_2 + \text{NADPH} \xrightarrow{\text{NADPH Oxidase}} 2O_2^{\bullet} + \text{NADP}^+ + H^+$$

$$2O_2^{\bullet} + 2H^{\bullet} \xrightarrow{\text{Superoxide Dismutase}} H_2O_2 + O_2$$

Further, more reactive species can be produced by the myeloperoxidase-halide-H_2O_2 system. The cytoplasmic granules of neutrophils possess an enzyme myeloperoxidase (MPO) which in the presence of chloride ions actively converts hydrogen peroxide into hypochlorous acid. Hypochlorous acid is a powerful oxidant and has antimicrobial potential (Aruoma 1999).

$$Cl^- + H_2O_2 + H^+ \xrightarrow{\text{Myeloperoxidase (MPO)}} \text{HOCl} + H_2O.$$

The production of these reactive oxygen species (ROS) is necessary for microbicidal activity. There are mechanisms present in the cell for balancing the oxidants and antioxidants in the cell, but sometime, this overgeneration of free radicals damages the immune cells themselves as well as the host tissues. High ROS generation can damage lipids, proteins and DNA. Oxidative damage is one of the most important pathological consequences of infections. It affects vital organs of the body manifesting in various degenerative damages. This suggests the potential for utilisation of ROS scavengers, or antioxidants, in control of certain aspects of infectious disease and reduction of host tissue damage.

16.3.2 Restricting the Pathogenesis of Infectious Diseases

Free radicals are also produced by numerous pathogenic agents which help in establishing their virulence and pathogenicity by damaging the cells of the immune system. *Pseudomonas aeruginosa*, a causative agent of pneumonia and various other life-threatening complications, produces the blue-green dye pyocyanin, a redox active toxin. Pyocyanin induces cytotoxicity in host cells by increasing the intracellular levels of ROS and induction of oxidative stress. The toxin inhibits phagocytosis in macrophages and causes premature senescence and apoptosis in neutrophils (Manago et al. 2015). The role of oxidative stress in pneumococcal meningitis and pathogenesis of *Helicobacter pylori* has also been documented. In *H. pylori* pathogenesis, bacterial colonisation is associated with apoptosis of mucosal cells and infiltration of immune cells, which causes tissue damage through production of ROS (Ding et al. 2007). *Streptococcus pneumoniae* (pneumococcus) is one of the leading causes of bacterial meningitis. The pneumococci are one of the few organisms that directly produce H_2O_2 (Regev-Yochay et al. 2006). Production of ROS is associated with cerebral oedema, neuronal and vascular damage and other associated complications.

The use of antioxidant therapy may also benefit in viral infections. Acquired immune deficiency syndrome (AIDS) caused by human immunodeficiency virus (HIV) has emerged as major global health threat. It is associated with secondary infections that are often mediated by production of ROS (Porter and Sutliff 2012). HIV-1 induces oxidative stress by deregulation of oxidative stress pathways resulting in enhanced ROS production and by inducing mitochondrial dysfunction. The increase in ROS production is also stimulated by various viral proteins like envelope protein Gp120 and reverse transcriptase (RT) (Isaguliants et al. 2013; Ivanov et al. 2016). Hepatitis C virus (HCV) infection is another example where free radicals are associated with disease pathogenesis. HCV infections can cause hepatitis, and cirrhosis and predispose to hepatocellular carcinoma. Besides the inflammatory response, during HCV infection, free radicals are also generated by viral proteins like HCV capsid, core protein and NS_3 by disrupting the electron transfer chain (Medvedev et al. 2016).

16.3.3 Enhancement in Immunity

Antioxidants like vitamins C, E, etc. can be beneficial in fighting infections by virtue of their ability to boost immunity. Increased proliferative response of T cells, augmentation of cytokine production and immunoglobulin synthesis have been associated with high vitamin C levels. Vitamin C is known to have an immune-enhancing role in improving activity of natural killer (NK) cells and proliferation of lymphocytes (Anderson 1984). It also maintains redox integrity of cell by protecting against reactive oxygen species produced during respiratory burst in response to infectious diseases (Beveridge et al. 2008)

Vitamin E plays an important role during infectious diseases. Immune responses mediated by vitamin E include increased lymphocyte proliferation, improved phagocytic function by alveolar macrophages and raised IL-2 production and NK cell activity. It also protects the lipid membranes from damage caused by ROS during infections, maintaining cell membrane integrity. It was reported that intake of vitamin E augmented immune response in mice immunised with tetanus toxoid and evoked the generation of cytokines that promote cell-mediated (TH1) immune response (Radhakrishnan et al. 2013). In chronic hepatitis B, high doses of vitamin E have a tendency to promote HBeAg (hepatitis B e-antigen) seroconversion (i.e. loss of HBeAg, undetectable levels of serum hepatitis B DNA), thereby generating a possibility to evaluate the use of vitamin E in further studies with this virus (Fiorino et al. 2017).

Vitamins have been found to have profound results while treating deadly disease like AIDS. Oxidative stress also contributes to deteriorating immune system and health in HIV patients. (Ivanov et al. 2016). Regular intake of multivitamins consisting of vitamins A and E, various B vitamins and folic acid by pregnant women with HIV infection boosted the immunity by improving CD4, CD3 and CD8 cell count, lowering the risk of premature birth, foetal death and low birth weight (Fawzi et al. 2003).

Vitamin A refers to any of the members of two different families of compounds: retinoids and carotenoids. Retinoids (preformed vitamin A) obtained from animal sources are the synthetic analogues of retinol and include all trans-retinolic acids and 13-cis-retinoic acids, while carotenoids (pro-vitamin A) obtained from plant source are converted to retinol in the body prior to utilisation. The essential role of vitamin A in modulating a broad range of immune processes, such as lymphocyte activation and proliferation, T helper cell differentiation, tissue-specific lymphocyte homing, the production of specific antibody isotypes and regulation of the immune response, is well reviewed (Mora et al. 2008). When the status of vitamin A retinoids lowers during epithelial damage then tissue repair, cell differentiation gets hampered, and chances of secondary infections are increased. Aibana et al. (2017) proposed that vitamin A could be an effective alternative among people at high risk for tuberculosis as it prevents progression from TB infection to TB disease.

16.3.4 Antimicrobial Agents

Several antioxidants in addition to reducing the oxidative stress also possess microbiocidal activity. These compounds can be used in conjunction with antibiotics or as alternative therapeutic strategies. Vitamin C is a strong reducing agent and has been studied for its antimicrobial activity. The significance of Vitamin C (ascorbic acid) in treatment of various bacterial infections is well recognised (Hemilä 2017). Antiviral activity is also exhibited by vitamin C. Ascorbic acid is reported to be an inhibitor of replication and infectivity of avian RNA tumour virus (Bissell et al. 1980). Uozaki et al. (2010) demonstrated that dehydroascorbic acid inhibits the multiplication of virus of three families, i.e. herpes simplex virus type 1 (HSV-1), influenza virus type A and poliovirus type 1. A clinical study revealed that the intravenous injection of high dose of vitamin C helped in viral antigen control by interfering with viral replication during Epstein-Barr (human herpesvirus 4/HHV-4) viral infection (Mikirova and Hunninghake 2014). In addition, intravenous administration also helps to reduce postherpetic neuralgia (PHN), thereby improving quality of life in infected patients (Kim et al. 2016). Many viral diseases like hepatitis affect the liver metabolism resulting in various biochemical disorders. Supplementation of diet with orange juice which is rich in vitamin C and antioxidants has been shown to have beneficial effects in patients under antiviral therapy (Gonçalves et al. 2017).

Phenolic compounds are the secondary metabolites that consist of an aromatic ring linked with one or more hydroxyl groups. Phenolics constitute major part of dietary components and are known for their free radical scavenging and antimicrobial activities. Antimicrobial potential of various dietary components like grapes, berries and walnuts which are rich in phenolics and flavonoids is well documented. Honey is also well known for its antimicrobial potential. These properties can be attributed to phenolics present in it. Estevinho et al. (2008) identified inhibitory potential of honey against various bacteria, viz. *Staphylococcus aureus*, *Bacillus subtilis*, *Klebsiella pneumonia* and *Escherichia coli*, which could be ascribed to flavonoids and phenolic acids, namely, p-hydroxybenzoic acid, naringenin, pinocembrin and chrysin. Mushrooms are good sources of phenolic compounds, and two phenolic compounds, namely, 2, 4-dihydroxybenzoic and protocatechuic acids, were found to possess good antibacterial activity against the majority of Gram-negative and Gram-positive bacteria. Methicillin-resistant *S. aureus* (MRSA) was found to be more susceptible to phenolic compounds of mushroom than methicillin-sensible *S. aureus* (MSSA). Docking studies revealed the importance of carboxylic acid (COOH), two hydroxyl (OH) groups in *para* and *ortho* positions of the benzene ring and a methoxyl (OCH3) group in the *meta* position for anti-MRSA activity (Alves et al. 2013).

The mode and site of action of flavonoids are unspecific. Intercalation and hydrogen bonding of B ring of flavonoids with nucleic acid bases can lead to inhibition of DNA and RNA synthesis in bacteria imparting them antibacterial activity. Antimicrobial activity against *Proteus vulgaris* and *S. aureus* is exhibited by some flavonoids by halting the synthesis of DNA and RNA, respectively (Mori et al. 1987). The inhibitory activity of phenolic compounds like quercetin, rutin, etc. on bacterial enzymes like topoisomerase IV and DNA gyrase has also been reported (Bernard et al. 1997; Ohemeng et al. 1993). Phenolics can also form complexes with proteins such as microbial adhesins, enzymes and cell envelope transport proteins and inactivate them. Their complexing properties have been proposed to inhibit bacterial enzyme glucosyltransferases isolated from *Streptococcus mutans* and to detoxify the cholera toxin isolated from *Vibrio cholera* (Borris 1996; Nakahara et al. 1993). Some lipophilic flavonoids can also break microbial membranes (Borris 1996; Cowan 1999). Flavonoids (naringenin and sophoraflavanone G) inhibited methicillin-resistant *S. aureus* (MRSA) and streptococci by altering the membrane fluidity. Flavonoids might reduce the fluidity of outer and inner layers of membranes by alteration in hydrophilic and hydrophobic regions (Tsuchiya and Iinuma 2000). Baicalein demonstrates antimicrobial activity against *S. aureus* by virtue of its ability to inhibit the quorum sensing system. It inhibited biofilm formation, destroyed biofilms and increased the permeability of vancomycin (Chen et al. 2016). Myricetin can inhibit *E. coli* DnaB helicase which is an essential enzyme in the replication and elongation of DNA (Griep et al. 2007).

Phenolic compounds also possess antiviral properties due to their ability to inhibit various enzymes involved in viral cycle. The enzyme inhibitory activity has been found to be dependent on structure of flavonoids. Studies have revealed the activities of flavonoids (robinetin and demethylated gardenin A) and catechins against enzymes of HIV-1; the former possessed ability to inhibit proteinase, and latter inhibited DNA polymerase (Cushnie and Lamb 2005). Flavonoids, viz. baicalein, quercetin, quercetagetin and myricetin, were found to be partial competitive and non-competitive inhibitors of reverse transcriptase of human immunodeficiency virus (HIV) with respect to template primer complex and triphosphate substrate, respectively (Ono et al. 1990). Flavonoids have also been documented for inhibition of HIV-1 penetration in cells having CD4 receptors and chemokine co-receptors. Williamson et al. (2006) documented the crucial role of epigallocatechin gallate (EGCG) in HIV transmission. It was found that EGCG blocked the attachment of CD4 receptor to glycoprotein 120 of HIV, thus lowering the transmission risk. Flavonoids such as chrysin, apigenin and acacetin are reported to block HIV-1 activation by a mechanism which might inhibit viral transcription (Critchfield et al. 1996).

Epigallocatechin gallate (EGCG) and chlorogenic acid isolated from green tea and coffee, respectively, were identified to inhibit the replication of HBV (Wang et al. 2009; He et al. 2011). Antiviral activities of various phenolic compounds have been observed against many other viruses such as herpes simplex virus (HSV), respiratory syncytial virus, polio virus and coxsackie B virus (Balde et al. 1990; Selway 1986; Suárez et al. 2010). Antioxidants such as quercetin, hesperetin and daidzein can affect different stages of DENV-2 (dengue virus type 2) infection and replication cycle (Zandi et al. 2011) leading to antiviral activity against it.

16.4 Conclusion

Dietary antioxidants not only serve as free radical scavengers but also possess antimicrobial properties. They can be used in conjunction with antibiotics or as alternative therapeutic strategy in infectious diseases. Antioxidants can provide promising leads to counter the challenge of growing drug resistance in microorganisms. However, the reactive oxygen species (ROS) produced in response to antigens are beneficial to counter the impact of infectious pathogens. ROS are also involved in various signalling pathways. Thus it is necessary that a cautionary approach should be followed while taking antioxidants as supplements. The inappropriate use of dietary supplements available commercially may lead to "antioxidative stress". It is crucial that a proportionate balance between free radicals generation and antioxidant intake is made as both extremes, oxidative and antioxidative stress, are damaging.

References

Aibana O, Franke MF, Huang CC, Galea JT, Calderon R, Zhang Z, Becerra MC, Smith ER, Ronnenberg AG, Contreras C, Yataco R (2017) Impact of vitamin A and carotenoids on the risk of tuberculosis progression. Clin Infect Dis 15(65):900–909

Alves MJ, Ferreira IC, Froufe HJ, Abreu RM, Martins A, Pintado M (2013) Antimicrobial activity of phenolic compounds identified in wild mushrooms, SAR analysis and docking studies. J Appl Microbiol 115:346–357

Anderson R (1984) The immunostimulatory, anti-inflammatory and anti-allergic properties of ascorbate. Adv Nutr Res 6:619–645

Aruoma OI (1999) Free radicals, antioxidants and international nutrition. Asia Pac J Clin Nutr 8:53–63

Balde AM, Van Hoof L, Pieters LA, Vanden Berghe DA, Vlietinck AJ (1990) Plant antiviral agents. VII. Antiviral and antibacterial proanthocyanidins from the bark of *Pavetta owariensis*. Phytother Res 4:182–188

Bernard FX, Sable S, Cameron B, Provost J, Desnottes JF, Crouzet J, Blanche F (1997) Glycosylated flavones as selective inhibitors of topoisomerase IV. Antimicrob Agents Chemother 41:99–98

Beveridge S, Wintergerst ES, Maggini S, Hornig D (2008) Immune-enhancing role of vitamin C and zinc and effect on clinical conditions. Proc Nutr Soc 1:67 (OCE)

Bissell MJ, Hatie C, Farson DA, Schwarz RI, Soo WJ (1980) Ascorbic acid inhibits replication and infectivity of avian RNA tumor virus. Proc Natl Acad Sci 77:2711–2715

Borris RP (1996) Natural products research: perspectives from a major pharmaceutical company. J Ethnopharmacol 41:587–590

Chen Y, Liu T, Wang K, Hou C, Cai S, Huang Y et al (2016) Baicalein inhibits *Staphylococcus aureus* biofilm formation and the quorum sensing system in vitro. PLoS One 11:e0153468

Cowan MM (1999) Plant products as antimicrobial agents. Clin Microbiol Rev 12:564–582

Critchfield JW, Butera ST, Folks TM (1996) Inhibition of HIV activation in latently infected cells by flavonoid compounds. AIDS Res Hum Retrovir 12:39–46

Cushnie TP, Lamb AJ (2005) Antimicrobial activity of flavonoids. Int J Antimicrob Agents 26:343–356

Ding SZ, Minohara Y, Fan XJ, Wang J, Reyes VE, Patel J et al (2007) *Helicobacter pylori* infection induces oxidative stress and programmed cell death in human gastric epithelial cells. Infect Immun 75:4030–4039

Elswaifi SF, Palmieri JR, Hockey KS, Rziqalinski BA (2009) Antioxidant nanoparticles for control of infectious disease. Infect Disord Drug Targets 9:445–452

Estevinho L, Pereira AP, Moreira L, Dias LG, Pereira E (2008) Antioxidant and antimicrobial effects of phenolic compounds extracts of Northeast Portugal honey. Food Chem Toxicol 31:3774–3379

Fawzi WW, Msamanga GI, Wei R, Spiegelman D, Antelman G, Villamor E, Manji K, Hunter D (2003) Effect of providing vitamin supplements to human immunodeficiency virus-infected, lactating mothers on the child's morbidity and $CD4^+$ cell counts. Clin Infect Dis 36:1053–1062

Fiorino S, Bacchi-Reggiani ML, Leandri P, Loggi E, Andreone P (2017) Vitamin E for the treatment of children with hepatitis B e antigen-positive chronic hepatitis: a systematic review and meta-analysis. World J Hepatol 9:333

Gonçalves D, Lima C, Ferreira P, Costa P, Costa A, Figueiredo W, Cesar T (2017) Orange juice as dietary source of antioxidants for patients with hepatitis C under antiviral therapy. Food Nutr Res 61(1):1296675

Griep MA, Blood S, Larson MA, Koepsell SA, Hinrichs SH (2007) Myricetin inhibits *Escherichia coli* DnaB helicase but not primase. Bioorg Med Chem 15:7203–7208

Hajhashemi V, Vaseghi G, Pourfarzam M, Abdollahi A (2010) Are antioxidants helpful for disease prevention? Res Pharm Sci 5:1–8

He W, Li LX, Liao QJ, Liu CL, Chen XL (2011) Epigallocatechin gallate inhibits HBV DNA synthesis in a viral replication – inducible cell line. World J Gastroenterol 17:1507–1514

Hemilä H (2017) Vitamin C and infections. Nutrients 9:339

Isaguliants M, Smirnova O, Ivanov AV, Kilpelainen A, Kuzmenko Y, Petkov S, Latanova A, Krotova O, Engstrom G, Karpov V, Kochetkov S, Wahren B, Starodubova E (2013) Oxidative stress induced by HIV-1 reverse transcriptase modulates the enzyme's performance in gene immunization. Hum Vaccines Immunotherapeutics 9:2111–2119

Ivanov AV, Valuey-Ellistov VT, Ivanova ON, Kochetkov SN, Starodubova ES, Bartosch B, Isaguliants MG (2016) Oxidative stress during HIV infection: mechanisms and consequences. Oxidative Med Cell Longev. https://doi.org/10.1155/2016/8910396

Kim MS, Kim DJ, Na CH, Shin BS (2016) A study of intravenous administration of vitamin C in the treatment of acute herpetic pain and postherpetic neuralgia. Ann Dermatol 28:677–683

Knight JA (2000) Review: free radicals, antioxidants and the immune system. Ann Clin Lab Sci 30:145–158

Manago A, Becker KA, Carpinteiro A, Wilker B, Soddemann M, Seitz AP, Edwards MJ, Grassme H, Szabo I, Gulbins E (2015) *Pseudomonas aeruginosa* pyocyanin induces neutrophil death via mitochondrial reactive oxygen species and mitochondrial acid sphingomyelinase. Antioxid Redox Signal 22(13):1097–1110. https://doi.org/10.1089/ars.2014.5979

Medvedev R, Ploen D, Hildt E (2016) HCV and oxidative stress: implications for HCV life cycle and HCV- associated pathogenesis. Oxidative Med Cell Longev. https://doi.org/10.1155/2016/9012580

Mikirova NA, Hunninghake R (2014) Effect of high dose vitamin C on Epstein-Barr viral infection. Med Sci Monit Int Med J Exp Clin Res 20:725–732

Mora JR, Iwata M, von Andrian UH (2008) Vitamin effects on the immune system: vitamins A and D take centre stage. Nat Rev Immunol 8:685–698

Mori A, Nishino C, Enoki N, Tawata S (1987) Antibacterial antibacterial activity and mode of action of plant flavonoids against *Proteus vulgaris* and *Staphylococcus aureus*. Phytochemistry 26:2231–2234

Nakahara KS, Kawabata S, Ono H, Oqura K, Tanaka T, Ooshima T, Hamada S (1993) Inhibitory effect of oolong tea polyphenols on glucosyltransferases of mutans *Streptococci*. Appl Environ Microbiol 59:968–973

Ohemeng KA, Schwender CF, Fu KP, Barrett JF (1993) DNA gyrase inhibitory and antibacterial activity of some flavones. Bioorg Med Chem Lett 3:225–230

Ono K, Nakane H, Fukushima M, Chermann JC, Barré-Sinoussi F (1990) Differential inhibitory effects of various flavonoids on the activities of reverse transcriptase and cellular DNA and RNA polymerases. Eur J Biochem 190:469–476

Porter KM, Sutliff RL (2012) HIV-1, reactive oxygen species and vascular complications. Free Radic Biol Med 53:142–159

Radhakrishnan AK, Mahalingam D, Selvaduray KR, Nesaretnam K (2013) Supplementation with natural forms of vitamin E augments antigen-specific Th1-type immune response to tetanus toxoid. Biomed Res Int 2013:782067,8

Regev-Yochay G, Trzcinski K, Thompson CM, Malley R, Lipsitch M (2006) Interference between *Streptococcus pneumoniae* and *Staphylococcus aureus*: in vitro hydrogen peroxide-mediated killing by *Streptococcus pneumoniae*. J Bacteriol 188:4996–5001

Selway JWT (1986) Antiviral activity of flavones and flavans. In: Cody V, Middleton E, Harborne JB (eds) Plant flavonoids in biology and medicine: biochemical, pharmacological, and structure–activity relationships. Alan R. Liss, New York 1986

Suárez B, Álvarez AL, García YD, Barrio YD, Lobo AP, Parra F (2010) Phenolic profiles, antioxidant activity and in vitro antiviral properties of apple pomace. Food Chem 120:339–342

Tsuchiya H, Iinuma M (2000) Reduction of membrane fluidity by antibacterial sophoraflavanone G isolated from *Sophora exigua*. Phytomedicine 7:161–165

Uozaki M, Ikeda K, Tsujimoto K, Nishide M, Yamasaki H, Khamsri B, Hajime Koyama A (2010) Antiviral effects of dehydroascorbic acid. Exp Ther Med 1:983–986

Wang GF, Shi LP, Ren YD, Liu QF, Liu HF, Zhang RJ, Li Z, Zhu FH, He PL, Tang W, Tao PZ, Li C, Zhao WM, Zuo JP (2009) Anti-hepatitis B virus activity of chlorogenic acid, quinic acid and caffeic acid in vivo and in vitro. Antivir Res 83:186–190

Williamson MP, McCormick TG, Nance CL, Shearer WT (2006) Epigallocatechin gallate, the main polyphenol in green tea, binds to the T-cell receptor, CD4: potential for HIV-1 therapy. J Allergy Clin Immunol 118:1369–1374

Young IS, Woodside JV (2001) Antioxidants in health and disease. J Clin Pathol 54:176–186

Zandi K, Teoh BT, Sam SS, Wong PF, Mustafa MR, Abubakar S (2011) Antiviral activity of four types of bioflavonoid against dengue virus type-2. Virol J 8:560

Chapter 17
Probiotic Lactobacilli, Infection, and Immunomodulation

Sumanpreet Kaur, Preeti Sharma, and Sukhraj Kaur

Abstract Immunomodulatory agents are potentially believed to play important roles in infectious, allergic, and autoimmune diseases. The probiotic bacteria such as lactobacilli have immunomodulatory properties and are generally regarded as safe. Thus, they can be used as adjuvants for the treatment of infectious and allergic/autoimmune diseases. The immunomodulatory properties of five lactobacilli species such as *Lactobacillus casei*, *L. rhamnosus*, *L. paracasei*, *L. gasseri*, and *L. acidophilus* have been well studied. This book chapter summarizes the various studies which have reported immunomodulation by *Lactobacillus* species. Further, the immunomodulatory molecules produced by lactobacilli have been discussed. The immunomodulatory effects of *Lactobacillus* species are strictly strain-specific and in some cases yielded contrasting results in different hosts. Thus, use of lactobacilli as immunomodulatory agent for therapeutic use should be strictly backed by human clinical trials. Further, some of the immunomodulatory molecules are known to play role(s) in immunopathogenesis of allergic diseases. Thus probiotic lactobacilli species to be used as therapeutic agents should be screened for their ability to secrete harmful metabolites.

Keywords Allergic disease · Cytokine · GABA · Immunomodulatory · Lactobacilli · SCFA

17.1 Introduction

Lactic acid bacteria (LAB) are Gram-positive, catalase-negative, acid-tolerant bacteria belonging to phylum *Firmicutes* (Fig. 17.1). They produce lactic acid as major metabolic end product of carbohydrate fermentation. The genus *Lactobacillus* comprises the largest group of rod-shaped, non-sporulating and facultative anaerobes that contains 154 species. Lactobacilli are ubiquitous in nature and found in milk, fermented food products, and beverages. They comprise almost 0.01% of the gut

S. Kaur · P. Sharma · S. Kaur (✉)
Department of Microbiology, Guru Nanak Dev University, Amritsar, India

P. P. Singh (ed.), *Infectious Diseases and Your Health*,
https://doi.org/10.1007/978-981-13-1577-0_17

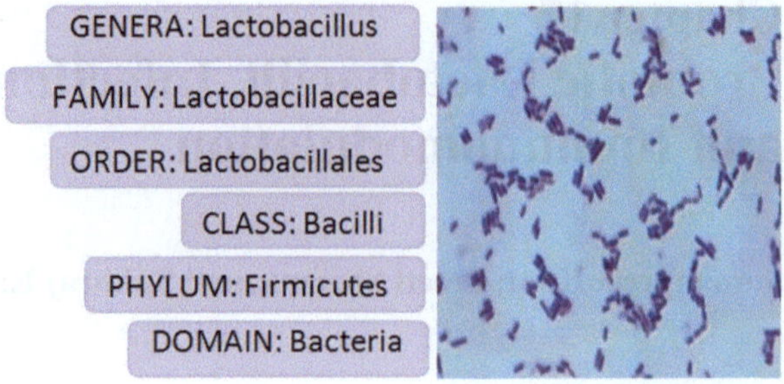

Fig. 17.1 The classification of *Lactobacillus* spp. and the Gram-stained microscopic view (1000×) of lactobacilli cells

microbiome of human and animal tracts (Finegold et al. 1983). The proportion of lactobacilli in the vagina and oral cavity is quite high probably because of their ability to persist under harsh physiological conditions such as at low pH and their ability to bind to the vaginal epithelial cells.

Lactobacilli have a long history of safe consumption by humans through fermented foods, and therefore they have been accorded GRAS status (generally regarded as safe) by the World Health Organization (FAO/WHO 2001). They are the most common bacterial genera that are used as probiotics. Probiotics are defined as live microbial food supplements of human origin beneficially influencing human health by improving the intestinal microbial balance (FAO/WHO 2001). As the probiotic potential of bacteria is strain-specific, every lactobacilli strain should be screened for some characteristics such as the ability to survive and adhere to the intestinal tract, form strong biofilms, and auto-aggregate and co-aggregate with pathogens (Havenaar et al. 1992). *Lactobacillus* strains are known to benefit the host due to their various functional properties such as acting as microbial barriers against gastrointestinal pathogens, causing competitive exclusion of pathogen binding, and producing inhibitory compounds, such as organic acids, e.g., lactic acid and acetic acid, hydrogen peroxide, and cationic peptides, e.g., bacteriocins (Bermudez-Brito et al. 2012). They are also known to strengthen both the innate and adaptive immune system against pathogens as they have immunomodulatory properties. In this chapter five *Lactobacillus* spp. whose immunomodulatory properties have been widely reported worldwide will be reviewed, and further the various compounds of the lactobacilli that have immunomodulatory potential shall be discussed.

17.2 Gut-Associated Lymphoid Tissue (GALT)

The lactobacilli form less than 1% of gut microflora. The various persistent species present in human feces are mostly *L. gasseri*, *L. crispatus*, *L. reuteri*, *L. salivarius*, and *L. ruminis*, and the sporadic species belong to *L. acidophilus*, *L. plantarum*, and *L. casei* group (*L. casei*, *L. paracasei*, and *L. rhamnosus*; Walter 2008). In the gut these lactobacilli species interact with the immune cells and are believed to cause immunomodulation. The gut mucosa consists of various immune cells that protect

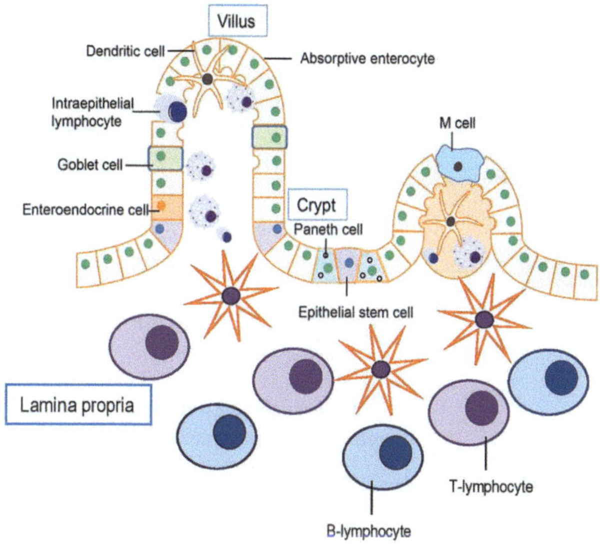

Fig. 17.2 The structure of small intestine mucosa along with associated immune cells

the host from the pathogens (Fig. 17.2). The epithelial cells of the gut contribute to innate immunity as they form tight junctions; secondly the mucus produced by goblet cells shields the epithelium from the action of gut enzymes, microorganisms, and other luminal contents. The intestinal epithelium contains a large population of lymphocytes known as intraepithelial lymphocytes (IELs), which recognize and eliminate infected epithelial cells and microorganisms. IELs are unique because they have high proportions of γδ T cells that do not require major histocompatibility complex (MHC) presentation of antigens for activation. Other important cells are Paneth cells located at the base of the crypts within the small intestine that secrete a vast array of antimicrobial proteins (Bevins and Salzman 2012) following microbial recognition via pattern recognition receptors, including Toll-like receptors (TLRs) and nucleotide-binding oligomerization domain (NOD)-like receptors (Kobayashi et al. 2005). In the lamina propria (LP), dendritic cells (DCs) extend processes between the epithelial cells and sample the contents of the gut lumen, following which they activate lymphocytes (Niess and Reinecker 2006). The LP contains large numbers of dendritic cells and effector lymphocytes such as immunoglobulin A (IgA)-producing plasma cells and cluster of differentiation (CD) 4+ T cells.

The commensals present in the gut are important for the health of gut immune system because they are both immunostimulatory and immunomodulatory in nature. Immunostimulatory functions involve development of immune system, maintenance of healthy epithelial cell lining, recruitment of immune cells to the epithelium, and stimulation of the production of antimicrobial peptides and other innate immune effector molecules by immune cells. Immunomodulatory functions involve influencing the type of immune response through modulating the differentiation of immune effector cells (Fig. 17.3; Ivanov and Honda 2012).

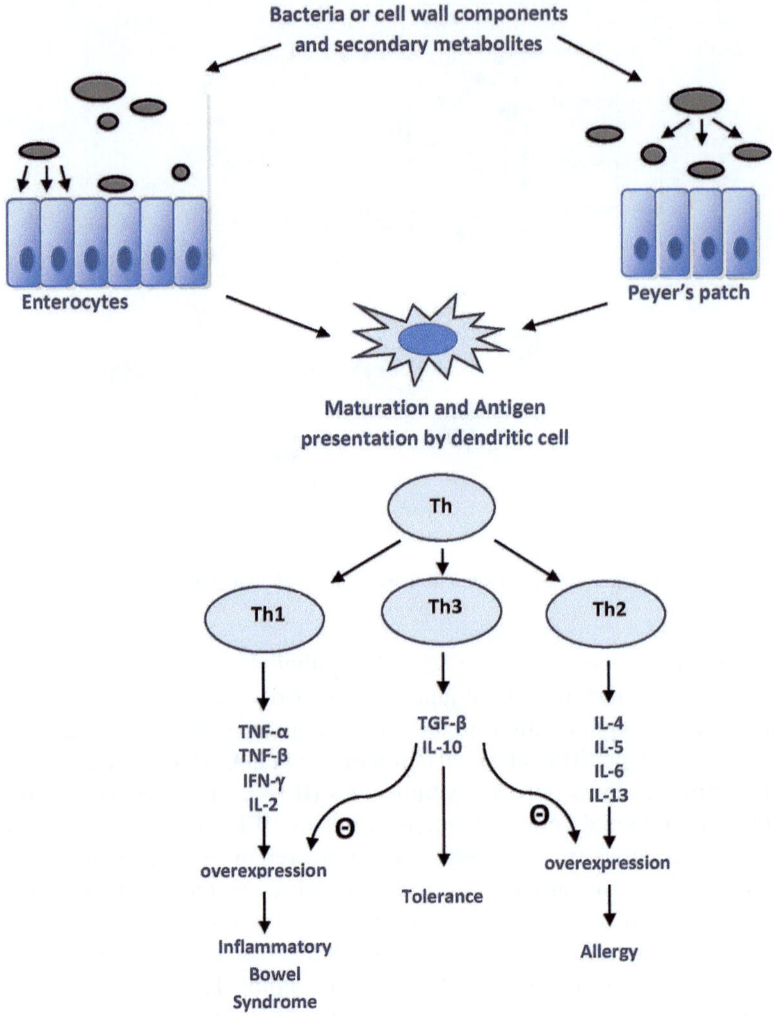

Fig. 17.3 Mechanism of bacterial immunomodulation of gut immune system

17.3 Immunomodulatory Role of Lactobacilli

Various reports have demonstrated that lactobacilli can modulate host's immune system because they are potential adjuvants triggering mucosal and systemic immune responses. Lactobacilli have been shown to interact with various cells like natural killer (NK) cells, enterocytes, dendritic cells, macrophages, and T helper (Th) 1 and Th2 and T regulatory (Treg) cells and may modulate the immune response either toward pro- or anti-inflammatory type. The immunomodulatory effects of lactobacilli include their potential to activate, cause maturation, and induce cytokine production through interaction with immune cells. The immunomodulatory properties of some lactobacilli such as *L. casei*, *L. rhamnosus*, *L. paracasei*, *L. gasseri*, and *L. acidophilus* have been very well studied. These strains have the potential to be used as probiotics for the treatment of allergic and inflammatory diseases as shown by various animal studies.

17.3.1 Lactobacillus casei

L. casei is a facultative heterofermentative, acid-resistant, lactic acid-producing bacterial strain found in the mouth and intestinal tract of humans. *L. casei* can be isolated from cheddar cheese during maturation process. *L. casei* Shirota (LcS) is commercially available in probiotic-fermented milk drink, Yakult. Both probiotic and immunomodulatory roles of L. casei have been well documented.

Some reports suggested that LcS enhanced the NK cell activity in vitro (Dong et al. 2010) as well as stimulated the cytokines such as interleukin (IL)-10, IL-12, and tumor necrosis factor (TNF)-α and interferon (IFN-γ) production (Shida et al. 2006). Another report showed that *L. casei* DN-114001 elevated the number of CD4 Fox P3 Tregs in the mesenteric lymph nodes and reduced the production of the pro-inflammatory cytokines TNF-α and IFN-γ (Zakostelska et al. 2011). Contrary findings were observed by Shida et al. (2002) in the food allergy mice model. Intraperitoneal injection of heat-killed LcS induced a rise in serum IL-12 levels, and the reduction in the levels of IgE and IgG by splenocytes was observed.

Kato et al. (1984) demonstrated that intraperitoneal application of *L. casei* in murine model activated macrophages by increasing their phagocytic and enzyme activities and also activated NK cells, which play an important role in tumor killing. In a human clinical trial, LcS-containing Yakult was orally administered to middle-aged and elderly individuals for 3 weeks, and the NK cell activity was determined. Although the numbers of NK cells and CD4+ and CD8+ T cells remained same, the NK cell activity was enhanced at 1 week, 3 weeks, and 6 weeks after the start of intake. However in elderly individuals, no effect on NK cell activity was observed (Takeda and Okumura 2007). In another similar report by Hashimoto et al. (1985), it was demonstrated in an in vitro assay that *L. casei* administration activated the Kupffer cells and immune cells associated with spleen, lung, and peritoneal macrophages. Perdigon et al. (1986) demonstrated that oral administration of *L. casei* led to lymphocytes and macrophage stimulation. Subsequent studies by Perdigon et al. (1990, 1991) showed that treatment with *L. casei*-activated cells in GALT even at low doses led to significantly higher production of secretory IgG in intestinal fluid, providing protection against infections such as *Salmonella*.

Herías et al. (2005) showed that LcS had positive effect in case of the prevention of experimental ulcerative colitis in a mice model. Treatment with LcS showed an increase in colonic epithelial regeneration in murine model of ulcerative colitis in the chronic stage. In vivo effects of oral administration of LcS were tested in skin allergy mice model and experimental autoimmune encephalomyelitis (EAE) rat model. LcS abrogated Th1 response and thus proved effective in mice allergy model, but it worsened the symptoms in EAE model (Baken et al. 2006). An ex vivo treatment of the gut mucosal cells isolated from the ileum of Crohn's disease patients with both *L. casei* and *L. bulgaricus* resulted in significant reduction in the numbers of CD4+ cells as well as TNF-α expression among IEL (Borruel et al. 2002).

17.3.2 Lactobacillus rhamnosus

L. rhamnosus is a heterofermentative, facultative anaerobe found in abundance in the genitourinary tract of healthy females and fecal samples. *L. rhamnosus* GG (LGG) is the most extensively studied lactobacilli strain. LGG was originally isolated from human feces of healthy adults by Sherwood Gorbach and Barry Goldwin (Doron et al. 2005).

Pena and Versalovic (2003) demonstrated that LGG specifically inhibited TNF-α production in murine macrophages. Balejko et al. (2015) demonstrated that LGG cells immobilized in capsules as well as LGG metabolites activated the release of cytokine IL-10 and tumor growth factor (TGF)-β1 in human peripheral blood mononuclear cells (PBMCs) and downregulated the production of IFN-γ. Kopp et al. (2008) demonstrated in an in vitro study that there was a significant increase in IL-10, IFN-γ levels in the supernatant of LGG-treated PBMCs. However, another study showed that LGG was least effective among other probiotic bacteria such as *Leuconostoc mesenteroides* and *Streptococcus* spp. in the induction of various cytokines such as TNF-α, IFN-γ, IL-12, and IL-10 in PBMC culture supernatants (Kekkonen et al. 2008).

Fong et al. (2015) studied the immunomodulatory effects of LGG on human dendritic cells (DCs), macrophages, and monocytes. Results showed that LGG downregulated the TLR2 mRNA levels of DCs and monocytes and both TLR2 and TLR8 mRNA levels of macrophages. The levels of IL-12, TNF-α, and IL-10 increased in both treated macrophages and monocytes, whereas the levels of Th2 cytokines, IL-4 and IL-25, decreased. Another report by Fong et al. (2016) showed immunomodulatory effect of LGG cells and LGG-derived soluble factors on human PBMCs + DC co-cultures. Both the treatments increased the TLRs on the surface of PBMCs and the pro-inflammatory Th1 and Th17 immune responses. In another study, immunomodulation of human DCs by *L. rhamnosus* Lcr35 was demonstrated (Evrard et al. 2011). Lcr35 treatment led to dose-dependent maturation of DCs that was associated with upregulation of membrane expression of CD86, CD83, human leukocyte antigen-antigen D related (HLA-DR), and TLR4. Further, Lcr35 induced strong dose-dependent increase of Th1/Th17 cytokine levels (TNF-α, IL-1β, IL-12, IL-23) but a low increase in IL-10 levels from DCs. In another study by Miettinen et al. (2000), LGG was shown to directly and very rapidly activate nuclear factor kappa B (NF-κB) in human macrophages that was not inhibited in the presence of protein synthesis inhibitor, cycloheximide.

To study the immunomodulatory effect of *Lactobacillus* spp. in mice, Kirjavainen et al. (1999) orally administered 10^9 colony-forming units (CFUs) of four different

Lactobacillus spp. (*L. acidophilus*, *L. casei*, *L. rhamnosus*, and *L. gasseri*) to mice for 7 days, *L. rhamnosus* along with *L. gasseri* and *L. casei* was shown to inhibit lipopolysaccharide (LPS)-mediated murine lymphocyte proliferation ex vivo. Wu et al. (2016) studied the effect of LGG in allergic murine model. The results showed that the levels of Th2 cytokines and IgE were significantly decreased in both serum and bronchoalveolar lavage fluid after treatment with LGG as compared to untreated OVA-sensitized mice.

The human clinical trials of LGG were also conducted. Kalliomaki et al. (2001, 2003, 2007) in a series of randomized placebo-controlled clinical trials demonstrated that the oral administration of LGG-containing capsules (10^{10} CFU daily) to mothers prenatally (2–4 weeks) and to postnatal pediatric subjects (6 months) in families with a history of atopic disease significantly lowered the risk of eczema at the age of 2, 4, and 7 years. However, allergic rhinitis and asthma tended to be more common in the LGG-treated group, and no significant differences were found in incidence of cow milk allergy. On the other hand, similar clinical trials with LGG repeated in Germany in families with atopic dermatitis did not yield any beneficial results (Fölster-Holst et al. 2006). The differences may be attributed to differences in the LGG strains and different population.

17.3.3 Lactobacillus paracasei

L. paracasei is a member of the normal human and animal gut microbiota. It is extensively used in the food industry as starter cultures for dairy products and also as probiotic (Marchand and Vandenplas 2000).

D'Arienzo et al. (2011) studied the immunomodulatory roles of five strains of *L. paracasei*. All the isolates had the ability to induce phenotypic maturation of DCs, increasing surface expression of CD11c and CD80 at levels comparable with LPS stimulation; however surface expression of CD86 was higher than that of LPS stimulation. The cytokine profile analysis showed variation among isolates. Isolate LMGP-17806 increased IL-12 production when co-administered with LPS, whereas in case of isolate ATCC334, IL-12 production was decreased. Similarly, IL-10 and IL-2 levels were increased when lactobacilli were co-administered with LPS. No significant differences were observed in case of TNF-α levels when lactobacilli were administered alone or along with LPS.

L. paracasei NCC2461 altered the cytokine profiles of murine CD4+ T lymphocytes. Dose-dependent inhibition of proliferative capacity of CD4+ T cells was seen after treatment with *L. paracasei*. Further, the in vitro stimulation of lymphocytes cultures with *L. paracasei* cells resulted in a dose-dependent increase in IL-10 and TGF-β levels, whereas the Th1 and Th2 effector cytokine production such as IL-4, IL-5, and IFN-γ decreased greatly (von de Weid et al. 2001).

In mice, oral administration of *L. paracasei* KW3110 increased IL-12 secretion and reduced IL-4 secretion from splenocytes, but no effect was seen in levels of IFN-γ (Fujiwara et al. 2004). A study conducted by Zhu et al. (2016) demonstrated the positive effect of oral administration of *L. paracasei* L9 on mouse systemic immunity by enhancing phagocytic activity of peritoneal macrophages and proliferation ratio of splenocytes, IgG levels in serum, and IgA levels in mucosa. L9 induced Th1-polarized immune response by elevating IFN-γ/IL-10 ratio in mucosa as well as induced IL-12 in macrophages. Increased expression of TLR-2 mRNA in mucosa was also observed.

Immunomodulatory activity of L. *paracasei* subsp. *paracasei* NTU101 in enterohemorrhagic *E.coli* O157:H7-infected BALB/c mice was investigated by Tsai et al. (2010). Oral administration of *L. paracasei* to mice resulted in weight gain and lowered the cumulative morbidity rates. The upregulation of dendritic cells, Th cell activation, and antibody production in post- and pre-treated mice were observed as compared to untreated mice. *L. paracasei* downregulated the expression of TLRs on macrophages and pro-inflammatory cytokines and chemokines in post- and prefeeding mice induced by *E. coli* infection, thus inhibiting the inflammation.

17.3.4 Lactobacillus gasseri

L. gasseri is a homofermentative rod-shaped bacterium which can be isolated from human mouth, gut, and vagina. *L. gasseri* is known for weight maintenance and providing protection against pathogens. Research suggested that *L. gasseri* speeds up the metabolism, therefore resulting in weight loss (Kang et al. 2013). The other in vivo effects of *L. gasseri* include cholesterol-lowering effects (Ooi et al. 2010), alleviating symptoms of allergic responses and asthma (Chen et al. 2010), and reduction in menstrual pain in women suffering from endometriosis (Itoh et al. 2011).

Luongo et al. (2013) demonstrated the immunomodulatory abilities of *L. gasseri* OLL2809 and L13-Ia. Direct incubation of murine DCs with irradiated *L. gasseri* cells induced the secretion of cytokines IL-12, IL-10, and TNF-α in the supernatant. Further they explored the cross talk between gut enterocytes and DCs that is influenced by probiotic bacteria such as *L. gasseri*. The treatment of DCs with the supernatant of bacterial-conditioned murine enterocyte cell line completely suppressed the expression of all the three cytokines.

Human clinical trials with different strains of *L. gasseri* have been conducted. Chen et al. (2010) conducted a randomized, double-blind, placebo-controlled study on the oral administration of *L. gasseri* A5 to the children suffering from asthma and allergic rhinitis. The study concluded that probiotic intake significantly improved the pulmonary function tests and the disease symptoms and significantly decreased TNF-α, IFN-γ, and IL-13 production by PBMCs was also observed. Another randomized, double-blind clinical trial conducted by Olivares et al. (2006) suggested that the consumption of the fermented product containing strains, *L. gasseri* CECT 5714 and *L. coryniformis* CECT 5711, boosted the immune system of healthy humans by increasing the numbers of phagocytic cells such as monocytes and neutrophils, as well as their phagocytic activity. The increase in proportion of NK cells,

IgA concentrations, and the levels of serum IL-10 and IL-4 was also observed after 2 weeks treatment. After 2 weeks of treatment, significant decrease was observed in serum IgE levels. IgE is involved in allergic responses; therefore decrease in IgE levels could be beneficial for allergic patients.

17.3.5 Lactobacillus acidophilus

L. acidophilus is a homofermentative bacterium that can be isolated from human vagina, gut, and mouth. As part of starter culture, it is used in the production of various dairy products such as acidophilus milk and cheese. *L. acidophilus* has been shown to inhibit the growth of *Candida albicans* by inhibiting the biofilm formation by *C. albicans* (Vilela et al. 2015).

The immunomodulatory role of *L. acidophilus* was studied in in vitro cultures. *L. acidophilus* was shown to bind to DC via DC-specific ICAM-3-grabbing nonintegrin (DC-SIGN) and activate concentration-dependent IL-10 production (Konstantinova et al. 2008). The binding of *L. acidophilus* cells with DCs with the mutant strain that lacked surface S-layer A protein and had dominant expression of S-layer protein B was significantly reduced. Also the DCs treated with mutant strain produced pro-inflammatory cytokines in higher amounts as compared to parental strain.

A study conducted by Gill et al. (2000) showed that mice fed on *L. acidophilus* (10^9 CFUs) had enhanced phagocytic activity of blood leukocytes and macrophages as compared to controls. The levels of sera antibodies were also significantly enhanced. Further, the spleen cells isolated from the treated mice produced significantly higher amounts of IFN-γ in response to concanavalin A, i.e., T-cell mitogen, as compared to control, whereas no significant increase was observed in IL-4 production. This study suggested that diet supplemented with these bacteria enhanced both cell-mediated and humoral immunity in healthy mice.

Maroof et al. (2012) showed that administration of *L. acidophilus* in breast cancer murine model can modulate immune responses and thereby cause reduction in tumor volume. The pro-inflammatory response due to enhanced production of IFN-γ and the decrease in IL-4 levels from splenocytes was observed.

17.4 Immunomodulatory Molecules

The immunomodulatory functions of different cell wall-associated and secreted molecules of lactobacilli have been studied. The prominent ones among cell wall-associated immunomodulatory molecules are exopolysaccharides (EPS) and among those secreted are secretory EPS, γ-aminobutyric acid (GABA), short-chain fatty acids (SCFA), and biogenic amines (Fig. 17.4).

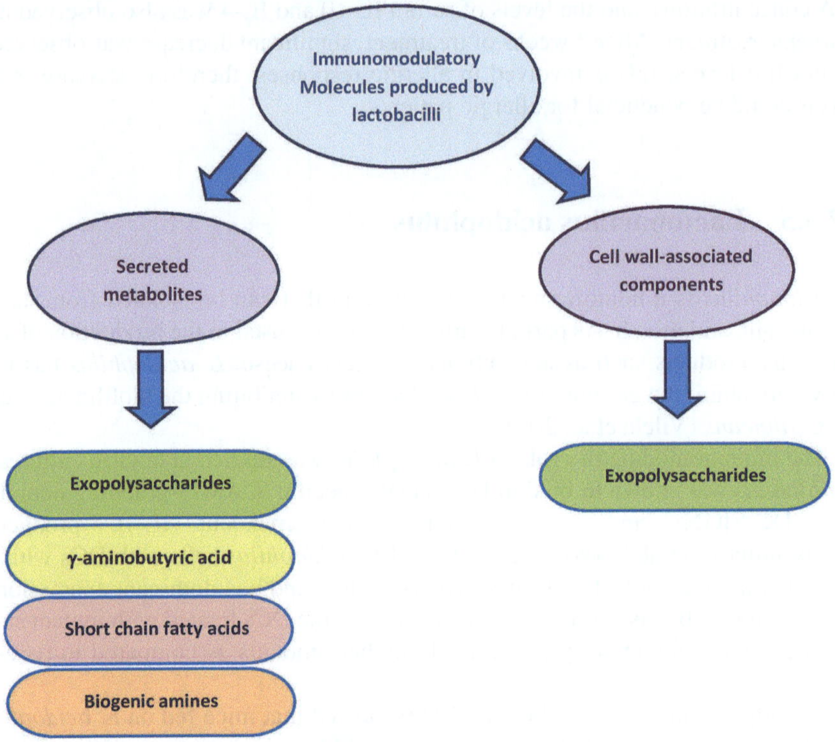

Fig. 17.4 The various immunomodulatory molecules of lactobacilli

17.4.1 EPS

EPS are high molecular weight molecules composed of sugar residues produced by microorganisms. They are important for the structural integrity of biofilms and also determine the physiochemical properties of the biofilm. They are made up of either homo- or heteropolysaccharides. The food applications of EPS are well known; for example, EPS have been commercially used as additives to improve texture as well as viscosity of naturally fermented milk products and prevented syneresis in fermented milk products. The lactobacilli-derived EPS have many health-benefiting potentials such as having antitumor, anti-ulcer, antioxidant, cholesterol-lowering, and immunostimulating activities (Patten and Laws 2015; Kim et al. 2010).

Few research studies have shown the immunomodulatory activities of purified EPS and EPS-producing bacteria. In in vitro study in murine macrophage cell line RAW264.7 cells, the EPS of *L. paracasei* subsp. *paracasei* NTU 101 and *L. plantarum* NTU 102 stimulated the dose-dependent increase in pro-inflammatory cytokine production (TNF-α, IL-6, and IL-1β). Further, EPS from both the cultures also increased the proliferative and phagocytic abilities of RAW264.7 cells (Liu et al. 2011). Similar results were obtained by Ciszek-Lenda et al. (2011), where they reported that EPS derived from *L. rhamnosus* KL37 enhanced the production of both pro-inflammatory (TNF-α, IL-6, IL-12) and anti-inflammatory (IL-10) cytokines from murine macrophages. They also demonstrated that EPS made cells tolerant to subsequent stimulation by same stimulus.

Gorska et al. (2014) demonstrated the differential immunomodulatory properties of two EPS – L900/2 and L900/3 produced by *L. rhamnosus* LOCK 0900. Exposure of mouse bone marrow-derived dendritic cells (BM-DC) to both the EPS did not trigger the production of cytokines; however, they differentially modulated the immune responses of BM-DC to *L. plantarum*. L900/2 along with *L. plantarum* cells induced the production of IL-10 levels, whereas L900/3 along with *L. plantarum* led to enhanced levels of IL-12 secretion by BM-DC. Further they showed in an experimental-induced allergy mouse model that L900/3 abrogated the ovalbumin allergen-induced IL-4, IL-5, IL-10, and IL-13 production in the spleen and mesenteric lymph nodes and thus has the potential to be used as therapeutic agent for the treatment of allergy (Gorska et al. 2017).

Another report by Gorska et al. (2016a) showed immunomodulatory properties of two different EPS 919/A and 919/B isolated from *L. casei* in BM-DC. Both the EPS did not induce the production of cytokines IL-10 and IL-12 in BM-DC; however coincubation of BM-DC with *L. plantarum* and the EPS induced the production of IL-10, but no changes in IL-12 levels were observed.

In another study by Gorska et al. (2016b), the EPS were purified from different lactobacilli species isolated from the gut of mice model of inflammatory bowel disease and from the gut of healthy mice. The EPS from all the isolates induced human mononuclear DCs to secrete cytokine weakly as compared to whole bacteria. The treatment of DCs with purified EPS did not alter the cytokine profile except *L. johnsonii* E142 which induced murine BM-DC to produce of IL-10 and IL-12 and human mononuclear DCs to secrete IL-6, IL-10, and TNF-α.

The in vivo effects of EPS of *L. kefiranofaciens* known as kefiran isolated from the fermented milk product kefir on the gut mucosal immunity were investigated (Vinderola et al. 2006). EPS in a dose-dependent manner induced the production of IgA in both the large and small intestine. It also induced the secretion of cytokines IL-4 and IL-12 in the intestinal fluid and slightly enhanced the number of IL4+, IL6+, and IL10+ T cells in the LP of the small and large intestine. In another study, oral administration of EPS of a particular strain of *L. delbrueckii* spp. *bulgaricus* OLL1073R-1 to mice induced the enhanced the NK cell activity, whereas, under in vitro conditions, it induced the mouse splenocytes to produce IFN-γ (Makino et al. 2006, 2016).

17.4.2 GABA

GABA is a nonprotein amino acid produced by decarboxylation of glutamate by action of enzyme, glutamic acid decarboxylase (GAD). Immune cells like macrophages, lymphocytes, dendritic cells, and monocytes express GABA receptors, through which GABA regulates their immune responses. There are two types of GABA receptors – $GABA_A$ and $GABA_B$ – expressed differentially on different immune cells. $GABA_A$ receptors are present on the cell surface of T cells and dendritic cells where they form functional channels through which they modulate proliferation, cytokine release, inflammatory response, and intracellular calcium concentrations.

GABA primarily functions as an inhibitor neurotransmitter in the brain of animals, but it is also known to have immunomodulatory functions in animals. Some of the reported immunomodulatory functions of GABA include downregulation of T-cell proliferation, reduction of the levels of IL-2, downregulation of Th1 proliferation, and therefore inhibition of delayed-type hypersensitivity response in vivo (Jin et al. 2013). GABA transporters (GAT) are also reported to be present on various immune cells. The deficiency of GAT in mice leads to increased T-cell proliferation and cytokine production. GABAergic agents, like topiramate and vigabatrin, lead to dose-dependent inhibition of cytokines IL-17 and IFN-γ produced by T cells and TNF, IL-6, and IL-10 produced by either T cells or DCs and macrophages (Bhat et al. 2009). Some *Lactobacillus* spp. are known to have gene for GAD enzyme and thus have been shown to produce GABA in the culture supernatant (Barrett et al. 2012; Dhakal et al. 2012), and thus theoretically the GABA-producing strains can modulate the immune response.

17.4.3 SCFA

Lactobacilli are known to carry out saccharolytic conversion of nondigestible carbohydrate to produce SCFA that are two- to six-carbon volatile acids (Pessione 2012). The prominent SCFA are acetate, propionate, and butyrate. Amino acid fermentation can also lead to production of SCFA mainly acetate and butyrate production. Lactobacilli are known to synthesize SCFA by fermenting pyruvate and by the phosphoketolase route under heterofermenting conditions (Pessione 2012). SCFA can easily pass into the blood stream through the gut epithelium. Butyrate and propionate have been shown to induce a strong anti-inflammatory response in the human monocyte-derived dendritic cells in vitro (Nastasi et al. 2015). Transcriptomic analysis showed that treatment of mature dendritic cells with butyrate upregulated 458 genes and downregulated 322 genes, whereas propionate treatment upregulated 230 genes and downregulated 41 genes in total. The prominent genes downregulated belong to pro-inflammatory chemokines such as chemokine (C-C motif) ligand (CCL)-3, CCL4, CCL5, chemokine (C-X-C motif) ligand (CXCL)-9, CXCL10, and CXCL11. Also butyrate and propionate inhibited the expression of LPS-induced cytokines such as IL-6 and IL-12p40. In another report, it was shown that feeding SCFA to germ-free mice enhanced the levels of colonic Tregs (Smith et al. 2013) and feeding butyrate enhanced the levels of peripheral Tregs in antibiotic-treated mice (Arpaia et al. 2013). Treatment of BM-DC with butyrate led to the downregulation of LPS-induced pro-inflammatory mediators, such as nitric oxide,

IL-6, and IL-12, via acting on histone deacetylases thus probably playing an important role in gut tolerance (Chang et al. 2014).

17.5 Biogenic Amines

Biogenic amines are molecules with one or more amine groups. They are generated by decarboxylation of amino acids or by transamination or amination of ketones or aldehydes. Biogenic amines include monoamines such as histamine, serotonin, and three catecholamines (epinephrine, norepinephrine, and dopamine) and polyamines such as cadaverine, putrescine, spermine, and spermidine (Fig. 17.5).

Several studies have suggested that histamine receptors (HRs) are expressed not only on mast cell and basophils but also on other immune cells such as lymphocytes, neutrophils, macrophages, and DCs, therefore modulating the function of these cells in the immune system. Histamine has a role in the immunopathogenesis of allergies and anaphylaxis as it is known to cause vasodilation, smooth muscle contraction, and mucus production. However, as it downregulates the proliferation of Th1 cells and upregulates the proliferation of Th2 cells, it again has a role in allergic disease and asthma. Histamine induces the secretion of Th2 cytokines such as IL-4, IL-5, IL-10, and IL-13 and inhibits the production of Th1 cytokines such as IL-2, IFN-γ, and IL-12 (Shahid et al. 2009). Lactobacilli are known to produce biogenic amines especially tyramine and putrescine (Lucas et al. 2007); however, the production of amines is strain-specific. There are reports that the absorption of biogenic amine in systemic circulation at high concentrations can result in toxicity due to enhanced release of adrenaline and noradrenaline that induces gastric acid secretion, increased blood glucose levels, and high blood pressure (Shalaby 1996) leading to hypertensive crisis ultimately causing end-organ damage in the heart or central nervous system (McCabe-Sellers et al. 2006; Blackwell 1963). Also increased levels of putrescine have been detected in gastric carcinomas caused by *Helicobacter pylori* (Shah and Swiatlo 2008).

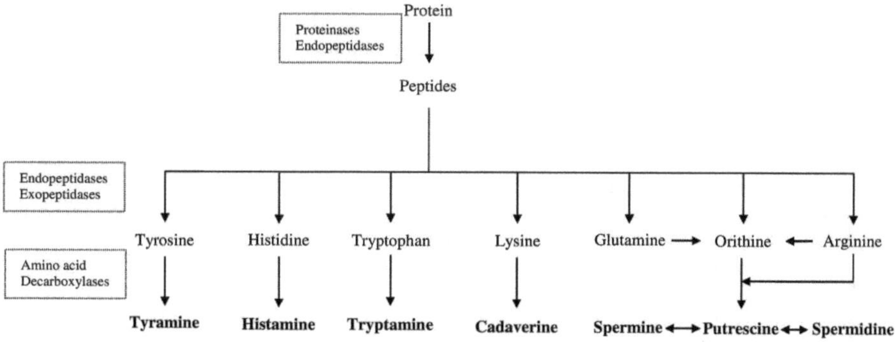

Fig. 17.5 Various biogenic amines and their precursors

17.6 Conclusions

Few *Lactobacillus* spp. have strong immunomodulatory abilities and thus can be potentially therapeutic in certain health conditions such as allergic, inflammatory autoimmune diseases, and infectious diseases. Further, due to their immunomodulatory nature, lactobacilli have been reported to enhance the effectiveness of several candidate mucosal vaccines against malaria, HIV, and infantile diarrhea as shown in animal models (Amdekar et al. 2010). However, the immunomodulation by lactobacilli depends on the bacterial strain and host, and it has been noted that at times in vitro studies do not correlate with in vivo results. Thus, human clinical trials of the therapeutic effectiveness of *Lactobacillus* spp. or its metabolites in patients are strongly recommended. Further, none of the clinical trials of *Lactobacillus* spp. in humans had shown any evidence of infection or side effects, thus, proving their GRAS status. Furthermore, the EPS of some *Lactobacillus* spp. have also shown immunomodulatory behavior; therefore they can be the safer alternatives as compared to live probiotic cells especially for the immunocompromised patients. As some lactobacilli strains are known to produce high amounts of biogenic amines, therefore probiotic lactobacilli should be screened for their ability to secrete biogenic amines.

Acknowledgment This work was supported by research grant (grant number: 42-478/2013 SR) sponsored by the University Grants Commission (UGC), New Delhi, India. Mrs. Sumanpreet Kaur is thankful to University of Potential for Excellence scheme of UGC for the fellowship. Ms. Preeti Sharma is thankful to UGC for the fellowship.

References

Amdekar S, Dwivedi D, Roy P, Kushwah S, Singh V (2010) Probiotics: multifarious oral vaccine against infectious traumas. FEMS Immunol Med Microbiol 58:299–306

Arpaia N, Campbell C, Fan X, Dikiy S, van der Veeken J, deRoos P et al (2013) Metabolites produced by commensal bacteria promote peripheral regulatory T cell generation. Nature 504:451–455

Baken KA, Ezendam J, Gremmer ER, Klerk R, Pennings JA, Matthee B et al (2006) Evaluation of immunomodulation by *Lactobacillus casei* Shirota: immune function, autoimmunity and gene expression. Int J Food Microbiol 112:8–18

Balejko E, Bogacka A, Balejko J, Kucharska E (2015) Immunomodulation effect of metabolites from *Lactobacillus rhamnosus* GG on interleukins release in vitro. J Food Nutr Res 3:297–302

Barrett E, Ross RP, O'Toole PW, Fitzgerald GF, Stanton C (2012) γ-aminobutyric acid production by culturable bacteria from the human intestine. J Appl Microbiol 113:411–417

Bermudez-Brito M, Plaza-Díaz J, Muñoz-Quezada S, Gómez-Llorente C, Gil A (2012) Probiotics mechanisms of action. Ann Nutr Metab 61:160–174

Bevins CL, Salzman NH (2012) Paneth cells, antimicrobial peptides and maintenance of intestinal homeostasis. Nat Rev Microbiol 9:356–368

Bhat R, Axtell R, Mitra A, Miranda M, Lock C, Tsien RW et al (2009) Inhibitory role for GABA in autoimmune inflammation. Proc Natl Acad Sci 107:2580–2585

Blackwell B (1963) Hypertensive crisis due to monoamine-oxidase inhibitors. Lancet 2:849–850

Borruel N, Carol M, Casellas F, Antolin M, de Lara F, Espin E et al (2002) Increased mucosal tumour necrosis factor a production in Crohn's disease can be downregulated ex vivo by probiotic bacteria. Gut 51:659–664

Chang PV, Hao L, Offermanns S, Medzhitov R (2014) The microbial metabolite butyrate regulates intestinal macrophage function via histone deacetylase inhibition. Proc Natl Acad Sci 111:2247–2252

Chen YS, Jan RL, Lin YL, Chen HH, Wang JY (2010) Randomized placebo-controlled trial of *Lactobacillus* on asthmatic children with allergic rhinitis. Pediatr Pulmonol 45:1111–1120

Ciszek-Lenda M, Nowak B, Srottek M, Gamian A, Marcinkiewicz J (2011) Immunoregulatory potential of exopolysaccharide from *Lactobacillus rhamnosus* KL37: effects on the production of inflammatory mediators by mouse macrophages. Int J Exp Pathol 92:382–391

D'Arienzo R, Bozzella G, Rossi M, De Bellis P, Lavermicocca P, Sisto A (2011) Distinct immunomodulatory properties of *Lactobacillus paracasei* strains. J Appl Microbiol 111:1482–1491

Dhakal R, Bajpai VK, Baek KH (2012) Production of GABA (γ – aminobutyric acid) by microorganisms: a review. Braz J Microbiol 43:1230–1241

Dong H, Rowland I, Tuohy KM, Thomasand LV, Yaqoob P (2010) Selective effects of *Lactobacillus casei* Shirota on T cell activation, natural killer cell activity and cytokine production. Clin Exp Immunol 161:378–388

Doron S, Snydman DR, Gorbach SL (2005) Lactobacillus GG: bacteriology and clinical applications. Gastroenterol Clin N Am 34:483–498

Evrard B, Coudeyras S, Dosgilbert A, Charbonnel N, Alamé J, Tridon A et al (2011) Dose-dependent immunomodulation of human dendritic cells by the probiotic *Lactobacillus rhamnosus* Lcr35. PLoS One 6:1–12

FAO/WHO (2001) Evaluation of health and nutritional properties of probiotics in food including powder milk with live lactic acid bacteria. Report of a joint Food Agricultural Organisation/ WHO expert consultation, Cordoba

Finegold SM, Sutter VL, Mathisen GE (1983) Normal indigenous intestinal flora. In: Hentges DJ (ed) Human intestinal microbiota in health and disease. Academic, New York, pp 3–31

Fölster-Holst R, Müller F, Schnopp N, Abeck D, Kreiselmaier I, Lenz T et al (2006) Prospective, randomized controlled trial on *Lactobacillus rhamnosus* in infants with moderate to severe atopic dermatitis. Br J Dermatol 155:1256–1261

Fong FLY, Kirjavainen P, Wong VHY, El-Nezami H (2015) Immunomodulatory effects of *Lactobacillus rhamnosus* GG on dendritic cells, macrophages and monocytes from healthy donors. J Funct Foods 13:71–79

Fong FL, Kirjavainen PV, El-Nezami H (2016) Immunomodulation of *Lactobacillus rhamnosus* GG (LGG)-derived soluble factors on antigen-presenting cells of healthy blood donors. Sci Rep 6:1–8

Fujiwara D, Inoue S, Wakabayashi H, Fujii T (2004) The anti-allergic effects of lactic acid bacteria are strain dependent and mediated by effects on both Th1/Th2 cytokine expression and balance. Int Arch Allergy Immunol 135:205–215

Gill HS, Rutherfurd KJ, Prasad J, Gopal PK (2000) Enhancement of natural and acquired immunity by *Lactobacillus rhamnosus* (HN001), *Lactobacillus acidophilus* (HN017) and *Bifidobacterium lactis* (HN019). Br J Nutr 83:167–176

Gorska S, Schwarzer M, Jachymek W, Srutkova D, Brzozowska E, Kozakova H (2014) Distinct immunomodulation of bone marrow-derived dendritic cell responses to *Lactobacillus plantarum* WCFS1 by two different polysaccharides isolated from *Lactobacillus rhamnosus* LOCK 0900. Appl Environ Microbiol 80:6506–6516

Gorska S, Hermanova P, Ciekot J, Schwarzer M, Srutkova D, Brzozowska E et al (2016a) Chemical characterization and immunomodulatory properties of polysaccharides isolated from probiotic *Lactobacillus casei* LOCK 0919. Glycobiology 26:1014–1024

Gorska S, Sandström C, Wojas-Turek J, Rossowska J, Pajtasz-Piasecka E, Brzozowska E et al (2016b) Structural and immunomodulatory differences among lactobacilli exopolysaccharides isolated from intestines of mice with experimentally induced inflammatory bowel disease. Sci Rep 6:1–16

Gorska S, Schwarzer M, Srutkova D, Hermanova P, Brzozowska E, Kozakova H et al (2017) Polysaccharides L900/2 and L900/3 isolated from *Lactobacillus rhamnosus* LOCK 0900 modulate allergic sensitization to ovalbumin in a mouse model. Microb Biotechnol 10:586–593

Hashimoto S, Seyama Y, Yokokura T, Mutai M (1985) Cytotoxic factor production by Kupffer cells elicited with *Lactobacillus casei* and *Corynebacterium parvum*. Cancer Immunol Immunother 20:117–121

Havenaar R, Brink BT, Huis in't Veld JHJ (1992) Selection of strains for probiotic use. In: Probiotics. Springer, Dordrecht, pp 209–224

Herías MV, Koninkx JF, Vos JG, Huis in't Veld JH, van Dijk JE (2005) Probiotic effects of *Lactobacillus casei* on DSS-induced ulcerative colitis in mice. Int J Food Microbiol 103:143–155

Itoh H, Uchida M, Sashihara T (2011) *Lactobacillus gasseri* OL2809 is effective especially on the menstrual pain and dysmenorrheal in endometriosis patients: randomized, double-blind, placebo-controlled study. Cytotechnology 63:153–161

Ivanov II, Honda K (2012) Intestinal commensal microbes as immune modulators. Cell Host Microbe 12:496–508

Jin Z, Mendu SK, Birnir B (2013) GABA is an effective immunomodulatory molecule. Amino Acids 45:87–94

Kalliomaki M, Salminen S, Arvilommi H, Kero P, Koskinen P, Isolauri E (2001) Probiotics in primary prevention of atopic disease: a randomised placebo-controlled trial. Lancet 357:1076–1079

Kalliomaki M, Salminen S, Poussa T, Arvilommi H, Isolauri E (2003) Probiotics and prevention of atopic disease: 4-year follow-up of a randomized placebo-controlled trial. Lancet 361:1869–1871

Kalliomaki M, Salminen S, Poussa T, Isolauri E (2007) Probiotics during the first 7 years of life: a cumulative risk reduction of eczema in a randomized, placebo- controlled trial. J Allergy Clin Immunol 119:1019–1021

Kang JH, Yun SI, Park MH, Park JH, Jeong SY, Park HO (2013) Anti-obesity effect of *Lactobacillus gasseri* BNR17 in high-sucrose diet-induced obese mice. PLoS One 8(1):1–8

Kato I, Yokokura T, Mutal M (1984) Augmentation of mouse natural killer cell activity by *Lactobacillus casei* and its surface antigens. Microbiol Immunol 28:209–217

Kekkonen RA, Kajasto E, Miettinen M, Veckman V, Korpela R, Julkunen I (2008) Probiotic *Leuconostoc mesenteroides sp. cremoris* and *Streptococcus thermophilus* induce IL-12 and IFN-gamma production. World J Gastroenterol 14(8):1192–1203

Kim Y, Oh S, Yun HS, Oh S, Kim SH (2010) Cell-bound exopolysaccharide from probiotic bacteria induces autophagic cell death of tumour cells. Lett Appl Microbiol 51:123–130

Kirjavainen PV, El-Nezami HS, Salminen SJ, Ahokas JT, Wright PA (1999) The effect of orally administered viable probiotic and dairy lactobacillus on mouse lymphocyte proliferation. FEMS Immunol Med Microbiol 26:131–135

Kobayashi KS, Chamaillard M, Ogura Y, Henegariu O, Inohara N, Nunez G et al (2005) Nod2-dependent regulation of innate and adaptive immunity in the intestinal tract. Science 307:731–734

Konstantinova SR, Smidta H, de Vosa WM, Bruijnsb SCM, Singh SK, Valence F et al (2008) S layer protein A of *Lactobacillus acidophilus* NCFM regulates immature dendritic cell and T cell functions. Proc Natl Acad Sci 105:19474–19479

Kopp MV, Goldstein M, Dietschek A, Sofke A, Heinzmann A, Urbanek R (2008) *Lactobacillus* GG has in vitro effects on enhanced interleukin-10 and interferon-γ release of mononuclear cells but no *in vivo* effects in supplemented mothers and their neonates. Clin Exp Allergy 38:602–610

Liu CF, Tseng KC, Chiang SS, Lee BH, Hsu WH, Pan TM (2011) Immunomodulatory and antioxidant potential of *Lactobacillus* exopolysaccharides. J Sci Food Agric 91:2284–2291

Lucas PM, Blancato VS, Claisse O, Magni C, Lolkema JS, Lonvaud-Funel A (2007) Agmatine deiminase pathway genes in *Lactobacillus brevis* are linked to the tyrosine decarboxylation operon in a putative acid resistance locus. Microbiology 153:2221–2230

Luongo D, Miyamoto J, Bergamo P, Nazzaro F, Baruzzi F, Sashihara T et al (2013) Differential modulation of innate immunity in vitro by probiotic strains of *Lactobacillus gasseri*. BMC Microbiol 13:298

Makino S, Ikegami S, Kano H, Sashihara T, Sugano H, Horiuchi H et al (2006) Immunomodulatory effects of polysaccharides produced by *Lactobacillus delbrueckii* ssp. *bulgaricus* OLL1073R-1. J Dairy Sci 89:2873–2881

Makino S, Goto A, Nakamura M, Ogawa M, Chiba Y, Hemmi J et al (2016) Enhanced natural killer cell activation by exopolysaccharides derived from yogurt fermented with *Lactobacillus delbrueckii* ssp. *bulgaricus*OLL1073R-1. J Dairy Sci 99:915–923

Marchand J, Vandenplas Y (2000) Micro-organisms administered in the benefit of the host: myths and facts. Eur J Gastroenterol Hepatol 12:1077–1088

Maroof H, Hassan ZM, Mobarez AM, Mohamadabadi MA (2012) Lactobacillus acidophilus could modulate the immune response against breast cancer in murine model. J Clin Immunol 32:1353–1359

McCabe-Sellers BJ, Staggs CG, Bogle ML (2006) Tyramine in foods and monoamine oxidase inhibitor drugs: a crossroad where medicine, nutrition, pharmacy, and food industry converge. J Food Compos Anal 19:58–65

Miettinen M, Lehtonen A, Julkunen I, Matikainen S (2000) Lactobacilli and Streptococci activate NF-kappa B and STAT signaling pathways in human macrophages. J Immunol 164:3733–3740

Nastasi C, Candela M, Bonefeld CM, Geisler C, Hansen M, Krejsgaard T et al (2015) The effect of short-chain fatty acids on human monocyte-derived dendritic cells. Sci Rep 5:1–10

Niess JH, Reinecker HC (2006) Dendritic cells in the recognition of intestinal microbiota. Cell Microbiol 8:558–564

Olivares M, Díaz-Ropero MP, Gómez N, Lara-Villoslada F, Sierra S, Maldonado JA et al (2006) The consumption of two new probiotic strains, *Lactobacillus gasseri* CECT 5714 and *Lactobacillus* coryniformis CECT 5711, boosts the immune system of healthy humans. Int Microbiol 9:47–52

Ooi LG, Ahmad R, Yuen KH, Liong MT (2010) *Lactobacillus gasseri* CHO220 and inulin reduced plasma total cholesterol and low-density lipoprotein cholesterol via alteration of lipid transporters. J Dairy Sci 93:5048–5058

Patten DA, Laws AP (2015) *Lactobacillus*-produced exopolysaccharides and their potential health benefits: a review. Benefic Microbes 6:457–471

Pena JA, Versalovic J (2003) Lactobacillus rhamnosus GG decreases TNF-a production in lipopolysaccharide-activated murine macrophages by a contact-independent mechanism. Cell Microbiol 5:277–285

Perdigon G, Nader de Maclas ME, Alverarez S, Medici M, Oliver G (1986) Effect of a mixture of *Lactobacillus casei* and *Lactobacillus acidophilus* administered orally on the immune system in mice. J Food Prot 49:986–989

Perdigon G, Macias ME, Alvarez S, Oliver G, Holgado AA (1990) Prevention of gastrointestinal infection using immunobiological methods with milk fermented with *Lactobacillus casei* and *Lactobacillus acidophilus*. J Dairy Res 57:255–264

Perdigón G, Alvarez S, Holgado AA (1991) Immunoadjuvant activity of oral *Lactobacillus* casei: influence of dose on the secretory immune response and protective capacity in intestinal infections. J Dairy Res 58:485–496

Pessione E (2012) Lactic acid bacteria contribution to gut microbiota complexity: lights and shadows. Front Cell Infect Microbiol 2:1–15

Shah P, Swiatlo E (2008) A multifaceted role for polyamines in bacterial pathogens. Mol Microbiol 68:4–16

Shahid M, Tripathi T, Sobia F, Moin S, Siddiqui M, Khan RA (2009) Histamine, histamine receptors, and their role in immunomodulation: an updated systematic review. Open Immunol J 2:9–41

Shalaby AR (1996) Significance of biogenic amines in food safety and human health. Food Res Int 29:675–690

Shida K, Takahashi R, Iwadate E, Takamizawa K, Yasui H, Sato T et al (2002) Lactobacillus casei strain Shirota suppresses serum immunoglobulin E and immunoglobulin G1 responses and systemic anaphylaxis in a food allergy model. Clin Exp Allergy 32:563–570

Shida K, Suzuki T, Kiyoshima-Shibata J, Shimada S, Nanno M (2006) Essential roles of monocytes in stimulating human peripheral blood mononuclear cells with *Lactobacillus casei* to produce cytokines and augment natural killer cell activity. Clin Vaccine Immunol 13:997–1003

Smith PM, Howitt MR, Panikov N, Michaud M, Gallini CA, Bohlooly-Y M et al (2013) The microbial metabolites, short chain fatty acids, regulate colonic Treg cell homeostasis. Science 341:569–573

Takeda K, Okumura K (2007) Effects of a fermented milk drink containing *Lactobacillus casei* strain shirota on the human NK-cell activity. J Nutr 137:791–793

Tsai YT, Cheng PC, Pan TM (2010) Immunomodulating activity of *Lactobacillus paracasei* subsp. *paracasei* NTU 101 in enterohemorrhagic *Escherichia Coli* O157:H7-infected mice. J Agric Food Chem 58:11265–11272

Vilela SF, Barbosa JO, Rossoni RD, Santos JD, Prata MC, Anbinder AL et al (2015) *Lactobacillus acidophilus* ATCC 4356 inhibits biofilm formation by *C. albicans* and attenuates the experimental candidiasis in *Galleria mellonella*. Virulence 6:29–39

Vinderola G, Perdigon G, Duarte J, Farnworth E, Matar C (2006) Effects of the oral administration of the exopolysaccharide produced by *Lactobacillus kefiranofaciens* on the gut mucosal immunity. Cytokine 36:254–260

von de Weid T, Bulliard C, Schiffrin E (2001) Induction by a lactic acid bacterium of a population of CD4+ T cells with low proliferative capacity that produce transforming growth factor β and interleukin-10. Clin Diagn Lab Immunol 8:695–701

Walter J (2008) Ecological role of lactobacilli in the gastrointestinal tract: implications for fundamental and biomedical research. Appl Environ Microbiol 74:4985–4996

Wu CT, Chen PJ, Lee YT, Ko JL, Ko HL (2016) Effects of immunomodulatory supplementation with *Lactobacillus rhamnosus* on airway inflammation in a mouse asthma model. J Microbiol Immunol Infect 49:625–635

Zakostelska Z, Kverka M, Klimesova K, Rossmann P, Mrazek J, Kopecny J et al (2011) Lysate of probiotic *Lactobacillus casei* DN-114 001 ameliorates colitis by strengthening the gut barrier function and changing the gut microenvironment. PLoS One 6:1–11

Zhu Y, Zhu J, Zhao L, Zhang M, Guo H, Ren F (2016) Effect of oral administration of *Lactobacillus paracasei* L9 on mouse systemic immunity and the immune response in the intestine. Arch Biol Sci 68:311–318

Part III
Viral Infections

Chapter 18
HIV-AIDS: An Unconquered Immune War

S. K. Arora and Gurleen Mehta

Abstract Acquired immune deficiency syndrome (AIDS) is a disease that has shaken the world affecting millions of people from all corners of the society. Syndrome involves the compromised state of the host defense system which results in impaired immunity leading to increased susceptibility of the host to opportunistic infections or predisposition to malignancies. The disease progression is monitored by CD4 cell counts and plasma virus loads in the blood of the infected patients. The disease progression in HIV infection is dependent on many factors, but most importantly, the disease progression depends upon how the viral factors and host factors interact with each other. Opportunistic infections (OIs) lead to increased mortality in the patients, and tuberculosis accounts for the most common OI. Early detection, diagnosis and appropriate treatment of OIs can slow down disease progression. The antiretroviral treatment leads to decrease in the viral loads and increase in the CD4 T lymphocyte count in the infected individuals. Strict adherence to treatment can significantly improve the quality of life of the patients infected with HIV.

Keywords AIDS – Acquired immune deficiency syndrome · ART – Antiretroviral treatment · CD – Cluster of differentiation · HIV – Human immunodeficiency virus · OI – Opportunistic infection

18.1 Introduction

Acquired immune deficiency syndrome (AIDS) is a disease that has shaken the world affecting millions of people from all corners of the society. Syndrome involves the compromised state of the host defense system which results in impaired immunity leading to increased susceptibility of the host to opportunistic infections or predisposition to malignancies. AIDS, as we know it better, is caused by human immunodeficiency virus (HIV) which is a lentivirus of the family *Retroviridae*. In 2015 there were 2.1 million (1.8 million to 2.4 million) new HIV infections

S. K. Arora (✉) · G. Mehta
Department of Immunopathology, Postgraduate Institute of Medical Education and Research, Chandigarh, India

P. P. Singh (ed.), *Infectious Diseases and Your Health*,
https://doi.org/10.1007/978-981-13-1577-0_18

worldwide, adding up to a total of 36.7 million (34.0 million to 39.8 million) people living with HIV (UNAIDS Global AIDS Update 2016). Two types of HIV have been characterized, namely, HIV-1 and HIV-2. HIV-1 is the virus that was initially termed as both lymphadenopathy-associated virus (LAV) and human T-lymphotropic virus-III (HTLV-III). It is more virulent and more infective than HIV-2 that makes it the cause of the majority of HIV infections globally.

18.2 Structure and Genome (Fig. 18.1)

HIV is complex retrovirus and is a member of the *Retroviridae* family. It is spherical in shape with a diameter of about 90–120 nm. The genome contains two copies of single-stranded positive-polarity RNA which is the basic genome of the virus. The two RNA strands are surrounded by a nucleocapsid which is composed of proteins. The nucleocapsid also contains the associated viral enzymes that are required for the replication of the virion such as reverse transcriptase (RT), protease, ribonuclease and integrase. The matrix is composed of the matrix protein or viral protein, i.e. p17, which contains the essential proteins and the nucleus. The virus envelope contains glycoprotein (gp) 120 and gp 41 in HIV-1 and gp 140 and gp 36 in HIV-2. HIV RNA genome encodes various genes which can be divided into three groups:

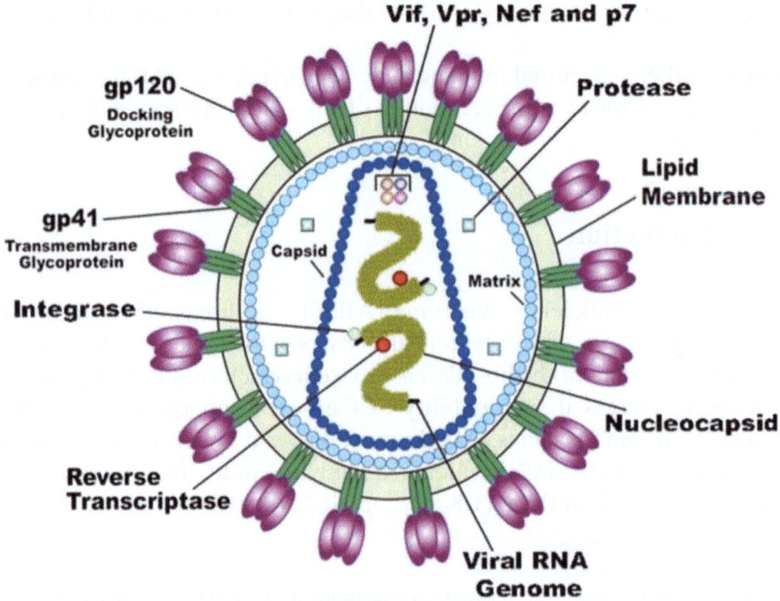

Fig. 18.1 HIV structure. (Image source: https://commons.wikimedia.org/wiki/File:HI-Virion-en.png)

1. The major structural proteins – Gag, Pol and Env
2. The regulatory proteins – Tat and Rev
3. The accessory proteins – Vpu, Vpr, Vif and Nef

18.3 Transmission

The HIV is generally transmitted by four processes only:

1. Unsafe/unprotected sexual (homo or hetero) exposure
2. Unsafe blood transfusions
3. Use of infected/unsterilized needles
4. From HIV infected mother to her child which can happen during pregnancy, labour or breastfeeding

Unprotected heterosexual route is the commonest and predominant mode of HIV transmission in India (Lal 2001). This mode of transmission amounts to high-risk activities. The virus may enter through cuts or sores on the penis, vagina or anus in case of an unprotected sexual intercourse. Unsafe blood transfusions, which have been a major concern in many countries, have been nearly eliminated in developed countries because of the mandatory routine screening of all blood donated units prior to each transfusion (Gayle and Hill 2001).

However in many developing countries, HIV transmission via the blood and blood products still poses a threat because screening of blood for HIV is still not in routine practice. The transmission of HIV via the infected blood or its products is still high with an infectivity rate of 2% in India (Steinbrook 2007). However, as per NACO Annual report, 2016–2017, the incidence of donor HIV sero-reactivity has declined from 1.2% to less than 0.14% in NACO supported Blood Banks.

The use of unsterilized needles and syringes has been a threat to the general public especially in the younger population which further increases the transmission of HIV infection. The intravenous drug users, who share the same needles, pose a major threat to the society. The transmission from an infected mother to her newly born child occurs during birth or just after in approximately 20–40% cases. The HIV may get transmitted through the genital secretions from the infected mother through maternal blood, during birth and also during breastfeeding as breast milk carries high content of the virus.

18.4 HIV Life Cycle/Infection Cycle

The primary site of HIV infection is the CD4 protein which is present on the surface of the T lymphocytes. The CD4+ T Lymphocyte is also known as T helper cell and the CD4 receptor is also present on the surface of some other cells of the immune system. HIV may infect other CD4-expressing cells such as dendritic cells, macrophages and microglial cells. For the entry of the virus in the host cell, besides a CD4 receptor, the HIV also requires a chemokine receptor as a co-receptor which may be CXCR4 or CCR5. These co-receptors are present on the target host cell membrane. The virus is phenotyped as a R5 (CCR5 tropic virus) or X4 (CXCR4 tropic virus) depending on the co-receptor usage.

18.5 The Lifecycle of the Virus May Be Described by Means of Various Stages (Fig. 18.2)

18.5.1 Binding and Fusion

This is the first encounter of the virus with the host cell. This step involves fusion of the virus via gp120 to a CD4 molecule present on the host cell. The binding of the glycoprotein triggers the conformational changes in the host cell membrane leading to exposure of the binding sites of either of the two co-receptors, i.e. CCR5 or CXCR4. The exposure of the binding sites of CCR5 or CXCR4 helps in the fusion or merging of the virus envelope and the cell membrane of the host, and this fusion initiates the transfer of the viral genome into the protoplasm of the target cell.

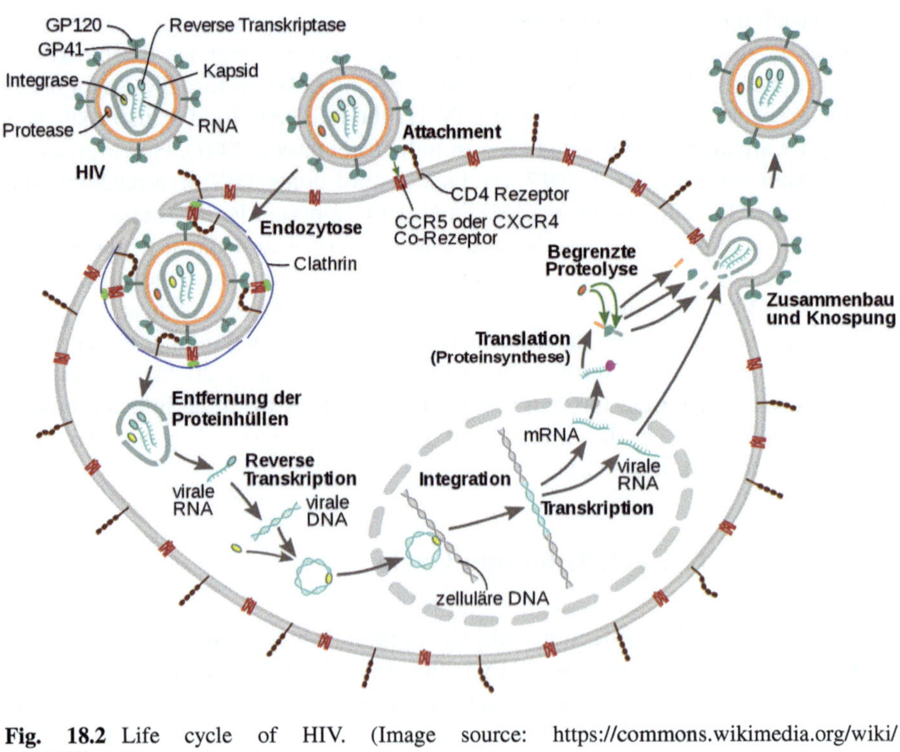

Fig. 18.2 Life cycle of HIV. (Image source: https://commons.wikimedia.org/wiki/File:HIV-replication-cycle-)

18.5.2 Reverse Transcription

Once the viral genome enters the protoplasm of the host CD4 cell, the single-stranded HIV RNA undergoes reverse transcription and forms a double-stranded complementary DNA (cDNA) by viral reverse transcriptase (RT) enzyme that subsequently translocates into the nucleus of the host cell.

18.5.3 Integration

The function of the HIV enzyme "integrase" which is present in the nucleus of the host cell is to integrate the viral DNA with the DNA of the infected/target CD4+ cell leading to the generation of provirus. This step is a nonreversible step. Once integration of the viral RNA with the host DNA happens, the host cell gets infected for rest of its life.

18.5.4 Transcription

The integrated proviral DNA undergoes transcription with the help of the enzyme RNA polymerase resulting in numerous copies of viral RNA and viral mRNAs which are then released out in the protoplasm of the host cell.

18.5.5 Protein Synthesis and Assembly

The next step is the translation of viral mRNAs into viral proteins which takes place in the protoplasm of the target cell. The HIV enzyme "protease" present in the protoplasm of the host cell facilitates the synthesis of smaller protein particles by cutting the long chains of HIV proteins. These HIV proteins along with the copies of HIV RNA generated by transcription now produce a new virus particle, leading to generation of numerous virus particles in the host cell.

18.5.6 Budding

Finally the newly generated virus particle assembly "buds" from the cell membrane. During the process of budding of the virus assembly, host membrane proteins and lipid bilayer are also acquired by the HIV envelope.

18.6 Immunopathogenesis

The rate of disease progression varies in different infected individuals. The disease progression in HIV infection is dependent on many factors, but most importantly, the disease progression depends upon how the viral factors and host factors interact with each other. The primary target of the HIV is the T lymphocytes or other cells like dendritic cells, macrophages and monocytes having CD4 receptor on their surface. The virus targets these cells leading to their destruction through multiple mechanisms including apoptosis (Paranjape 2005).

The HIV virus resides inside the infected CD4 cells and keeps replicating that manipulates the host's immune system. There are two important subsets of T lymphocytes which are the principal cells involved in the HIV infection. These are CD4 T (helper) cells and CD8 T (cytotoxic) cells which express different T cell receptors (TCRs) on their surface. The T cell receptors enable these principal cells to bind to antigen-presenting cells. The important antigen-presenting cells in the host are the dendritic cells (DCs), follicular dendritic cells (FDCs), macrophages and B cells. Besides these, HIV can also infect monocytes, microglia, oligodendrocytes and retinal cells and several other cells in the body.

18.7 Cellular Immune Response

The transmission of HIV occurs through various routes that include sexual intercourse (vaginal, anal or oral), through blood or blood products (IV drug use, unsafe blood transfusion or organ transplantation) and from infected mother to child (during pregnancy, child birth or breastfeeding). HIV is acquired through any route of transmission as described above and targets the T lymphocyte (or other cells) having CD4 receptor on the surface and a co-receptor. The glycoprotein present on the membrane of HIV-1, i.e. gp120, binds to the CD4 receptors present on the T lymphocytes along with a co-receptor which is either X4 (CXCR4) or R5 (CCR5). This interaction of the virus and the cell leads to activation of the T lymphocytes on account of the infection caused by the virus, and HIV starts replicating rapidly in these activated T cells. This virus multiplies and spreads throughout the lymphoid tissue; thus there is an increase in virus-infected cells in the plasma of the infected individuals. The viral load increases exponentially during the first 2–3 weeks of infection. The host immune response is activated at this stage, but as the virus keeps multiplying rapidly in the activated T lymphocytes, it leads to destruction of CD4 cells and effector memory T cells by various mechanisms such as apoptosis. Another important mechanism is continuous budding of the viruses from infected cells that cause membrane disruption leading to increased permeability of the host cell membrane resulting in death of the cell. Initially the virus is M-tropic and infects the CD4-positive monocytes by utilizing CCR5 co-receptor. The CXCR4 tropic or X4 tropic viruses are found in the later stages of HIV infection. These X4 viruses mostly infect T cells and form giant multinucleate cells called syncytium by inducing plasma membrane fusion between adjacent cells which are generally seen in the late stages of HIV infection. During advanced disease, both the HIV-infected and the bystander cells participate in the formation of syncytium which results in cell destruction (Alimonti et al. 2003).

18.8 Humoral Immune Response

During the active infection, both effective cell-mediated immune response and humoral response come into play. HIV antibodies are detectable by IgM ELISA or third-generation rapid serological tests typically 3–4 weeks after infection. There is typically a low titer of antibodies during the earlier weeks of infection which is not effective against the virus population. The time period from which the virus enters the host until detectable levels of HIV-specific antibodies appear in the blood of the infected individual is called the "window period" or "acute infection phase". During this period, although the antibodies are undetectable, the individual is highly infectious as the level of viral load is very high because of the continuous replication of the virus in the host. During this phase of the infection, as the antibody levels are undetectable in the infected individuals, this may be falsely interpreted as sero-negative, i.e. tests for detecting anti-HIV antibodies are negative.

18.9 Disease Progression and Mechanisms of Immune Cell Destruction (Fig. 18.3)

The disease progression in HIV characteristically has three stages, i.e. acute infection phase, chronic asymptomatic phase and AIDS.

18.10 Acute Infection Phase/Primary Infection

This phase of primary infection is the primary stage of infection which develops within 2–4 weeks of infection with the virus. The symptoms in the acute phase may vary in different individuals which may be flu-like symptoms, such as fever, headache and rash. In the acute stage of infection, HIV multiplies rapidly and disseminates throughout the body. The viral load is highest at this stage which leads to destruction of the infection-fighting CD4 cells also known as T helper (Th) cells of the immune system. The chance of transmission of HIV from an infected individual to an uninfected individual is high during this particular stage because of the high level of circulating viral load in the body of the infected individual.

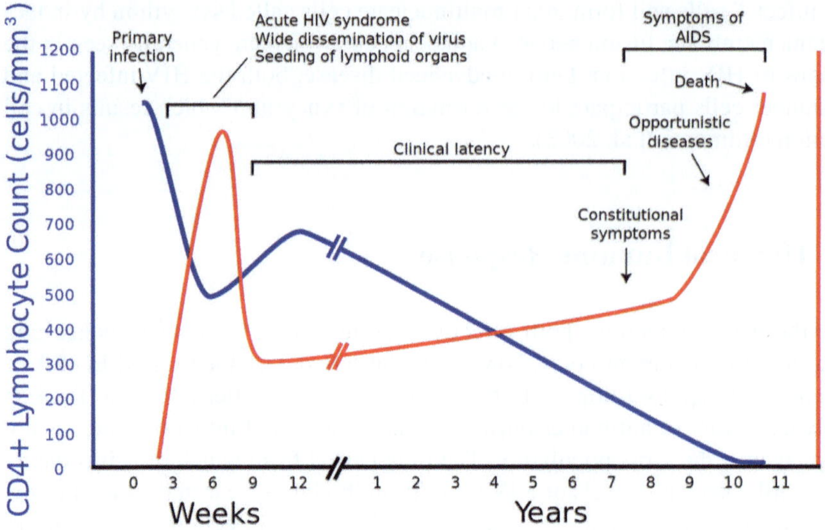

Fig. 18.3 Graph showing HIV copies and CD4 counts in a human over the course of a treatment-naive HIV infection. (Image source: https://tl.wikipedia.org/wiki/HIV)

During the initial phase of infection, the virus disseminates in the host and causes the destruction of the CD4 helper T cells. As viraemia increases, the CD4 T cell populations gets depleted due to destruction of the cells by the virus using various mechanisms. The immune system activates in response to the infection, leading to the activation of the humoral immune response causing a decrease in viraemia. This decrease in viraemia may be because the cytotoxic T cells are able to block active HIV replication through non-cytolytic virus-suppressive mechanisms. The soluble factors have the capability of suppressing primary HIV isolates that are both CCR5 tropic and CXCR4 tropic resulting in the blockage of HIV replication (Walker et al. 1986).

The two cell populations, plasmacytoid dendritic cells (PDC) and natural killer (NK) cells, play predominant role in innate cellular immune responses. This phase is followed by activation of adaptive immune response to HIV.

18.11 Chronic Asymptomatic Phase

The next stage of HIV infection is the chronic but asymptomatic phase of HIV infection. This stage is also called clinical latency period. During this phase, the viral load in the body decreases due to the activation of the immune response leading to increase in the CD4 cells, but HIV still keeps replicating at very low levels. During this stage of infection, the patient may not show any HIV-related symptoms, but they can still spread HIV to others as the virus is replicating in the body of the infected individual. It becomes extremely important to monitor the disease progression during this stage as it is asymptomatic. The CD4 count is monitored every 6 months till the CD4 number remains stable. As per latest guidelines, all the individuals who are detected with HIV infection are started with the antiretroviral treatment (ART) irrespective of CD4 count. The patient while on ART may remain on this viral suppressed stage for several years, but slowly the viral load starts going up, and the patient progresses to AIDS.

Although the activated host immune response lowers the viraemia, HIV is never completely eliminated from the host or the infected individual and slowly leads to the progression of the HIV disease to the chronic phase.

The hallmark of this phase is the progressive depletion of the CD4 pool in the host. The disease progression to acquired immune deficiency syndrome (AIDS) depends on several factors that include host's susceptibility to the virus, genetic mechanisms like HLA, chemokine receptors and MHC and the presence of co-infection(s)/opportunistic infections. The viral "set point" is an important determinant of how the disease progresses in a certain individual and is determined between 6 and 12 months after the acute infection and is established by the host's immune response. The virus largely spreads to the lymphoid tissue and replicates in the tissue resulting in the destruction of HIV infected as well as bystander CD4 T cells through various immune mechanisms (Finkel et al. 1995). The lymphoid architecture now becomes extremely disrupted because of this increased replication and infection of the various host cells. The host immune response gets downregulated and the plasma virus burden gets upregulated which causes the disease progression towards AIDS. The chronic infection phase may last for 7–10 years and may vary in different individuals.

18.12 Acquired Immune Deficiency Syndrome (AIDS)

This is the last and most critical stage of HIV infection. At this stage, HIV has completely damaged the immune system, and the patient becomes immune-compromised along with the activation of the opportunistic infections. AIDS may be defined in an infected individual when the CD4 count becomes less than 200 cells/mm^3. The most common symptoms of AIDS include chills, fever, sweats, swollen lymph glands, weakness and unexplained weight loss. At this stage, the host has a high viral load and becomes very infectious (https://www.cdc.gov/actagainstaids/basics/whatishiv.html).

This stage is particularly characterized by increased levels of plasma viral load. The downregulation of the immune system occurs because of the impaired function of the T lymphocytes and other cells of the immune system like dendritic cells (DCs), natural killer (NK) cells and macrophages. There is immense immune suppression leading to opportunistic infections and malignancies. The CD4 count drops down and keeps declining progressively. This stage may last for 1–3 years and ultimately causes the death of the infected individuals (National AIDS Control Organization Manual 2007).

18.13 Disease Monitoring and Treatment

The detection of HIV infection is characterized by active viral replication leading to very high levels of viral antigens and viral proteins such as p24 being found in the HIV-infected patients (Daar et al. 1991). Subsequent to the activation of the immune system, the viraemia falls as the cytotoxic killing of the infected cells leads to a decline in the CD4+ T cell counts. Although the amount of virus in blood falls with the activated immune response, yet the replication continues steadily in the lymphoid organs. Subsequently there is a gradual decrease in the CD4 cell count; the rate of this decline varies in different individuals based on many factors. Hence monitoring of CD4 cell count and HIV-1 viral load are the two important parameters of disease progression in HIV-infected individuals.

The CD4 cell count is an immunological indicator and is currently used to assess the response to antiretroviral treatment. The status of CD4 cell reconstitution is an important indicator for first-line failure cases which is further confirmed by viral load estimation. However viral load is now considered a better indicator of treatment failure as it reflects a change at an early time point, so monitoring by sequential viral load estimation is being considered to be implemented soon as a national program for monitoring the progression of HIV disease and response to treatment. As per latest policy of "test and treat", all the HIV-infected individuals are supposed to be started with the antiretroviral treatment (ART) irrespective of the CD4 count (WHO 2016).

The ART regimen as per the national guidelines is given in Table 18.1 (National AIDS Control Organization Manual 2013).

18.14 Summary

HIV infection primarily causes the killing of the CD4-positive cells causing immune suppression and increased susceptibility to opportunistic infections. The disease progression is monitored by CD4 cell counts and plasma viral loads in the blood of the infected patients. Opportunistic infections (OIs) lead to increased mortality in the patients, and tuberculosis accounts for the most common OI. Other commonly reported OIs are candidiasis, cryptosporidiosis, toxoplasmosis and pneumocystis pneumonia. However early detection, diagnosis and appropriate treatment of OIs

Table 18.1 Antiretroviral therapy guidelines for HIV-infected adults and adolescents (Reproduced from NACO Manual 2013)

Regimen I	Zidovudine + Lamivudine + Nevirapine	First-line regimen for patients with Hb ≥9 gm/dl and not on concomitant ATT
Regimen I (a)	Tenofovir + Lamivudine + Nevirapine	First-line regimen for patients with Hb <9 gm/dl and not on concomitant ATT
Regimen II	Zidovudine + Lamivudine + Efavirenz	First line Regimen for patients with Hb ≥9 gm/dl and on concomitant ATT
Regimen II (a)	Tenofovir + Lamivudine + Efavirenz	First-line regimen for patients with Hb <gm/dl and on concomitant ATT first line for all patients with hepatitis B and/ or hepatitis C co-infection First-line regimen for pregnant women, with no exposure to sd-NVP in the past
Regimen III	Zidovudine + Lamivudine + Atazanavir/ ritonavir	Regimen for patients on AZT Containing first line regimen, who develop toxicity to both NVP and EFV Also Second line regimen for those who are on TDF containing first line regimen if Hb≥9 gm/dl
Regimen III (a)	Zidovudine + Lamivudine + Lopinavir/ ritonavir	For patients of regimen III who develop severe atazanavir toxicity, first-line regimen for patients with HIV-2 infection with Hb ≥9 gm/dl
Regimen IV	Tenofovir + Lamivudine+ Atazanavir/ ritonavir	Second-line regimen for those who are on AZT/d4T containing regimen in the first line and also for patients on TDF containing first-line regimen who develop toxicity to both NVP and EFV
Regimen IV (a)	Tenofovir + Lamivudine+ Lopinavir/ ritonavir	For patients on regimen IV who develop severe atazanavir toxicity, first-line regimen for patient with HIV-2 infection with Hb <9 gm/dl and first-line regimen for all women exposed to sd-NVP in the past
Regimen V	Stavudine+ Lamivudine+ Atazanavir/ ritonavir	Second line for those who are on TDF containing regimen in the first line if Hb <9 gm/dl
Regimen V (a)	Stavudine+ Lamivudine+ Lopinavir/ ritonavir	For patients on regimen V who develop severe atazanavir toxicity

can slow down disease progression. The antiretroviral treatment leads to decrease in the viral loads and increase in the CD4 T lymphocyte count in the infected individuals. Strict adherence to treatment can significantly improve the quality of life of the patients infected with HIV.

References

Alimonti JB, Blake Ball T, Fowke KR (2003) Mechanisms of CD4+ T lymphocyte cell death in human immunodeficiency virus infection and AIDS. J Gen Virol 84:1649–1661
Daar ES, Moudgil T, Meyer RD, Ho DD (1991) Transient high levels of viremia in patients with primary human immunodeficiency virus type 1 infection. N Engl J Med 324:961–964
Finkel TH, Tudor-Williams G, Banda NK, Cotton MF, Curiel T, Monks C et al (1995) Apoptosis occurs predominantly in bystander cells and not in productively infected cells of HIV- and SIV-infected lymph nodes. Nat Med 1:129–134
Gayle HD, Hill GL (2001) Global impact of human immunodeficiency virus and AIDS. Clin Microbiol Rev 14:327–335
https://www.cdc.gov/actagainstaids/basics/whatishiv.html
Lal S (2001) Surveillance of HIV/AIDS in India (editorial). Indian J Community Med 27:3–9
National AIDS Control Organization (2007) Manual for quality standards for HIV Testing Laboratory India
National AIDS Control Organization (2013) Antiretroviral therapy guidelines for HIV-Infected Adults and Adolescents India
Paranjape RS (2005) Immunopathogenesis of HIV infection. Indian J Med Res 121:240–255
Steinbrook R (2007) HIV in India-a complex epidemic. N Engl J Med 356:1089–1093
UNAIDS Global AIDS Update (2016)
Walker CM, Moody DJ, Stites DP, Levy JA (1986) CD8+ lymphocytes can control HIV infection in vitro by suppressing virus replication. Science 234:1563–1566
World Health Organization (2016) Progress report

Chapter 19
Chikungunya Fever: Where Are We Today?

Ritu Garg and Varsha Gupta

Abstract Chikungunya fever is a mosquito-borne viral disease caused by Chikungunya virus. Chikungunya virus (CHIKV) is an RNA virus belonging to family *Togaviridae*. Mosquitoes primarily responsible for transmission of Chikungunya are *Aedes aegypti*, *A. albopictus* and *A. polynesiensis*, which breed in clean water collections like in tanks, disposable items and other scrap material from domestic and peri-domestic sites. It is a re-emerging viral disease characterized by sudden onset of fever with severe arthralgia followed by rash and constitutional symptoms lasting for 1–7 days. It is more or less a self-limiting disease, but long-term sequelae of the disease like persisting arthralgia/arthritis (arthralgia/joint stiffness plus joint swelling) are the most frequently encountered. In India several outbreaks were documented during 1963–1973. After then, no outbreak was reported between the period of 1973 and 2005 from most parts of the world, except for a few sporadic cases occurring in various parts of the world including India. After dormancy of almost three decades, Chikungunya virus re-emerged in India in the states of Karnataka, Andhra Pradesh, Maharashtra, Tamil Nadu and Madhya Pradesh since December 2005. Recently in 2016 a big upsurge due to Chikungunya was reported from all over India including the capital city of Delhi and other states and UTs. Mumbai reported a 12.5% seroprevalence rate in 2016. Laboratory diagnosis is very essential to validate the clinical diagnosis of suspected cases which can be done by virus isolation, serological tests and molecular methods by polymerase chain reaction. By establishing the exact diagnosis, specific public health measures can be initiated timely. Appropriate surveillance is, thus, compulsory to minimize re-emergence and in controlling the future outbreaks.

Keywords *Aedes aegypti* · Chikungunya virus · Mosquitoes · Re-emergence

---

R. Garg
Department of Microbiology, Maharishi Markandeshwar Institute of Medical Sciences & Research, Mullana, India

V. Gupta (✉)
Department of Microbiology, Government Medical College Hospital, Chandigarh, India

P. P. Singh (ed.), *Infectious Diseases and Your Health*,
https://doi.org/10.1007/978-981-13-1577-0_19

19.1 Introduction

Chikungunya fever is a mosquito-borne viral disease caused by Chikungunya virus. Chikungunya virus is an enveloped, spherical and single-stranded positive-sense RNA alpha virus belonging to family *Togaviridae*. The genome size is approximately 12 kb and cleaved into four nonstructural proteins and five structural proteins (C, E3, E2, 6 K and E1). The glycoproteins E1 and E2 both play an important role in viral replication. The E1 glycoprotein is important for membrane fusion, and the E2 glycoprotein allows the virus to enter the cell through endocytosis (Wahid et al. 2017; Ananthanarayan and Paniker 2013).

19.2 Vector Responsible for Chikungunya

Mosquitoes primarily responsible for transmission of Chikungunya are *Aedes aegypti*, *A. albopictus* and *A. polynesiensis*, which breed in clean water collections like in tanks, disposable items and other scrap material from domestic and peri-domestic sites. These mosquitoes can also breed in natural habitats like tree holes and plantations. Chikungunya transmission is linked to rainfall and temperature like in case of dengue. A rise in the number of cases has been observed in the recent years during monsoon and post-monsoon periods. During post-monsoon period, high vector density amplifies virus transmission. In India, the main vector for transmission of Chikungunya is *A. aegypti*, but in some areas *A. albopictus* has also been found. These mosquitoes are day biters, and their flight range is also less (NVBCP 2016).

19.3 Genotypes of Chikungunya

There are three genotypes of Chikungunya virus:

1. West African genotype
2. East/Central/South African (ECSA) genotype
3. Asian genotypes.

Most of the Indian cases before 1973 were due to Asian genotypes of E1 gene. It was prevalent in Thailand, Malaysia and Indonesia. However, Reunion outbreaks during 2006 was caused due to mutated strain which was closely related to East/Central/South African genotype from Kenya and is responsible for most of the current outbreaks in India as well as in other parts of the world (NVBCP 2016; Sastry and Bhat 2016).

19.4 History of Chikungunya

The name Chikungunya was derived from the "Makonde" word "Kungunyala" (Makonde is a local language spoken in Tanzania) meaning that which bends up or gets folded which is mention to the stooped posture developed as a result of the severe joint pain that occurs during the course of illness (Sastry and Bhat 2016).

The first case of Chikungunya virus was reported from serum of a febrile patient during the dengue epidemic that occurred in Newala District, Tanzania, Africa, in 1953. It was then introduced to Asia and Bangkok, Thailand, in 1958 and caused several outbreaks in various African and Southeast Asian countries. Viral transmission continued till 1964. In India several outbreaks were documented during 1963–1973. In Kolkata it was in 1963 and in South India, i.e. Puducherry, Chennai-Vellore, in 1964. Afterwards, a small outbreak was reported from Barsi, Solapur District, Maharashtra, in 1973. After that, no outbreak was reported between the period of 1973 and 2005 from most parts of the world, except for a few sporadic cases occurring in various parts of the world including India. After dormancy of almost three decades, Chikungunya virus re-emerged in India in the states of Karnataka, Andhra Pradesh, Maharashtra, Tamil Nadu and Madhya Pradesh since December 2005. There were cases reported from Gujarat, Rajasthan and Kerala. The disease was thought to be dengue initially, but the debilitating arthralgia raised the doubt, and ultimately in January 2006, the outbreak was confirmed as Chikungunya with the help of laboratory findings. Consequently, WHO also confirmed re-occurrence of Chikungunya fever in India. The outbreak of Chikungunya fever had an attack rate of 4–45% (NVBCP 2016; Sastry and Bhat 2016).

19.5 Reasons for Re-emergence

Re-emergence of Chikungunya in 2005 is believed to occur due to a novel mutation in virus, i.e. alanine in the 226 position of E1 glycoprotein gene is replaced by valine (E1-A226V), and therefore due to this new mutation, there was shift of vector preference. With this mutation, now Chikungunya virus was found to be 100 times more infective to *A. albopictus* than to *A. aegypti* (Sastry and Bhat 2016; Singh et al. 2012).

19.6 Past and Present Status of Chikungunya in India

The first outbreak occurred in Kolkata in 1963, followed by a number of other outbreaks in Maharashtra, Andhra Pradesh, Tamil Nadu and Barsi from 1964 to 1973. The virus reappeared in 2006 and badly hit 13 Indian states including Gujarat, Kerala, Tamil Nadu, Andhra Pradesh, Madhya Pradesh, Maharashtra and Karnataka.

About 2994 individuals out of a total 60,777 suspected Chikungunya cases lost their lives. Lakshadweep experienced a small outbreak in 2007. One lac people were again infected with CHIKV in 2008 in Kerala. During subsequent years, several other large outbreaks occurred in Maharashtra, Andaman and Nicobar Islands, West Bengal, Orissa, Rajasthan and Puducherry. In 2010, the seroprevalence rate was 9.91% in the National Capital Region of India. In the current scenario, Chikungunya is endemic in a number of states in India. Tamil Nadu, Andhra Pradesh and West Bengal have reported higher number of cases. Karnataka accounted for the maximum number of cases in the year 2013 and 2014. A total of 27,553 clinically suspected cases of Chikungunya have been reported from 22 states and 3 UTs during 2015. Recently in 2016 a big upsurge due to Chikungunya was reported from all over India including the capital city of Delhi and other states and UTs. Mumbai reported a 12.5% seroprevalence rate in 2016 (Wahid et al. 2017; NVBCP 2016; Sastry and Bhat 2016; Chabbra et al. 2008).

19.7 Transmission Cycle

Chikungunya virus is most commonly transmitted to humans through the bite of infected mosquitoes of genus *Aedes*. *A. aegypti* and *A. albopictus* bite during daytime and mostly in the morning and late afternoon. Rare routes of transmission are by vertical transmission or blood transmission. Vertical transmission of CHIKV has been observed in a recent outbreak at La Reunion Island (Robillard et al. 2006). CHIKV is maintained in the environment through urban cycle and sylvatic transmission cycle.

Urban cycle: Transmission occurs between humans and *Aedes aegypti* mosquito.

Sylvatic cycle: It occurs in African forests. Monkeys act as reservoir, and forest species of *Aedes* such as *Aedes furcifer* and *Aedes taylori* serve as vectors (Sastry and Bhat 2016; Thiberville et al. 2013).

19.8 Clinical Manifestations of Chikungunya

Chikungunya virus causes an acute febrile illness of abrupt onset signalled by fever and severe arthralgia followed by other constitutional symptoms like headache and myalgia and rash lasting for a period of 1–7 days. Incubation period is usually 2–3 days with a range of 1–12 days. Clinical manifestations may be mild to moderate and severe, and most of the symptoms may subside within 3 weeks from the onset of illness. Some of the symptoms may persist for 3 months or even longer. Generally 10–15% of the patients suffered from severe Chikungunya that progresses to chronic phase.

Abrupt rise of fever repeatedly touching 39–40 °C is accompanied by intermittent shivering or chills. The temperature may remit for 1–2 days, after the gap of 4–10 days, resulting in saddle back fever curve, i.e. 3–5 days of high temperature are followed by 3–5 days of low temperature, occurring in a number of waves before resolution. Arthralgia is polyarticular, which is migratory and primarily affects the small joints of the hands, wrists, ankles and feet with lesser involvement of larger joints like knee and shoulder joints. Patients in acute stage complain of intense pain on movement, so patients prefer to lie still in the attitude of flexion, i.e. "walking bent over". Pain on movement is worse in the morning which gets better with mild physical activity, but it is exacerbated by strenuous exercise. Joints with previous trauma or degenerating changes are more prone to early or more significant involvement. Transient maculopapular rash is seen in up to 50% of patients. The trunk and limbs are commonly involved, but the face, palms and soles may also show lesions. Petechiae occur either alone or in association with rash. Photophobia, retro-orbital pain and conjunctival redness are present in some patients (Sastry and Bhat 2016; Chabbra et al. 2008).

Chikungunya in Children Clinical manifestations in children can vary from asymptomatic to severe. Children may have lymphadenopathy, swelling of eyelids, minor haemorrhages and pharyngitis. Infrequent clinical features include seizures, altered sensorium, acute flaccid paralysis and blindness due to retrobulbar neuritis. Symptoms of Chikungunya in infants can be lethargy, irritability, excessive crying and sometimes watery stools in addition to fever. Although rare, the infection can result in meningoencephalitis especially in newborns.

Chikungunya in Elderly Chikungunya is more severe and dangerous and takes longer time for recovery. This is because of weak immune system of the body. There could be cerebral problems like dementia, paralysis and kidney disorders. Patients with co-morbidities may have more complications and psychological sequelae than patients belonging to other age groups.

Chikungunya-Dengue Coinfection The presence of dual infection of Chikungunya and dengue has been reported from various states of India. In these reports both the viruses were reported simultaneously from the same sera, and both the viruses can cause the disease at the same time. Symptoms are often confusing with that of dengue. In general, Chikungunya is less severe and less acute, and haemorrhages are rare compared to dengue. This is not very unusual because both the viruses are arboviruses and also transmitted by same vector *Aedes aegypti* mosquitoes. Clinical signs and symptoms of Chikungunya and dengue are given in Table 19.1 (Sastry and Bhat 2016; Chabbra et al. 2008).

High-Risk Group Chikungunya infections worsened and are also accompanied with adverse outcomes of the disease in patients with co-morbidities like hypertension, diabetes mellitus, heart disease and pregnancy and also in patients with co-infections of tuberculosis, enteric fever, HIV, pneumonia and malaria (Sastry and Bhat 2016; Chabbra et al. 2008).

Table 19.1 Clinical signs and symptoms of Chikungunya and dengue

Symptoms	Chikungunya	Dengue
Fever	Common	Common
Polyarthritis	Common	Rare
Rash	Appears in 1–4 days	Appears in 3–7 days
Myalgia	Possible	Common
Leucopenia	Rare	Common
Thrombocytopenia	Rare	Common
Retro-orbital pain	Rare	Common
Hypotension	Possible	Common
Haemorrhages	Rare	Common

Sequelae of Chikungunya Generally Chikungunya is a self-limiting disease, but long-term sequelae of the disease in terms of persisting signs and symptoms have been known for long. In old patients the presence of osteoarthritis and pre-existing severe joint pains increases the sequelae of Chikungunya fever. The most frequent problem encountered is persisting arthralgia/arthritis, i.e. arthralgia/joint stiffness plus joint swelling, which is associated too with female gender, older age, some co-morbidities and the severity of the acute phase. Other sequelae according to literature could be alopecia and depression. Quality of life can be reduced for months to years after the acute phase of Chikungunya (Van Aalst et al. 2017).

19.9 Laboratory Diagnosis of Chikungunya Fever

Chikungunya fever should be suspected when an epidemic occurs with the characteristics of sudden onset of fever, arthralgia and myalgia, with or without rash. Therapeutic actions for control of the diseases and symptomatic treatment of the suspected Chikungunya fever cases should be started immediately on the basis of clinical diagnosis of the disease. The laboratory diagnosis helps to validate the clinical diagnosis of suspected cases and is very essential because clinical manifestations of Chikungunya fever look a lot like dengue fever and other arboviruses. Dengue fever and Chikungunya both causes outbreaks at a same time in India. Laboratory investigations should be done to establish the exact diagnosis; depending upon the diagnosis, specific public health measures can be initiated timely.

Definitive diagnosis of Chikungunya can be done by virus isolation, serological tests and molecular methods by polymerase chain reaction.

Sample Collection, Storage and Transportation Samples are usually blood or serum and CSF in meningoencephalitic patients. Blood sample should be collected within the 5 days of the onset of illness. Convalescent or paired sample should be collected after 10–14 days after the first sample. Then transport the sample to the laboratory at 2–8 °C as soon as possible. Do not freeze the sample as haemolysis may interfere with serological results. If delay of 24 h is expected for transportation, then serum should be separated from red blood cells and stored frozen. For virus isolation and RT-PCR, samples should be transported to reference laboratory in cold or preferably frozen within 48 h of collection of sample.

Virus Isolation It provides the most definitive diagnosis. But it is time consuming, and it should be carried out in biosafety level III laboratories to reduce the risk of viral transmission.

Serological diagnosis: This approach is the most commonly used. Seroconversion is demonstrated by comparing acute and convalescent phase sera in the haemagglutination inhibition, serum neutralization or complement fixation test.

Serum Antibody Detection IgM appears after 4 days of infection and lasts for 3 months, while IgG appears late, i.e. after 2 weeks, and lasts for years. So detection of Chikungunya-specific IgM levels or fourfold rise in IgG titre between acute and convalescent sera is more significant.

Immunoglobulin M antibody (IgM) capture enzyme-linked immunosorbent assay (MAC-ELISA): It is the best format available showing excellent sensitivity and specificity with very little cross-reactivity with other alphaviruses and dengue.

In India, Chikungunya IgM ELISA test kits are provided to the identified laboratories (Apex Referral Laboratories) through the National Institute of Virology, Pune, since 2007. NIV, Pune, is the only institute in the country which prepares reagents for laboratory diagnosis of CHIKV. Cost is borne by the government of India.

RT-PCR: Reverse transcriptase (RT)-PCR technique using nested primer pairs is used to amplify several Chikungunya-specific genes from whole blood; it is the best method for early diagnosis (0–7 days). RT-PCR can also be used to quantify the viral load in the blood. Using RT-PCR, diagnostic results can be available in 1–2 days.

Biological markers: IL-1 β and IL-6 are increased, and RANTES (regulated on activation, normal T cell expressed and secreted) levels are decreased in Chikungunya virus infection. Leucopenia with leucocyte predominance is the common observation. Thrombocytopenia is rare. ESR and CRP are increased during acute period. Some of the patients may be rheumatoid factor positive during and after clinical episode.

NIV, Pune, a WHO collaborating centre for arboviral diseases, is involved in outbreak investigations, diagnosis and preparation of reagents for diagnosis of arboviral infections. The National Institute of Communicable Diseases, Delhi, has assisted state governments in the investigation of the CHIKV through multidisciplinary teams. NICD laboratory is fully equipped for confirmation of diagnosis. For improvement of trained manpower, training courses on diagnosis of arboviral infections are conducted as WHO biennium activity (NVBCP 2016; Sastry and Bhat 2016; Kawle et al. 2017).

Management The disease is commonly self-limiting as there is no specific treatment against Chikungunya virus. The treatment of Chikungunya is purely symptomatic. Supportive care with rest and healthy nutrition is indicated during acute phase of the disease. Pain reliever, antipyretics and fluid supplementation are important aspects of management. Movement and mild exercise are likely to improve stiffness and morning arthralgia. Heavy exercise may exacerbate rheumatic symptoms. As nonaspirin and non-steroidal anti-inflammatory drugs (NSAID) are commonly recommended, in unresolved arthritis which is refractory to NSAID, chloroquine at 250 mg was evidenced to be useful. In high-risk group, patients' proper management of co-morbidities and co-infections is advised. Self-medication mainly antibiotics, steroids, other painkillers, aspirin and hot fomentation is to be avoided as in acute stages it can worsen the joint symptom. Infective persons should be protected from further mosquito exposure so that they cannot contribute to the transmission cycle. Promptly refer the case to higher centres when fever persisting for more than 5 days, altered sensorium and any bleeding under the skin or through any orifice are present. All the high-risk patients should be managed at higher centres (NVBCP 2016; Chabbra et al. 2008).

Public Health Measures No vaccine and specific treatment are available against Chikungunya infection. Vector control is of utmost importance in controlling or preventing CHIKV transmission. *Aedes aegypti* and *Aedes albopictus* are vectors for Chikungunya which breed in clean water containers (metallic, plastic, rubber, cement, earthen materials, etc.). Elimination of breeding sites or source reduction is an effective method of control.

19.10 Control of Mosquito Breeding (Vector Control)

The *Aedes aegypti* and *Aedes albopictus* mosquito should be the foremost target of control activities. It will be possible only with vigorous involvement of community to keep water storage containers free of culprit mosquitoes and also to eliminate the other breeding places of mosquitoes in and around houses and lodgings:

- Take away stagnant water from all scrap items lying around household areas like old tyres, tin cans, buckets, drums, bottles and coolers or from other mosquito breeding places.
- Change the water frequently in plant pots, at least twice a week. Pets' water bowls should be emptied daily.
- Weeds and tall grass should be cut short as adult mosquitoes look for these shady places to rest during the hot daylight hours.
- The organophosphorus insecticide, Abate, is used as a larvicide. It can prevent breeding for up to 3 months when applied on sand granules, does not harm man and does not affect the taste of water. Anti-larval measures can prevent an epidemic, but do not give immediate results when an epidemic has already broken out. In such cases, anti-larval measures alone can bring about a rapid interruption of transmission.
- Larvivorous fish (e.g. gambusia, guppy) need to be introduced in aquaria, garden pools, etc.

- Aerosol spray of ultra-low volume quantities of malathion or Sumithion (250 ml/ha) has been found to be effective in interrupting transmission and stopping epidemics.

19.11 Protection from Mosquito Bites

- Insecticide-treated mosquito nets should be used even while sleeping during daytime.
- Insecticide spraying should be done to kill mosquitoes. Fogging with 2% pyrethrum space spray is strongly recommended in high-risk villages/wards where clustering of cases has been reported.
- Use an insect repellent with registered active ingredient on exposed skin.
- Wear pants and clothes with long sleeves.
- Use protected screens on windows and doors to keep mosquitoes out.

So the high prevalence of Chikungunya fever in India suggests its continuance as a major health threat in the present scenario. A multi-diagnostic approach is proposed for the timely detection of CHIKV infection and other vector-borne diseases which should be given utmost attention that will in turn help in the prediction, prevention and control of awaiting and sporadic outbreaks in developing countries. Proper surveillance is, therefore, necessary to minimize re-emergence and in controlling these future and sporadic outbreaks.

References

Ananthanarayan A, Paniker CKJ. (2013) Arboviruses. In: Textbook of microbiology, 9th edn. Universities Press, p 519

Chabbra M, Mittal V, Bhattacharya D, Rana UVS, Lal S (2008) Chikungunya fever: a re-emerging viral fever. Indian J Med Microbiol 26:5–12

Kawle AP, Nayak AR, Bhullar SS, Borkar SR, Patankar SD, Daginawala HF et al (2017) J Vector Borne Dis 54:35–43

NVBCP (2016) National guidelines for clinical management of Chikungunya

Robillard PY, Boumahni B, Gerardin P, Michault A, Fourmaintraux A, Shuffenceker I et al (2006) Vertical maternal fetal transmission of the chikungunya virus. Presse Med 35:785–788

Sastry AS, Bhat SK. (2016) Arboviruses. In: Essentials of medical microbiology, 1st edn. JAYPEE Publications, pp 486–487

Singh P, Mittal V, Rizvi AM, Bhattacharya D, Chabbra M, Rawat DS et al (2012) Northward movement of East Central South African genotype of chikungunya virus causing an epidemic between 2006 – 2010 in India. J Infect Dev Ctries 6:563–571

Thiberville SD, Moyen N, Maguiraga LD, Nougairede A, Gould EA, Roques P et al (2013) Chikungunya fever: epidemiology, clinical syndrome, pathogenesis and therapy. Antivir Res 99:345–370

Van Aalst M, Nelen CM, Goorhuis A, Stijnis C, Grobusch MP (2017) Long-term sequelae of chikungunya virus disease: a systematic review. Travel Med Infect Dis 15:8–22

Wahid V, Ali A, Rafique S, Idrees M (2017) Global expansion of chikungunya virus: mapping the 64 year history. Int J Infect Dis 58:69–76

Chapter 20
Epigenetics and Infectious Pathogens: Interactions, Ploy and Perspectives

Nitu Saha and Raghuvir Singh Tomar

Abstract Since the inception of life, organisms have enormously diversified, from simple unicellular to complex multicellular. As many life forms coexist in an environment, they constantly interact with each other via beneficial or harmful interactions. Humans, being part of the same environment, have confronted many hostile organisms partly comprising of disease-causing pathogens. Pathogens use humans as host for completion of their life cycle or nutrition. Regardless of their purpose, pathogens inflict profound harm to health and well-being of humankind. To name a few, millions of people have died due to the havoc caused by diseases such as plague, tuberculosis, cholera, Spanish flu, Ebola, etc. Revolution in the field of medicine has offered multiple drugs/medications and vaccinations against these pathogens, but owing to their misusage and the increasing anthropogenic rooted climate change, multidrug-resistant species have evolved, capable of attacking more smartly on the host. It has now been established that many such harmful organisms induce epigenetic modifications in the host to suppress host immunity, maintain their own latency, etc. in the course of establishing themselves. To curb this problem, epigenetic modifiers have now been formulated into drugs. These drugs have demonstrated promising results, paving way towards novel cure. This chapter is an attempt to introduce epigenetics and its modifications in host, mediated by pathogens, with an emphasis on bacteria and viruses. Finally, it gives an overview of different novel epigenetic approaches to combat these pathogens.

Keywords Epigenetics · Epigenetic drugs · Host · Interactions · Pathogen

N. Saha · R. S. Tomar (✉)
Laboratory of Chromatin Biology, Department of Biological Sciences, Indian Institute of Science Education and Research, IISER Bhopal, Bhopal, MP, India
e-mail: nitu16@iiserb.ac.in; rst@iiserb.ac.in

P. P. Singh (ed.), *Infectious Diseases and Your Health*,
https://doi.org/10.1007/978-981-13-1577-0_20

20.1 Introduction

The ever-increasing human population has led to the present scenario where earth is inhabited by more than seven billion people. Needless to say that such a growth has led to scarcity of resources for consumption by the humans. This demand has led to ruthless exploitation of environment, changing climate patterns, and more urbanization. These anthropogenic changes have caused an increase in the number of infectious diseases by creating opportunities for climate-dependent active infiltration of pathogens (Bloom 2011; Vitousek et al. 1997).

Infectious diseases have had dramatic influence on human population, and enough evidences depict that from time to time they have caused massive wipeout of individuals. For example, millions of people yielded to Spanish flu in the early twentieth century. Now, with increased facility of medications and lifestyle, the death trolls have decreased, but the pathogens are finding new ways to evade and re-emerge (Johnson and Mueller 2002).

The emerging diseases extend to a bigger geographical area, increasing the number of incidences and infecting new populations. Thus, these diseases pose major threat to humans. Of all the pathogens capable of causing infectious diseases, viruses have succeeded the most. They have evolved to exhibit high mutation and replication rates (Taylor et al. 2001; Medicine IO 2015). Next in the line are bacteria known to cause havoc and re-emerging episodes of disease in a given area, for example, the recurrence of plague and cholera in a country like India. Bacteria exhibit high mutation rates like viruses and are also capable of lateral transfer of genetic material (Lan and Reeves 1996; Ochman et al. 2000). Below is the description of different pathogens and the diseases they cause:

20.1.1 Bacteria

Bacteria [light micrograph of bacteria; refer to Fig. 20.1, source: (Dumler et al. 2005)] are known to inhabit some of the most diverse niches around the world. They appeared long back in the geological time scale. This long history is responsible for the diverse environment they have confronted over time. Thus, they have evolved with a remarkable capacity to respond as per the environmental needs (Persat et al. 2015). Bacteria are classified into at least two classes based on Gram's staining, viz. gram-positive bacteria and gram-negative bacteria. Gram-negative bacteria are particularly notorious and are known to cause variety of diseases (Anwar and Choi 2014).

A description of few bacteria and the associated diseases are as follows:

Mycobacterium tuberculosis: The bacterium is an obligate pathogen responsible for tuberculosis. Infection is manifested in the form of coughing of blood and severe wasting. It may also infect the central nervous system causing meningitis, the digestive tract, the urogenital tract, or the cutaneous region causing lupus vul-

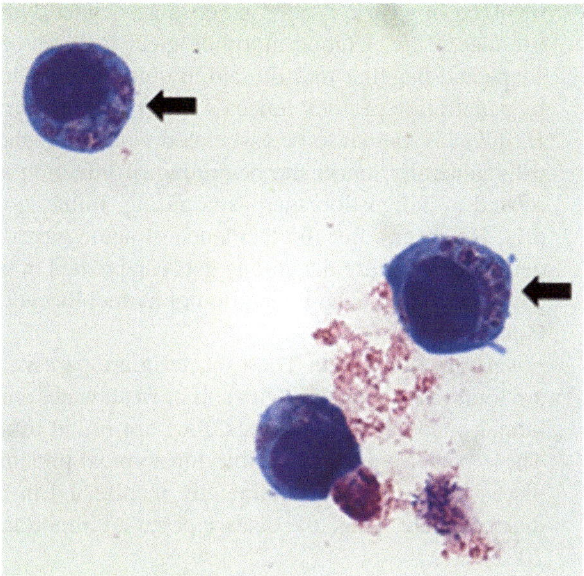

Fig. 20.1 An image of HL-60 (human promyelocytic) cell line containing A. phagocytophilum indicated via arrows © 2014 Kim KH. Published in Kim et al., 2014. Available from: https://dx.doi.org/10.3201/eid2010.131680 (Kim et al., 2014)

garis. The disease is still a major concern of different health organizations (Smith 2003).

Neisseria meningitidis: These are gram-negative, fastidious, and encapsulated bacteria. These organisms are responsible for septicaemia and meningitis in humans. Severity of disease is dependent on virulence of infecting organism, susceptibility factors of host, and the environmental conditions. The disease can lead to potent problems in hearing, behaviour, education, limb loss, and cognitive dysfunction. Disease may be asymptomatic or may cause local inflammation. Many reports of this disease have been reported in Asia, Latin America, Africa, etc. (Rouphael and Stephens 2012).

Salmonella enterica: These rod-shaped, gram-negative bacteria account for major infections in homeothermic organisms. One of the most distinguishing features is the abundance of different serovars within *S. enterica* which define the subspecies. These serovars exhibit different host specificity and virulence which is probably dependent upon their differential gene expression. They mainly infect the intestinal tract of host organisms such as humans. The infected organisms excrete out these bacteria in faeces which are then carried by insects to water sources and into the organisms drinking it. Severity of infection in humans is dependent on the microbial load and host immunity level. Collectively, these serovars are responsible for both typhoid and non-typhoidal infections causing millions of death across the globe, so many measures are being taken by scientists to understand the basis of these infections (Porwollik et al. 2004; Andino and Hanning 2015).

Helicobacter pylori: It is a gram-negative bacterium responsible for chronic gastritis infection, affecting a large number of people all over the world. It is also involved in gastric cancer, mucosa-associated lymphoid tissue lymphoma, gastric ulcers, etc. Unique morphological features of these bacteria such as spiral shape and flagellar motility aid in quick movement towards the neutral pH area by penetration of thick mucus which provides favourable conditions for growth. *H. pylori* is known to be associated with acute and chronic gastritis. Acute gastritis generally marks the beginning of infection and is exhibited by symptoms related to indigestion such as vomiting, fullness, nausea, etc. and hypochlorhydria. However, after the incidence of acute gastritis, the bacterial colony might get cleared. In case the colony gets established in the host, it gives rise to chronic gastritis leading to augmentation of hypochlorhydria (Kusters et al. 2006; Garza-Gonzalez et al. 2014).

Legionella pneumophila: These bacteria are parasitic or commensal in nature and are found in association with soil or fresh water amoebae in nature. In the case of human-made aquatic habitats, they are found in the form of complex biofilms. These bacteria are responsible for atypical pneumonia known as Legionnaire's disease. This disease is clinically manifested in the form of headache, cough, diarrhoea, etc. Thus, the disease poses an important health problem (Fields et al. 2002).

Chlamydia trachomatis: These bacteria are ovoid or spherical in shape and are obligate intracellular pathogens. These are responsible for multiple inflammations like urethritis, cervicitis, and endometritis in both men and women. Additionally, it may also cause tubal factor infertility, pelvic inflammatory disease, ectopic pregnancy, etc. Though the disease is curable, it makes the infected person more prone to acquisition or transmission of HIV and contributes to the development of cancer in the cervix region. Varieties of symptoms are associated with the disease such as vaginal discharge, high fever, severe pain in the abdominal region, prolonged menstrual cycles, etc. (Malhotra et al. 2013).

Anaplasma phagocytophilum: It is the causal organism of human granulocytic anaplasmosis (HGA). The pathogen is known to employ many mechanisms and adaptations to infect its niche, the neutrophil cells present in human body. HGA is transmitted by infected ticks during their blood meal and is manifested by multiple symptoms such as leukopenia, myalgia, malaise, fever, headache, etc. The incidence of this disease is particularly high in the areas of Europe, California, New England, etc. *Anaplasma* is one of the few organisms which infect neutrophils, so understanding the mechanism of interactions between the two would aid in drug designing against this deadly disease (Dumler et al. 2005).

Ehrlichia chaffeensis: It causes human monocytic ehrlichiosis or human monocytotropic ehrlichiosis (HME). The bacterium is obligately intracellular in nature and exists in the form of two morphologically different reticulate and dense-cored forms. While the reticulate cells have nucleoid DNA fibrils and uniform distribution of ribosomes, the dense-cored cells have both their nucleoid DNA and ribosomes condensed in the centre. The bacterium is introduced into the human body as and when an infected tick bites. Though it mainly targets monocytes, there have been cases of infections in lymphocytes, segmented neutrophils, etc. Symptoms of the infection develop with time and include low back pain, gastrointestinal symptoms, vomiting, abdominal pain, nausea, malaise, etc. The disease has been found in regions of Oklahoma, Georgia, Arkansas, Maryland, etc. raising concern among the people. Thus, development of potent drugs against the disease is important (Paddock and Childs 2003).

Listeria monocytogenes: It is a gram-positive and facultative anaerobic bacterium causing a food-borne infection called listeriosis which might lead to meningitis and sepsis. Listeriosis is associated with the central nervous system leading to endocarditis, meningitis leading to malaise, vomiting, headache, etc. During pregnancy it has been shown to be associated with abortions. The bacterium is present in soil, water, air, sewage, faeces, etc. (Farber and Peterkin 1991).

Shigella flexneri: One of the most communicable bacterial dysenteries, shigellosis, is caused by the bacterium *S. flexneri*. The bacterium affects rectal and colonic epithelial cells and leads to chronic inflammation and destruction of the epithelia which is exhibited by severe pain in the abdomen and diarrhoea leading to bloody mucoid stool. The disease, if not treated, may further give rise to pneumonia, septicaemia, etc. The pathogen is very infectious, and the incidence of the disease increases due to malnutrition and poor sanitation in developing countries. Transmission of the disease is through faecal oral route and personal contact (Jennison and Verma 2004).

20.1.2 Viruses

These are simple, obligate intracellular parasites which are infectious in nature [electron micrograph of viruses, refer to Fig. 20.2a, b, source: (Barreto-Vieira and Barth 2015; Kawase 2013)]. The genome of viruses contains either RNA or

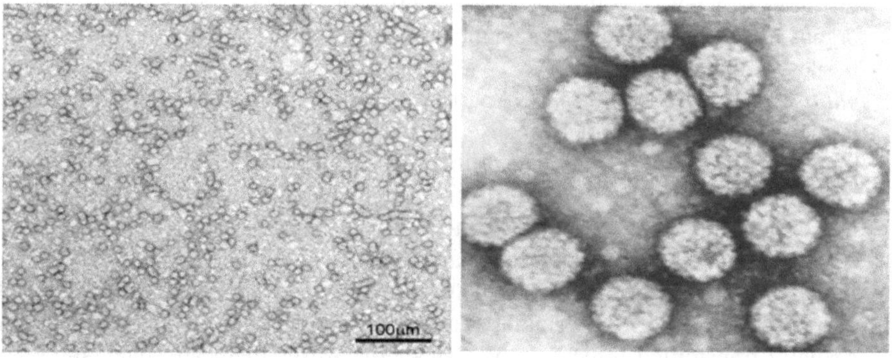

A B

Fig. 20.2 HBV as seen under electron microscope © 2015 ferreira Barreto-Vieira D, Barth OM. Published in (Barreto-Vieira and Barth, 2015) under CC BY 3.0 license. Available from: http://dx.doi.org/10.5772/60511 (**a**) HPV5 virions as seen under electron microscope © 2013 Kawase M. Published in (Kawase, 2013) under CC BY 3.0 license. Available from: http://dx.doi.org/10.5772/55340 (**b**)

DNA. The main function of the genome is to hijack the host cellular machinery to synthesize viral components. Once the components form, viruses assemble to form the progeny viruses called virions which break open the host cells to release and repeat the process of infecting new cells. As the viruses cannot multiply on their own, they invade host cells to grow and flourish themselves and in the process give rise to multiple diseases (Flint et al. 2009).

Description of few viruses and the associated diseases are as under:

Cytomegalovirus (CMV): It is one of the most prevalent causes of congenial viral infection. The herpesvirus is ubiquitously found and is transmitted through interpersonal contact via blood, genital secretions, breast milk, or urine. Infection of this virus is generally mildly symptomatic or asymptomatic. However, in case of immunocompromised hosts, it leads to morbidity, and premature babies exhibit multiple symptoms associated with respiratory status, septic appearance, etc. (Swanson and Schleiss 2013).

Kaposi's sarcoma herpesvirus/human herpesvirus-8 (KSHV/HHV8): The virus causes Kaposi's sarcoma (KS), the most common cancer in HIV-positive untreated patients. KS is a type of vascular tumour which gives rise to lesions. These KS lesions have infiltration of inflammatory agents. Endothelial cells infected by the virus exhibit defective vascularization or angiogenesis and aberrant differentiation (Gramolelli and Schulz 2015; Mesri et al. 2010).

Epstein-Barr virus (EBV): It is ubiquitously found virus and infects mostly during early childhood or infancy. Though infection during early childhood leads to non-specific or no symptoms at all, infection in the adult stage is associated with sore throat, fever, lymphadenopathy, etc. Major concern lies in the fact that the virus is associated with multiple lymphomas, carcinoma, etc. Chronic active EBV is prevalent across continents of South America and Asia and is associated with NK cells or T-cells, while in the Unites States, it is associated with B-cells. Chronic active EBV is characterized by alarmingly elevated antibodies in response to EBV infection or elevated level of the viral DNA, infiltration of organs by EBV-infected cells, and the presence of viral nucleic acid or protein in tissues (Cohen 2009).

Hepatitis B virus (HBV): It is a DNA virus but exhibits properties of retroviruses. It is the causal organism of liver disease and liver cancer. Infection may not be manifested or may be exhibited in the form of hepatitis or liver inflammation. The virus may also lead to other liver diseases like cirrhosis, chronic hepatitis, etc. Proper and correct diagnosis of this disease requires multiple biochemical, serological, and histological tests (Liang 2009).

Human papillomavirus (HPV): These DNA tumour viruses are associated with anal cancer, cervix cancer, and genital warts. HPV causes one of the most common sexually transmitted infections. Within these viruses there are many subtypes each categorized into high-risk or oncogenic types and low-risk or non-oncogenic types. High-risk HPV has been shown to be involved in cervical cancer and dysplasia. The low-risk types are involved in the formation of genital warts which generally cause morbidity or embarrassment among those who have it (Braaten and Laufer 2008).

Simian virus 40 (SV40): Originally the virus was isolated from rhesus monkey, but it gradually got introduced to the human system, and it is now suspected to be involved in causing cancer. Introduction of SV40 to humans occurred accidently during the administration of Salk and Sabin polio vaccines. These vaccine preparations were contaminated with the virus as they were obtained from primary kidney cell cultures of rhesus monkey which most of the time were SV40 infected. Thus, it could be possible that the viruses, over years, have become pathogen requiring humans as host (Garcea and Imperiale 2003; Vilchez and Butel 2004).

Foot-and-mouth disease virus (FMDV): The virus is from *Picornaviridae* family and is the causal agent for one of the most contagious diseases called the foot-and-mouth disease. Symptoms include the presence of vesicles on the foot and mouth of the cattle. This disease causes huge economic losses as the affected cattle and the ones nearby it are to be slaughtered to prevent any further infections in a given area. The virus is highly notorious as it exhibits high mutation rates and high levels of heterogeneity within a single host, so protection against one serotype does not confer resistance against all. Thus, drugs which can effectively eradicate the virus are to be designed using a novel approach (Salguero et al. 2005).

Thus, novel approaches are required to fight against these pathogens. A new ray of hope is now emerging with epigenetic studies of these pathogens and their interactions with different hosts thus paving way for novel drug designing.

In the 1950s, it was Conrad Waddington who had put forth meaning of the term 'epigenetics', but the history and idea of it dates back to the discovery of chromosomes by Flemming following which Muller performed experiments to produce a class of mutations in drosophila which depended on rearrangements in chromosomes (Berger et al. 2009; Holliday 2006). Now, decades later, we know that these epigenetic changes are responsible for regulating biological processes in a body. To sum up, epigenetics deals with the study of heritable changes in gene function without any change in actual DNA sequence. In other words, it supplies an additional layer of transcriptional regulation which aids in modulating the expression pattern of genes. Thus, it has multifaceted roles in gametogenesis, embryogenesis, reorganization of genome, and cell differentiation. Under the effect of non-coding RNAs (ncRNAs) and regulatory proteins, histone post-translational alteration and DNA methylation occur which aid in the rearrangement into heterochromatin, euchromatin, and compartmentalization of nucleus (Moosavi and Motevalizadeh Ardekani 2016). This burgeoning field has been providing greater insights into understanding of diseases or disorders associated with humans. Human genome has approximately 23,000 genes, the expression of which is tightly regulated based on the precise requirement of cells. Control over gene expression is achieved by wrapping of ~147 bp DNA around an octamer of histones containing two copies of H2A, H2B, H3, and H4 each. This unit is known as nucleosome, and many such units are packed

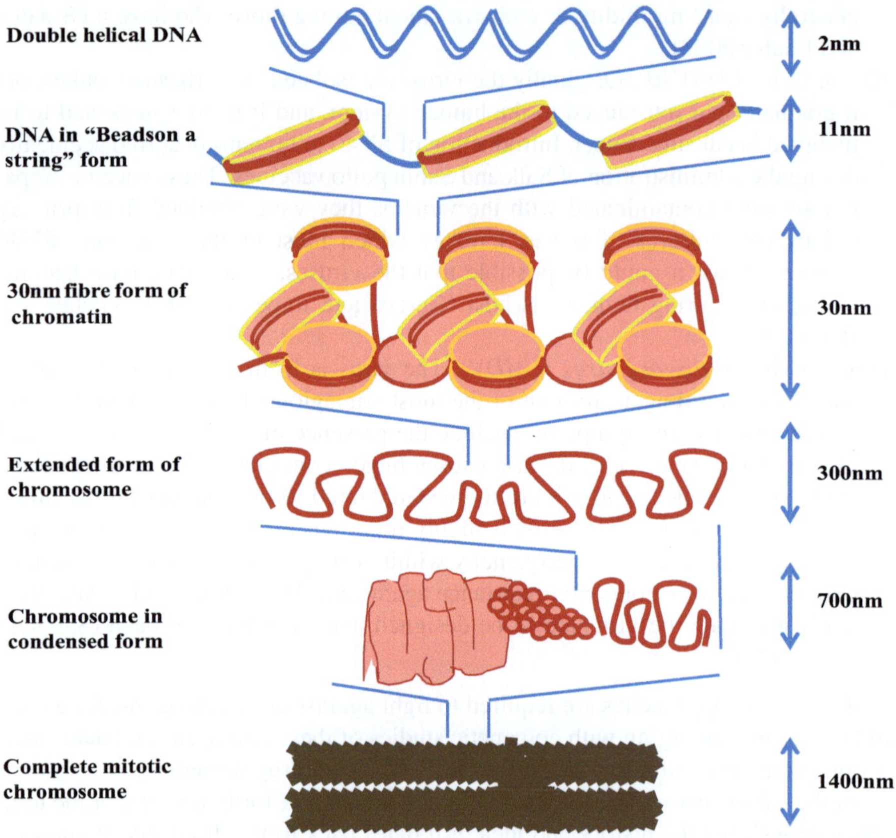

Fig. 20.3 Model depicting multiple levels of chromatin packing. Nucleosome forms the structural unit of chromatin which together gives an appearance of beads on a string. The nucleosomes associate together to form more compact solenoid structure via linker histones H1. The solenoid structure further compacts to frm looped dna alsp known as 700nm and ultimately 1400nm chromosome is formed. The DNA compresses about 10,000 times to form highly condensed structures

at multiple levels to form chromatin or chromosome. Nucleosome units arrange themselves in 'beads on a string fashion' which then form 30 nm chromatin and finally the condensed chromosome (Fig. 20.3; Kouzarides 2007; Alberts et al. 2013). Histone protein exhibits globular structure except the N-terminal tail of all and C-terminal tails of H2A and H2B which have undefined structure and protrude out of octamer to interact with DNA (Luger et al. 1997). Besides, the core histones, viz. H2A, H2B, H3, and H4, different variants of histones exist as well. These variants are involved in modulation of gene expression and are associated with particular chromatin forms (Buschbeck and Hake 2017; Brazel and Vernimmen 2016). For example, an H3 variant named CENP-A has been shown to be associated with a centromere-linked event during interphase (Figueroa et al. 1998). Another variant, H2A.Z, has been implicated in regulation of gene expression as it does not form condensed chromatin under in vitro conditions (Fan et al. 2002). Also, previous studies have suggested that macroH2A, a histone variant, is involved in inactivated X chromosome of mammalian female (Costanzi et al. 2000). H2AX, another histone variant, plays a rather important role in repair of DNA damage (detailed under phosphorylation). Finally, H3.3, a variant of H3 has been shown to be involved in plausible exchange with H3 to enable activation of transcription (Ahmad and Henikoff 2002). There are at least 60 different residues on histones which are known to be modified. Different modifications may define an outcome dictated by signalling within the cell (Jenuwein and Allis 2001). Also, a given residue may show different levels of modifications. For example, methylation on a histone residue could be mono (one), di (two), tri (three), etc., based on which activation or repression of the associated gene occurs. Certain common histone modifications have come to light based on global analyses done in yeast. The actively transcribing genes have the following common characteristics:

Acetylation is found at the 5′-end of the coding region and the promoter region. Initiation site is flanked by a couple of nucleosomes enriched in Htz1, a variant of H2A, and also has few hypoacetylated lysine residues. Coding region exhibits enrichment of trimethylation and three widely known methylations, viz. H3K4, H3K36, and H3K79 show specific pattern. Thus, these all characteristics point to a basic pattern of modifications necessary for the functioning of a cell. An understanding of all the histone modifications is therefore required. There are, however, technical limitations to the studies in detail. For example, presently, a method like ChIP on CHIP is used, but it cannot give information about modifications on different histones at a given time, a major limitation. Another method which can be harnessed involves the use of mass spectrometry, but it also has a major limitation as the protein in question needs to be digested. So, a modified method comprising of knowledge about the protein following its digestion might provide better insights into the global histone modifications. Next challenge is to get a dynamic picture of histone modifications. The proposed method would provide a static idea of global histone modifications, but these modifications are known to rapidly change under different stimuli. Thus, examination of global histone modifications can only be partially done. Even so, there are problems associated with that too, like availability of specific antibodies, the absence of proper controls as it is almost impossible in mammalian cells to create a mutation that would render the residue completely inactive, and last the inhibition of the binding to a histone residue due to a neighbouring residue which would lead to misleading results. With mass spectrometry the problem lies in the non-uniform coverage of peptides in different sections of a protein which would result in inaccuracy.

It is also difficult to pinpoint the fact that a given histone modification caused by an enzyme leads to a specific function as the histone-modifying enzymes are known to have multiple substrates. So, a function shown may take place via an unknown different molecule. Also, redundant pathways may exist in cellular milieu, masking the effect of inactivation of enzyme so that the function remains unaltered. A double check, therefore, is required by mutating the residue and showing the same effect on the function though it is not possible in a mammalian system as they have multiple genes for histone. Thus, it may be said that an absolute answer cannot be derived using present techniques but based on how rigorous experiments have been performed which a level of certainty can be reached.

Majority of histone modifications are deemed dynamic, i.e. modification on a residue by one enzyme is removed by another enzyme. However, for modifications of arginine methylation, any direct demethylating enzyme is not known; instead the removal of methylation seems to be tied with the deimination. It has also been found that specificity of a histone-modifying enzyme to a residue on nucleosome or free histones and the levels of modification of a residue as mono, di, and tri is dependent on other associated molecules or proteins (Kouzarides 2007; Steward et al. 2006).

20.2 Types of Epigenetic Modifications

Modification of histone leads to two things: First, it helps to 'open' the nucleosome by disrupting the interaction between histone and DNA. Second, it facilitates the binding of a group of molecules or occlusion of them from a chromatin. These molecules have varied enzymatic activities to bring about more changes in a chromatin according to an external or internal stimulus which in turn may result in gene activation due to an active chromatin or gene inactivation due to a condensed chromatin (Badeaux and Shi 2013). Thus, there is a need to bring different proteins in sequence of their involvement in multistep processes like transcription, repair, etc. (Kouzarides 2007).

The switch between condensed and active states of a chromatin requires epigenetic signatures like DNA methylation and histone modifications via enzymes called epigenetic modifiers. There are different classes of epigenetic modifiers which have been named on the basis of modifications they cause. These enzymes are as under:

20.2.1 DNA Methyltransferases (DNMT)

These enzymes are subdivided into de novo DNA methyltransferases which are enzymes involved in DNA methylation during embryogenesis there by establishing them, e.g. DNMT3A and DNMT3B, and maintenance DNA methyltransferases which are enzymes involved in copying methylation to newly replicated strand, e.g. DNMT1. DNMT1 is part of a complex which is involved in recognition of hemimethylated DNA followed by methyl group addition to the non-methylated daughter DNA strand during replication. Thus, methylation is maintained in a reciprocal manner during replication cycles (Bhutani et al. 2011; Chen and Riggs 2011; Bierne et al. 2012).

20.2.2 Histone Modifiers

It includes various enzymes involved in post-translational modification of multiple residues of histones (Fig. 20.4). Post-translational modifications and their importance are listed below:

20.2.3 Acetylation

Histone acetylation refers to addition of acetyl group to arginine (R) and lysine (K) residues of histone. Acetyl group is added to the ε-amino group of lysine residues. The acetyl group donor is charged CoA or acetylated CoA. Initially, it was shown that treatment of Friend erythroleukaemic cells by n-butyrate differentiated them into haemoglobin-synthesizing normoblast-like cells (Riggs et al. 1977). Later experiments confirmed that n-butyrate acts as histone deacetylase inhibitor or HDACi. Thus, the experiment uniquely defined the role of small molecule like n-butyrate leading to differentiation of cancer cells via inhibition of deacetylation on histone. Prescient experiments by Allfrey et al. laid the basis of histone

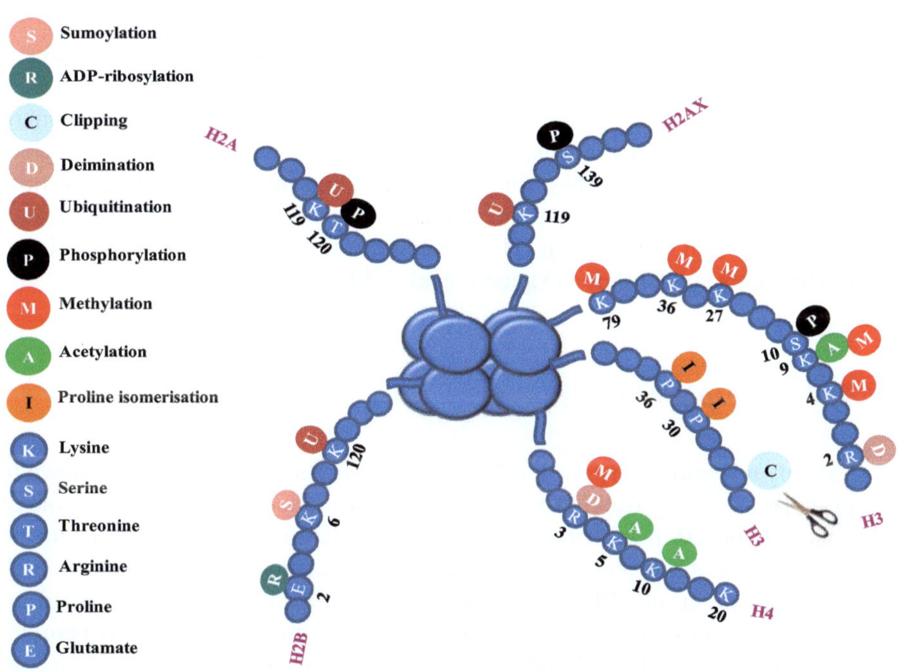

Fig. 20.4 Summary of different modifications on histone amino acid residues. Four types of core histones H2A, H2B, H3, H4, and histone variant H2AX are epigenetically modified through attachment of various chemical groups and/or proteins

acetylation and deacetylation governing the activated and inhibited RNA synthesis, respectively. Afterwards scientists could correlate the acetylation of core histone with the transcriptional activation of genes. It was thus shown that the modifications on histone could bring epigenetic changes. With the basis of an understanding of acetylation, scientists began to speed up the search of possible acetylating and/or deacetylating enzymes and their roles in histone modifications (Kouzarides 2007). Acetylation is associated mostly with activation of gene. So far, most of the acetylations have been described for residues on N-terminal tail of histone as it is the most accessible part. Although, recently, acetylation on H3K56 residue has also been reported which is localized at the core region. H3K56Ac is laid on newly formed histones during the S-phase of cell division. If there is no damage, then the modification is removed. In the presence of DNA damage, however, the modification is retained indicating its significance during DNA damage (Celic et al. 2006; Maas et al. 2006). Another enzyme called Hat1 has been shown to acetylate the lysine-12 on H4 (H4K12) at the sites of DNA repair. Studies suggest that acetylation of H4 has a role to play in the fixation of replication origins and initiation of S-phase (Qin and Parthun 2006; Kouzarides 2007).

20.2.4 Deacetylation

Deacetylation encompasses the reversal of acetylation. Histone deacetylation is involved in the regulation of multiple pathways, and they are part of multiple repression complexes. They are divided into different classes and are called as classical or nonclassical based on their dependence on Zn^{2+} or nicotinamide adenine dinucleotide (NAD^+), respectively. They are known for aiding in chromatin condensation. For example, in vitro studies have shown that sirtuin type2 (SirT2) exhibits specificity to acetylation of lysine-16 on H4 (H4K16Ac) and causes deacetylation to bring about chromatin condensation (Vaquero et al. 2006).

20.2.5 Phosphorylation

A relatively unexplored histone modification is phosphorylation which is known to occur on serine (S), tyrosine (Y), or threonine (T) residues on histone. The role of phosphorylation has mainly been explored in response to DNA damage wherein the serine-139 on H2AX is phosphorylated (in case of mammalian cells) (Celeste et al. 2003; Rogakou et al. 1998) and the modified histone variant is called as ϒH2AX. Yeast does not have H2AX, but parallel to mammalian cells, phosphorylation occurs at H2A serine-129 (Downs et al. 2000; Redon et al. 2003). In yeast, mutation of this residue to non-phosphorylatable alanine makes them hypersensitive to DNA-damaging agents such as methyl methanesulphonate (MMS), highlighting its importance in double-strand break repair (Downs et al. 2000). In addition, this modification distributes over the broad region around the DNA damage probably creating a unique platform to recruit DNA damage repair proteins (Jungmichel and Stucki 2010; Stucki et al. 2005).

Phosphorylation of serine-1 in H4 (H4S1P) has been shown to be induced both in DNA damage repair by negatively regulating acetylation on H4 and during the transcriptional activation of genes. This seems to be contradictory as transcription generally requires hyperacetylation. One hypothesis derived from studies point that phosphorylation of this residue helps to stabilize nucleosome after the ribosome moves in order to prevent inappropriate initiation of transcription within coding sequence of genes which are active (Utley et al. 2005; Cheung et al. 2008; Li et al. 2007). Another phosphorylation is found on serine-47 of H4 (H4S47) which is shown to help in incorporating H3.3-H4 into the nucleosome by associating with a chaperone histone cell cycle regulation defective homolog A (HIRA) specific for H3.3 (Kang et al. 2011).

20.2.6 Dephosphorylation

Like phosphorylation, dephosphorylation also plays a significant role in the regulation of cell functioning and is brought about by class of enzymes known as phosphatase. Studies show that a phosphatase complex, viz. HTP-C, helps in the recovery from DNA damage checkpoint by dephosphorylating phoshorylated serine-129 of H2A (H2AS129P). Similarly, defect in H2AXY142 dephosphorylation recruits pro-apoptotic molecule instead of repair setup to ϒH2AX (Rossetto et al. 2012; Kouzarides 2007).

20.2.7 Methylation

It occurs on various amino acid residues of histone as lysine, arginine, and histidine (Byvoet et al. 1972; Murray 1964; Fischle et al. 2008). Lysine residues get methylated on their ε-amine group, and it can either be mono-, di-, or trimethylation (Murray 1964; Hempel et al. 1968). Arginine residues get methylated on their guanidinyl group, and it can be mono- or dissymmetric or asymmetric dimethylation. Histidine residues are known to be monomethylated (Borun et al. 1972; Gershey et al. 1969). Few types of methylation on H3 and H4 have been studied extensively, viz. H3K4, H3K9, H3K27, H3K36, H3K79, and H4K20. Arginine methylations which include H3R2, H3R8, H3R17, H3R26, and H4R3 have also been reported. Recent studies using mass spectrometry have revealed the presence of methylated basic residues (Greer and Shi 2012).

Histone methyltransferases use methyl groups from S-adenosylmethionine to methylate the histone residues. They belong to three different families: SET domain containing proteins (Rea et al. 2000), Dotl-like proteins (Feng et al. 2002), and ariginine N-methyltransferase (PRMT) proteins (Bannister and Kouzarides 2011).

H3K4 trimethylation (H3K4me3) in general is linked to active transcription or with poised genes which are ready for activation. H3K4 monomethylation (H3K4me) is often linked to function of enhancer. On the contrary, H3K27 trimethylation (H3K27me3) is a mark of repressed form of chromatin. Cell cycle is regulated by H3K79 dimethylation (H3K79me2) (Bernstein et al. 2002; Santos-Rosa et al. 2002; Heintzman et al. 2007; Schulze et al. 2009). Though methylation on specific residues, in general, is associated with a state of chromatin, it is not always the case. For instance, H3K4 di- and trimethylation (H3K4me2 and H3K4me3) are deemed to be activation marks, but they may be involved in repression. H3K4 di- and trimethylation associate with an inhibitor of growth family member 2 (ING2) protein and help in repression via stabilization of histone deacetylase complex (Shi et al. 2006). Histone methylation has been implicated in the transcriptional regulation, ageing, and intellectual disability (Bernstein et al. 2002; Santos-Rosa et al. 2002; Noma et al. 2001; Pollina and Brunet 2011). Aberrant methylation pattern is found in many types of cancer (Chi et al. 2010).

20.2.8 Demethylation

Methylation mark on histone residues is removed by enzymes called demethylases which are categorized into one of the two families, viz. jumonji-C (JmjC) domain-containing dioxygenases which depend on iron and amine oxidases (Shi et al. 2004; Tsukada et al. 2006; Whetstine et al. 2006; Cloos et al. 2006; Klose et al. 2006a).

One of the lysine demethylases, lysine-specific histone demethylase 1 (LSD1), acts to demethylate H3K4 and repress transcription, but when it combines with androgen receptor, it helps in demethylation of H3K9 and activates transcription (Shi et al. 2006; Metzger et al. 2005). Jumonji-C (JMJC) domain-containing histone demethylases 1,3A, i.e. JHDM1, JMJC domain protein 2A (JMJD2A)/JHDM3A, and JMJC domain protein 2C (JMJD2C)/GASC1, have been reported to be involved in the removal of H3K36 methylation (Tsukada et al. 2006; Whetstine et al. 2006; Klose et al. 2006b; Cloos et al. 2006). Thus, like methyltransferases, demethylases are important to bring about changes in chromatin state (Kouzarides 2007).

20.2.9 Ubiquitination

It is defined as addition of one or more ubiquitin molecule to histone residues. Most abundantly ubiquitinated histones are H2A and H2B. Monoubiquitinated H2A and H2B are the most widespread and take place on Lys-119 for H2A and Lys-123 for H2B. Enzymes which help in the ubiquitination of histone are known as histone ubiquitin ligases (Goldknopf et al. 1975; West and Bonner 1980). For example,

H2AK119 ubiquitination is catalysed by ring finger protein1B (RING1B) (Wang et al. 2004; Cao et al. 2005). Polyubiquitination of H2A and H2AX, a histone variant at K63, has been shown to play an important role in DNA damage repair (Stewart et al. 2009).

20.2.10 Deubiquitination

Like other histone modifications, ubiquitination is also reversible. Enzymes that aid in the removal of ubiquitin from histones are known as deubiquitinating enzymes. Different deubiquitinases as ubiquitin-specific peptidase 16 (USP16), BRCA1-associated protein 1 (BAP1), Myb like, SWIRM and MPN domains 1 (MYSM1) alias 2A-DUB, and USP21 are specific for H2A. They have been shown to play roles in transcription, DNA damage response, etc. (Joo et al. 2007; Shanbhag et al. 2010; Cao and Yan 2012).

20.2.11 Proline Isomerization

Proline (P) isomerization encompasses a non-covalent histone modification which converts peptidyl-proline residues in between *cis* and *trans* form. This causes changes in polypeptide secondary structure which in turn affects histone methylation and therefore gene expression. An enzyme involved in proline isomerization is known as proline isomerase, peptidylprolyl isomerase FPR4 (Fpr4). Studies have implicated a crosstalk in between H3P38 isomerization and H3K36 methylation. It has been shown that H3P38 isomerization to its *cis* form changes the secondary structure of H3 tail so that H3K36 does not fit into Set2 methyltransferase active site. In case of active transcription, the movement of RNA polymerase through genes disturbs the nucleosome and reveals H3K36 to be modified by Set2. The modified H3K36 trimethylation inhibits activity of Fpr4 and maintains an active chromatin state (Nelson et al. 2006).

20.2.12 ADP-Ribosylation

It is a reversible post-translational modification of proteins (Hottiger et al. 2010). Like other modifications, proteins can be both mono- or poly-ADP-ribosylated. During the modification, there is transfer of one ADP-ribose moiety from NAD^+ to acceptor protein amino acid, and thus mono-ADP-ribosylation occurs. The ADP-ribosylated protein may further be ADP-ribosylated (Messner and Hottiger 2011). In general, residues like lysine, glutamate (E), cysteine (C), asparagines (N), phosphor-serine, aspartate (D), and arginine have been shown to undergo this modification. ADP-ribosylation is catalysed by enzymes called ADP-ribosyltransferases (ARTs) belonging to three different classes, viz. ARTDs (where D stands for diphtheria toxin-like), ARTCs (where C stands for clostridial toxin-like), and NAD^+-dependent protein deacetylases called sirtuins (Hottiger et al. 2010; Hassa et al. 2006; Koch-Nolte et al. 2008; Milne and Denu 2008). This modification has been shown to mediate nucleosome structure dynamics by interacting with 'super beads' made of eight to ten nucleosomes. When NAD^+ is absent, binding of ARTD1 to nucleosome encourages compaction of chromatin, whereas in the presence of NAD^+, automodification of ARTD1 takes place leading to a relax state of chromatin. Poly-ADP-ribosylated chromatin shows a distinct structure and is more sensitive and accessible to treatment of nuclease. It has also been shown that increased levels of mono- and poly-ADP-ribosylated proteins are found in case of DNA damage induction. Besides, ADP-ribosylation level is shown to increase in case of regions which are transcriptionally active. ARTD1 and histone H1 show a reciprocal pattern with respect to chromatin binding. ARTD1 is found to be enriched at transcriptionally active gene promoter (Messner and Hottiger 2011).

20.2.13 ADP-Ribosylation Removal

Poly-ADP-ribosylation of histone is reversible in nature, and the enzymes responsible for removal of this modification are categorized into two classes, viz. PAR glycohydrolases (PARGs) and ADP-ribosylhydrolases (ARHs). The importance of these enzymes lies in their role of primary ADP-ribosyl group removal and maintenance of PAR groups in a cell (Messner and Hottiger 2011).

20.2.14 Histone Clipping

Histone clipping is defined as cleavage of histones. Unlike other histone modifications, this modification is irreversible. All the core histones can undergo cleavage, but H3 clipping has been of great interest because of abundant cleavage sites on the tail, and also it has been found to be cleaved during sporulation, infection, differentiation, ageing, and spermatogenesis (Mandal et al. 2014). Besides, clipping of histone H3 has been found to be involved in alleviation of cytotoxicity or in increasing it. Thus, targeting the clipping of H3 histone might enable improved survival by controlling inflammation. Various enzymes have been shown to be associated with H3 clipping. For example, cathepsin L has been shown to cleave H3 at specific sites Ala21-Thr22 and Lys27-Ser28. Similarly, glutamate dehydrogenase GDH has been shown to target H3 on two sites Lys23-Ala24 and Lys27-Ser28 (Adams-Cioaba et al. 2011; Mandal et al. 2012, 2013; Zhou et al. 2014).

20.2.15 *Deimination*

It is a modification involving arginine conversion to citrulline. Deamination might play a role in antagonizing the arginine methylation-induced activation as its conversion to citrulline essentially prevents the methylation. One of the enzymes required for this conversion is peptidylarginine deiminase 4 (PADI4), and it has been shown that deimination is specific to monomethylated arginine (Kouzarides 2007).

20.2.16 *Sumoylation*

Sumoylation involves addition of small ubiquitin-related modifier (SUMO) proteins to lysine residues on proteins. Sumoylation occurs through multistep reactions which involves conversion of precursor SUMO protein into its active form via hydrolysis. Activation leads to exposed gly-gly motif in the SUMO protein following which it gets conjugated to SUMO-activating enzyme E1 via thioester bond. Finally, ubiquitin-conjugating enzyme 9 (UBC9) covalently attaches SUMO to lysine residues. Studies have shown that the modification results in repression as both ubiquitination and acetylation are oppressed by it (Maejima and Sadoshima 2014; Kouzarides 2007).

20.3 Epigenetics and Pathogenic Interactions

Once pathogens invade their host, they cause a variety of epigenetic modifications which enable them to thrive and evade the host defence measures (Chen et al. 2014).

20.3.1 *Role of Epigenetics in Bacterial Pathogenicity*

Bacterial invasion on host triggers changes in epigenetic marks such as DNA methylation, microRNAs (miRNAs), and histone post-translational modifications (Table 20.1). Bacterial lipopolysaccharide (LPS) has been shown to induce miRNA levels. Different bacterial strains bring about different epigenetic modifications in host (Yaseen et al. 2015) (Fig. 20.5).

M. tuberculosis has been one of the most notorious strains of bacteria causing havoc in wide population. Because of its evasion mechanisms, it has become difficult to tackle. Evidence from recent studies suggest that it secretes a protein named Rv1988, a methyltransferase which methylates the histone H3 arginine 42, i.e.

Table 20.1 Summary of epigenetic modifications induced by bacterial pathogens

Organism	Epigenetic modification	Effector (E); target (T)	Functional consequence	Refs.
Bacteria				
Chlamydia	Methylation of histones	NUE (E); histones in mammals (T)	Target genes unknown	Pennini et al. (2010)
Plasma phagocyophilum	Deacetylation of H3 at CYBB locus	AnkA (E); CYBB locus (T)	Repression of CYBB which affects survival of organism	Garcia-Garcia et al. (2009)
Ehrlichia chaffeensis	NA	P200 (E); Alu-Sx elements (T)	Possibly involved in global gene transcription	Zhu et al. (2009)
Lysteria monocytogenes	Histone acetylation	LntA (E); BAHD1 (T)	Enhancement of ISG expression	Bierne et al. (2009)
Shigella	Prevention of phosphorylation	OspF (E); H3 (T)	Inhibition of MAPK	Arbibe et al. (2007)
Shigella	Ubiquitinylation	IpaH9.8 (E); U2AF (T)	Degradation of splicing factor U2AF	Bierne et al. (2012)
Listeria monocytogenes	H3 dephosphorylation, H4 deacetylation	Listeriolysin O (E); H3, H4 (T)	Reduced transcriptional activity of key immunity genes and helps the bacteria to survive in host	Hamon et al. (2007)
Helicobacter pylori	Aberrant methylation of DNA	NA (E); tumour suppressor genes, DNA repair genes (T)	Hypermethylation at tumour suppressor genes, DNA repair genes	Bierne et al. (2012)

H3R42, and thus aids in the repression of defence mechanism employed by the host against *Mycobacteria* (Yaseen et al. 2015).

One of the causal organisms for bacterial meningitidis is *N. meningitidis*. Two potent virulence factors secreted by this organism are meningococcal serine protease A (MspA) and adhesion and penetration protein (App). These two factors cause death of dendritic cells by caspase-dependent apoptosis following proteolytic cleavage of H3 (Khairalla et al. 2015).

Pathogens also employ epigenetics to maintain varied levels of gene transcription. For example, serovars of *S. enterica* have varied levels of DNA methylation which may lead to difference in virulence. Another example is furnished by *H. pylori*, which, after infecting the gastric cells, causes the dephosphorylation of H3S10 and decrease in H3K23ac. H3S10 dephosphorylation is possibly associated with cytotoxin-associated gene A pathogenicity island (cagPAI), as cagPAI deletion leads to restoration of phosphorylation in H3S10. Furthermore, it has also been shown that these changes lead to upregulation of c-Jun, an oncogene, and

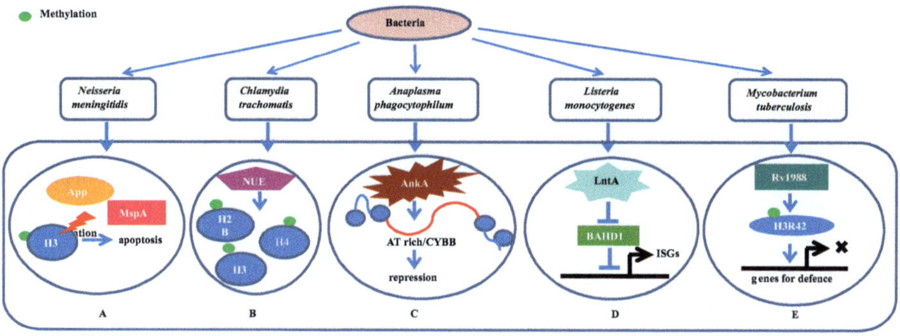

Fig. 20.5 Schematic of epigenetic modifications in bacterial infections. Neisseria meningitides releases App and MspA into the host cells which enter the nucleus to cleave H3 histone and trigger apoptosis (**a**); Chlamydia trachomatis releases NUE which enters inside host cell and methylate the histones (**b**); Anaplasma phagocytophilum secretes AnkA which recognizes AT rich sequence in chromatin and specifically downregulates CYBB which affects survival of the bacteria (**c**); Listeria monocytogenes release LntA which alleviate the interaction of repressor BAHD1 from ISG gene in an attempt to open the chromatin (**d**); Mycobacterium tuberculosis methylates the H3R42 residue which suppresses the expression of genes involved in defence against the bacteria (**e**)

downregulation of heat shock protein 70 (Hsp 70), implicating the possibility of tumour development besides inflammation.

L. pneumophilia employs a regulator of methylation A (Rom A), type 4 secretion system effector, Suvar3-9, enhancer of zeste, and trithorax (SET) domain inhabiting methyltransferase involved in H3K14me3 increase and H3K14ac reduction leading to switching off of gene transcription.

C. trachomatis secretes a nuclear effector (NUE) which enters inside the host cell nuclei and gets associated to the chromatin. NUE methylates histones H2B, H3, and H4 though the advantage of this modification is still not clear.

A. phagocytophilum encodes Ank-containing (AnkA) protein which shows preference for AT-rich chromatin region. AnkA is known to repress CYBB which encodes cytochrome b-245, part of phagocyte oxidase, and is known to play a crucial role in survival of the bacteria.

E. chaffeensis encodes p200 protein which binds to chromatin at specific elements called Alu-Sx. Thus, it might have a role in affecting global gene transcription.

Listeria monocytogenes codes for listeria nuclear targeted protein A (LntA), a nuclear targeted factor. LntA has been shown to interact with BAH domain-containing protein 1 (BAHD1), a factor involved in formation of heterochromatin. BAHD1 forms a complex along with other chromatin factors, viz. heterochromatin protein 1 (HP1), SETDB1, KRAB-associated protein 1 (KAP1), histone deacetylases (HDACs), and methyl-CpG-binding domain protein (MBD1) all of which help in silencing expression of genes. Depending on the signal in the cell and the cell types, the BAHD1 complex represses the expression of genes. BAHD1 is involved in the interferon-stimulated gene (ISG) repression in epithelial cells. An unknown signal triggers expression of *lntA* gene which then enters the nucleus and releases binding of BAHD1 to ISG promoters thus helping in upregulation of ISG expression. LntA most probably facilitates unwinding of chromatin by restricting BAHD1 recruitment to ISG promoters. Thus, the interplay of LntA and BAHD1 helps in modulation of interferon response.

In another example, bacteria *S. flexneri* have been shown to target the chromatin via modulation of transcription factor activity. OspF of *S. flexneri* is a phosphothreonine lyase which converts the phosphothreonine residue of MAPK into dehydrobutyrine. Dehydrobutyrine cannot be phosphorylated, so the MAPK remains in inactive form, resulting in MAPK signalling inhibition. Inhibition of MAPK signalling deters phosphorylation of H3 at promoters regulated by nuclear factor kappa-light-chain-enhancer of activated B-cells (NF-*k*B) and ultimately blocks pro-inflammatory gene expression. Studies show that OspF and OspB interact with Rb, a human retinoblastoma protein which binds to many chromatin remodelling factors. Thus, *Shigella* uses both OspF and OspB to suppress host immunity via alteration of specific genes. Another factor which aids in pathogenesis of *Shigella* is IpaH9.8, an E3 ubiquitin ligase which binds to host proteins and channels them for proteasomal degradation. IpaH9.8 affects activity of U2A, an mRNA splicing factor and NF-*k*B pathway, and therefore modulates the host response (Bierne et al. 2012; Kaul et al. 1997).

20.3.2 Role of Epigenetics in Viral Pathogenicity

Viruses have developed numerous ways to attack the host cells, and using epigenetic modifications to disarm host has not remained untouched (Fig. 20.6) (Table 20.2). For example, cytomegalovirus (CMV) replication induces H3K79 dimethylation (H3K79me2) which is associated with disruptor of telomeric silencing (DOT1). The absence of DOT1L leads to decreased CMV replication. Also, CMV replication is associated with an increase in H3K27 monomethylation (H3K27me) and H3K36 dimethylation (H3K36me2) and a decrease in H4K16 acetylation (H4K16ac). Like with other epigenetic modifications, sirtuin 1 (SIRT1) aids in latency of human herpes-8 virus via H3K27me3, a repressive mark on the viral replication and transcription activator. Thus, these epigenetic modifications limit the spread of the virus. Knockdown of SIRT1 led to a decrease of H3K27me3 and an increase in the H3K4me3 resulting in lytic *infection* (Cole et al. 2016).

KSHV is involved in B-cell lymphoma. It encodes a protein named latency-associated nuclear antigen (LANA) which interacts with DNMTs in the cell. Studies have implicated an association of LANA with H-cadherin (CDH13), a tumour suppressor gene which is methylated in variety of cancers. Also, it has been shown that LANA binds to DNMT1, DNMT3A, and DNMT3B. Thus, it is likely that LANA helps in de novo DNA methylation via recruitment of DNMT3A in B-cell lymphomagenesis. Besides, LANA is shown to bind to transforming growth factor (TGF)-β type II receptor (TGF-βRII) gene promoter and inhibiting its transcription. TGF

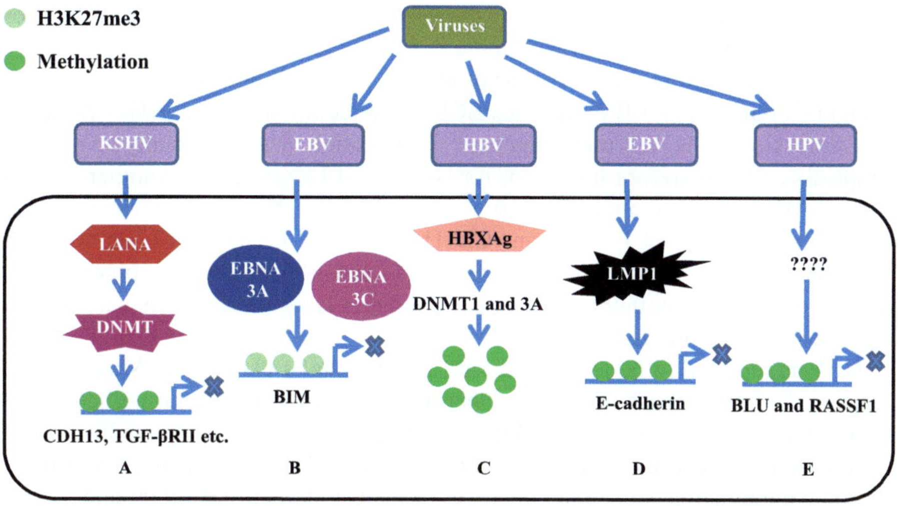

Fig. 20.6 KSHV employs LANA which interact with DNMT to methylate H-cadherin, TGF-βRII in cancerous cells (**a**); EBV uses latency associated molecules to negativelyregulate BIM, a tumor suppressor by methylation of its gene in associated cancer (**b**); HBV codes for HBV X antigen and induce DNMT1 and 3A to methylate E-cadherin and p16INK4A tumor suppressor in cancer (**c**); EBV encodes LMP1 to negatively regulate E-cadherin via methylation (**d**); HPV induces tumor formation by negatively regulating BLU and RASSF1, where RASSF1 is a tumor suppressor (**e**)

Table 20.2 Summary of epigenetic modifications induced by viral pathogens

Organism	Epigenetic modification	Effector (E); target (T)	Functional consequence	Refs.
Kaposi's sarcoma associated virus	Methylation of promoter region of genes	Latency-associated nuclear antigen (E); transforming growth factor (TGF)-β type II receptor (TGF-βRII) (T)	Downregulation of transcription of TGF- βRII and possibly aiding maintenance of latency of virus and lymphoma development	Di Bartolo et al. (2008)
Simian vacuolating virus	Methylation of gene promoter	Large T-antigen (Tag) (E); RASSF1 (T)	Methylation of promoter RASSF1 and associated with pathogenesis of cancer	Paschos and Allday (2010)
Human immunodeficiency virus	Gene promoter methylation	NA (E); GNE (T)	Methylation of GNE gene	Paschos and Allday (2010)
Human adenovirus	Blockage of H2B monoubiquitination	HAdV E1A (E); hBre1 complex (T)	Evasion of interferon (IFN) response	Fonseca et al. (2014)

pathway has pro-apoptotic and/or anti-proliferative role in B-cells, so its inhibition would lead to increased survival of cells.

EBV has been found to be associated with B-cell, T-cell, and some forms of Hodgkin's lymphoma. EBV mediates BCL-2-interacting mediator (*BIM*) transcription repression via H3K27me3 epigenetic modification. *BIM* is crucially involved in apoptosis and is therefore important for survival of lymphocytes. Another way by which EBV aids in tumour formation is by latency membrane protein LMP1-assisted methylation of E-cadherin (CDH1) promoter. E-cadherin is an adhesion molecule which controls tumour invasiveness and is found to be epigenetically repressed in many carcinomas. Thus, EBV induces lymphomagenesis via different epigenetic modifications.

HBV is reported to be associated with hepatocellular carcinoma. Many studies indicate the involvement of HBV in aberrant DNA methylation during hepatocellular carcinoma. A protein named HBV X antigen (HBVXAg) may mediate methylation as it has been shown to induce DNMT1 and DNMT3A. Thus, EBV triggers aberrant methylation via induction of epigenetic modifiers (Paschos and Allday 2010).

HPV like other viruses has been implicated to DNA methylation in cervical carcinoma cells. Studies show that HPV is linked to enhanced DNA methylation of putative RAS association domain family protein 1 (RASSF1) and BLU (tumour suppressor) gene promoter (Lai et al. 2007; Dammann et al. 2000).

SV40 has similarly been shown to enhance the level of DNMTs and promoter DNA methylation of tumour suppressor gene like RASSF1. Adenoviruses have been reported to block H3K18Ac globally via an oncoprotein e1a thereby repressing transcription of many genes. Human immunodeficiency virus has been shown to enable promoter methylation of glucosamine (UDP-N-acetyl)-2-epimerase/N-acetylmannosamine kinase (*GNE*) gene and repress its transcription. This may lead to disruption of sialylation of cell surface required for cell-cell recognition and may have effects on lymphocyte trafficking (Paschos and Allday 2010).

Studies have shown that FMDV cleaves H3 in a process known as H3 clipping. This modification is irreversible and is catalysed by FMDV 3C protease. Thus, FMDV causes epigenetic modulation of the host resulting in a change of transcriptional regulation of several genes (Falk et al. 1990).

20.4 Future Perspectives

Over several decades now, scientific community has been facing problems with the tackling of bacterial, viral, and other infectious diseases. These organisms exhibit high mutation rates and are capable of lateral gene transfer, rendering them resistant to multiple drugs. The field of epigenetics has given a new scope for fending off these pathogens.

Initially, a number of epigenome modulators were designed to alter the modifications imposed by the pathogens or to let pathogens come out of their latent states so that drugs can act on them and prevent further chances of recurrence from reactivation of their latent forms. The designed drugs span from the generic bromodomain inhibitors and HDAC inhibitors (HDACis) to specific clustered regularly interspersed palindromic repeats (CRISPR)/CRISPR-associated protein 9 (Cas9) coupled to acetylases and methyltransferases (Heerboth et al. 2014; Hilton et al. 2015). The most extensively used class of epigenetic drug are HDAC inhibitors (HDACis). In humans there are approximately 18 different kinds of HDACis and have been classified into 4 classes on the basis of their homology with that of yeast HDACis. The major concern of these drugs is their off-target effects. They have been shown to alleviate problem associated with infections but at the same time have disturbed the normal functioning of immune system, for example, by impairing macrophage function. To overcome this disadvantage, several site-specific modifiers have been designed such as tubastatin A, HDAC6 inhibitor, etc. (Vishwakarma et al. 2013). When taken for trial, these HDAC6 inhibitors which target more specifically yielded more positive results. Thus these drugs could be used to target epigenetic pathways individually.

Another class of bromodomain inhibitors are known to inhibit bromodomain and extraterminal domain (BET) proteins. These BET proteins are involved in associating the histone acetylation with transcription (Smale et al. 2014). They recognize the acetyl group on chromatin and then help in its association with RNA polymerase II for transcription. Also, the ET domains of BET proteins interact with several proteins which can possibly influence remodelling of chromatin. Thus, the specific BET inhibitors could modulate the response to lipopolysaccharide to alter levels of interleukins IL-6 and IL-12 and nitric oxide (NO) which in turn be utilized for bringing down pathological inflammation while maintaining the robust immune response during chronic infections (Smale et al. 2014). A bromodomain inhibitor, JQ1, has been used to alter the c-Myc (a transcription factor) functioning in myeloma (Delmore et al. 2011).

A more targeted approach has been under an intensive research which exploits the use of CRISPR technology to specifically modulate epigenetic changes (Ledford 2015). Cas9, a DNA endonuclease enzyme, is used in general for specificity to a particular site. In one of the studies, scientists used an inactive Cas-9 and acetyltransferase (p300) fusion construct to acetylate H3K27 at specific sites leading to targeted gene activation (Hilton et al. 2015). In another instance, inactivated Cas9 was fused with LSD1 to activate specific genes which resulted in a map of 'enhancer' sequences at genetic level, giving insight of their locations (Kearns et al. 2015).

In one study, it has been shown that bromodomain inhibitor could act as a potential therapeutic candidate for diseases associated with human T-cell leukaemia virus 1 (HTLV-1). The viral infection causes prolonged activation of NF-*k*B and its targeted gene expression. Acetylation of NF-*k*B subunit and RelA and recruitment of Brd4 are required following which expression of NF-*k*B targeted genes occur. The acetylation of RelA and the binding of Brd4 to acetylated RelA are induced by Tax molecule. The presence of JQ1, a BET inhibitor, disrupts the interaction between bromodomain-containing protein 4 (Brd4) and acetylated REL-associated protein (RelA) and thus suppresses Tax, an oncogenic protein-mediated tumorigenesis of HTLV-1-infected cells. Thus, such inhibitors can be utilized for cancer treatment (Wu et al. 2013). JQ1 has also been shown to be effective in the reactivation of latent HIV-1 and aiding in the suppression of T-cell activation gene C-X-C chemokine receptor type 4 (CXCR4) and cluster of differentiation CD3 and CD28 thereby minimizing the T-cell proliferation. Thus, it can be of therapeutic use for treatment of viral infection (Banerjee et al. 2012).

Bacterial and viral infections have long been known to be associated with cancer of different types such as head and neck, liver, cervical, and gastric cancers. One of the common characteristics of these cancers is the presence of aberrant methylation of DNA. An interesting fact about DNA methylation is that the level of aberrant methylation pattern correlates with cancer development risk. Bacterial or viral infections are known to cause severe inflammatory response which gives rise to abnormal pattern of DNA methylation and which culminates in cancer. Now, according to one study, administration of 5-aza-2′-deoxycytidine (5-aza-dC), a demethylating agent, has been shown to decrease the incidence of gastric cancers (Niwa et al. 2013).

Influenza virus infections have become difficult to deal with as they have evolved to multiple antigenic variants and drug-resistant varieties. Researchers have developed C646, a histone acetyltransferase (HAT) inhibitor which binds to p300/cAMP response element-binding protein (CREB) and affects multiple stages of virus life cycle thus helping in suppression of influenza A virus infection (Zhao et al. 2015). Furthermore, it has been proposed that the combined highly active antiretroviral therapy (HAART) and inhibitor of histone methyltransferase (HMT) could be used for inducing HIV-1 recovery. It was shown earlier that H3K9 methylation was required for the HIV-1 promoter. Thus, it can lead to powerful remedy towards the cure of HIV (Bouchat et al. 2012).

HDACis have been shown to aid in the clearance of intracellular bacteria like *E.coli* and *Salmonella typhimurium* of macrophage by increasing the production of mitochondrial reactive oxygen species generated by these cells. HDACis have also been known to inhibit hypoxia-inducible factor, HIF-1α, thereby abrogating the pro-inflammatory cytokine secretion (Ariffin et al. 2015).

Thus, a combination of different epigenetic modifiers can be employed to eradicate different multidrug-resistant pathogenic species.

Acknowledgements We thank all the members of our lab for their immense patience, discussions, and repeated corrections which helped us in writing the chapter. NS acknowledges IISER Bhopal for her fellowship support. SERB and DBT are acknowledged for the funding support to RST.

References

Adams-Cioaba MA, Krupa JC, Xu C, Mort JS, Min J (2011) Structural basis for the recognition and cleavage of histone H3 by cathepsin L. Nat Commun 2:197

Ahmad K, Henikoff S (2002) The histone variant H3.3 marks active chromatin by replication-independent nucleosome assembly. Mol Cell 9:1191–1200

Alberts B, Bray D, Hopkin K, Johnson A, Lewis J, Raff M et al (2013) Essential cell biology. Garland Science, New York

Andino A, Hanning I (2015) Salmonella enterica: survival, colonization, and virulence differences among serovars. Sci World J 2015:520179

Anwar MA, Choi S (2014) Gram-negative marine bacteria: structural features of lipopolysaccharides and their relevance for economically important diseases. Mar Drugs 12:2485–2514

Arbibe L, Kim DW, Batsche E, Pedron T, Mateescu B, Muchardt C et al (2007) An injected bacterial effector targets chromatin access for transcription factor NF-kappaB to alter transcription of host genes involved in immune responses. Nat Immunol 8:47–56

Ariffin JK, Das Gupta K, Kapetanovic R, Iyer A, Reid RC, Fairlie DP et al (2015) Histone deacetylase inhibitors promote mitochondrial reactive oxygen species production and bacterial clearance by human macrophages. Antimicrob Agents Chemother 60:1521–1529

Badeaux AI, Shi Y (2013) Emerging roles for chromatin as a signal integration and storage platform. Nat Rev Mol Cell Biol 14:211–224

Banerjee C, Archin N, Michaels D, Belkina AC, Denis GV, Bradner J et al (2012) BET bromodomain inhibition as a novel strategy for reactivation of HIV-1. J Leukoc Biol 92:1147–1154

Bannister AJ, Kouzarides T (2011) Regulation of chromatin by histone modifications. Cell Res 21:381–395

Barreto-Vieira DF, Barth OM (2015) In: Shah MM (ed) Negative and positive staining in transmission electron microscopy for virus diagnosis. InTech, Rijeka

Berger SL, Kouzarides T, Shiekhattar R, Shilatifard A (2009) An operational definition of epigenetics. Genes Dev 23:781–783

Bernstein BE, Humphrey EL, Erlich RL, Schneider R, Bouman P, Liu JS et al (2002) Methylation of histone H3 Lys 4 in coding regions of active genes. Proc Natl Acad Sci U S A 99:8695–8700

Bhutani N, Burns DM, Blau HM (2011) DNA demethylation dynamics. Cell 146:866–872

Bierne H, Tham TN, Batsche E, Dumay A, Leguillou M, Kerneis-Golsteyn S et al (2009) Human BAHD1 promotes heterochromatic gene silencing. Proc Natl Acad Sci U S A 106:13826–13831

Bierne H, Hamon M, Cossart P (2012) Epigenetics and bacterial infections. Cold Spring Harb Perspect Med 2:a010272

Bloom DE (2011) 7 billion and counting. Science 333:562–569

Borun TW, Pearson D, Paik WK (1972) Studies of histone methylation during the HeLa S-3 cell cycle. J Biol Chem 247:4288–4298

Bouchat S, Gatot JS, Kabeya K, Cardona C, Colin L, Herbein G et al (2012) Histone methyltransferase inhibitors induce HIV-1 recovery in resting CD4(+) T cells from HIV-1-infected HAART-treated patients. AIDS 26:1473–1482

Braaten KP, Laufer MR (2008) Human papillomavirus (HPV), HPV-related disease, and the HPV vaccine. Rev Obstet Gynecol 1:2–10

Brazel AJ, Vernimmen D (2016) The complexity of epigenetic diseases. J Pathol 238:333–344

Buschbeck M, Hake SB (2017) Variants of core histones and their roles in cell fate decisions, development and cancer. Nat Rev Mol Cell Biol 18:299–314

Byvoet P, Shepherd GR, Hardin JM, Noland BJ (1972) The distribution and turnover of labeled methyl groups in histone fractions of cultured mammalian cells. Arch Biochem Biophys 148:558–567

Cao J, Yan Q (2012) Histone ubiquitination and deubiquitination in transcription, DNA damage response, and cancer. Front Oncol 2:26

Cao R, Tsukada Y, Zhang Y (2005) Role of Bmi-1 and Ring1A in H2A ubiquitylation and Hox gene silencing. Mol Cell 20:845–854

Celeste A, Fernandez-Capetillo O, Kruhlak MJ, Pilch DR, Staudt DW, Lee A et al (2003) Histone H2AX phosphorylation is dispensable for the initial recognition of DNA breaks. Nat Cell Biol 5:675–679
Celic I, Masumoto H, Griffith WP, Meluh P, Cotter RJ, Boeke JD et al (2006) The sirtuins hst3 and Hst4p preserve genome integrity by controlling histone h3 lysine 56 deacetylation. Curr Biol CB 16:1280–1289
Chen ZX, Riggs AD (2011) DNA methylation and demethylation in mammals. J Biol Chem 286:18347–18353
Chen R, Kang R, Fan XG, Tang D (2014) Release and activity of histone in diseases. Cell Death Dis 5:e1370
Cheung V, Chua G, Batada NN, Landry CR, Michnick SW, Hughes TR et al (2008) Chromatin- and transcription-related factors repress transcription from within coding regions throughout the *Saccharomyces cerevisiae* genome. PLoS Biol 6:e277
Chi P, Allis CD, Wang GG (2010) Covalent histone modifications – miswritten, misinterpreted and mis-erased in human cancers. Nat Rev Cancer 10:457–469
Cloos PA, Christensen J, Agger K, Maiolica A, Rappsilber J, Antal T et al (2006) The putative oncogene GASC1 demethylates tri- and dimethylated lysine 9 on histone H3. Nature 442:307–311
Cohen JI (2009) Optimal treatment for chronic active Epstein-Barr virus disease. Pediatr Transplant 13:393–396
Cole J, Morris P, Dickman MJ, Dockrell DH (2016) The therapeutic potential of epigenetic manipulation during infectious diseases. Pharmacol Ther 167:85–99
Costanzi C, Stein P, Worrad DM, Schultz RM, Pehrson JR (2000) Histone macroH2A1 is concentrated in the inactive X chromosome of female preimplantation mouse embryos. Development 127:2283–2289
Dammann R, Li C, Yoon JH, Chin PL, Bates S, Pfeifer GP (2000) Epigenetic inactivation of a RAS association domain family protein from the lung tumour suppressor locus 3p21.3. Nat Genet 25:315–319
Delmore JE, Issa GC, Lemieux ME, Rahl PB, Shi J, Jacobs HM et al (2011) BET bromodomain inhibition as a therapeutic strategy to target c-Myc. Cell 146:904–917
Di Bartolo DL, Cannon M, Liu YF, Renne R, Chadburn A, Boshoff C et al (2008) KSHV LANA inhibits TGF-beta signaling through epigenetic silencing of the TGF-beta type II receptor. Blood 111:4731–4740
Downs JA, Lowndes NF, Jackson SP (2000) A role for *Saccharomyces cerevisiae* histone H2A in DNA repair. Nature 408:1001–1004
Dumler JS, Choi KS, Garcia-Garcia JC, Barat NS, Scorpio DG, Garyu JW et al (2005) Human granulocytic anaplasmosis and Anaplasma phagocytophilum. Emerg Infect Dis 11:1828–1834
Falk MM, Grigera PR, Bergmann IE, Zibert A, Multhaup G, Beck E (1990) Foot-and-mouth disease virus protease 3C induces specific proteolytic cleavage of host cell histone H3. J Virol 64:748–756
Fan JY, Gordon F, Luger K, Hansen JC, Tremethick DJ (2002) The essential histone variant H2A.Z regulates the equilibrium between different chromatin conformational states. Nat Struct Biol 9:172–176
Farber JM, Peterkin PI (1991) *Listeria monocytogenes*, a food-borne pathogen. Microbiol Rev 55:476–511
Feng Q, Wang H, Ng HH, Erdjument-Bromage H, Tempst P, Struhl K et al (2002) Methylation of H3-lysine 79 is mediated by a new family of HMTases without a SET domain. Curr Biol CB 12:1052–1058
Fields BS, Benson RF, Besser RE (2002) Legionella and Legionnaires' disease: 25 years of investigation. Clin Microbiol Rev 15:506–526
Figueroa J, Saffrich R, Ansorge W, Valdivia MM (1998) Microinjection of antibodies to centromere protein CENP-A arrests cells in interphase but does not prevent mitosis. Chromosoma 107:397–405

Fischle W, Franz H, Jacobs SA, Allis CD, Khorasanizadeh S (2008) Specificity of the chromodomain Y chromosome family of chromodomains for lysine-methylated ARK(S/T) motifs. J Biol Chem 283:19626–19635

Flint SJ, Racaniello VR, Enquist LW, Skalka AM (2009) Principles of virology, volume 2: pathogenesis and control. ASM press, Washington, DC

Fonseca GJ, Cohen MJ, Mymryk JS (2014) Adenovirus E1A recruits the human Paf1 complex to enhance transcriptional elongation. J Virol 88:5630–5637

Garcea RL, Imperiale MJ (2003) Simian virus 40 infection of humans. J Virol 77:5039–5045

Garcia-Garcia JC, Rennoll-Bankert KE, Pelly S, Milstone AM, Dumler JS (2009) Silencing of host cell CYBB gene expression by the nuclear effector AnkA of the intracellular pathogen Anaplasma phagocytophilum. Infect Immun 77:2385–2391

Garza-Gonzalez E, Perez-Perez GI, Maldonado-Garza HJ, Bosques-Padilla FJ (2014) A review of Helicobacter pylori diagnosis, treatment, and methods to detect eradication. World J Gastroenterol 20:1438–1449

Gershey EL, Haslett GW, Vidali G, Allfrey VG (1969) Chemical studies of histone methylation. Evidence for the occurrence of 3-methylhistidine in avian erythrocyte histone fractions. J Biol Chem 244:4871–4877

Goldknopf IL, Taylor CW, Baum RM, Yeoman LC, Olson MO, Prestayko AW et al (1975) Isolation and characterization of protein A24, a "histone-like" non-histone chromosomal protein. J Biol Chem 250:7182–7187

Gramolelli S, Schulz TF (2015) The role of Kaposi sarcoma-associated herpesvirus in the pathogenesis of Kaposi sarcoma. J Pathol 235:368–380

Greer EL, Shi Y (2012) Histone methylation: a dynamic mark in health, disease and inheritance. Nat Rev Genet 13:343–357

Hamon MA, Batsche E, Regnault B, Tham TN, Seveau S, Muchardt C et al (2007) Histone modifications induced by a family of bacterial toxins. Proc Natl Acad Sci U S A 104:13467–13472

Hassa PO, Haenni SS, Elser M, Hottiger MO (2006) Nuclear ADP-ribosylation reactions in mammalian cells: where are we today and where are we going? Microbiol Mol Biol Rev 70:789–829

Heerboth S, Lapinska K, Snyder N, Leary M, Rollinson S, Sarkar S (2014) Use of epigenetic drugs in disease: an overview. Genet Epigenet 6:9–19

Heintzman ND, Stuart RK, Hon G, Fu Y, Ching CW, Hawkins RD et al (2007) Distinct and predictive chromatin signatures of transcriptional promoters and enhancers in the human genome. Nat Genet 39:311–318

Hempel K, Lange HW, Birkofer L (1968) Epsilon-N-trimethyllysine, a new amino acid in histones. Naturwissenschaften 55:37

Hilton IB, D'Ippolito AM, Vockley CM, Thakore PI, Crawford GE, Reddy TE et al (2015) Epigenome editing by a CRISPR-Cas9-based acetyltransferase activates genes from promoters and enhancers. Nat Biotechnol 33:510–517

Holliday R (2006) Epigenetics: a historical overview. Epigenetics 1:76–80

Hottiger MO, Hassa PO, Luscher B, Schuler H, Koch-Nolte F (2010) Toward a unified nomenclature for mammalian ADP-ribosyltransferases. Trends Biochem Sci 35:208–219

Jennison AV, Verma NK (2004) Shigella flexneri infection: pathogenesis and vaccine development. FEMS Microbiol Rev 28:43–58

Jenuwein T, Allis CD (2001) Translating the histone code. Science 293:1074–1080

Johnson NP, Mueller J (2002) Updating the accounts: global mortality of the 1918–1920 "Spanish" influenza pandemic. Bull Hist Med 76:105–115

Joo HY, Zhai L, Yang C, Nie S, Erdjument-Bromage H, Tempst P et al (2007) Regulation of cell cycle progression and gene expression by H2A deubiquitination. Nature 449:1068–1072

Jungmichel S, Stucki M (2010) MDC1: the art of keeping things in focus. Chromosoma 119:337–349

Kang B, Pu M, Hu G, Wen W, Dong Z, Zhao K et al (2011) Phosphorylation of H4 Ser 47 promotes HIRA-mediated nucleosome assembly. Genes Dev 25:1359–1364

Kaul R, Hoang A, Yau P, Bradbury EM, Wenman WM (1997) The chlamydial EUO gene encodes a histone H1-specific protease. J Bacteriol 179:5928–5934
Kawase M (2013) Genetics of epidermodysplasia verruciformis. InTech, Rijeka
Kearns NA, Pham H, Tabak B, Genga RM, Silverstein NJ, Garber M et al (2015) Functional annotation of native enhancers with a Cas9-histone demethylase fusion. Nat Methods 12:401–403
Khairalla AS, Omer SA, Mahdavi J, Aslam A, Dufailu OA, Self T et al (2015) Nuclear trafficking, histone cleavage and induction of apoptosis by the meningococcal App and MspA autotransporters. Cell Microbiol 17:1008–1020
Kim K-H, Yi J, Won Sup O, Kim N-H, Choi SJ, Choe PG, Kim N-J, Lee J-K, Myoung-don O (2014) Human granulocytic Anaplasmosis, South Korea, 2013. Emerg Infect Dis 20(10):1708–1711
Klose RJ, Kallin EM, Zhang Y (2006a) JmjC-domain-containing proteins and histone demethylation. Nat Rev Genet 7:715–727
Klose RJ, Yamane K, Bae Y, Zhang D, Erdjument-Bromage H, Tempst P et al (2006b) The transcriptional repressor JHDM3A demethylates trimethyl histone H3 lysine 9 and lysine 36. Nature 442:312–316
Koch-Nolte F, Kernstock S, Mueller-Dieckmann C, Weiss MS, Haag F (2008) Mammalian ADP-ribosyltransferases and ADP-ribosylhydrolases. Front Biosci 13:6716–6729
Kouzarides T (2007) Chromatin modifications and their function. Cell 128:693–705
Kusters JG, Van Vliet AH, Kuipers EJ (2006) Pathogenesis of Helicobacter pylori infection. Clin Microbiol Rev 19:449–490
Lai HC, Lin YW, Chang CC, Wang HC, Chu TW, Yu MH et al (2007) Hypermethylation of two consecutive tumor suppressor genes, BLU and RASSF1A, located at 3p21.3 in cervical neoplasias. Gynecol Oncol 104:629–635
Lan R, Reeves PR (1996) Gene transfer is a major factor in bacterial evolution. Mol Biol Evol 13:47–55
Ledford H (2015) Epigenetics: the genome unwrapped. Nature 528:S12–S13
Li B, Carey M, Workman JL (2007) The role of chromatin during transcription. Cell 128:707–719
Liang TJ (2009) Hepatitis B: the virus and disease. Hepatology 49:S13–S21
Luger K, Mader AW, Richmond RK, Sargent DF, Richmond TJ (1997) Crystal structure of the nucleosome core particle at 2.8 A resolution. Nature 389:251–260
Maas NL, Miller KM, Defazio LG, Toczyski DP (2006) Cell cycle and checkpoint regulation of histone H3 K56 acetylation by Hst3 and Hst4. Mol Cell 23:109–119
Maejima Y, Sadoshima J (2014) SUMOylation: a novel protein quality control modifier in the heart. Circ Res 115:686–689
Malhotra M, Sood S, Mukherjee A, Muralidhar S, Bala M (2013) Genital Chlamydia trachomatis: an update. Indian J Med Res 138:303–316
Mandal P, Azad GK, Tomar RS (2012) Identification of a novel histone H3 specific protease activity in nuclei of chicken liver. Biochem Biophys Res Commun 421:261–267
Mandal P, Verma N, Chauhan S, Tomar RS (2013) Unexpected histone H3 tail-clipping activity of glutamate dehydrogenase. J Biol Chem 288:18743–18757
Mandal P, Verma N, Azad GK, Singh V, Golla U, Tomar RS (2014) Epigenetics: role of histone proteases in cellular functions and diseases. Springer, New York
Medicine IO (2015) Emerging viral diseases: the one health connection: workshop summary. The National Academies Press, Washington, DC
Mesri EA, Cesarman E, Boshoff C (2010) Kaposi's sarcoma and its associated herpesvirus. Nat Rev Cancer 10:707–719
Messner S, Hottiger MO (2011) Histone ADP-ribosylation in DNA repair, replication and transcription. Trends Cell Biol 21:534–542
Metzger E, Wissmann M, Yin N, Muller JM, Schneider R, Peters AH et al (2005) LSD1 demethylates repressive histone marks to promote androgen-receptor-dependent transcription. Nature 437:436–439
Milne JC, Denu JM (2008) The Sirtuin family: therapeutic targets to treat diseases of aging. Curr Opin Chem Biol 12:11–17

Moosavi A, Motevalizadeh Ardekani A (2016) Role of epigenetics in biology and human diseases. Iran Biomed J 20:246–258

Murray K (1964) The occurrence of epsilon-N-methyl lysine in histones. Biochemistry 3:10–15

Nelson CJ, Santos-Rosa H, Kouzarides T (2006) Proline isomerization of histone H3 regulates lysine methylation and gene expression. Cell 126:905–916

Niwa T, Toyoda T, Tsukamoto T, Mori A, Tatematsu M, Ushijima T (2013) Prevention of Helicobacter pylori-induced gastric cancers in gerbils by a DNA demethylating agent. Cancer Prev Res (Philadelphia, Pa.) 6:263–270

Noma K, Allis CD, Grewal SI (2001) Transitions in distinct histone H3 methylation patterns at the heterochromatin domain boundaries. Science 293:1150–1155

Ochman H, Lawrence JG, Groisman EA (2000) Lateral gene transfer and the nature of bacterial innovation. Nature 405:299–304

Paddock CD, Childs JE (2003) Ehrlichia chaffeensis: a prototypical emerging pathogen. Clin Microbiol Rev 16:37–64

Paschos K, Allday MJ (2010) Epigenetic reprogramming of host genes in viral and microbial pathogenesis. Trends Microbiol 18:439–447

Pennini ME, Perrinet S, Dautry-Varsat A, Subtil A (2010) Histone methylation by NUE, a novel nuclear effector of the intracellular pathogen Chlamydia trachomatis. PLoS Pathog 6:e1000995

Persat A, Nadell CD, Kim MK, Ingremeau F, Siryaporn A, Drescher K et al (2015) The mechanical world of bacteria. Cell 161:988–997

Pollina EA, Brunet A (2011) Epigenetic regulation of aging stem cells. Oncogene 30:3105–3126

Porwollik S, Boyd EF, Choy C, Cheng P, Florea L, Proctor E et al (2004) Characterization of Salmonella enterica subspecies I genovars by use of microarrays. J Bacteriol 186:5883–5898

Qin S, Parthun MR (2006) Recruitment of the type B histone acetyltransferase Hat1p to chromatin is linked to DNA double-strand breaks. Mol Cell Biol 26:3649–3658

Rea S, Eisenhaber F, O'Carroll D, Strahl BD, Sun ZW, Schmid M et al (2000) Regulation of chromatin structure by site-specific histone H3 methyltransferases. Nature 406:593–599

Redon C, Pilch DR, Rogakou EP, Orr AH, Lowndes NF, Bonner WM (2003) Yeast histone 2A serine 129 is essential for the efficient repair of checkpoint-blind DNA damage. EMBO Rep 4:678–684

Riggs MG, Whittaker RG, Neumann JR, Ingram VM (1977) n-Butyrate causes histone modification in HeLa and Friend erythroleukaemia cells. Nature 268:462–464

Rogakou EP, Pilch DR, Orr AH, Ivanova VS, Bonner WM (1998) DNA double-stranded breaks induce histone H2AX phosphorylation on serine 139. J Biol Chem 273:5858–5868

Rossetto D, Avvakumov N, Cote J (2012) Histone phosphorylation: a chromatin modification involved in diverse nuclear events. Epigenetics 7:1098–1108

Rouphael NG, Stephens DS (2012) Neisseria meningitidis: biology, microbiology, and epidemiology. Methods Mol Biol (Clifton N.J.) 799:1–20

Salguero FJ, Sanchez-Martin MA, Diaz-San Segundo F, De Avila A, Sevilla N (2005) Foot-and-mouth disease virus (FMDV) causes an acute disease that can be lethal for adult laboratory mice. Virology 332:384–396

Santos-Rosa H, Schneider R, Bannister AJ, Sherriff J, Bernstein BE, Emre NC et al (2002) Active genes are tri-methylated at K4 of histone H3. Nature 419:407–411

Schulze JM, Jackson J, Nakanishi S, Gardner JM, Hentrich T, Haug J et al (2009) Linking cell cycle to histone modifications: SBF and H2B monoubiquitination machinery and cell-cycle regulation of H3K79 dimethylation. Mol Cell 35:626–641

Shanbhag NM, Rafalska-Metcalf IU, Balane-Bolivar C, Janicki SM, Greenberg RA (2010) ATM-dependent chromatin changes silence transcription in cis to DNA double-strand breaks. Cell 141:970–981

Shi Y, Lan F, Matson C, Mulligan P, Whetstine JR, Cole PA et al (2004) Histone demethylation mediated by the nuclear amine oxidase homolog LSD1. Cell 119:941–953

Shi X, Hong T, Walter KL, Ewalt M, Michishita E, Hung T et al (2006) ING2 PHD domain links histone H3 lysine 4 methylation to active gene repression. Nature 442:96–99

Smale ST, Tarakhovsky A, Natoli G (2014) Chromatin contributions to the regulation of innate immunity. Annu Rev Immunol 32:489–511

Smith I (2003) Mycobacterium tuberculosis pathogenesis and molecular determinants of virulence. Clin Microbiol Rev 16:463–496

Steward MM, Lee JS, O'Donovan A, Wyatt M, Bernstein BE, Shilatifard A (2006) Molecular regulation of H3K4 trimethylation by ASH2L, a shared subunit of MLL complexes. Nat Struct Mol Biol 13:852–854

Stewart GS, Panier S, Townsend K, Al-Hakim AK, Kolas NK, Miller ES et al (2009) The RIDDLE syndrome protein mediates a ubiquitin-dependent signaling cascade at sites of DNA damage. Cell 136:420–434

Stucki M, Clapperton JA, Mohammad D, Yaffe MB, Smerdon SJ, Jackson SP (2005) MDC1 directly binds phosphorylated histone H2AX to regulate cellular responses to DNA double-strand breaks. Cell 123:1213–1226

Swanson EC, Schleiss MR (2013) Congenital cytomegalovirus infection: new prospects for prevention and therapy. Pediatr Clin N Am 60:335–349

Taylor LH, Latham SM, Woolhouse ME (2001) Risk factors for human disease emergence. Philos Trans R Soc Lond Ser B Biol Sci 356:983–989

Tsukada Y, Fang J, Erdjument-Bromage H, Warren ME, Borchers CH, Tempst P et al (2006) Histone demethylation by a family of JmjC domain-containing proteins. Nature 439:811–816

Utley RT, Lacoste N, Jobin-Robitaille O, Allard S, Cote J (2005) Regulation of NuA4 histone acetyltransferase activity in transcription and DNA repair by phosphorylation of histone H4. Mol Cell Biol 25:8179–8190

Vaquero A, Scher MB, Lee DH, Sutton A, Cheng HL, Alt FW et al (2006) SirT2 is a histone deacetylase with preference for histone H4 Lys 16 during mitosis. Genes Dev 20:1256–1261

Vilchez RA, Butel JS (2004) Emergent human pathogen simian virus 40 and its role in cancer. Clin Microbiol Rev 17:495–508, table of contents

Vishwakarma S, Iyer LR, Muley M, Singh PK, Shastry A, Saxena A et al (2013) Tubastatin, a selective histone deacetylase 6 inhibitor shows anti-inflammatory and anti-rheumatic effects. Int Immunopharmacol 16:72–78

Vitousek PM, Mooney HA, Lubchenco J, Melillo JM (1997) Human domination of Earth's ecosystems. Science 277:494–499

Wang H, Wang L, Erdjument-Bromage H, Vidal M, Tempst P, Jones RS et al (2004) Role of histone H2A ubiquitination in Polycomb silencing. Nature 431:873–878

West MH, Bonner WM (1980) Histone 2B can be modified by the attachment of ubiquitin. Nucleic Acids Res 8:4671–4680

Whetstine JR, Nottke A, Lan F, Huarte M, Smolikov S, Chen Z et al (2006) Reversal of histone lysine trimethylation by the JMJD2 family of histone demethylases. Cell 125:467–481

Wu X, Qi J, Bradner JE, Xiao G, Chen LF (2013) Bromodomain and extraterminal (BET) protein inhibition suppresses human T cell leukemia virus 1 (HTLV-1) tax protein-mediated tumorigenesis by inhibiting nuclear factor kappaB (NF-kappaB) signaling. J Biol Chem 288:36094–36105

Yaseen I, Kaur P, Nandicoori VK, Khosla S (2015) Mycobacteria modulate host epigenetic machinery by Rv1988 methylation of a non-tail arginine of histone H3. Nat Commun 6:8922

Zhao D, Fukuyama S, Sakai-Tagawa Y, Takashita E, Shoemaker JE, Kawaoka Y (2015) C646, a novel p300/CREB-binding protein-specific inhibitor of histone acetyltransferase, attenuates influenza A virus infection. Antimicrob Agents Chemother 60:1902–1906

Zhou P, Wu E, Alam HB, Li Y (2014) Histone cleavage as a mechanism for epigenetic regulation: current insights and perspectives. Curr Mol Med 14:1164–1172

Zhu B, Nethery KA, Kuriakose JA, Wakeel A, Zhang X, Mcbride JW (2009) Nuclear translocated Ehrlichia chaffeensis ankyrin protein interacts with a specific adenine-rich motif of host promoter and intronic Alu elements. Infect Immun 77:4243–4255

Part IV
Fungal Infections

Chapter 21
An Update on Sexually Transmitted Infections: An Indian Context

Priya Datta and Varsha Gupta

Abstract Sexually transmitted infections (STIs) are an important public health problem in the young reproductive age group worldwide. The exact prevalence of STIs in India is not known because of insufficient reporting and privacy regarding these diseases. There is a change in epidemiology of various STIs in India. The incidence of bacterial diseases like syphilis, gonorrhea, and chancroid is decreasing, and the viral STIs are increasing. Various newer methods like NAAT (nucleic acid amplification technique) and point-of-care (POC) testing have increased the sensitivity of the diagnosis and made it more rapid. Various newer methods like real-time PCR, strand displacement amplification, and transcription-mediated amplification have been approved for the diagnosis of gonorrhea, chlamydial, and trichomonal infections causing STIs using different primers. There has been isolation of strains of *Neisseria gonorrhoeae* having high MIC for cefixime and ceftriaxone from France, Spain, and Japan. The spread of this highly resistant strain of *N. gonorrhoeae* would be catastrophic. Government of India advocates syndromic management of STIs wherein treatment for sexually transmitted infections is given according to the syndrome the patient is presenting with like vaginal discharge, urethral discharge, genital ulcer, etc.

Keywords Diagnosis · India · Sexually transmitted infections

21.1 Introduction

There is a great burden placed on the healthcare system of India due to sexually transmitted infections (STIs). These infections typically affect young reproductive age group. There is a changing epidemiology of various STIs in India as bacterial diseases like syphilis, gonorrhea, and chancroid are decreasing and the viral STIs are increasing. Various diagnostic tests like nucleic acid amplification technique (NAAT) and point-of-care (POC) test kit have increased the sensitivity of the

P. Datta (✉) · V. Gupta
Department of Microbiology, Government Medical College Hospital, Chandigarh, India

P. P. Singh (ed.), *Infectious Diseases and Your Health*,
https://doi.org/10.1007/978-981-13-1577-0_21

diagnosis of various STIs and made it more rapid. These include real-time PCR, strand displacement amplification (SDA), and transcription-mediated amplification (TMA) especially for the diagnosis of gonorrhea, chlamydial, and trichomonas infections causing STIs using different primers. The problem of STIs has been further compounded with the isolation of strains of *Neisseria gonorrhoeae* having high MIC for cefixime and ceftriaxone from France, Spain, and Japan, leaving very few treatment options. This multidrug resistant bacteria has not been detected in India till now and spread of this highly resistant strain of *N. gonorrhoeae* would be disastrous. Therefore, continuous monitoring, surveillance, and reporting of various STIs are essential to control the spread of this deadly disease. Syndromic management is the mainstay of treatment of STIs in India, wherein treatment for sexually transmitted infections is given according to the syndrome the patient.

The STIs are communicable diseases transmitted by sexual contact. The various causative agents can be bacterial, viral, protozoal, fungal, or rarely ectoparasitic. STIs differ from STDs, as STDs include infection resulting in clinical diseases that involve genitalia and other parts of the body participating in sexual infection, whereas STIs include infection that may cause clinical diseases of genital area but are transmitted by sexual interaction, e.g., all STDs, hepatitis B, and HIV.

STIs are an important public health problem worldwide as well as in India. These have a profound effect on sexual and reproductive health worldwide. STIs are an important cause of infertility in both women and men. The consequence of STIs in pregnancy leads to still birth, abortions, low birth weight, prematurity, sepsis, congenital deformity, and neonatal death (Park 2015). In recent years, drug resistance in patients with gonorrhea has become a cause of concern globally. Most of the cases of STIs are curable and preventable (Government of India 2014. National guidelines on prevention, management, and control of reproductive tract infections and sexually transmitted infections). A prevalence study done in 2002–2003 by Indian Council Medical Research among the community population in India showed that around 6% of adult population are having one or more STI/year (Ray et al. 2006).

The various agents which cause STIs are bacterial, viral, fungal, and parasitic agents. The important bacterial agents are *N. gonorrhoeae*, *Chlamydia trachomatis*, *Treponema pallidum*, *Haemophilus ducreyi*, *Mycoplasma hominis*, *Ureaplasma urealyticum*, *Calymmatobacterium granulomatis*, *Shigella* spp., *Campylobacter* spp., *Group B streptococcus*, and bacterial vaginosis-associated organisms. The important viral causes of STIs are *herpes simplex virus 1* and 2, *Cytomegalovirus*, *Hepatitis B virus* , *human papilloma viruses*, *Molluscum contagiosum virus*, and *human immunodeficiency virus.* The protozoal and fungal agents implicated in causing STIs are *Entamoeba histolytica*, *Giardia lamblia*, *Trichomonas vaginalis*, and *Candida albicans*, respectively. Sometimes ectoparasites like *Pthirus pubis* and *Sarcoptes scabiei* are also linked to STIs (Thappa and Kaimal 2007).

21.2 Epidemiology of STI in India

As there is no reporting system for STIs in India, data regarding the prevalence of these diseases are not readily available. This is because of inadequate reporting and secrecy. But few studies done on major STI prevalent in India have studied the prevalence. In a study done in 2012, there were 97,180 cases of gonorrhea (31,564 male and 65,616 female) and 33,570 cases (18,081 male and 15,489 female) of syphilis reported in India. LGV (lymphogranuloma venereum) is more prevalent in southern states of Tamil Nadu, Andhra Pradesh, Maharashtra, and Karnataka. Donovanosis or granuloma inguinale is endemic in Tamil Nadu, Andhra Pradesh, Orissa, Karnataka, and Maharashtra.

Current data available from various STI control program from India shows major decline in bacterial STIs (syphilis and gonorrhea). Additionally the prevalence of chancroid has decreased significantly and is on the verge of disappearing. On the other hand, the incidence of viral STIs like herpes and genital warts has increased (Park 2015).

21.3 Clinical Spectrum

21.3.1 Gonococcal Infection

N. gonorrhoeae, the causative agent of gonorrhea, is transmitted by sexual contact or during childbirth. It is characterized by purulent infection of genital mucous surfaces, i.e., urethra, cervix, and rectum. The main symptoms in female's gonorrhea consist of vaginal discharge, mild lower abdominal pain, dysuria, dyspareunia, and intermenstrual bleeding. Untreated infection can progress and lead to pelvic inflammatory disease (PID). Long-term sequelae of PID are increased risk of ectopic pregnancy, infertility, and chronic pelvic pain. In males, the major genitourinary symptoms of gonorrhea include urethritis, acute epididymitis, urethral strictures, and rectal infection. The most common laboratory methods for diagnosis include culture on selective media, followed by NAAT – nucleic acid amplification technique. NAAT is more sensitive for pharyngeal and rectal samples. The various FDA-approved NAATs for *N. gonorrhoeae* include real-time PCR, strand displacement amplification (SDA), and transcription-mediated amplification (TMA). The various targets are 16SrRNA, DR 9, *opa genes*, cytosine DNA methyltransferase gene, and pilin. The advantage of culture for diagnosis of gonorrhea is that it is more sensitive, highly specific in optimized circumstances, comparatively inexpensive, and permits antimicrobial susceptibility. The criterion standard for diagnosis is a swab taken from all potential sites of gonococcal infection. Additionally culture can be beneficial in patients where clinical diagnosis is unclear, in treatment-failure patients, difficult contact tracing, and medicolegal cases. The drug of choice in uncomplicated urogenital, anorectal, and pharyngeal gonococcal cases are ceftriaxone, cefixime, azithromycin, ciprofloxacin, doxycycline, or spectinomycin.

Lately, the strains of *N. gonorrhoeae* being isolated have shown resistance to first-line antimicrobials like penicillin, tetracycline, and fluoroquinolones; thereby, the only alternative left is the third-generation expanded-spectrum cephalosporins like ceftriaxone and cefixime. This problem has been further compounded by appearance of extensively drug-resistant (XDR) gonococcal strains which have high degree of resistance to ceftriaxone and cefixime from France, Spain, and Japan. Therefore, testing of *N. gonorrhoeae* to monitor antimicrobial sensitivity is of paramount importance.

21.3.2 Syphilis

The spirochete *Treponema pallidum* causes syphilis which is an infectious venereal disease. The various ways of spread of syphilis include sexual contact, blood and blood products, from mother to child, and occasionally through direct contact with infectious lesions. The clinical course of the disease, if left untreated, progresses through four stages: primary, secondary, latent, and tertiary.

The diagnosis of syphilis is complex because *T. pallidum* cannot be cultivated in vitro and cannot be seen under the light microscope. Therefore, in all the stages of syphilis, serological tests are considered the standard method of laboratory diagnosis. In a suspected acquired syphilis, traditionally screening using the non-specific tests like venereal disease research laboratory (VDRL) or rapid plasma reagin (RPR) is done first. Confirmation is done with specific treponemal test for any positive or equivocal nontreponemal test result to rule out false-positive results. The various treponemal tests include the *T. pallidum* hemagglutination (TPHA), microhemagglutination assay *T. pallidum* (MHA-TP), fluorescent treponemal antibody-absorption (FTA-ABS), and treponemal enzyme immunoassay (EIA) for IgG and IgM.

After positive VDRL or RPR test, FTA-ABS is a frequently used confirmatory test. It has a sensitivity of 84% in diagnosing primary syphilis infection and almost 100% sensitivity for diagnosing syphilis infection in other stages (WHO 2013, laboratory diagnosis of sexually transmitted infections, including human immunodeficiency virus). Its specificity is 96%. In evaluating clinical cases of syphilis presenting with moist cutaneous lesions such as the chancre of primary syphilis or the condyloma lata of secondary syphilis, dark-field microscopy can be used. An alternative to dark-field microscopy is direct immunofluorescence staining of fixed smears (direct fluorescent antibody *T. pallidum* [DFA-TP]) in these clinical cases. Since the VDRL test for CSF (VDRL-CSF) is highly specific but has low sensitivity, therefore the diagnosis of neurosyphilis can be challenging. The combination of CSF cell count, CSF protein, and clinical manifestations with or without a reactive VDRL-CSF is used for the diagnosis of neurosyphilis. The antibiotics used to treat syphilis are penicillin, doxycycline, and erythromycin.

In India, RPR/VRDL is not available at most peripheral healthcare centers. Thus, the government of India has introduced rapid point-of-care testing for syphilis. These rapid POC (immunochromatographic) tests are stored at 4–30 °C for up to a year, don't require any special equipment, and can be done by a trained paramedical staff (Government of India 2014. National guidelines on prevention, management, and control of reproductive tract infections and sexually transmitted infections).

21.3.3 Chlamydial Infection

Chlamydiae are obligate, gram-negative intracellular pathogens that specifically infect squamocolumnar epithelial cells. *Chlamydia trachomatis* causes two types of genital infections on the basis of different serovars – serovars D–K cause genital tract infections, and serovars L1–L3 cause lymphogranuloma venereum (LGV). Genital chlamydiasis is the common STIs worldwide. The main site of infection by *C. trachomatis* infection includes the cervix, urethra, salpinges, uterus, urethra, epididymis, and nasopharynx. In this, nearly 80% of infected female and 50% of infected males are asymptomatic. In males, the most common clinical manifestation includes urethritis, epididymitis, proctitis, and conjunctivitis, whereas in females, mucopurulent cervicitis, endometritis, and salpingitis are the common sites of infection. The cause of PID, chronic pelvic pain, and perinephritis is primarily due to ascending infection. Also some serotypes like *C. trachomatis* serotype C have 6.5 times more risk of developing cervical cancer. The various samples for diagnosis of *C. trachomatis* include endocervical, urethral, rectal, or oropharyngeal specimens. In males, voided urine sample, whether midstream or first void, is an ideal sample as it successfully captures the chlamydial organism for nucleic acid amplification testing (NAAT). For all these specimens, NAATs are the most sensitive tests and the recommended method for diagnosis of *C. trachomatis* infection. Additionally the sensitivity and specificity of self-collected vaginal swab specimens are equivalent to those collected by a clinician. Various FDA-approved NAATs include real-time PCR, SDA, and TMA, and the various target regions are chlamydial cryptic plasmids and 23S rRNA. Several POC (point-of-care) tests have been developed for the diagnosis of *C. trachomatis* infection based on principle of lateral flow and antigen membrane capture on immunochromatographic strips. The antimicrobial agents used for chlamydia infection are doxycycline, azithromycin, or tetracycline.

21.3.4 Trichomoniasis

Trichomoniasis is a parasitic infection by *Trichomonas vaginalis* of multiple sites, i.e., vaginal epithelium, Skene's glands, Bartholin's glands, and the urethra. In females and males, usually there are no symptoms. Occasionally females can present with vaginitis and vaginal discharge. There is major difference between *T.*

vaginalis infections and that of other causes of vaginal discharge. Firstly, *T. vaginalis* infection occurs in higher age group like 40–50 years (as compared to 20–30 years in other infective causes of vaginal discharge), and secondly, *T. vaginalis* infects females more than males (4:1). In pregnancy, there are chances of adverse effects like low birth weight and premature rupture of membranes. There are four major diagnostic methods available – wet preparation microscopy, antigen detection, culture, and NAATs. Additionally various POC tests are also available. The treatment option is tinidazole or metronidazole.

21.3.5 Chancroid

H. ducreyi that causes chancroid is a bacterial STI. The patient presents with painful necrotizing genital ulcers along with inguinal lymphadenopathy. *H. ducreyi* penetrates the skin of the external genitalia subsequently colonizing the subcutaneous tissue and leading to tissue damage and ulceration. Currently no rapid laboratory testing method is available for the diagnosis of chancroid (Levis 2000). Culture of *H. ducreyi* on special media (owing to the fastidious nature of the organism) is needed for the definitive diagnosis of chancroid. The drawback of culture is that the sensitivity of culture from lesions is less than 80% and culture is not readily available in many centers. Therefore, the role of NAAT in rapid detection of *H. ducreyi* is encouraging and may surpass culture in diagnosis (Orle et al. 1996). The antibiotics used for treatment are ciprofloxacin, erythromycin, ceftriaxone, and azithromycin.

21.3.6 Lymphogranuloma Venereum

Lymphogranuloma venereum (LGV) is an ulcerative STI caused by *C. trachomatis*. The other various causative agents of genital ulcer diseases include HSV-2, syphilis, and chancroid. In LGV, the clinical manifestation is self-limited genital papules or ulcers followed by painful inguinal and/or femoral lymphadenopathy (Mabey and Peeling 2002). In patients participating in receptive anal intercourse, the presentation may be rectal pain, discharge, and bleeding due to rectal ulcerations and proctocolitis. The complications of LGV, if not treated, may be enlargement of the external genitalia, disfiguring ulceration, and subsequent lymphatic obstruction. Laboratory diagnosis depends on detecting *C. trachomatis* – specific DNA – and then genotyping to classify serovars L1, L2, or L3 found in LGV. For culture, the sample of choice is needle aspiration of an involved bubo. But the culture of *C. trachomatis* is technically challenging and expensive, and the sensitivity is only 30% (Alexander et al. 2008). Serologic testing is tricky because of cross-reactivity of the many different serotypes.

In LGV complement-fixation titer, greater than 1:64, along with the proper clinical situation, is considered diagnostic. The problem with chlamydial urethritis, cervicitis, or conjunctivitis is that antibodies titer is rarely greater than 1:16. Additionally cross-reactivity between various chlamydial infections occurs on complement-fixation testing. Other tests like polymerase chain reaction (PCR) and immunofluorescent testing with monoclonal antibodies are effective but are limited in availability. Therefore, currently the diagnosis is based primarily on clinical findings, along with positive nucleic acid amplification tests (NAATs). Commercially available NAATs, using urine as specimen for demonstration of *C. trachomatis*, have demonstrated an increased sensitivity and specificity, i.e., 96–100% and 99.1–100%, respectively (WHO 2013, laboratory diagnosis of sexually transmitted infections, including *human immunodeficiency virus*). The antibiotics used are doxycycline, erythromycin, and tetracycline. Surgical operation may be used in cases of extensive elephantiasis or deformity (White 2009).

21.3.7 Donovanosis/Granuloma Inguinale

Granuloma inguinale is a chronic STI caused by *Klebsiella granulomatis*. This disease is characterized by Donovan bodies which are intracellular inclusions in macrophages. The clinical presentation of this STI is nodular lesions involving the skin and mucous membranes in the genital region, which progress into ulcers. The ulcers gradually enlarge and cause local necrosis. The culture of *K. granulomatis* is difficult because the organism is exceedingly fastidious and culture is beyond the skill of most laboratories. The commonest and easiest method of diagnosis is to visualize the organism within the cytoplasm of histiocytes in smears made from the base of the ulcer. These typically exhibit bipolar staining, giving safety-pin appearance, and are referred to as Donovan bodies (O'Farrrell 2002). Additionally stains like Wright-Giemsa, Warthin-Starry, toluidine blue, or Leishman stain are used to demonstrate the Donovan bodies in tissue biopsy specimens. The use of PCR techniques in diagnosis is limited to research though it has good sensitivity. Recently, *K. granulomatis* has been isolated from feces using monocyte co-culture system and a modified *Chlamydia* culture (Carter and Kemp 2000). Antibiotics used are azithromycin, doxycycline, tetracycline, or trimethoprim-sulfamethoxazole.

21.3.8 Mycoplasma

Mycoplasmas are very small, cell wall-deficient, free-living bacteria. Due to the absence of cell wall, mycoplasma is resistant to beta-lactams. The urogenital mycoplasma includes *M. genitalium* and *M. hominis*. *M. genitalium* has shown strong association in patients with non-gonococcal urethritis (NGU) especially in persistent or recurrent NGU. This could be because of reduced microbiologic treatment efficacy of tetracyclines. The other clinical entities associated with urogenital mycoplasma include cervicitis, preterm birth, endometritis, and infertility

(Taylor-Robinson and Jensen 2011). The culture of mycoplasma has many disadvantages – i.e., takes long (several months), difficult, and insensitive. The sample of choice is first-void urine from men and vaginal swabs from women as these contain the highest load of bacteria. The best method diagnosis of *M. genitalium* uses NAAT, as it is sensitive and practical method. In NAATs for the diagnosis of *M. genitalium*, the MgPa adhesin gene and 16S rRNA gene are used.

21.3.9 Genital Herpes

Herpes simplex viruses cause a wide variety of disease and are present ubiquitously. There are two types: *Herpes simplex virus* type 1 (HSV-1) and type 2 (HSV-2). These are closely related but differ in epidemiology. HSV-1 commonly causes orofacial disease, while HSV-2 is commonly associated with genital disease. Classical genital herpes are identified by the presence of typical popular lesions that progress to multiple blisters and ulcers. HSV-2 infections are typically lifelong and recurrent ulcerative episodes keep occurring. The gold standard of diagnosis of *herpes simplex virus* (HSV) infection is isolation of the virus in tissue culture. The disadvantage of tissue culture is that it is operator-dependent, but it can produce positive results within 48 h of inoculation. The characteristic of cytopathic effect associated with HSV is ballooning of cells and cell death of the entire monolayer of cells. Also this characteristic cytological changes induced by HSV can be demonstrated in Tzanck smear. To distinguish HSV type 1 and 2, immunofluorescent staining of the tissue culture cells can be done. Rapid diagnosis, i.e., within an hour, is based on the histological appearance of the lesion (Le Goff et al. 2014).

In the herpetic lesions, there occur multinucleated giant cells and epithelial cells with characteristic eosinophilic intranuclear inclusion bodies. In lesions infected with bacteria and fungi, punch biopsy provides more reliable material for histological examination. Detection of HSV DNA in clinical specimens by PCR techniques is more sensitive than culture, and additionally PCR is preferred test for ocular and CNS infections. In ocular lesions, cells scraped from ulcer bases can be also stained with a direct fluorescent antibody. This procedure can usually be performed within 2–3 h. Oral antiviral drugs like acyclovir, valacyclovir, and famciclovir are effective in reducing the severity and duration of genital herpes (Domeika et al. 2010).

21.3.10 Human Papillomavirus

HPV are small ds-DNA that infects epithelium. Based on the genetic sequence of the outer capsid protein L1, more than 120 types have been identified. There are about 40 types which infect the mucosal epithelium. Infection with low-risk type such as type 6 and 11 can cause low-grade cervical cell abnormality and genital warts. High-risk types (16, 18) are linked to cervical and anogenital cancer. In India, the most common oncogenic type is HPV 16 and 18 (98%), with HPV 16 exclusively causing 80–90% of cancer. The epithelial tumors of the skin and mucous membranes are caused by HPV (Dunne and Markowitz 2006). According to anatomic area involved, the clinical history and presentation of HPV infection vary. Due is because certain viral genotypes have a predilection for infecting certain epidermal sites. The various conditions associated with HPV include the following:

1. Anogenital warts (condylomata acuminata): They are found near the perianal area, vagina, labia, and vulva. These lesions are not painful but are associated with pruritus or bleeding.
2. Cervical disease.
3. Anal cancer: In 50% of men who are homosexual, there occur anorectal warts and can progress to anal squamous cell carcinoma (SCC).

The clinical diagnosis of cutaneous, external genital warts can be made with application of acetic acid and biopsy. When genital intraepithelial neoplasia is suspected, the careful inspection and colposcopy are done to determine the extent of disease. HPV DNA testing is done for detection of HPV and posttreatment follow-up of cervical intraepithelial neoplasia. HPV DNA testing is done by hybrid capture II or PCR assay. For prevention of HPV, infection vaccines are available, and suggested vaccination schedules have been drawn. Recently, two virus-like particles (VLPs) based prophylactic HPV vaccine. Gardasil and Cervarix have been introduced in India. Gardasil is a quadrivalent (HPV16/18/6/11) vaccine to be given in 0.5 ml intramuscular dose i/m 0, 2 and 6 months. Cervarix is a bivalent (HPV16/18) vaccine to be given in 0.5 ml intramuscular dose i/m 0, 1 and 6 months (Bhardwaj et al. 2009).

21.4 Syndromic Approach to STI

The traditional method of diagnosis of STI is by phenotypic and genotypic laboratory tests. Since 1990, WHO has recommended syndromic management of STIs in patients presenting with recognized signs and symptoms. The syndromic approach is a scientifically derived efficient approach, using flow charts which offer accessible, cost-effective, and immediate treatment to the patient. The advantage of syndromic diagnosis is that it leads to fast treatment for all the most significant likely causative agents. This is imperative because mixed infections occur regularly in STIs. The basis of syndromic approach is that the treatment for each syndrome is toward the most common organisms which caused that syndrome.

The management is decided according to the clinical management flowchart for the symptom as majority of patients with STIs present with symptoms such as urethral discharge/dysuria, genital ulcer/s, vaginal discharge, and lower abdominal pain. Also there are partner management, management among pregnant group, condom demonstration and provision, standardized drug regimen, and directly observed treatment strategies (to enhance compliance) wherever possible.

The syndromic case management of STIs has been adopted universally and is applicable at all levels of healthcare system. In India, this ensures access to the general population to standardized STI management. Additionally guidance is provided to address STI among high-risk group such as female sex workers, men who have sex with men (MSM), transsexual and transgender, and intravenous drug users.

However, the disadvantage of syndromic management is that it does not address asymptomatic and/or subclinical infections, often seen in women. If laboratory facilities are available, appropriate tests should be done to assist the healthcare provider in arriving at an etiological diagnosis and known drug resistance pattern. Appropriate treatment of patients with STI is important to prevent the development of complications and sequelae (Government of India 2014. National guidelines on prevention, management, and control of reproductive tract infections and sexually transmitted infections).

References

Alexander S, martin IM, Iscon C (2008) Comparison of two methods for the diagnosis of LGV. J Med Microbiol 57:962–965

Bhardwaj N, Hussain S, Nasare V, Das BC (2009) HPV and HPV vaccination: issues in developing countries. Indian J Med Res 130:327–333

Carter JS, Kemp DJ (2000) A calorimetric detection system of Calymmatobacterium granulomatis. Sex Transm Infect 76:134–136

Domeika M, Bashmakova M, Savicheva A, Kolomiec N, Sokolovskiy E, Hollen A (2010) Guidelines the laboratory diagnosis of genital herpes in Eastern European countries. Eurosurveillance 15:pii19730

Dunne E, Markowitz L (2006) Genital human papillomavirus infection. Clin Infect Dis 43:624–629

Government of India (2014) National guidelines on prevention, management and control of reproductive tract infections and sexually transmitted infections. Department of AIDS Control, Ministry of Health and Family Welfare, New Delhi

Le Goff J, Pere H, Belec L (2014) Diagnosis of genital herpes simplex virus infection in the clinical laboratory. Virol J 11:83–100

Levis DA (2000) Diagnostic tests for chancroid. Sex Transm Infect 76:137–141

Mabey D, Peeling RW (2002) Lymphogranuloma venereum. Sex Transm Dis 78:90–92

O'Farrell N (2002) Donovanosis Sex Transm Infect 78:452–457

Orle KA, Gates CA, Martin DH, Body BA, Weiss JB (1996) Simultaneous PCR detection of Haemophilus ducreyi, Treponema pallidum and herpes simplex virus type 1 and 2 from genital ulcers. J Clin Microbiol 34:49–54

Park K (2015) Epidemiology of communicable disease. In: Park's textbook of preventive and social medicine. Bhanot Publishers, Jabalpur, pp 330–341

Ray K, Bala M, Gupta M, Khunger N, Puri P, Muralidhar S et al (2006) Changing trends in sexually transmitted infections at a regional STD centre in Northern India. Indian J Med Res 124:559–568

Taylor-Robinson D, Jensen JS (2011) Mycoplasma genitalium: from chrysalis to multicolored butterfly. Clin Microbiol Rev 24:498–514

Thappa DM, Kaimal S (2007) Sexually transmitted infections: current status except HIV. Indian J Dermatol 52:78–82

White JA (2009) Manifestations and management of lymphogranuloma venereum. Curr Opin Infect Dis 22:55–59

WHO (2013) Laboratory diagnosis of sexually transmitted infections including human immunodeficiency viruses. WHO, Geneva

Chapter 22
Candida: Friend and Foe of Humans

Priyanka Bhakt, Anamika Battu, and Rupinder Kaur

Abstract Opportunistic fungal infections pose a serious public health challenge. *Candida* species are the most common cause of hospital-acquired bloodstream fungal infections and are associated with a mortality rate of about 40%. Majority of these infections are caused by the yeast *Candida albicans*, which lives in the mouth, intestines, and urinary and reproductive tracts in healthy individuals but causes various infections in patients with weakened immune systems. *C. albicans*, which is multimorphic in nature, has no known terrestrial life cycle. The spectrum of diseases caused by *C. albicans* includes superficial infections of the skin (thrush), vagina (yeast infections), mouth (oral thrush), and life-threatening bloodstream infections (candidemia). Candidemia, which is diagnosed by fungal culture of the blood, usually results in the multiple organ failure and patient death. *Candida* infections commonly arise from the microflora of the patient but can spread in health-care settings due to contaminated beds, medical equipment, surfaces, and hands of health-care workers. Recent global surveillance programs have revealed significant contribution of four other species of *Candida*, *C. glabrata*, *C. tropicalis*, *C. parapsilosis*, and *C. krusei*, to bloodstream *Candida* infections. We will discuss salient features and prevalence of *Candida* species, the spectrum of diseases caused, and their prevention, diagnosis, and treatment in this chapter.

Keywords Antifungals · *Candida albicans* · Fungal infections · Oral and vaginal thrush · Sepsis

---

P. Bhakt · A. Battu
Laboratory of Fungal Pathogenesis, Centre for DNA Fingerprinting and Diagnostics, Hyderabad, India

Graduate Studies, Manipal Academy of Higher Education, Manipal, India
e-mail: priyankabhakt@cdfd.org.in; anamikab@cdfd.org.in

R. Kaur (✉)
Laboratory of Fungal Pathogenesis, Centre for DNA Fingerprinting and Diagnostics, Hyderabad, India
e-mail: rkaur@cdfd.org.in

P. P. Singh (ed.), *Infectious Diseases and Your Health*,
https://doi.org/10.1007/978-981-13-1577-0_22

22.1 Introduction

Fungi are eukaryotic organisms which lack the green pigment chlorophyll, obtain nutrients from their environment through absorption, and are able to reproduce sexually as well as asexually (Stajich et al. 2009). For a common man, fungi may simply imply mold or mushrooms. However, the fungal kingdom, which is estimated to be represented by more than 2.0 million species, is very diverse (Hawksworth and Lücking 2017). Fungi are important for human health, agriculture, ecosystem sustenance, and industry. The best characterized fungal species is *Saccharomyces cerevisiae*, also known as baker or brewer's yeast (Duina et al. 2014). It has been used in baking and fermentation (conversion of sugar into alcohol) industry for ages (Duina et al. 2014). Although fungi are primarily known as a source of food, alcohol, and pharmaceuticals worldwide, several fungal species have also been reported to live in or on human body as pathogens or commensals (Huffnagle and Noverr 2013). In the commensal state, fungi cause no harm to the human host and stay in harmony with other microbes in the body. The switch from the commensal state to the pathogenic stage is primarily triggered by defects in the immune system of the human host and can be detrimental to human health (Huffnagle and Noverr 2013).

The term "mycoses" is generally used to refer to diseases caused by fungi which range from the most common superficial infection of the skin and nails (caused by dermatophytes) to life-threatening bloodstream infections (BSI) that are mainly caused by species of *Aspergillus*, *Cryptococcus*, and *Candida* genera (Pfaller and Diekema 2007; Brown et al. 2012). Fungal BSIs are associated with a mortality rate of up to 95% (Brown et al. 2012). Besides being a clinical challenge, fungal BSIs also pose a serious economic threat owing to the high cost of associated medical care (Pfaller and Diekema 2007). Fungal pathogens are considered as hidden killers, as the number of deaths due to top 10 invasive fungal diseases is similar to those from tuberculosis or malaria per year (Brown et al. 2012). *Candida* species (spp.) are the most common cause of opportunistic bloodstream fungal infections and account for 8–10% of all nosocomial (hospital-acquired) BSIs (Pfaller and Diekema 2007; Brown et al. 2012). Furthermore, a mortality rate of 40% has been associated with *Candida* BSIs (Pfaller and Diekema 2007).

22.1.1 Candidiasis

Candida spp. reside on or inside the human body as commensal (Moran et al. 2012). This coexistence requires a healthy relationship between the fungus and the human host and/or with other human microbial flora. Any factor that negatively affects the host immune response, either innate or adaptive, may result in *Candida* overgrowth leading to fungal diseases. Of more than 150 known species of *Candida*, at least 30

Table 22.1 Risk factors associated with infections caused by *Candida* species

Disease type	Major risk factors	*Candida* spp.
Oropharyngeal candidiasis	Steroid or tobacco consumption Smoking Radiotherapy or chemotherapy of the head or neck Neutropenia Acquired immunodeficiency syndrome (AIDS)	*C. albicans* *C. krusei* *C. glabrata* *C. tropicalis*
Esophageal candidiasis	Solid and hematological malignancy Antibiotics and corticosteroids therapy Old age Diabetes mellitus AIDS	*C. albicans* *C. glabrata* *C. tropicalis* *C. krusei* *C. parapsilosis*
Vulvovaginal candidiasis	Pregnancy Diabetes mellitus Oral contraceptives Broad-spectrum antibiotics Immunodeficiency Intrauterine devices Corticosteroid therapy	*C. albicans* *C. glabrata* *C. krusei* *C. tropicalis* *C. parapsilosis*
Urinary tract infections	Long ICU stay Diabetes mellitus Increased age Urinary catheters Prolonged exposure to antimicrobial agents	*C. albicans* *C. glabrata* *C. tropicalis* *C. parapsilosis*
Bloodstream infections	Broad-spectrum antimicrobial agents Neutropenia Organ malignancy Parenteral nutrition and central venous catheter Intra-abdominal surgery Organ transplantation Immunosuppressive treatments	*C. albicans* *C. glabrata* *C. tropicalis* *C. parapsilosis* *C. krusei*

have been reported to cause infections in humans (Pfaller and Diekema 2007; Gabaldón et al. 2016; Chowdhary et al. 2017). Infections caused by *Candida* spp. are collectively known as "candidiasis" and range from common superficial infections, known as mucosal candidiasis, to severe disseminated bloodstream infections (Moran et al. 2012; Revankar and Sobel 2012). Mucosal candidiasis involves oropharyngeal, esophageal, vaginal, and urinary tract infections. Risk factors and predominant *Candida* spp. associated with different forms of candidiasis are listed in Table 22.1.

22.1.2 Oropharyngeal Candidiasis

Oropharyngeal candidiasis (OPC), infection of the mouth and the throat, is the most common opportunistic *Candida* infection in human immunodeficiency virus (HIV)-infected individuals (Sangeorzan et al. 1994; Goldman et al. 2005). Clinically, OPC primarily consists of four types: (1) thrush (pseudomembranous) which affects the buccal cavity, throat, and tongue and is recognized by appearance of curd-like white patches; (2) erythematous wherein red patches are found on the soft and hard palate, the inner part of the tongue, and the buccal mucosa; (3) hyperplastic which is typified by strongly adhered white patches on the palate, tongue, and buccal mucosa; and (4) denture-induced stomatitis, also known as denture sore mouth, is represented by red lesions at the mouth corners and is prevalent in denture wearers (Williams and Lewis 2011). Although not life-threatening, OPC causes a great deal of discomfort and adversely affects the quality of life. OPC is currently being diagnosed via fungal detection through either culturing of swabs collected from lesions or microscopic examination of lesion biopsies (Williams and Lewis 2011).

22.1.3 Esophageal Candidiasis

Esophageal candidiasis (EC), the *Candida* infection of the food pipe, esophagus, commonly occurs in immunodeficient patients including HIV-infected individuals and solid organ transplant patients (Kliemann et al. 2008; Takahashi et al. 2015). Occasionally, EC has also been observed in healthy individuals (Choi et al. 2013). EC patients usually suffer from oral lesions, nausea, vomiting, weight loss, and painful swallowing (Vazquez 2010). EC is clinically characterized by distinct plaque-like lesions in the esophagus and primarily diagnosed by endoscopy and histopathology analysis (Revankar and Sobel 2012).

22.1.4 Vulvovaginal Candidiasis

Vulvovaginal candidiasis (VVC), infection of the external female genitalia, vulva, affects both healthy and immunocompromised women and ranges from a commonly occurring to a severe disease which may also be recurrent in some cases (Achkar and Fries 2010). Certain conditions including pregnancy, diabetes mellitus, and uptake of oral contraceptives, antimicrobials, or corticosteroids are associated with increased prevalence of VVC (Cassone 2015). The symptoms associated with VVC are irritation and vaginal itching, soreness, reddening, edema, and discharge (Achkar and Fries 2010; Revankar and Sobel 2012). VVC is diagnosed by microscopic examination or culturing of the vaginal discharge and/or by the polymerase chain reaction (PCR) method (Revankar and Sobel 2012).

22.1.5 Candiduria

Candiduria, the urinary tract infection (UTI) caused by *Candida* spp., is a common hospital-acquired infection (Achkar and Fries 2010; Sobel et al. 2011). Mostly candidal UTIs are asymptomatic; however, these can clinically be manifested as infection of the urinary bladder and renal parenchyma and urinary tract fungus balls (Fisher et al. 2011). *Candida* UTIs are primarily diagnosed through urine culture, though ultrasonography of the kidneys and bladder is also conducted in high-risk patients (Achkar and Fries 2010; Fisher et al. 2011).

22.1.6 Candidemia

Candidemia refers to hospital-acquired bloodstream *Candida* infections in immunocompromised and critically ill patients (Pfaller and Diekema 2007). It is also known as invasive candidiasis (IC) and associated with significant mortality and morbidity (Delaloye and Calandra 2014). IC symptoms are mostly non-specific; may include high fever, chills, fast heart rate, rapid breathing, abdominal pain, and low blood pressure; and are clinically represented by sepsis (blood poisoning) (Delaloye and Calandra 2014). Sepsis is a life-threatening illness arising from the response of the human body to an infection wherein the infection can reach into sterile body fluids and tissues, thereby, resulting in multiple organ failure. Candidemia is primarily diagnosed by fungal culturing of the patient blood sample and detecting *Candida* in the blood. Methods involving detection of circulating *Candida* cell wall components, metabolism products, and DNA have also been used for rapid diagnosis (Delaloye and Calandra 2014). Recently, the matrix-assisted laser desorption ionization time of flight (MALDI-TOF) mass spectrometry and T2 magnetic resonance (T2MR) systems have been used to identify *Candida* directly from patient blood samples without culturing it on the medium in the clinical laboratory (Mery et al. 2016; Pfaller et al. 2016).

22.1.7 Predisposing Conditions

Major risk factors for IC can be broadly categorized into two groups, host-related factors and health-care-associated factors. Health-care-associated risk factors include long stay in ICUs, immunosuppressive drug usage, the use of catheter and other medical interventions, and prolonged use of broad-spectrum antibiotics, while major host-related factors are presence of any immunodeficient disorder, neutropenia (low levels of neutrophils in the blood), diabetes, and advanced age (Pfaller and Diekema 2007). Although *Candida* infections commonly arise from the microflora of the patient, they can spread in health-care settings due to contaminated beds, medical equipment, surfaces, and hands of health-care workers (Pfaller and Diekema 2007; Chowdhary et al. 2017).

22.1.8 Epidemiology of Candidiasis

The World Health Organization (WHO; www.who.int/topics/epidemiology/en/) defines epidemiology as "the study of the distribution and determinants of health-related states or events (including disease), and the application of this study to the control of diseases and other health problems." Epidemiological studies have revealed *Candida* spp. to be the fourth leading cause of all hospital-acquired BSIs (Wisplinghoff et al. 2004; Pfaller and Diekema 2007; Chakrabarti et al. 2015). Among infectious *Candida* spp., *C. albicans* is the most prevalent, and it, in conjunction with four other species, *C. glabrata*, *C. parapsilosis*, *C. tropicalis*, and *C. krusei*, accounts for more than 90% of *Candida* BSIs (Pfaller et al. 2014; Pappas et al. 2016). Further, candidemia due to the emergence of the multidrug-resistant species, *C. auris*, has also recently been reported and poses a grave threat to human health (Chowdhary et al. 2017).

Based on the geographical location, the frequency and distribution of Candida species in *Candida* BSIs varies, and an increase in the occurrence of non-albicans *Candida* (NAC) species in many countries has been reported (Pfaller and Diekema 2007; Falagas et al. 2010; Pfaller et al. 2014; Chakrabarti et al. 2015; Kaur and Chakrabarti 2017). The precise reason for the recent emergence of NAC species is yet to be determined, though the intrinsic or acquired resistance against existing antifungal agents in these *Candida* spp. is thought to play some role.

Morphologically, *Candida* spp. either grow in the single-celled yeast form (unicellular) or multicellular pseudohyphal and hyphal forms (Turner and Butler 2014). *Candida* spp. can replicate sexually as well as asexually and are assigned to the class Ascomycetes (Gabaldón et al. 2016). Many medically important *Candida* spp. belong to the phylogenetic subgroup, the CTG clade, wherein the CTG codon codes for the serine amino acid instead of the leucine amino acid during protein synthesis (Gabaldón et al. 2016). Salient features of the five most commonly found *Candida* spp. in BSIs, *C. albicans*, *C. glabrata*, *C. tropicalis*, *C. parapsilosis*, and *C. krusei*, are described below.

Candida albicans: C. albicans is the most prevalent candidiasis-causing agent and accounts for about 60% of the total invasive candidiasis cases (Pfaller and Diekema 2007; Falagas et al. 2010). It resides as a harmless commensal on the mucosal layer of various organs like the intestine, oral cavity, urinary tract, and vagina of healthy individuals and becomes pathogenic in individuals with weakened immune system (Turner and Butler 2014). *C. albicans* is a diploid yeast and multimorphic in nature, i.e., it grows in three different morphological forms, viz., yeast, pseudohyphae, and true hyphae (Turner and Butler 2014). It has no known terrestrial life cycle. The hyphal form of *C. albicans* is considered as the invasive form, as it is more effective in attacking the host cell epithelium and immune system resulting

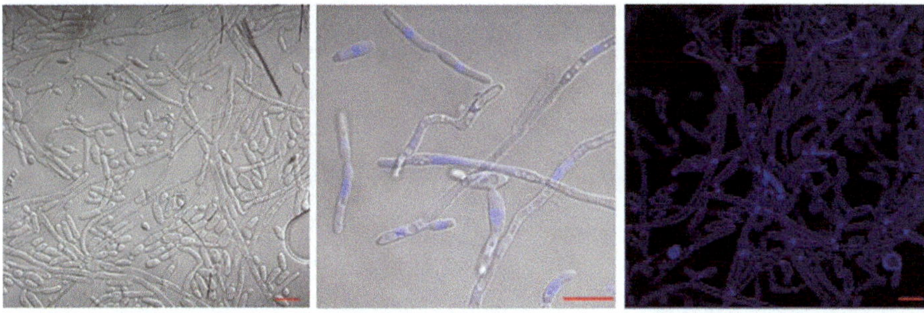

Fig. 22.1 Confocal microscopy images of *C. albicans* cells. The middle and right panel depict *C. albicans* cells stained with DAPI (4′,6-diamidino-2-phenylindole) and calcofluor white (CFW) fluorescent dye, respectively. DAPI binds to DNA and stains nuclei, while CFW binds to chitin in the cell wall and indicates the septum, which separates two cells in filamentous forms. Scale bar = 10 μm

in superficial and bloodstream infections (Turner and Butler 2014). Figure 22.1 shows different morphological forms of *C. albicans* and their stained nuclei (contains the hereditary material DNA) and cell wall (the first point of contact with the host).

Factors that help the pathogen to establish an infection and cause disease are called virulence factors. The major virulence traits of *C. albicans* (listed in Table 22.2) include the ability to undergo morphological switching (from yeast to pseudohyphae and hyphae) and reversible colony switching (from white to opaque), form biofilms on inert surfaces and strongly adhere to, and invade, host tissues through expression of cell surface adhesins and secretion of enzymes that hydrolyze host proteins (Turner and Butler 2014).

Candida glabrata: *C. glabrata*, earlier known as *Torulopsis glabrata*, is generally the second most common cause of invasive candidiasis (Pfaller and Diekema 2007). It is also the second most common cause of UTIs after *C. albicans* (Achkar and Fries 2010). *C. glabrata* is a haploid non-dimorphic fungus which only exists in the yeast form and divides by budding (Silva et al. 2012). Figure 22.2 shows cell wall-stained and unstained budding *C. glabrata* cells. Instead of the CTG clade, *C. glabrata* belongs to the Nakaseomyces clade containing many nonpathogenic yeasts (Gabaldón et al. 2016). Its known virulence factors include survival and proliferation in macrophages, melanin production, biofilm formation and expression of families of adhesins and cell surface-associated proteases (Silva et al. 2012; Gabaldón et al. 2016). *C. glabrata* is inherently less susceptible to widely used azole antifungals, is prevalent in diabetic and elderly patients, and can be associated with a high mortality rate (Silva et al. 2012; Gabaldón et al. 2016).

Candida parapsilosis: *C. parapsilosis* is a diploid *Candida* species, which largely exists either in the yeast or pseudohyphal form, and does not produce true hyphae (Silva et al. 2012; Turner and Butler 2014). It is a frequent cause of infections in

Table 22.2 Key virulence factors of *C. albicans*

Virulence factors	Function/s	Schematic representation
Morphological plasticity (ability to exist in more than one morphological forms)	Facilitates host invasion and host killing	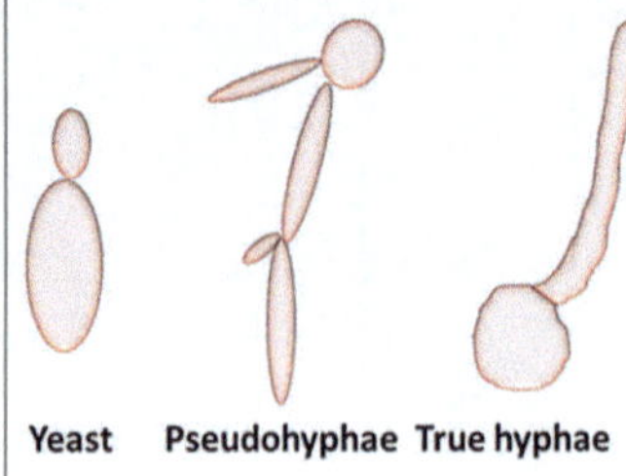
Secretory proteases (breaks down proteins)	Involved in nutrient acquisition and protection from, and evasion of, the host defense	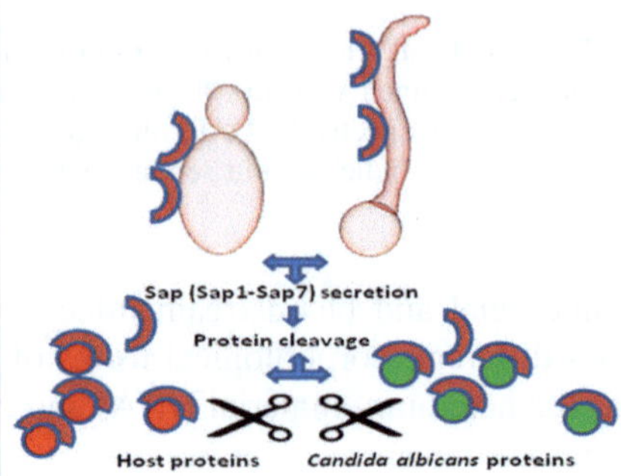
Adhesins (cell surface proteins with capability to stick to surfaces)	Promotes adhesion to host cells as well as to abiotic surfaces	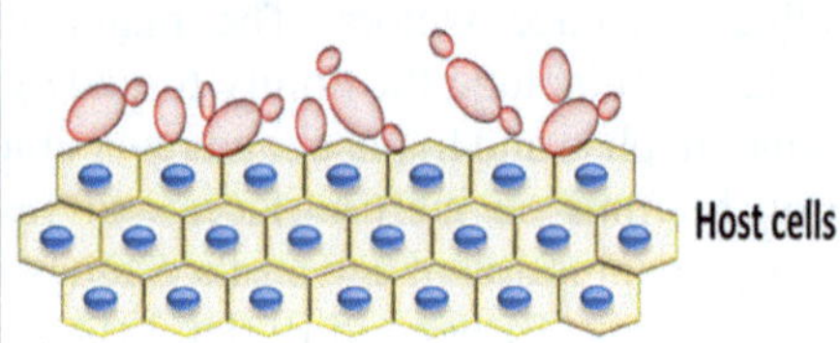
Phenotypic switching (ability to exist in more than one colony forms)	Facilitates mating and helps in adaptation to different environmental conditions	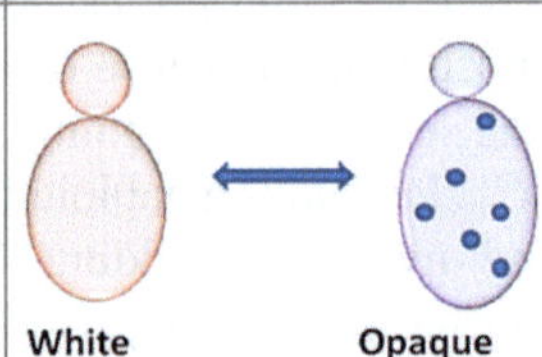
Biofilm formation (ability to form a city of cells that attach to a surface and are nested in a protective matrix)	Provides resistance to antimicrobial agents and host immune factors	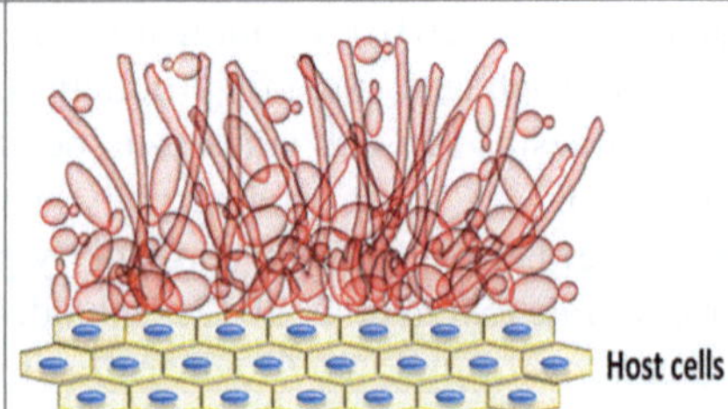

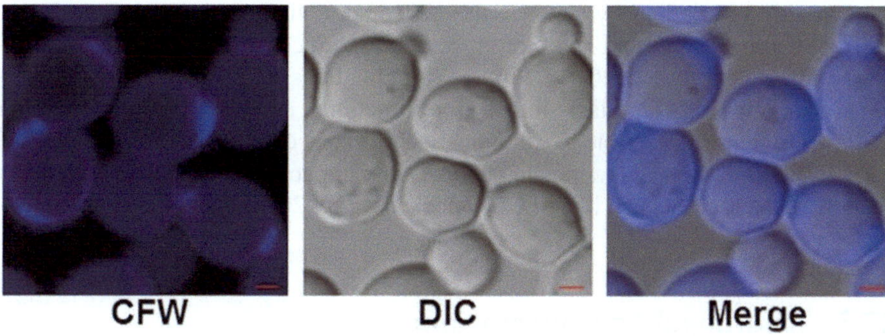

Fig. 22.2 Confocal microscopy images of calcofluor white (CFW)-stained *C. glabrata* cells. CFW binds to chitin in the cell wall and indicate bud scars. *DIC* differential interference contrast. Scale bar = 1 μm

infants and neonates (babies born prematurely) (Silva et al. 2012; Turner and Butler 2014). Hand carriage of the health-care workers is thought to be the primary reason for spread of *C. parapsilosis* infections in the hospital environment (Silva et al. 2012). The pathogenesis of *C. parapsilosis* is attributed to adherence to host cells, biofilm formation, and secretion of hydrolytic enzymes (Silva et al. 2012; Turner and Butler 2014).

Candida tropicalis: *C. tropicalis* is a common invasive fungal pathogen in neutropenic patients (Silva et al. 2012; Turner and Butler 2014). It is diploid and exists in the budding yeast, pseudohyphal and hyphal shapes (Silva et al. 2012). Biofilm formation and secretory proteases are known virulence attributes with clinical isolates of *C. tropicalis* displaying robust biofilm-forming ability (Silva et al. 2012).

Candida krusei: *C. krusei* is a haploid *Candida* species which has not been extensively studied (Turner and Butler 2014). *C. krusei* is intrinsically resistant to commonly used azole antifungals. Despite its low prevalence in healthy humans, *C. krusei* infections pose a serious challenge owing to limited treatment options (Turner and Butler 2014).

22.2 Host Immune Response to Candidiasis

Immune system of the human body, which protects us from infections, consists of two types: the non-specific innate and the pathogen-specific-acquired system. Compared to other *Candida* species, immune response toward *C. albicans* infections is well-studied (Netea et al. 2015). It is known that both innate and acquired defense mechanisms are pivotal to keep *C. albicans* in the commensal state. Any vulnerability in the human body defense system can trigger the commensal to pathogenic switch which may result in a full-blown disease (Netea et al. 2015). Receptors on immune cells recognize *C. albicans* primarily through molecules present in its cell wall (Smeekens et al. 2013; Netea et al. 2015). Hence, individuals with any kind of deficiency either in these receptors, viz., dectin 1, or their downstream effectors, Card9, are more prone to develop mucosal and/or systemic candidiasis (Smeekens et al. 2013). Similarly, individuals with monogenic primary immunodeficiency, which lack one or more functional components of the immune system, are more susceptible to candidiasis (Smeekens et al. 2013). Human genetic variations (alterations in the human genome) that impact susceptibility to candidiasis have also recently been identified (Smeekens et al. 2013; Kumar et al. 2014).

22.3 Antifungal Armamentarium

The last two decades have seen a shift in the epidemiology of candidemia with increased non-albicans *Candida* infections (Falagas et al. 2010; Pfaller et al. 2014; Chakrabarti et al. 2015; Kaur and Chakrabarti 2017). This epidemiological shift has been postulated to be due to the widespread use of antifungals leading to the emergence of *Candida* species that are intrinsically resistant to currently available antifungals (Perlin et al. 2015). Antifungal arsenal, in general, is limited owing to the eukaryotic nature of fungi, that results in largely conserved processes between the human host and the fungal pathogen, and very few attackable targets specific to the pathogen. As with other antimicrobials, antifungal drug resistance (AFDR) is emerging as an important clinical issue (Lewis 2011). AFDR can be of two types: intrinsic and acquired. Intrinsic drug resistance is the primary resistance, which is inherent in nature, and does not emerge from prior exposure to antifungals (Perlin et al. 2015). Examples of *Candida* spp. with intrinsic resistance include the azole antifungal-resistant *C. krusei* and recently emerged multi-antifungal-resistant *C. auris* (Perlin et al. 2015). Acquired drug resistance, as the name indicates, arises after treatment with antifungals and is secondary in nature (Perlin et al. 2015). *C. glabrata* is well-known for displaying acquired resistance to echinocandin antifungals (Perlin et al. 2015). Following four major classes of antifungals represent current anti-*Candida* therapies in clinical settings (Pianalto and Alspaugh 2016).

22.3.1 Azoles

Azoles are the most frequently used antifungal drugs which target the fungal cell membrane via impediment of the ergosterol biosynthetic pathway, with ergosterol being the major sterol of the *Candida* cell membrane (Ghannoum and Rice 1999). Azoles inhibit the cytochrome P450-dependent lanosterol 14α-demethylase, a core enzyme in ergosterol biosynthesis, which is encoded by the *ERG11* gene (Ghannoum and Rice 1999). Methylated lanosterol is converted, by cellular enzymes, to other sterol molecules, which are toxic for *Candida* growth (Ghannoum and Rice 1999; Lewis 2011). Antifungal action of azoles is attributed to reduced ergosterol levels in

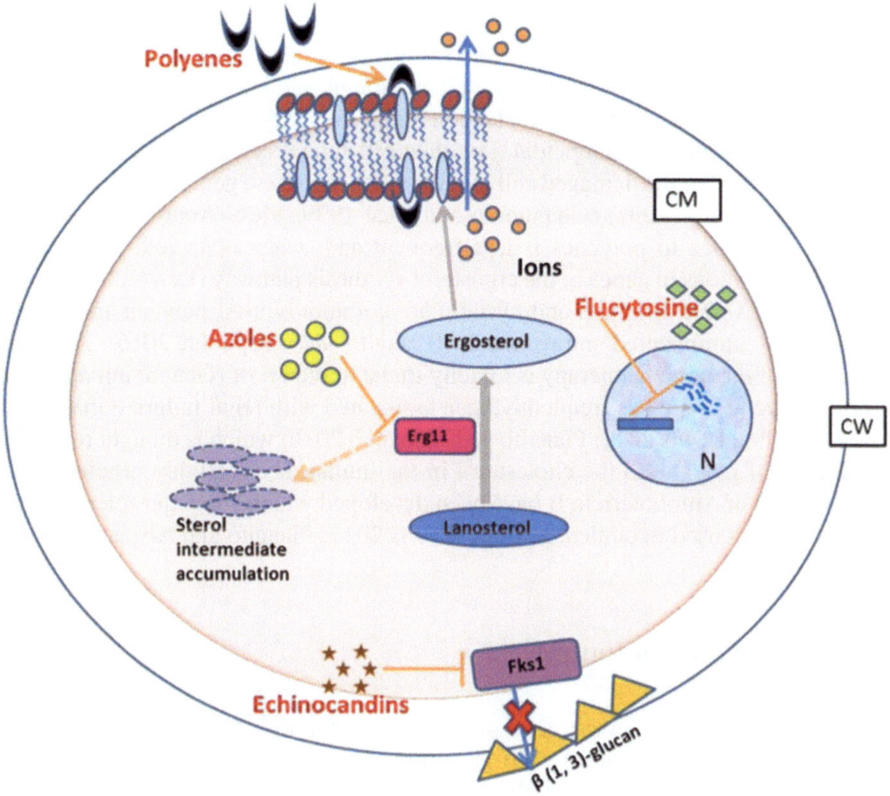

Fig. 22.3 A schematic illustrating mechanisms of action of antifungal drugs. Azoles inhibit ergosterol biosynthesis by targeting the Erg11 enzyme, while polyenes bind to ergosterol present in the cell membrane and create pores. Echinocandins inhibit the β-(1, 3)-glucan synthase enzyme, Fks1. Flucytosine impedes nucleic acid (DNA and RNA) and protein synthesis. *CM* cell membrane, *CW* cell wall, *N* nucleus

the plasma membrane and buildup of toxic sterols in the cell (Fig. 22.3). Azoles are fungistatic, i.e., they inhibit fungal growth but do not kill fungal cells. Hence, resistance to azole antifungals is common and is largely due to overexpression of multidrug efflux pumps, alteration or overexpression of the Erg11 enzyme, and mutations in the enzymes involved in ergosterol biosynthesis (Marie and White 2009; Lewis 2011; Perlin et al. 2015). These mechanisms lead to reduced intracellular drug accumulation, weak azole binding to the Erg11 enzyme, amplification of the drug target, and diminished generation of toxic sterols, respectively (Marie and White 2009; Lewis 2011; Perlin et al. 2015). Azole antifungals are available in both formulations: creams/lotions/tablets and oral suspensions/intravenous injections (Lewis 2011). Commonly used azole drugs are fluconazole, voriconazole, posaconazole, miconazole, ketoconazole, and itraconazole (Pianalto and Alspaugh 2016).

22.3.2 Polyenes

Polyene drugs bind to ergosterol in the cell membrane and form pores through which intracellular contents including metal ions leak out (Ghannoum and Rice 1999; Fig. 22.3). Polyenes are fungicidal, i.e., they kill fungal cells, and their cidality is largely attributed to the damaged cell membrane as well as to generated free reactive oxygen species (oxidants) (Ghannoum and Rice 1999; Mesa-Arango et al. 2014). Overall, resistance to polyenes is less frequent and, when observed, is generally owing to alterations in genes of the ergosterol synthesis pathway (Lewis 2011; Perlin et al. 2015). Amphotericin B and nystatin are commonly used polyene antifungals and usually administered intravenously (Pianalto and Alspaugh 2016). A major drawback of the polyene therapy is toxicity to the human host (Ghannoum and Rice 1999). Polyene usage has frequently been associated with renal failure (Ghannoum and Rice 1999; Lewis 2011; Pianalto and Alspaugh 2016) which is thought to be due to binding of polyene to the cholesterol in the human body. Highly efficient lipid formulations of Amphotericin B have been developed which are better tolerated and currently being used in clinical settings (Lewis 2011; Pianalto and Alspaugh 2016).

22.3.3 Echinocandins

Echinocandins are the newest class of antifungals and target an important component of the fungal cell wall, β-(1,3)-glucan (Marie and White 2009; Lewis 2011). They non-competitively inhibit the β-(1,3)-glucan synthase enzyme (Fig. 22.3) which is encoded by the *FKS1* gene (Marie and White 2009; Lewis 2011). Echinocandins are fungicidal due to the fragile fungal cell wall, as an intact cell wall is pivotal to maintenance of cellular integrity (Marie and White 2009; Lewis 2011). Caspofungin, micafungin, anidulafungin, commonly used echinocandins, have little oral bioavailability, and are, thus, given intravenously (Lewis 2011; Pianalto and Alspaugh 2016). Echinocandin resistance is being increasingly observed among clinical isolates of *Candida* species and is exclusively associated with mutations in the *FKS* genes (Marie and White 2009; Lewis 2011; Perlin et al. 2015). An alarming trend in hospitals worldwide is the cross resistance of azole-resistant *C. glabrata* isolates to echinocandins, thereby, rendering treatment of *C. glabrata* infections difficult (Pfaller et al. 2012; Perlin et al. 2015).

22.3.4 Flucytosine

Flucytosine, also known as 5-fluorocytosine (5FC), targets nucleic acid and protein synthesis in the cell (Fig. 22.3) (Ghannoum and Rice 1999; Marie and White 2009). It is taken up by the cytosine permease enzyme and is first converted to 5-fluorouracil (5FU) by the cytosine deaminase enzyme. Further conversion of 5FU leads to its incorporation into RNA, eventually inhibiting protein synthesis. Modified 5FU is also an inhibitor of the thymidylate synthase enzyme which is required for DNA synthesis (Ghannoum and Rice 1999; Marie and White 2009). Resistance to 5FC is common and has been attributed to mutations in genes encoding cytosine permease and cytosine deaminase enzymes (Marie and White 2009; Lewis 2011). Thus, 5FC is often used in combination with either azole or polyene antifungals (Marie and White 2009; Lewis 2011).

Allylamines are another class of antifungals, which inhibit an early enzyme, squalene epoxidase, of the ergosterol biosynthetic pathway, and are primarily used to treat superficial cutaneous mycoses (Ghannoum and Rice 1999). Terbinafine is a widely used allylamine drug whose toxicity is attributed to the intracellular accumulation of the squalene (Ghannoum and Rice 1999; Lewis 2011). Griseofulvin is a tubulin-binding drug which impedes cell division and is generally used to treat infections of the skin and nails caused by dermatophytic fungi (Lewis 2011). Overall, due to the availability of intravenous formulations, azoles, polyenes, and echinocandins are the first-choice drugs to treat *Candida* bloodstream infections (Lewis 2011).

Finally, owing to emerging resistance of *Candida* spp. to current antifungals, efforts, at present, are also being directed toward identification of novel and/or combinatorial antifungal therapeutic strategies.

22.4 Human Mycobiome

The human mycobiome refers to all fungal organisms present on or within the body. Previously, microbial culture-based methods were used to identify fungal community of humans (Huffnagle and Noverr 2013). With the advent of high-throughput next-generation sequencing technologies, which do not require culturing of human samples, efforts have been made to identify the human mycobiome (Seed 2014). Recent studies have revealed high occurrence of *Candida* spp. in the oral cavity, gut, fecal, and vaginal samples of healthy individuals (Seed 2014). These findings indicate that *Candida* spp. are an integral component of the human microbiome (all microorganisms present in or on the body) and may stay in a mutually beneficial relationship with the human host (Seed 2014). Studies are currently being carried out to decipher the precise role; *Candida* spp. play in both homeostasis of the human metabolism and keeping other microbiota in check.

22.5 Conclusion

Candida bloodstream infections pose a serious global health and economic challenge due to high mortality rates and medical costs. With the ever-increasing prevalence of predisposing risk factors, the invasive candidiasis burden is likely to rise.

Strict implementation of preventive measures, including frequent hand hygiene among health-care providers, regular surveillance and early detection in high-risk patients, and appropriate antifungal therapy, may result in better control of *Candida* infections. Lastly, research and sizeable funding toward development of early and species-specific diagnostic markers and novel or combinatorial antifungal drugs will help clinicians treat life-threatening *Candida* bloodstream infections more successfully in future.

Acknowledgments We thank Reshma Chowdary Alokam for the help with confocal imaging. Research in RK laboratory is supported by the Wellcome Trust/DBT India Alliance Senior Fellowship [IA/S/15/1/501831] to RK, and grants from the Department of Biotechnology, Government of India [BT/PR7388/MED/29/650/2012 and BT/PR7123/BRB/10/1149/2012] to RK. PB is a recipient of Junior and Senior Fellowship sponsored by the University Grants Commission, New Delhi, India. AB is a recipient of Junior Fellowship sponsored by the Department of Biotechnology, Government of India.

References

Achkar JM, Fries BC (2010) Candida infections of the genitourinary tract. Clin Microbiol Rev 23:253–273

Brown GD, Denning DW, Gow NA, Levitz SM, Netea MG, White TC (2012) Hidden killers: human fungal infections. Sci Transl Med 4:165rv13

Cassone A (2015) Vulvovaginal *Candida albicans* infections: pathogenesis, immunity and vaccine prospects. Br J Obstet Gynaecol 122:785–794

Chakrabarti A, Sood P, Rudramurthy SM, Chen S, Kaur H, Capoor M et al (2015) Incidence, characteristics and outcome of ICU-acquired candidemia in India. Intensive Care Med 41:285–295

Choi JH, Lee CG, Lim YJ, Kang HW, Lim CY, Choi JS (2013) Prevalence and risk factors of esophageal candidiasis in healthy individuals: a single center experience in Korea. Yonsei Med J 54:160–165

Chowdhary A, Sharma C, Meis JF (2017) *Candida auris*: a rapidly emerging cause of hospital-acquired multidrug-resistant fungal infections globally. PLoS Pathog 13:e1006290

Delaloye J, Calandra T (2014) Invasive candidiasis as a cause of sepsis in the critically ill patient. Virulence 5:161–169

Duina AA, Miller ME, Keeney JB (2014) Budding yeast for budding geneticists: a primer on the *Saccharomyces cerevisiae* model system. Genetics 197:33–48

Falagas ME, Roussos N, Vardakas KZ (2010) Relative frequency of albicans and the various non-albicans *Candida* spp. among candidemia isolates from inpatients in various parts of the world: a systematic review. Int J Infect Dis 14:e954–e966

Fisher JF, Sobel JD, Kauffman CA, Newman CA (2011) Candida urinary tract infections–treatment. Clin Infect Dis 52(Suppl. 6):S457–S466

Gabaldón T, Naranjo-Ortíz MA, Marcet-Houben M (2016) Evolutionary genomics of yeast pathogens in the *Saccharomycotina*. FEMS Yeast Res 16. pii: fow064

Ghannoum MA, Rice LB (1999) Antifungal agents: mode of action, mechanisms of resistance, and correlation of these mechanisms with bacterial resistance. Clin Microbiol Rev 12:501–517

Goldman M, Cloud GA, Wade KD, Reboli AC, Fichtenbaum CJ, Hafner R et al (2005) A randomized study of the use of fluconazole in continuous versus episodic therapy in patients with advanced HIV infection and a history of oropharyngeal candidiasis: AIDS Clinical Trials Group Study 323/Mycoses Study Group Study 40. Clin Infect Dis 41:1473–1480

Hawksworth DL, Lücking R (2017) Fungal diversity revisited: 2.2 to 3.8 million species. Microbiol Spectr 5(4):1-17 (FUNK-0052–2016)

Huffnagle GB, Noverr MC (2013) The emerging world of the fungal microbiome. Trends Microbiol 21:334–341

Kaur H, Chakrabarti A (2017) Strategies to reduce mortality in adult and neonatal candidemia in developing countries. J Fungi 3:41

Kliemann DA, Pasqualotto AC, Falavigna M, Giaretta T, Severo LC (2008) Candida esophagitis: species distribution and risk factors for infection. Rev Inst Med Trop Sao Paulo 50:261–263

Kumar V, Cheng SC, Johnson MD, Smeekens SP, Wojtowicz A, Giamarellos-Bourboulis E et al (2014) Immunochip SNP array identifies novel genetic variants conferring susceptibility to candidaemia. Nat Commun 5:4675

Lewis RE (2011) Current concepts in antifungal pharmacology. Mayo Clin Proc 86:805–817

Marie C, White TC (2009) Genetic basis of antifungal drug resistance. Curr Fungal Infect Rep 3:163–169

Mery A, Sendid B, François N, Cornu M, Poissy J, Guerardel Y et al (2016) Application of mass spectrometry technology to early diagnosis of invasive fungal infections. J Clin Microbiol 54:2786–2797

Mesa-Arango AC, Trevijano-Contador N, Román E, Sánchez-Fresneda R, Casas C, Herrero E et al (2014) The production of reactive oxygen species is a universal action mechanism of Amphotericin B against pathogenic yeasts and contributes to the fungicidal effect of this drug. Antimicrob Agents Chemother 58:6627–6638

Moran G, Coleman D, Sullivan D (2012) An introduction to the medically important Candida species. In: Calderone RA, Clancy CJ (eds) Candida and candidiasis. ASM Press, Washington, DC, pp 11–25

Netea MG, Joosten LA, van der Meer JW, Kullberg BJ, van de Veerdonk FL (2015) Immune defence against Candida fungal infections. Nat Rev Immunol 15:630–642

Pappas PG, Kauffman CA, Andes DR, Clancy CJ, Marr KA, Ostrosky-Zeichner L et al (2016) Clinical practice guideline for the management of candidiasis: 2016 update by the Infectious Diseases Society of America. Clin Infect Dis 62:e1–e50

Perlin DS, Shor E, Zhao Y (2015) Update on antifungal drug resistance. Curr Clin Microbiol Rep 2:84–95

Pfaller MA, Diekema DJ (2007) Epidemiology of invasive candidiasis: a persistent public health problem. Clin Microbiol Rev 20:133–163

Pfaller MA, Castanheira M, Lockhart SR, Ahlquist AM, Messer SA, Jones RN (2012) Frequency of decreased susceptibility and resistance to echinocandins among fluconazole-resistant bloodstream isolates of *Candida glabrata*. J Clin Microbiol 50:1199–1203

Pfaller MA, Andes DR, Diekema DJ, Horn DL, Reboli AC, Rotstein C et al (2014) Epidemiology and outcomes of invasive candidiasis due to non-albicans species of Candida in 2,496 patients: data from the Prospective Antifungal Therapy (PATH) registry 2004–2008. PLoS One 9:e101510

Pfaller MA, Wolk DM, Lowery TJ (2016) T2MR and T2Candida: novel technology for the rapid diagnosis of candidemia and invasive candidiasis. Future Microbiol 11:103–117

Pianalto KM, Alspaugh AJ (2016) New horizons in antifungal therapy. J Fungi 2:26

Revankar SG, Sobel JD (2012) Mucosal candidiasis. In: Calderone RA, Clancy CJ (eds) Candida and candidiasis. ASM Press, Washington, DC, pp 419–427

Sangeorzan JA, Bradley SF, He X, Zarins LT, Ridenour GL, Tiballi RN et al (1994) Epidemiology of oral candidiasis in HIV-infected patients: colonization, infection, treatment, and emergence of fluconazole resistance. Am J Med 97:339–346

Seed PC (2014) The human mycobiome. Cold Spring Harb Perspect Med 5:a019810

Silva S, Negri M, Henriques M, Oliveira R, Williams DW, Azeredo J (2012) *Candida glabrata*, *Candida parapsilosis* and *Candida tropicalis*: biology, epidemiology, pathogenicity and antifungal resistance. FEMS Microbiol Rev 36:288–305

Smeekens SP, van de Veerdonk FL, Kullberg BJ, Netea MG (2013) Genetic susceptibility to Candida infections. EMBO Mol Med 5:805–813

Sobel JD, Fisher JF, Kauffman CA, Newman CA (2011) Candida urinary tract infections – epidemiology. Clin Infect Dis 52(Suppl. 6):S433–S436

Stajich JE, Berbee ML, Blackwell M, Hibbett DS, James TY, Spatafora JW et al (2009) The fungi. Curr Biol 19:R840–R845

Takahashi Y, Nagata N, Shimbo T, Nishijima T, Watanabe K, Aoki T et al (2015) Long-term trends in esophageal candidiasis prevalence and associated risk factors with or without HIV infection: lessons from an endoscopic study of 80,219 patients. PLoS One 10:e0133589

Turner SA, Butler G (2014) The Candida pathogenic species complex. Cold Spring Harb Perspect Med 4:a019778

Vazquez JA (2010) Optimal management of oropharyngeal and esophageal candidiasis in patients living with HIV infection. HIV AIDS (Auckland, N.Z.) 2:89–101

Williams D, Lewis M (2011) Pathogenesis and treatment of oral candidosis. J Oral Microbiol 3:5771

Wisplinghoff H, Bischoff T, Tallent SM, Seifert H, Wenzel RP, Edmond MB (2004) Nosocomial bloodstream infections in US hospitals: analysis of 24,179 cases from a prospective nationwide surveillance study. Clin Infect Dis 39:309–317

MIX
Papier aus verantwortungsvollen Quellen
Paper from responsible sources
FSC® C105338

If you have any concerns about our products,
you can contact us on
ProductSafety@springernature.com

In case Publisher is established outside the EU,
the EU authorized representative is:
Springer Nature Customer Service Center GmbH
Europaplatz 3, 69115 Heidelberg, Germany

Printed by Libri Plureos GmbH
in Hamburg, Germany